Student Workbook

ANATOMY & PHYSIOLOGY

SECOND EDITION

ANATOMY & PHYSIOLOGY

Student Workbook

First edition: 2009
Second edition: 2013
Seventh printing

ISBN: 978-1-927173-57-2

Published by **BIOZONE International Ltd**

Printed by REPLIKA PRESS PVT LTD using paper produced from renewable and waste materials

Front cover photographs:
X-ray of skull: iStockphoto © 2009 David Marchal
Artist's impression of neurons: iStockphoto © 2009 ktsimage
istockphotos (www.istockphoto.com)

About the Writing Team

Tracey Greenwood joined the staff of BIOZONE at the beginning of 1993. She has a Ph.D in biology, specialising in lake ecology, and taught undergraduate and graduate biology at the University of Waikato for four years.

Lissa Bainbridge-Smith worked in industry in a research and development capacity for eight years before joining BIOZONE in 2006. Lissa has an M.Sc from Waikato University.

Kent Pryor has a BSc from Massey University majoring in zoology and ecology. He was a secondary school teacher in biology and chemistry for 9 years before joining BIOZONE as an author in 2009.

Richard Allan has had 11 years experience teaching senior biology at Hillcrest High School in Hamilton, New Zealand. He attained a Masters degree in biology at Waikato University, New Zealand.

Purchases of this workbook may be made direct from the publisher:

USA & CANADA:

BIOZONE Corporation,
18801 E. Mainstreet
Suite 240, Parker, CO, 80134-3445
United States
Toll FREE phone: 1-855-246-4555
Toll FREE fax: 1-855-935-3555
Email: sales@thebiozone.com
Website: www.the**BIOZONE**.com

UNITED KINGDOM & EUROPE:

BIOZONE Learning Media (UK) Ltd.
Unit 5/6, Greenline Business Park,
Wellington Street, Burton-on-Trent,
DE14 2AS, United Kingdom
Telephone: +44 1283 530 366
Fax: +44 1283 530 961
Email: sales@biozone.co.uk
Website: www.**BIOZONE**.co.uk

AUSTRALIA:

BIOZONE Learning Media Australia
P.O. Box 2841, Burleigh BC,
QLD 4220,
Australia
Telephone: +61 7 5535 4896
Fax: +61 7 5508 2432
Email: sales@biozone.com.au
Website: www.**BIOZONE**.com.au

NEW ZEALAND & REST OF WORLD:

BIOZONE International Ltd.
P.O. Box 5002,
Hamilton 3242,
New Zealand
Telephone: +64 7-856 8104
Fax: +64 7-856 9243
Email: sales@biozone.co.nz
Website: www.the**BIOZONE**.com

ANATOMY & PHYSIOLOGY
Student Workbook

Biozone's *Anatomy and Physiology Student Workbook* provides students with a set of comprehensive guidelines and highly visual worksheets through which to explore anatomy and physiology. *Anatomy & Physiology* is the ideal companion for students of the life sciences, encompassing the basic principles of cell biology, functional anatomy, physiological processes, and the impact of disease and aging on the body's systems. Homeostasis is the unifying theme throughout, and contextual examples provide the student with a relevant framework for their knowledge. This workbook comprises twelve chapters corresponding to each of the eleven body systems and an introductory chapter covering basic cells and tissues. The material is explained through a series of activities, usually of one or two pages, each of which explores a specific concept (e.g. joints or cell signaling). *Anatomy & Physiology* is a student-centered resource and is complemented by the ***Anatomy & Physiology Presentation Media CD-ROM*** and the ***Student Review Series***, downloadable to iPod and iPad. Students completing the activities, in concert with their other classroom and practical work, will consolidate existing knowledge and develop and practise skills that they will use throughout their course. This workbook may be used in the classroom or at home as a supplement to a standard textbook. Some activities are introductory in nature, while others may be used to consolidate and test concepts already covered by other means. Key features include:

- Concept maps introduce each major section of the workbook with a particular focus on how disease, medicine and technology, aging, and exercise affect each body system. There is also a focus on the interaction of body systems, encouraging students to consider how the body systems work together.
- An easy-to-use chapter introduction summarizing learning objectives in a list to be completed by the student. Chapter introductions also include a list of key terms and a summary of key concepts.
- An emphasis on acquiring skills in scientific literacy. Each chapter includes a literacy activity, and the appendix includes references for works cited as *Periodicals* throughout the text.
- *Related Activities* help students to locate related material within the workbook, and *Weblinks* provide support material (in the form of URLs) to support the material provided on activity pages.

Acknowledgements & Photo Credits

Biozone's authors acknowledge the generosity of all those who have kindly provided information and photographs for this edition: • Dept of Biological Science, University of Delaware • L. Howard and K Connolly, Dartmouth College • Wadsworth Center, New York State Dept of Health • Danny Wann, Carl Albert State College • Wellington Harrier Club • Helen Hall • David Fankhauser, Uni. of Cincinnati, Clermont College • Dan Butler • LocalFitness Pty Ltd • Charles Golderby (UCSD School of Medicine) • Dr David Wells, AgResearch • Wintec Academy of Sport • Deborahripley.com • J. Armstrong • Roger LeMoyne © UNICEF • Ed Uthman

We also acknowledge the photographers that have made their images available through **Wikimedia Commons** under Creative Commons Licences 2.0, 2.5. or 3.0 • Y tambe • Michael Berry • Emmanuelm • Department of Histology, Jagiellonian University Medical College • Johnmaxmena • Graham Crumb • Ildar Sagdejev • RM Hunt • DEA (drug enforcement agency) • Lyn Bry (Mad Science) • Bill Rhodes • Georgetown University Hospital • Stevenfruitsmaak • pan Pavel Recyl • Dan Ferber • UC Regents Davis Campus • Roadnottaken • Onderwijsgek • Mikael Hagstrom • Jpogi • Lucien Monfils • Jacoplane • Clinical Cases • USGS • Gleam • Pöllö • Dr D. Cooper, University of California, San Francisco

Coded credits are: **BF**: Brian Finerran (University of Canterbury), **CA**: Clipart.com, **CDC**: Centers for Disease Control and Prevention, Atlanta, USA, **DS**: Digital Stock, **EII**: Education Interactive Imaging, **Eyewire**: Eyewire Inc © 1998-2001, www.eyewire.com, **JDG**: John Green, **NASA**: National Aeronautics and Space Administration **NIH**: National Institutes of Health, **RA**: Richard Allan, **RCN**: Ralph Cocklin, **TG**: Tracey Greenwood, **WMU**: Waikato Microscope Unit, **WIKI**: Wikimedia Commons, **WMRCVM**: VA Maryland Regional Veterinary College of Medicine

Royalty free images, purchased by Biozone International Ltd, are used throughout this manual and do not have on-page credits. They have been obtained from: istockphotos (www.istockphoto.com) • Corel Corporation from various titles in their Professional Photos CD-ROM collection; ©Hemera Technologies Inc, 1997-2001; © 2005 JupiterImages Corporation www.clipart.com; PhotoDisc®, Inc. USA, www.photodisc.com; and ©Digital Vision. 3D models were created using Poser IV, Curious Labs and Bryce 5.5, PEIR Digital Library.

Modular Titles

Skills in Biology

Health & Disease

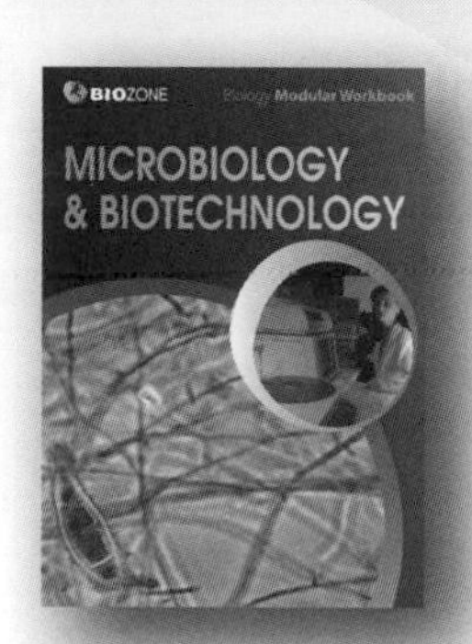

Microbiology & Biotechnology

Cell Biology & Biochemistry

Contents

CODES: △ **Significant changes** in this edition ★ **New** this edition **Activity** is marked: ☐ to be done; ☑ when completed

Contents

The Cardiovascular System

The Lymphatic System & Immunity

The Respiratory System

The Digestive System

The Urinary System

Reproduction and Development

Concepts in Anatomy and Physiology

Cells and Tissues

Cell structure and function
- Cellular membranes and organelles
- Cellular transport processes
- Cell division and specialization

Tissues are made up of cells with different roles
- Epithelial tissues
- Connective tissues
- Muscle tissue
- Nervous tissue

Organs are made up of different tissues. Organ systems have different roles.
- Exchanges with the environment
- Support and movement
- Control and coordination
- Internal transport
- Internal defense
- Reproduction and development
- Excretion and fluid balance

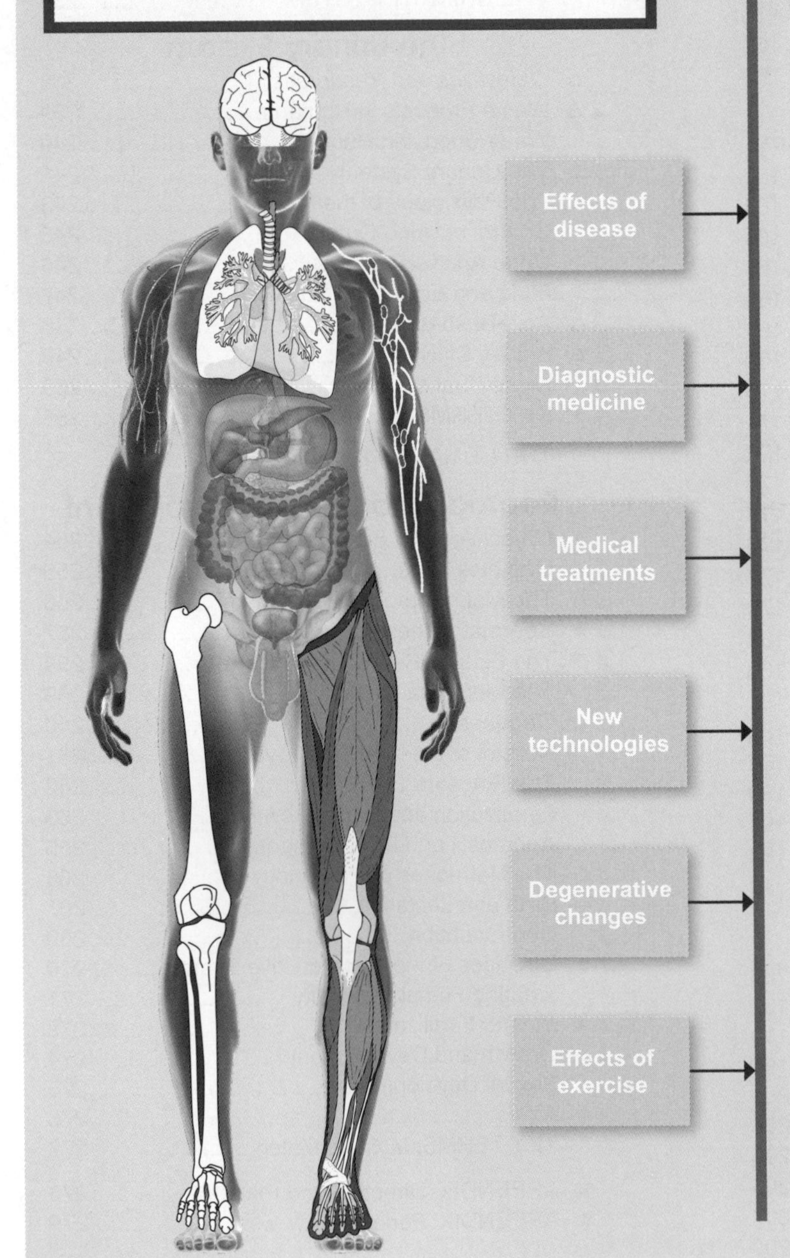

Effects of disease

Diagnostic medicine

Medical treatments

New technologies

Degenerative changes

Effects of exercise

The Integumentary System
The skin and its accessory structures

The Skeletal System
The bones, cartilage, and ligaments

The Muscular System
Smooth, cardiac, and skeletal muscle

The Nervous System
Neurons, glial cells, sensory receptors, and sense organs

The Endocrine System
Endocrine glands and hormones, including the hypothalamus

The Cardiovascular System
The heart, blood vessels, and blood

The Lymphatic System
The lymphoid tissues and organs, the leukocytes

The Digestive System
The digestive tract and accessory organs, including the liver

The Respiratory System
The lungs and associated structures

The Urinary System
The kidneys, bladder, and accessory structures

The Reproductive System
The sex organs and associated structures

Getting The Most From This Resource

This workbook is designed as a resource that will help to increase your understanding of the content and skill requirements of your course, and reinforce and extend the ideas developed by your teacher. This workbook includes many useful features outlined below to help you locate activities and information.

Constructing New Ideas: The Five Es

Engage:	Object, event, or question used to engage students.
Explore:	Objects and phenomena are explored.
Explain:	Student explains their understanding of concepts and processes.
Elaborate:	Student can apply concepts in contexts, and build on or extend their understanding and skills.
Evaluate:	Students assess their knowledge, skills and abilities.

The important key ideas in this chapter. You should have a thorough understanding of the concepts summarized here.

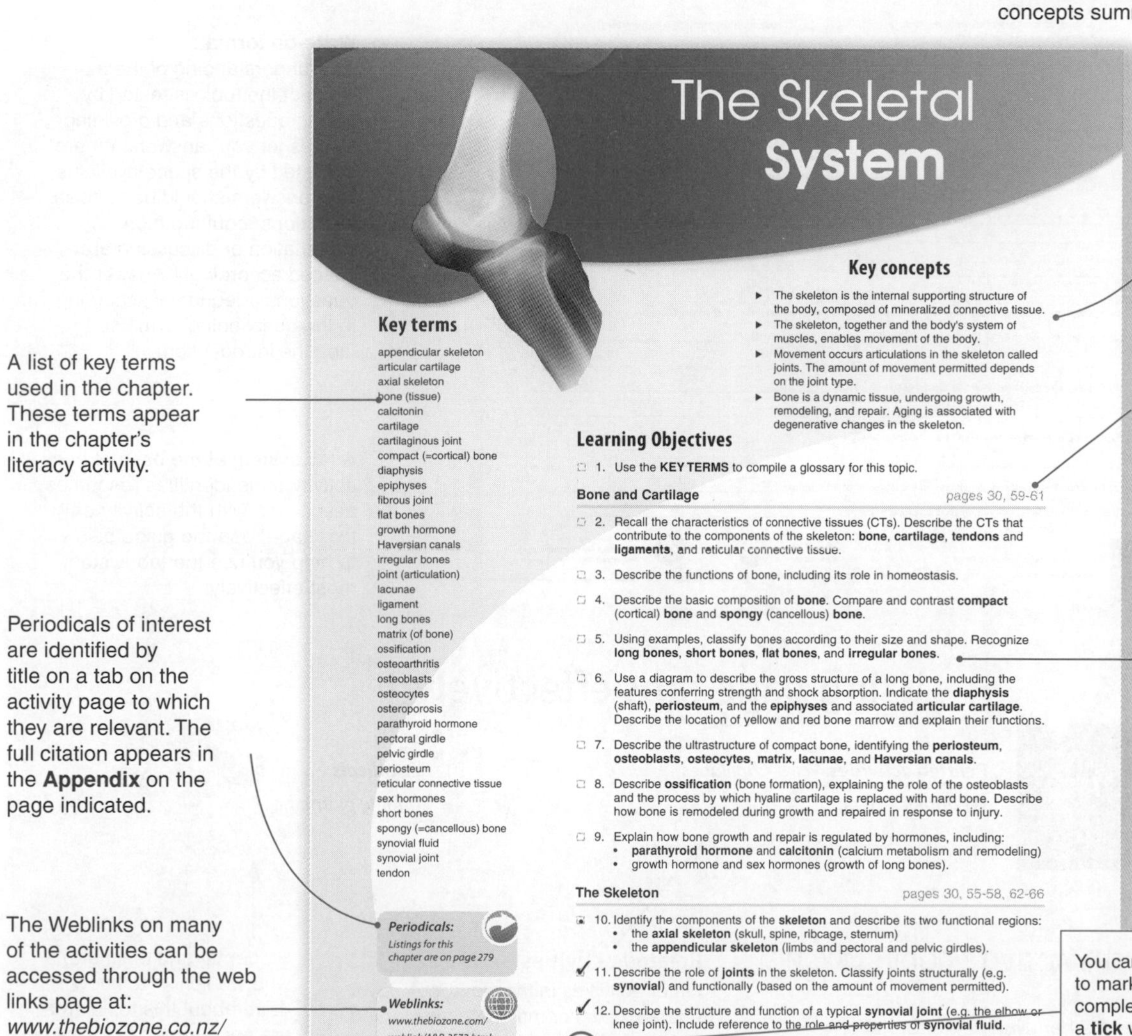

The Skeletal System

Key concepts

- The skeleton is the internal supporting structure of the body, composed of mineralized connective tissue.
- The skeleton, together and the body's system of muscles, enables movement of the body.
- Movement occurs articulations in the skeleton called joints. The amount of movement permitted depends on the joint type.
- Bone is a dynamic tissue, undergoing growth, remodeling, and repair. Aging is associated with degenerative changes in the skeleton.

Key terms

appendicular skeleton
articular cartilage
axial skeleton
bone (tissue)
calcitonin
cartilage
cartilaginous joint
compact (=cortical) bone
diaphysis
epiphyses
fibrous joint
flat bones
growth hormone
Haversian canals
irregular bones
joint (articulation)
lacunae
ligament
long bones
matrix (of bone)
ossification
osteoarthritis
osteoblasts
osteocytes
osteoporosis
parathyroid hormone
pectoral girdle
pelvic girdle
periosteum
reticular connective tissue
sex hormones
short bones
spongy (=cancellous) bone
synovial fluid
synovial joint
tendon

Periodicals: Listings for this chapter are on page 279

Weblinks: www.thebiozone.com/weblink/A&P-3572.html

BIOZONE APP: Student Review Series The Skeletal & Muscular Systems

Learning Objectives

1. Use the **KEY TERMS** to compile a glossary for this topic.

Bone and Cartilage pages 30, 59-61

2. Recall the characteristics of connective tissues (CTs). Describe the CTs that contribute to the components of the skeleton: **bone**, **cartilage**, **tendons** and **ligaments**, and reticular connective tissue.
3. Describe the functions of bone, including its role in homeostasis.
4. Describe the basic composition of **bone**. Compare and contrast **compact** (cortical) **bone** and **spongy** (cancellous) **bone**.
5. Using examples, classify bones according to their size and shape. Recognize **long bones**, **short bones**, **flat bones**, and **irregular bones**.
6. Use a diagram to describe the gross structure of a long bone, including the features conferring strength and shock absorption. Indicate the **diaphysis** (shaft), **periosteum**, and the **epiphyses** and associated **articular cartilage**. Describe the location of yellow and red bone marrow and explain their functions.
7. Describe the ultrastructure of compact bone, identifying the **periosteum**, **osteoblasts**, **osteocytes**, **matrix**, **lacunae**, and **Haversian canals**.
8. Describe **ossification** (bone formation), explaining the role of the osteoblasts and the process by which hyaline cartilage is replaced with hard bone. Describe how bone is remodeled during growth and repaired in response to injury.
9. Explain how bone growth and repair is regulated by hormones, including:
   - **parathyroid hormone** and **calcitonin** (calcium metabolism and remodeling)
   - growth hormone and sex hormones (growth of long bones).

The Skeleton pages 30, 55-58, 62-66

10. Identify the components of the **skeleton** and describe its two functional regions:
    - the **axial skeleton** (skull, spine, ribcage, sternum)
    - the **appendicular skeleton** (limbs and pectoral and pelvic girdles).
11. Describe the role of **joints** in the skeleton. Classify joints structurally (e.g. **synovial**) and functionally (based on the amount of movement permitted).
12. Describe the structure and function of a typical **synovial joint** (e.g. the elbow or knee joint). Include reference to the role and properties of **synovial fluid**.
13. Describe the degenerative changes in the skeleton that occur with increasing age, including a reduction in the rate of bone remodeling, accelerated rates of bone loss, **osteoporosis**, and **osteoarthritis**.

A list of key terms used in the chapter. These terms appear in the chapter's literacy activity.

The page numbers for the activities covering the material in this subsection of objectives.

Periodicals of interest are identified by title on a tab on the activity page to which they are relevant. The full citation appears in the **Appendix** on the page indicated.

The objectives provide a point by point summary of what you should have achieved by the end of the chapter.

The Weblinks on many of the activities can be accessed through the web links page at: *www.thebiozone.co.nz/weblink/AnaPhy-3572.html*
See page 5 for more details.

You can use the check boxes to mark objectives to be completed (a **dot** to be done; a **tick** when completed).

Student Review Series provide color review slides for purchase. Download via the free BIOZONE App, available on the App Store.

Using the Activities

The activities make up most of the content of this book. Your teacher may use the activity pages to introduce a topic for the first time or you may use them to revise ideas already covered by other means. You can use the activities in the classroom, as homework exercises and topic revision, and for self-directed study and personal reference. You may wish to read the related material in your textbook before you attempt the activities or use simpler activities as an introduction to your textbook reading.

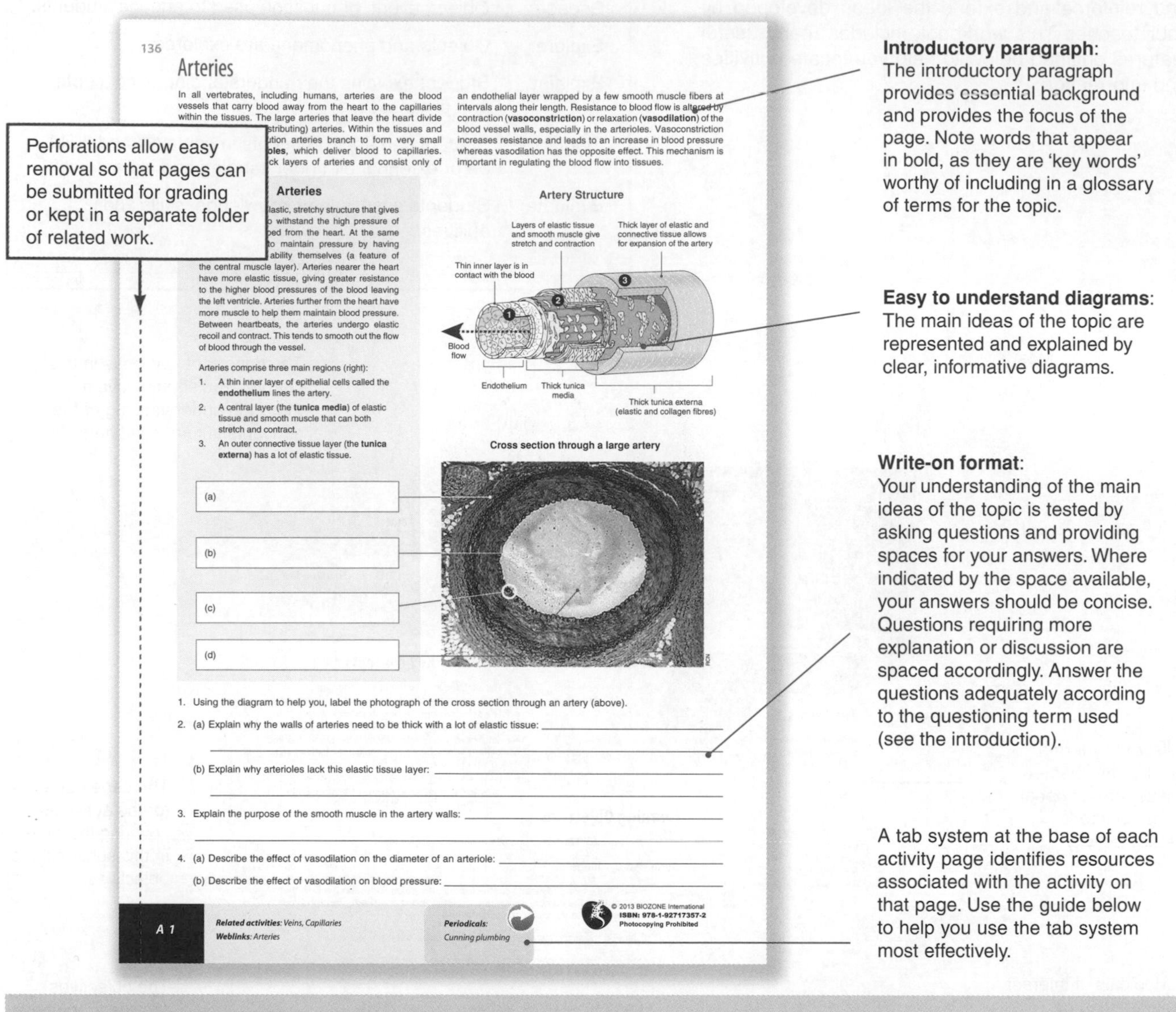

136

Arteries

In all vertebrates, including humans, arteries are the blood vessels that carry blood away from the heart to the capillaries within the tissues. The large arteries that leave the heart divide ... stributing) arteries. Within the tissues and ... ution arteries branch to form very small ... **oles**, which deliver blood to capillaries. ... ck layers of arteries and consist only of an endothelial layer wrapped by a few smooth muscle fibers at intervals along their length. Resistance to blood flow is altered by contraction (**vasoconstriction**) or relaxation (**vasodilation**) of the blood vessel walls, especially in the arterioles. Vasoconstriction increases resistance and leads to an increase in blood pressure whereas vasodilation has the opposite effect. This mechanism is important in regulating the blood flow into tissues.

Arteries

... lastic, stretchy structure that gives ... o withstand the high pressure of ... ped from the heart. At the same ... to maintain pressure by having ... ability themselves (a feature of the central muscle layer). Arteries nearer the heart have more elastic tissue, giving greater resistance to the higher blood pressures of the blood leaving the left ventricle. Arteries further from the heart have more muscle to help them maintain blood pressure. Between heartbeats, the arteries undergo elastic recoil and contract. This tends to smooth out the flow of blood through the vessel.

Arteries comprise three main regions (right):

1. A thin inner layer of epithelial cells called the **endothelium** lines the artery.
2. A central layer (the **tunica media**) of elastic tissue and smooth muscle that can both stretch and contract.
3. An outer connective tissue layer (the **tunica externa**) has a lot of elastic tissue.

Artery Structure

Cross section through a large artery

1. Using the diagram to help you, label the photograph of the cross section through an artery (above).
2. (a) Explain why the walls of arteries need to be thick with a lot of elastic tissue: ______

   (b) Explain why arterioles lack this elastic tissue layer: ______
3. Explain the purpose of the smooth muscle in the artery walls: ______
4. (a) Describe the effect of vasodilation on the diameter of an arteriole: ______

   (b) Describe the effect of vasodilation on blood pressure: ______

A 1 | *Related activities: Veins, Capillaries* | *Weblinks: Arteries* | *Periodicals: Cunning plumbing* | © 2013 BIOZONE International ISBN: 978-1-92717357-2 Photocopying Prohibited

Perforations allow easy removal so that pages can be submitted for grading or kept in a separate folder of related work.

Introductory paragraph: The introductory paragraph provides essential background and provides the focus of the page. Note words that appear in bold, as they are 'key words' worthy of including in a glossary of terms for the topic.

Easy to understand diagrams: The main ideas of the topic are represented and explained by clear, informative diagrams.

Write-on format: Your understanding of the main ideas of the topic is tested by asking questions and providing spaces for your answers. Where indicated by the space available, your answers should be concise. Questions requiring more explanation or discussion are spaced accordingly. Answer the questions adequately according to the questioning term used (see the introduction).

A tab system at the base of each activity page identifies resources associated with the activity on that page. Use the guide below to help you use the tab system most effectively.

Using page tabs effectively

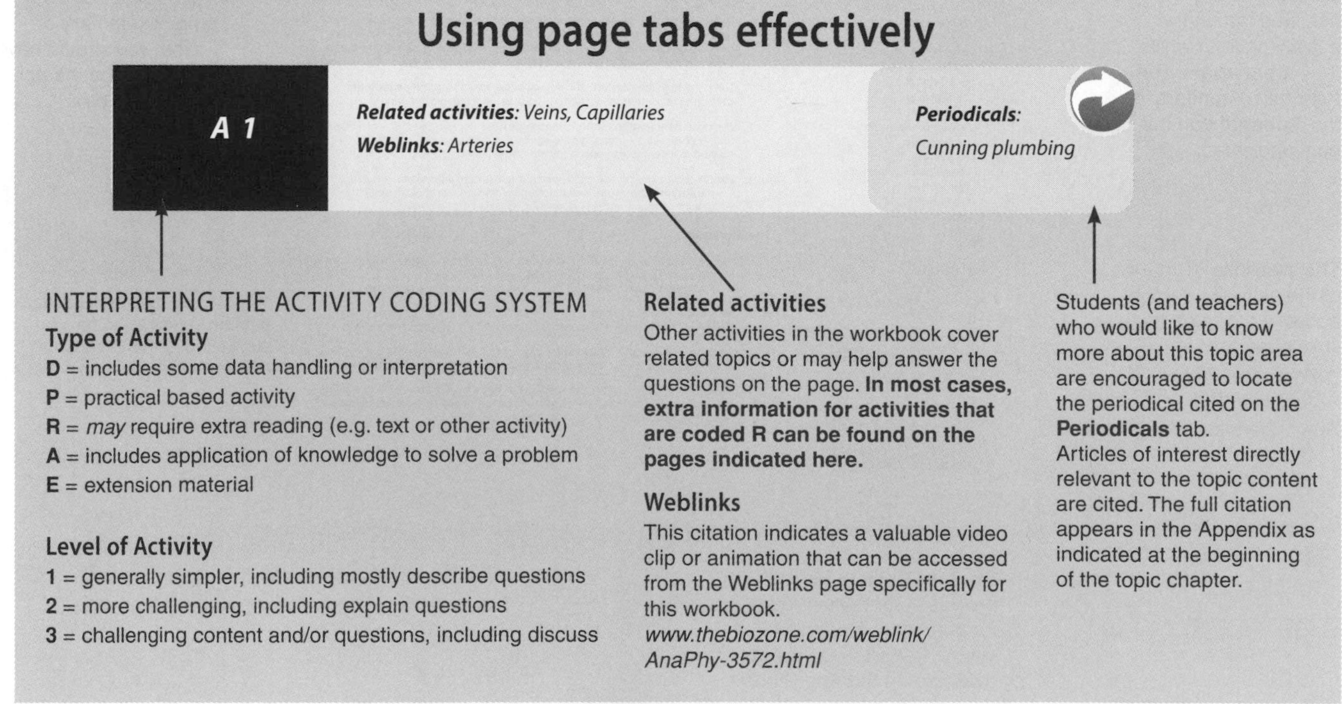

INTERPRETING THE ACTIVITY CODING SYSTEM

Type of Activity

D = includes some data handling or interpretation
P = practical based activity
R = *may* require extra reading (e.g. text or other activity)
A = includes application of knowledge to solve a problem
E = extension material

Level of Activity

1 = generally simpler, including mostly describe questions
2 = more challenging, including explain questions
3 = challenging content and/or questions, including discuss

Related activities

Other activities in the workbook cover related topics or may help answer the questions on the page. **In most cases, extra information for activities that are coded R can be found on the pages indicated here.**

Weblinks

This citation indicates a valuable video clip or animation that can be accessed from the Weblinks page specifically for this workbook. *www.thebiozone.com/weblink/AnaPhy-3572.html*

Students (and teachers) who would like to know more about this topic area are encouraged to locate the periodical cited on the **Periodicals** tab. Articles of interest directly relevant to the topic content are cited. The full citation appears in the Appendix as indicated at the beginning of the topic chapter.

Chapter Summary and Contexts

Each chapter (or in some cases two chapters) in *Anatomy & Physiology* is preceded by a two page summary of homeostatic interactions and contextual examples. The first of the two pages provides an overview of how the specific body system (in this case, the respiratory system) interacts with the other body systems to maintain homeostasis. A lower panel summarizes the general functions of the system (for example, the respiratory system's functional role is in gas exchange). Homeostasis is a unifying theme throughout *Anatomy & Physiology*. The second of the two pages continues this theme by showing how the selected body system can be affected by disease, aging, and exercise, and how our current medical knowledge can be applied to disorders of homeostasis. These contexts provide a relevant framework for understanding the subject matter.

A Contextual Framework

Interacting Systems:
The purpose of this page is to summarize the interactions of the body system under study (in this case the respiratory system) with all other body systems in turn. This summary describes the way in which systems work together to maintain homeostasis.

The intersecting regions of the center panel of the context map highlight topics of focus within each context. These are specifically addressed within the workbook.

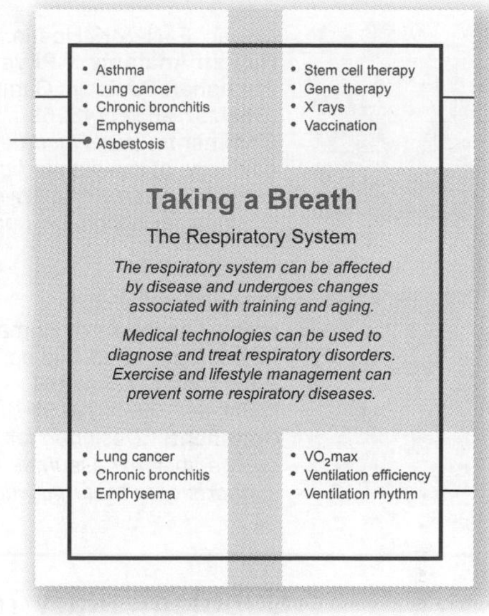

Most systems are treated singly, although those systems that operate very closely (e.g. muscular and skeletal) are mapped together.

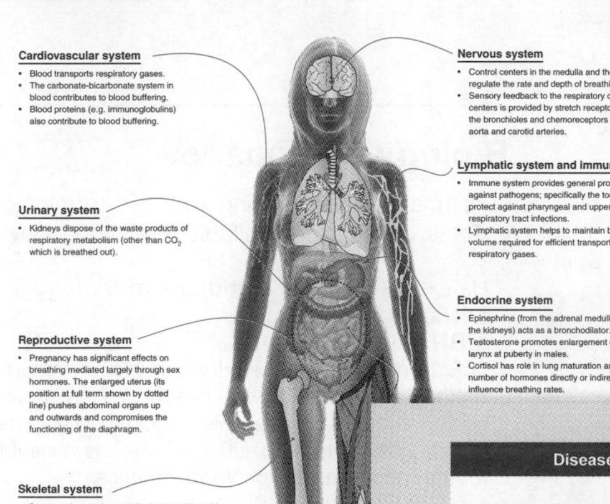

Disease:
A summary of some of the diseases affecting the body system. These provide a good context for examining departures from homeostasis.

Medicine and Technology:
A summary of how medicine and technology are used to study the chosen body system, and how new technologies can be used to diagnose and treat specific diseases. An awareness of how technology is applied in a medical context is essential to the study of anatomy and physiology.

Four-panel Focus:
Each of the four panels on this page focuses on one context to which you can apply your knowledge and understanding of the topic's content.

Exercise:
Exercise has different effects on different body systems. Some of the physiological effects of exercise to be considered in the workbook are summarized here.

The Effects of Aging:
Degenerative changes in the body system are summarized in this panel. The effects of aging provide another context for considering disruptions to homeostasis.

Resources Information

Your set textbook should always be a starting point for information, but there are also many other resources available. A list of readily available resources is provided below. Please note that our listing of any product in this workbook does not denote Biozone's endorsement of it.

Comprehensive Texts

We have listed two commonly used texts for anatomy and physiology below. There are many other excellent texts available also.

Marieb, E. N. & K. Hoehn, 2012
Human Anatomy & Physiology, 9 ed., 1264 pp.
Publisher: Benjamin Cummings.
ISBN: 978-0321743268
Comments: *Well illustrated and clearly explained coverage of the human body. 'The Essentials of Human Anatomy and Physiology', also by Marieb is a popular option for a one-semester course.*

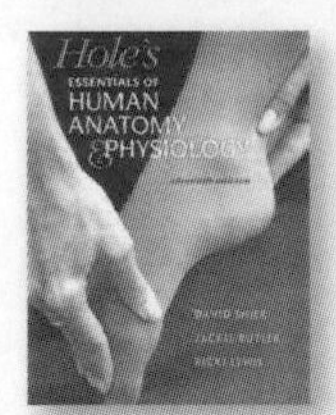

Shier, D.N *et.al*, 2012
Hole's Essentials of Human Anatomy & Physiology, 11 ed. 640 pp.
Publisher: McGraw-Hill
ISBN: 978-0073378152
Comments: *Designed for a one-semester A&P course, this text assumes no prior knowledge and supports core material with clinical examples.*

Supplementary Texts

Morton, D. & J.W. Perry, 1997
Photo Atlas for Anatomy & Physiology, 160 p.
Publisher: Brooks Cole. **ISBN**: 0-534-51716-1
Comments: *An excellent photographic guide to lab work. Includes a good section on histology, with representations of tissue types and clear images for comparison with textbook drawings.*

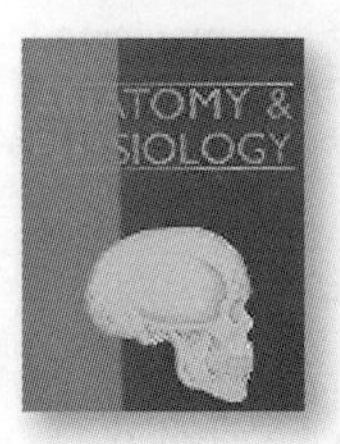

Rowett, H.G.Q, 1999
Basic Anatomy & Physiology, 4 ed. 132 pp.
Publisher: John Murray
ISBN: 0-7195-8592-9
Comments: *A revision of a well established reference book for the basics of human anatomy and physiology. Accurate coverage of required AS/A2 content with clear, informative diagrams.*

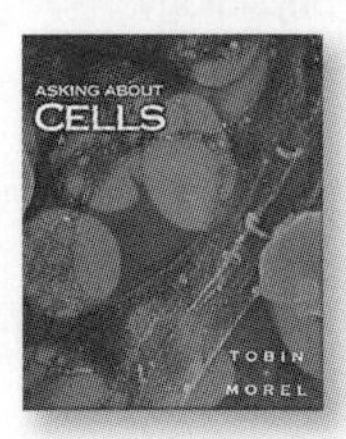

Tobin, A.J. and R.E Morel, 1997
Asking About Cells, 698 pp (paperback)
Publisher: Thomson Brooks/Cole
ISBN: 0-030-98018-6
Comments: *An introduction to cell biology, cellular processes and specialisation, DNA and gene expression, and inheritance. The focus is on presenting material through inquiry.*

Periodicals, Magazines, & Journals

Biological Sciences Review: *An informative quarterly publication for biology students.* Enquiries: **UK**: Philip Allan Publishers **Tel**: 01869 338652 **Fax**: 01869 338803 **E-mail**: sales@philipallan.co.uk **Australasia**: **Tel**: 08 8278 5916, **E-mail**: rjmorton@adelaide.on.net

New Scientist: *Widely available weekly magazine with research summaries and features.* Enquiries: Reed Business Information Ltd, 51 Wardour St. London WIV 4BN **Tel**: (UK and intl):+44 (0) 1444 475636 **E-mail**: ns.subs@qss-uk.com *or subscribe from their web site.*

Scientific American: *A monthly magazine containing specialist features. Articles range in level of reading difficulty and assumed knowledge.* Subscription enquiries: 415 Madison Ave. New York. NY10017-1111 **Tel**: (outside North America): 515-247-7631 **Tel**: (US& Canada): 800-333-1199

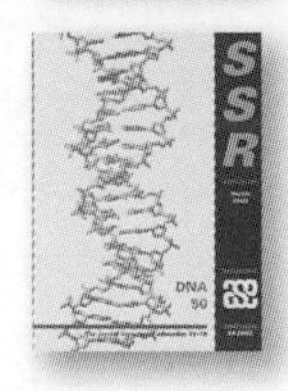

School Science Review: *A quarterly journal which includes articles, reviews, and news on current research and curriculum development. Free to Ordinary Members of the ASE or available on subscription.* Enquiries: **Tel**: 01707 28300 **Email**: info@ase.org.uk *or visit their web site.*

The American Biology Teacher: *The peer-reviewed journal of the NABT. Published nine times a year and containing information and activities relevant to biology teachers.* Contact: NABT, 12030 Sunrise Valley Drive, #110, Reston, VA 20191-3409 **Web**: www.nabt.org

Biology Dictionaries

A good dictionary is useful when dealing with the terminology of human biology. Some of the titles available are listed below.

Hale, W.G. **Collins: Dictionary of Biology** 4 ed. 2005, 528 pp. Collins.
ISBN: 0-00-720734-4.
Updated to take in the latest developments in biology and now internet-linked. (§ This latest edition is currently available only in the UK. The earlier edition, ISBN: 0-00-714709-0, is available though amazon.com in North America).

Henderson, I.F, W.D. Henderson, and E. Lawrence. **Henderson's Dictionary of Biological Terms**, 1999, 736 pp. Prentice Hall.
ISBN: 0582414989
An updated edition, rewritten for clarity, and reorganized for ease of use. An essential reference and the dictionary of choice for many.

Thain, M. **Penguin Dictionary of Human Biology**, 2009, 768 pp. Penguin Global.
ISBN: 978-0140514827
An essential dictionary for those studying human biology, medicine, or nursing. With a focus on human anatomy and physiology, this dictionary (in paperback) would be a good choice for students of the life sciences.

Using BIOZONE's Website

Access the **BIOLINKS** database of web sites directly from the homepage of our new website. Biolinks is organized into easy-to-use sub-sections relating to general areas of interest. It's a great way to quickly find out more on topics of interest.

Contact us with questions, feedback, ideas, and critical commentary. We welcome your input.

Use Google to search for websites of interest. The more precise your search words are, the better the list of results. Be specific, e.g. "biotechnology medicine DNA uses", rather than "biotechnology".

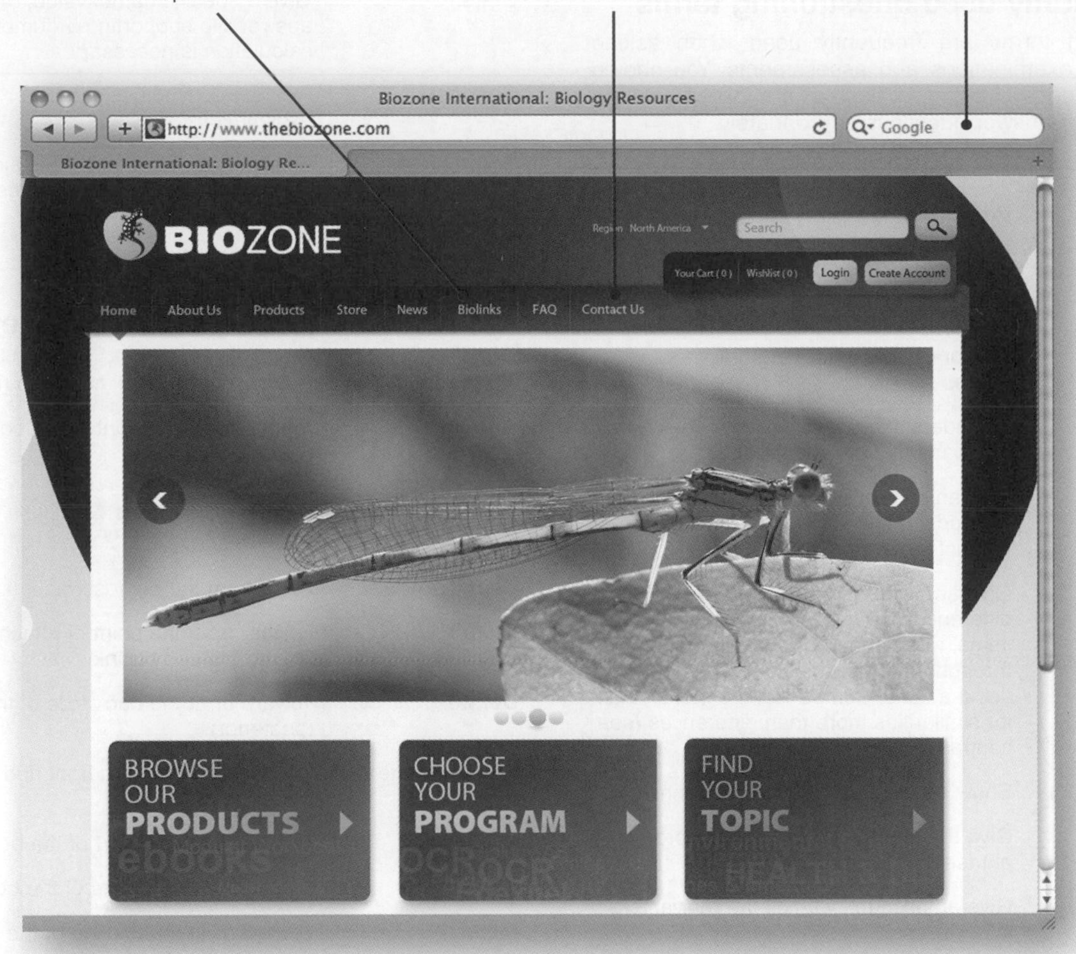

Weblinks: www.thebiozone.com/weblink/AnaPhy-3572.html

BOOKMARK WEBLINKS BY TYPING IN THE ADDRESS: IT IS NOT ACCESSIBLE DIRECTLY FROM BIOZONE'S WEBSITE

Throughout this workbook, some pages make reference to websites that have particular relevance to the activity by providing an explanatory animation or video clip. They are easy to use and a very useful supplement to the activity.

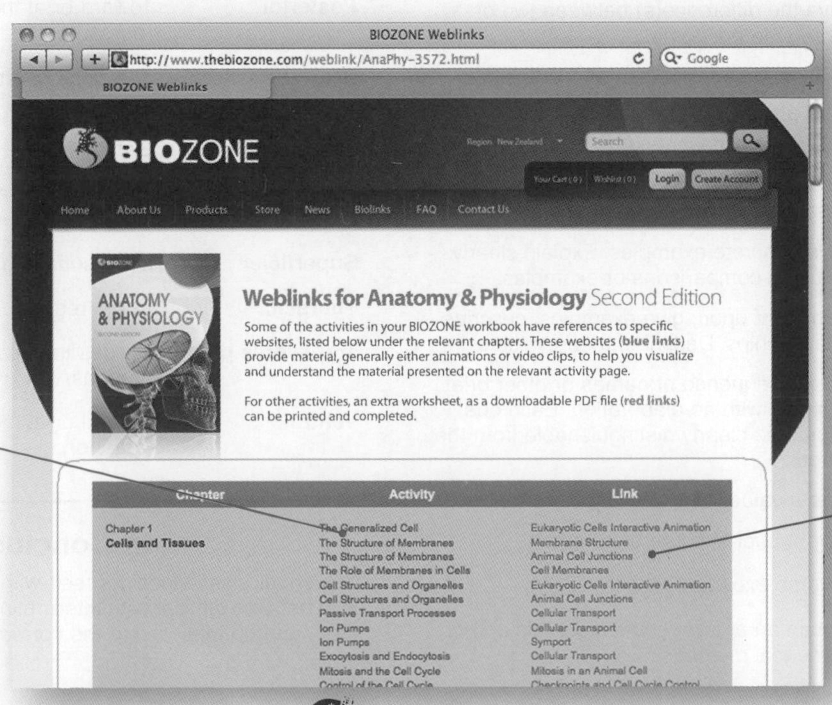

Activity reference
The activity on which the weblink is cited.

Weblink
Provides a link to an **external web site** with supporting information for the activity. If there are any additional activities, these are also listed here as links.

Terms in Anatomy and Physiology

The study of anatomy and physiology requires a good mental map of the body's different regions and a sound understanding of the terms used to describe the location of structures on the body. As well as this, when answering questions, you should have a good knowledge of what specific questioning terms mean and phrase your answers accordingly. The following short guide, which is by no means exhaustive, lists some of the basic questioning terms, and directional and regional terms that you will come across when you study anatomy and physiology. Refer to as frequently as you see fit.

Commonly used questioning terms

The following terms are frequently used when asking questions in examinations and assessments. You should have a clear understanding of each of the following terms and use this to answer questions appropriately.

Account for: Provide a satisfactory explanation or reason for an observation.

Analyze: Interpret data to reach stated conclusions.

Annotate: Add **brief** notes to a diagram, drawing, or graph.

Appreciate: To understand the meaning or relevance of a particular situation.

Calculate: Find an answer using mathematical methods. Show the working unless instructed not to.

Compare: Give an account of similarities and differences between two or more items, referring to both (or all) of them throughout. Comparisons can be given using a table. Comparisons generally ask for similarities more than differences (see contrast).

Contrast: Show differences. Set in opposition.

Define: Give the precise meaning of a word or phrase as concisely as possible.

Describe: Give an account, with all the relevant information.

Discuss: Give an account including, where possible, a range of arguments, assessments of the relative importance of various factors, or comparison of alternative hypotheses.

Distinguish: Give the difference(s) between two or more different items.

Evaluate: Assess the implications and limitations.

Explain: Give a clear account including causes, reasons, or mechanisms.

Identify: Find an answer from a number of possibilities.

Illustrate: Give concrete examples. Explain clearly by using comparisons or examples.

Interpret: Comment upon, give examples, describe relationships. Describe, then evaluate.

List: Give a sequence of names or other brief answers with no elaboration. Each one should be clearly distinguishable from the others.

Measure: Find a value for a quantity.

Outline: Give a brief account or summary.

Predict: Give an expected result.

Solve: Obtain an answer using algebraic and/or numerical methods.

State: Give a specific name, value, or other answer. No supporting argument or calculation is necessary.

Suggest: Propose a hypothesis or other possible explanation.

Summarize: Give a brief, condensed account. Include conclusions and avoid unnecessary details.

Commonly used anatomical terms

Abdominal: Anterior body trunk inferior to ribs.

Anterior: Toward or at the front of the body, (ventral).

Cephalic: Head region.

Cervical: Neck region.

Deep: Away from the body surface, (internal).

Distal: Farther from the point of attachment of a limb to the body trunk.

Dorsal: Toward or at the back side of the body, (posterior).

Frontal plane: Divides the body into front and back portions.

Inferior: Toward the lower part of the body.

Lateral: Away from the midline of the body.

Medial: Toward or at the midline of the body.

Median plane: A vertical plane through the midline of the body; divides the body into right and left halves, (midsagittal).

Pelvic: Area overlying the pelvis.

Posterior: Toward or at the back side of the body, (dorsal).

Proximal: The point of attachment of a limb to the body trunk.

Pubic: Genital region.

Superior: Toward the head or upper part of the body.

Superficial: Body surface, (external).

Thoracic: Chest region.

Transverse plane: Divides the body into top and bottom portions.

Ventral: Toward or at the front of the body, (anterior).

In Conclusion

You should familiarize yourself with this list of terms. The aim is to become familiar with interpreting questions and answering them appropriately using the correct anatomical terminology.

Cells and **Tissues**

Key terms

active transport
anaphase
cell
cellular differentiation
chromosome
connective tissue
cytokinesis
diffusion
DNA
endocytosis
epithelial tissue
exocytosis
facilitated diffusion
hypertonic
hypotonic
interphase
ion pumps
isotonic
light microscope
metaphase
mitosis
muscle tissue
nervous tissue
organ
organ system
organelle
osmosis
partially permeable membrane
passive transport
phospholipids
prophase
surface area: volume ratio
telophase
tissue

Periodicals:
Listings for this chapter are on page 279

Weblinks:
www.thebiozone.com/weblink/AnaPhy-3572.html

BIOZONE APP:
Student Review Series
Cells and Tissues

Key concepts

- Cells are the basic units of life. Microscopy can be used to understand cellular structure.
- Cellular metabolism depends on the transport of substances across cellular membranes.
- Cell size is limited by surface area to volume ratio.
- New cells arise through cell division.
- Cellular diversity arises through specialization from stem cell progenitors.

Learning Objectives

☐ 1. Use the **KEY TERMS** to compile a glossary for this topic.

Features of Cells

pages 8-17

☐ 2. Describe a **generalized cell**, identifying and describing the following **organelles**:
- plasma membrane, nucleus, nuclear envelope, nucleolus
- mitochondria, rough/smooth endoplasmic reticulum, ribosomes,
- Golgi apparatus, lysosomes, peroxisomes
- cytoplasm, cytoskeleton (of microtubules), centrioles, cilia (if present)

☐ 3. Recognize features of a generalized cell in light and electron micrographs.

☐ 4. Describe the interrelationship between the organelles involved in the production and secretion of proteins.

☐ 5. PRACTICAL: Using prepared slides of animal tissues, demonstrate an ability to correctly use a **light microscope** to locate material and focus images.

☐ 6. Describe the **fluid mosaic model** of membrane structure, including the role of **phospholipids**, cholesterol, glycolipids, proteins, and glycoproteins. Describe the functions of membranes (including the plasma membrane) in cells.

Cellular Transport

pages 19-22

☐ 7. Explain **passive transport** across membranes by **diffusion** and **osmosis**. Explain the terms **hypotonic**, **isotonic**, and **hypertonic** with reference to water fluxes in cells.

☐ 8. Describe **facilitated diffusion** (facilitated transport) involving carrier or channel proteins. Identify when and where facilitated diffusion might occur in a cell.

☐ 9. Distinguish between passive and **active transport** mechanisms. Using examples, explain active transport in cells, including **ion pumps**, **endocytosis**, and **exocytosis**.

☐ 10. Explain the role of **surface area: volume ratio** in limiting the size of cells. Describe factors affecting the rates of transport processes in cells.

Cell Division and Cellular Differentiation

pages 23-32

☐ 11. Describe the cell cycle in eukaryotes such as humans. Include reference to **mitosis**, **growth** (G1, G2), and **DNA replication** (S). Explain how the cell cycle is regulated.

☐ 12. Recall the general structure of nucleic acids. Describe the structure and role of the **nucleus** and its contents, including the **DNA** and **chromosomes**.

☐ 13. Recognize phases in mitosis: prophase, metaphase, anaphase, and telophase. Define **cytokinesis**, and distinguish between nuclear division and division of the cytoplasm. .

☐ 14. Describe the role of mitosis in growth and repair. Explain how carcinogens can upset the normal controls regulating cell division.

☐ 15. Describe how a zygote undergoes mitotic cell division and **differentiation** to produce an adult. Giving examples, explain what is meant by a **specialized cell**.

☐ 16. Recognize the hierarchy of organization in multicellular organisms (including humans). Appreciate the role of cooperation between **cells**, **tissues**, **organs**, and **organ systems** in the structure and function of human body.

☐ 17. With reference to specific examples, explain how cells are organized into tissues. Recognize structural and functional diversity in the cells that make up human tissues. Recognize the characteristic features and functional roles of the four main tissue types in humans.

The Biochemical Nature of the Cell

Water is the main component of organisms, and provides an equable environment in which metabolic reactions can occur. Apart from water, most other substances in cells are compounds of **carbon**, **hydrogen**, **oxygen**, and **nitrogen**. The combination of carbon atoms with the atoms of other elements provides a huge variety of molecular structures, collectively called **organic molecules**. The organic molecules that make up living things can be grouped into four broad classes: carbohydrates, lipids, proteins, and nucleic acids. In addition, a small number of **elements** and **inorganic ions** are also essential for life as components of larger molecules or extracellular fluids.

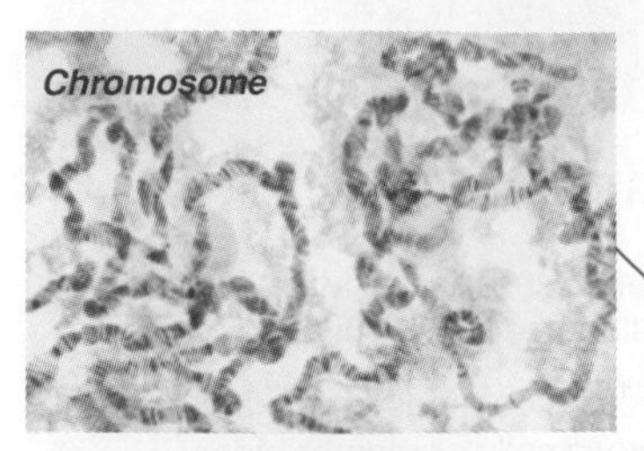

Nucleotides and nucleic acids
Nucleic acids (DNA and RNA) encode information for the construction and functioning of an organism. Most of a eukaryotic cell's DNA is found in the nucleus. The nucleotide, ATP, is the energy currency of the cell.

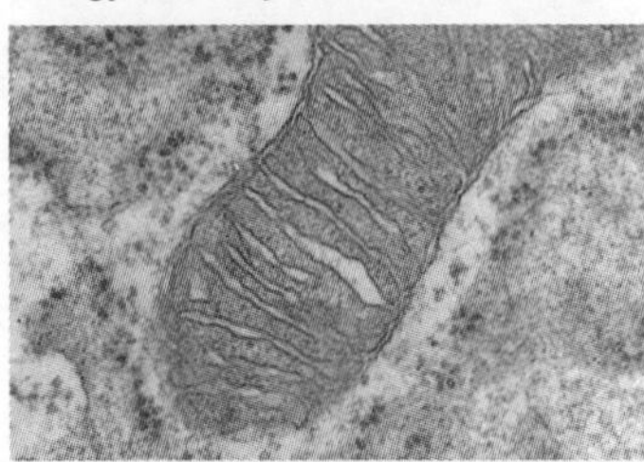

Lipids provide insulation and a concentrated source of energy. Phospholipids are a major component of **cellular membranes**, including the membranes of organelles.
Above: Mitochondrion

Water is a major component of cells: many substances dissolve in it, metabolic reactions occur in it, and it provides support and turgor.

Carbohydrates act as energy stores (e.g. glycogen, arrowed above), are involved in cellular recognition, cell signalling, and membrane stability (as glycoproteins and glycolipids).

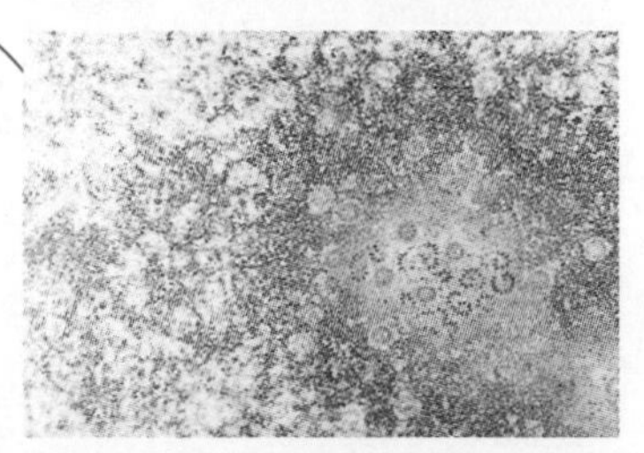

Proteins may be catalytic (enzymes), structural (e.g. collagen in skin, proteins in ribosomes), or they may be involved in movement, message signaling, internal defense and transport, or storage.
Above: ribosomes in translation

Certain elements and inorganic ions are important for the structure and metabolism of all living organisms. An ion is simply an atom (or group of atoms) that has gained or lost one or more electrons. Many of these ions are soluble in water. Some of these elements and inorganic ions required by organisms and their biological roles are listed in the table on the right.

Ion or Element	Name	Example of Biological Roles
Ca^{2+}	Calcium	Component of bones and teeth, required for muscle contraction.
NO_3^-	Nitrate	Component of amino acids.
Fe^{2+}	Iron (II)	Component of hemoglobin and cytochromes.
S	Sulfur	Component of the thiol (-SH) functional group and part of many organic molecules, e.g. coenzyme A and some amino acids.
P	as phosphate	Component of phospholipids, and nucleotides, including ATP.
Na^+	Sodium	Component of extracellular fluid and needed for nerve function.
K^+	Potassium	Important intracellular ion, needed for heart and nerve function.
Cl^-	Chloride	Component of extracellular fluid in multicellular organisms.

1. Describe the biologically important roles of each of the following molecules:

   (a) Lipids: ______________________

   (b) Carbohydrates: ______________________

   (c) Proteins: ______________________

   (d) Nucleic acids: ______________________

2. Identify the most four most common chemical elements in living organisms: ______________________

3. Giving examples, describe the roles of some of the less common elements in living organisms: ______________________

____

____

ISBN: 978-1-92717357-2

Related activities: Basic Cell Structure

Basic Cell Structure

Cells have a similar basic structure, although they may vary tremendously in size, shape, and function. Features common to almost all **eukaryotic cells** include the **nucleus** (often near the cell's center), surrounded by a watery **cytoplasm**, which is itself enclosed by the **plasma membrane**. Animal cells do not have a regular shape, and some (such as phagocytes) are quite mobile. The diagram below shows the ultrastructure of a **liver cell** (hepatocyte). It contains organelles common to most relatively unspecialized human cells. Hepatocytes make up 70-80% of the liver's mass. They are metabolically active, with a large central nucleus, many mitochondria, and large amounts of rough endoplasmic reticulum. Thin, cellular extensions called microvilli increase surface area of the cell, increasing its capacity for absorbing nutrients.

The Structure of a Liver Cell

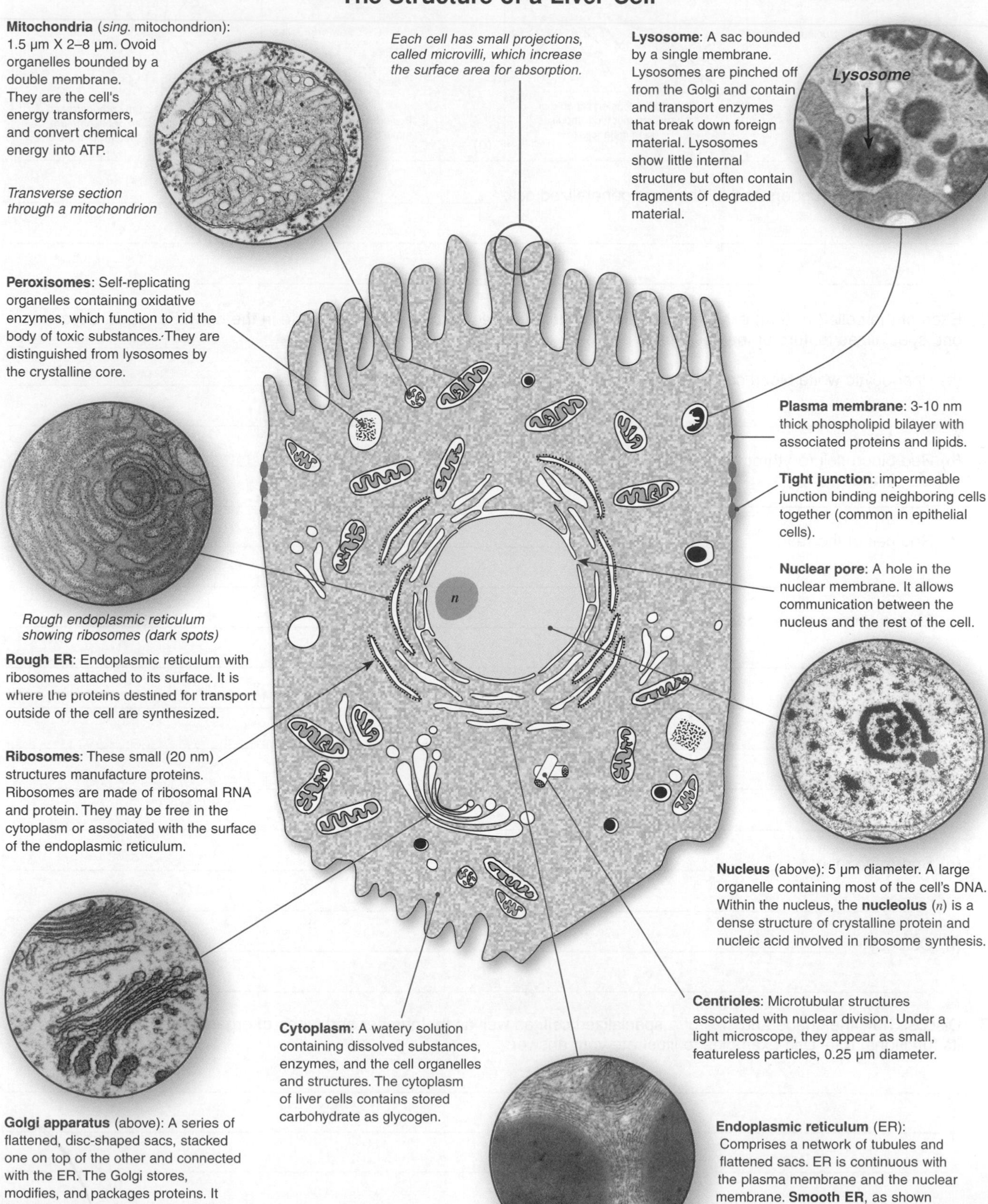

ISBN: 978-1-92717357-2

Related activities: *Cell Structures and Organelles, Cell Processes*
Weblinks: *Eukaryotic Cells Interactive Animation*

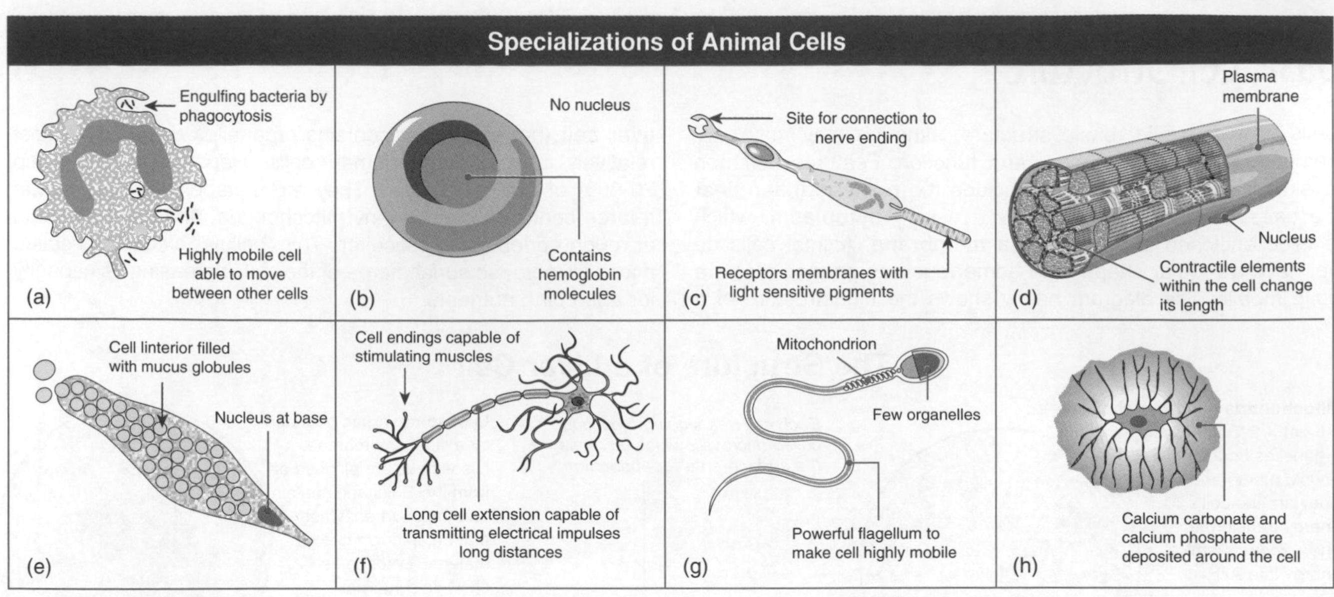

1. Explain what you understand by the term generalized cell: ______

______

______

2. Each of the cells (a) to (h) exhibits **specialized features** specific to its **functional role** in the body. For each, describe one specialized feature of the cell and its purpose:

(a) Phagocytic white blood cell: ______

______

(b) Red blood cell (erythrocyte) ______

______

(c) Rod cell of the retina: ______

______

(d) Skeletal muscle fiber (part of): ______

______

(e) Intestinal goblet cell: ______

______

(f) Motor neuron: ______

______

(g) Spermatozoon: ______

______

(h) Osteocyte: ______

______

3. Discuss how the shape and size of a specialized cell, as well as the number and types of organelles it has, is related to its functional role. Use examples to illustrate your answer:

______

______

______

______

______

ISBN: 978-1-92717357-2

The Structure of Membranes

All cells have a **plasma membrane** forming the outer limit of the cell. Cellular membranes are also found inside eukaryotic cells as part of **organelles**, such as the endoplasmic reticulum. Present day knowledge of membrane structure has been built up as a result of many observations and experiments. The now-accepted model of membrane structure is the **fluid-mosaic model** (below). The plasma membrane is more than just a passive envelope; it is a dynamic structure actively involved in cellular activities. Specializations of the plasma membrane, including microvilli and membrane junctions (e.g. desmosomes and tight junctions), are particularly numerous in epithelial cells, which line hollow organs such as the small intestine.

The Fluid Mosaic Model of Membrane Structure

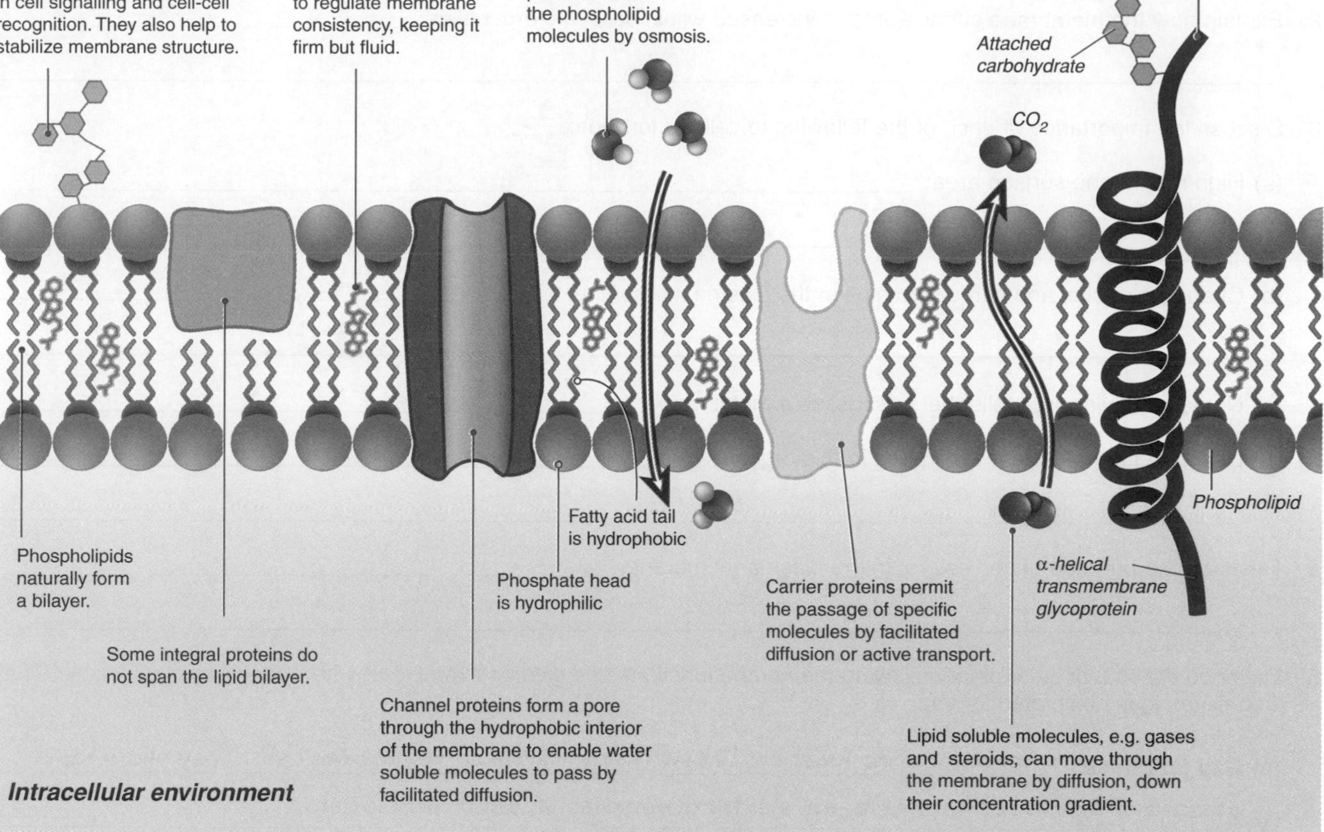

Based on a diagram in Biol. Sci. Review, Nov. 2009, pp. 20-21

Membrane Specializations

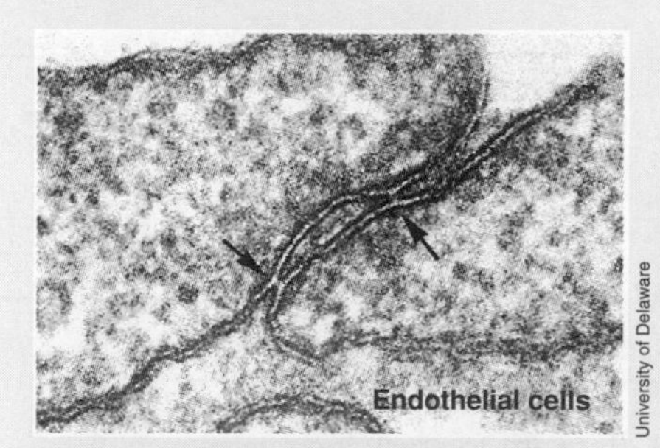

University of Delaware

Tight junctions bind the membranes of neighboring cells together to form a virtually impermeable barrier to fluid. Tight junctions prevent molecules passing through the spaces between cells.

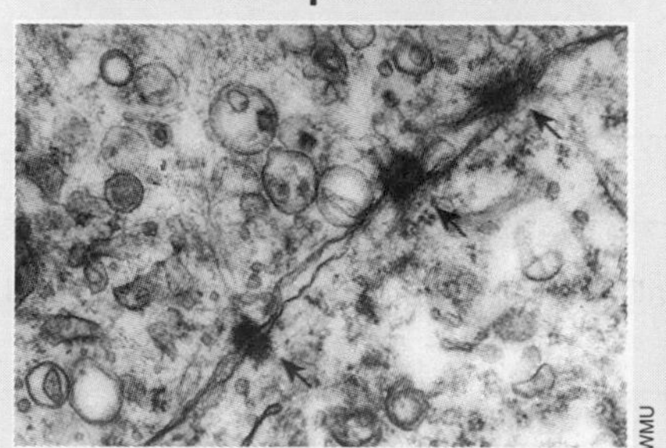

WMU

Desmosomes (arrowed) are anchoring junctions that allow cell-to-cell adhesion. Desmosomes help to resist shearing forces in tissues subjected to mechanical stress (such as skin cells).

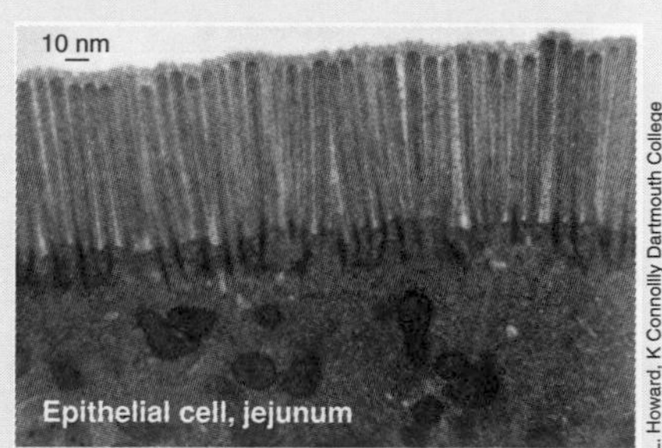

L. Howard, K Connolly Dartmouth College

Microvilli are microscopic protrusions of the plasma membrane that increase the surface area of cells. Microvilli are involved in a wide variety of functions, including absorption (e.g. in the intestine).

1. (a) Explain how phospholipids organize themselves into a bilayer in an aqueous environment:

ISBN: 978-1-92717357-2

Periodicals: *Border control,* *The fluid mosaic model*

Related activities: *The Role of Membranes in Cells*
Weblinks: *Membrane Structure, Animal Cell Junctions*

(b) Explain how the fluid mosaic model accounts for the observed properties of cellular membranes:

Molecular model showing how phospholipid molecules naturally orientate to form a bilayer.

2. Explain how the membrane surface area is increased within cells and organelles:

3. Discuss the importance of each of the following to cellular function:

(a) High membrane surface area:

(b) Channel proteins and carrier proteins in the plasma membrane:

4. (a) Name a cellular organelle that possesses a membrane:

(b) Describe the membrane's purpose in this organelle:

5. Describe the purpose of cholesterol in the plasma membrane:

6. Describe the role of each of the following membrane junctions and give an example of where they commonly occur. The first example is completed for you:

(a) **Gap junctions**: Communicating junctions linking the cytoplasm of neighboring cells. They allow rapid passage of signals between cells, e.g. electrical messages in cardiac muscle cells.

(b) **Tight junctions**:

(c) **Desmosomes**:

7. Explain why tight junctions are especially abundant in epithlelial cells, e.g. in the skin and intestine:

8. Use the symbol for a phospholipid molecule (below) to draw a **simple labelled diagram** to show the structure of a plasma membrane (include features such as lipid bilayer and various kinds of proteins):

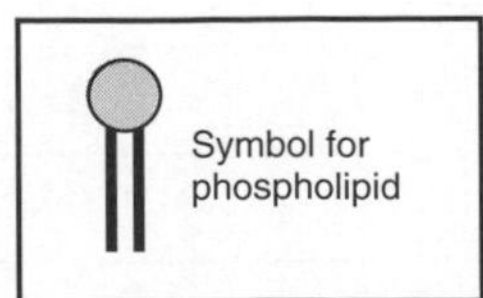

ISBN: 978-1-92717357-2

The Role of Membranes in Cells

Many of the important structures and organelles in cells are composed of, or are enclosed by, membranes. These include the endoplasmic reticulum, mitochondria, nucleus, Golgi apparatus, lysosomes, peroxisomes, and the plasma membrane itself. All membranes within eukaryotic cells share the same basic structure as the plasma membrane that encloses the entire cell. They perform a number of critical functions in the cell: serving to compartmentalize regions of different function within the cell, controlling the entry and exit of substances, and fulfilling a role in recognition and communication between cells. Some of these roles are described below and electron micrographs of the organelles involved are on the following page.

What Membranes Do In a Cell

Isolation of enzymes
Membrane-bound lysosomes contain enzymes for the destruction of wastes and foreign material. Peroxisomes are the site for destruction of the toxic and reactive molecule, hydrogen peroxide (formed as a result of some cellular reactions).

Role in lipid synthesis
The smooth ER is the site of lipid and steroid synthesis.

Containment of DNA
The nucleus is surrounded by a nuclear envelope of two membranes, forming a separate compartment for the cell's genetic material.

Role in protein and membrane synthesis
Some protein synthesis occurs on free ribosomes, but much occurs on membrane-bound ribosomes on the rough endoplasmic reticulum. Here, the protein is synthesized directly into the space within the ER membranes. The rough ER is also involved in membrane synthesis, growing in place by adding proteins and phospholipids.

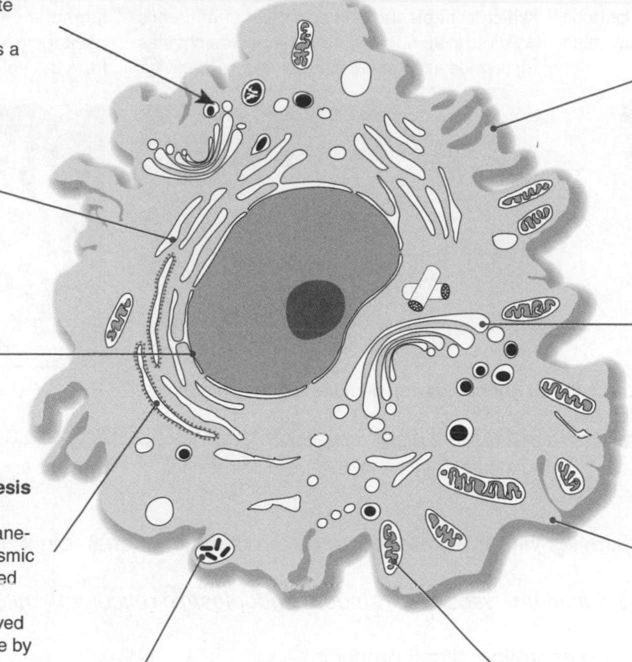

Cell communication and recognition
The proteins embedded in the membrane act as receptor molecules for hormones and neurotransmitters. Glycoproteins and glycolipids stabilize the plasma membrane and act as cell identity markers, helping cells to organize themselves into tissues, and enabling foreign cells to be recognized.

Packaging and secretion
The Golgi apparatus is a specialized membrane-bound organelle which produces lysosomes and compartmentalizes the modification, packaging and secretion of substances such as proteins and hormones.

Transport processes
Channel and carrier proteins are involved in selective transport across the plasma membrane. The level of cholesterol in the membrane influences permeability and transport functions.

Entry and export of substances
The plasma membrane may take up fluid or solid material and form membrane-bound vesicles (or larger vacuoles) within the cell. Membrane-bound transport vesicles move substances to the inner surface of the cell where they can be exported from the cell by exocytosis.

Energy transfer
The reactions of cellular respiration (and photosynthesis in plants) take place in the membrane-bound energy transfer systems occurring in mitochondria and chloroplasts respectively. See the example explained below.

Cisternae (membrane-bound sacs)

Vesicles transporting material to the Golgi

Vesicles leaving the Golgi

Compartmentation within Membranes

Membranes play an important role in separating regions within the cell (and within organelles) where particular reactions occur. Specific enzymes are therefore often located in particular organelles. The reaction rate is controlled by controlling the rate at which substrates enter the organelle and therefore the availability of the raw materials required for the reactions.

The Golgi *(diagram left and TEM right) modifies, sorts, and packages macromolecules for cell secretion. Enzymes within the cisternae modify proteins by adding carbohydrates and phosphates. To do this, the Golgi imports the substances it needs from the cytosol.*

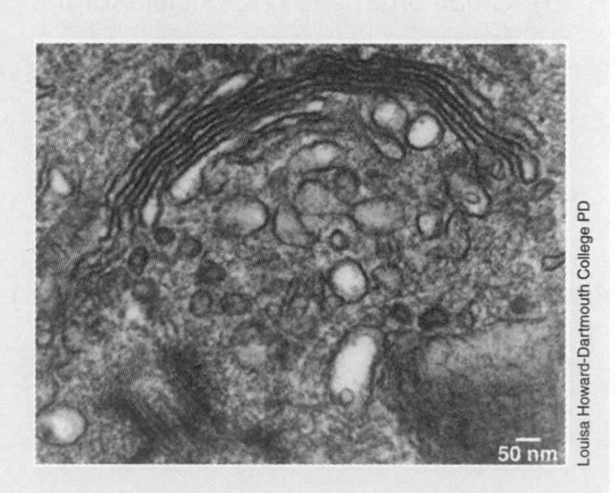

Louisa Howard-Dartmouth College PD

1. Discuss the importance of membrane systems and organelles in providing compartments within the cell:

ISBN: 978-1-92717357-2

Functional Roles of Membranes in Cells

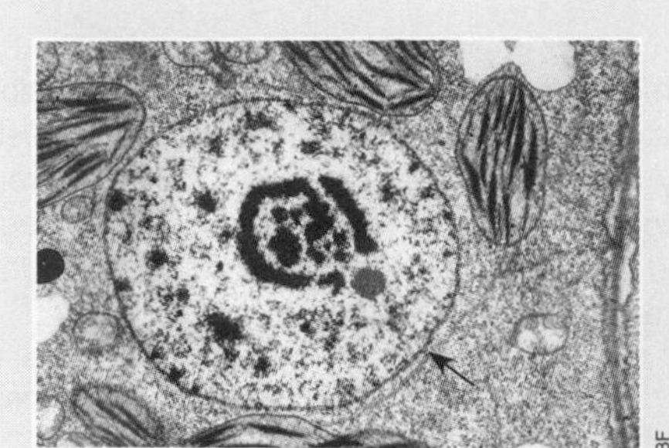

The **nuclear membrane**, which surrounds the nucleus, regulates the passage of genetic information to the cytoplasm and may also protect the DNA from damage.

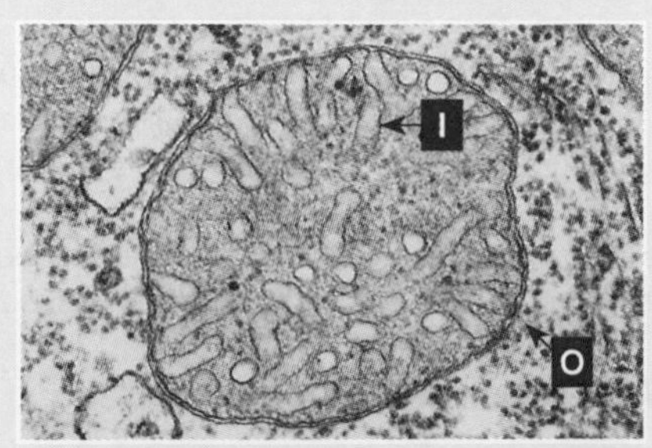

Mitochondria have an outer membrane (**O**) which controls the entry and exit of materials involved in aerobic respiration. Inner membranes (**I**) provide attachment sites for enzyme activity.

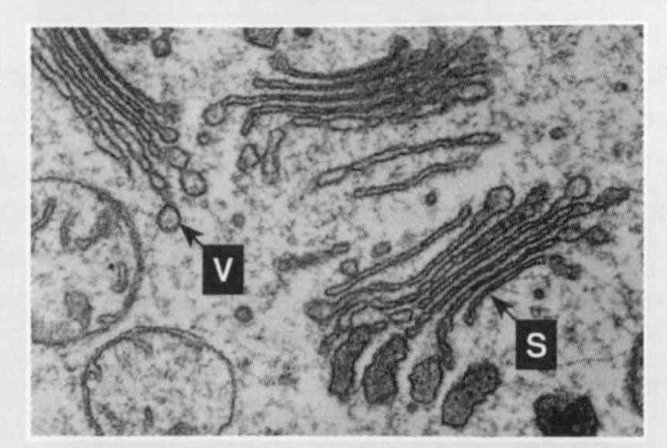

The **Golgi apparatus** comprises stacks of membrane-bound sacs (**S**). It is involved in packaging materials for transport or export from the cell as secretory vesicles (**V**).

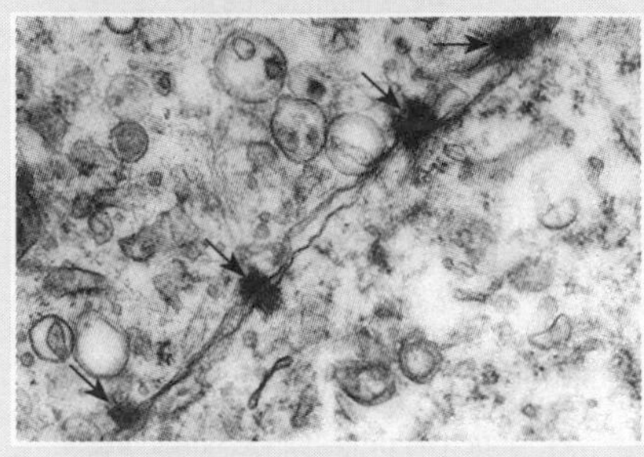

The **plasma membrane** surrounds the cell. In this photo, intercellular junctions called **desmosomes**, which connect neighbouring cells, are indicated with arrows.

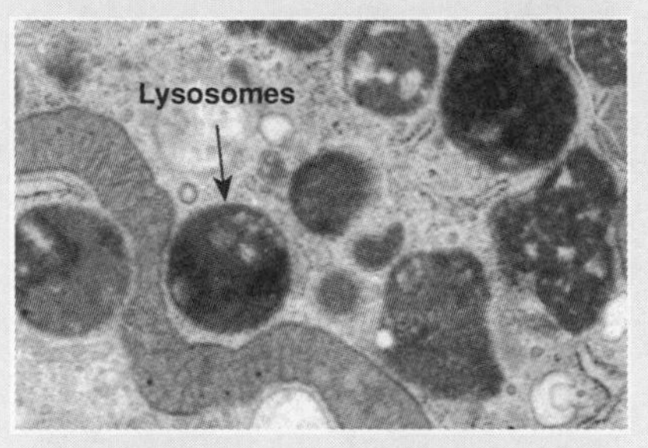

Lysosomes are membrane-bound organelles containing enzymes capable of digesting worn-out cellular structures and foreign material. They are abundant in phagocytes.

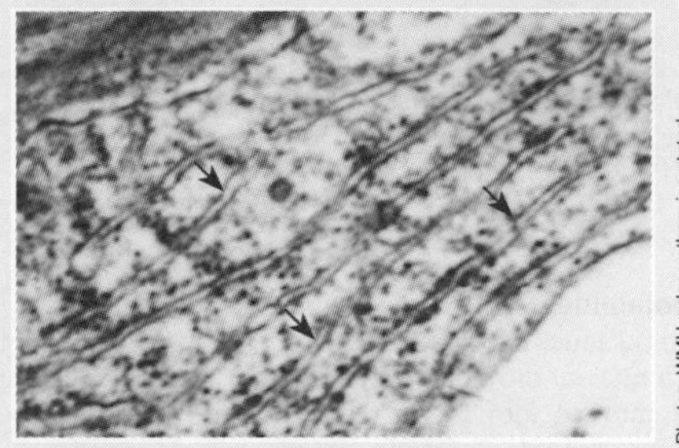

This EM shows stacks of rough endoplasmic reticulum (arrows). The membranes are studded with ribosomes, which synthesize proteins into the intermembrane space.

Photos: WMU unless otherwise stated.

2. Match each of the following organelles with the correct description of its functional role in the cell:

   peroxisome, rough endoplasmic reticulum, lysosome, smooth endoplasmic reticulum, mitochondrion, Golgi apparatus

   (a) Active in synthesis, sorting, and secretion of cell products: ______________________

   (b) Digestive organelle where macromolecules are hydrolyzed: ______________________

   (c) Organelle where most cellular respiration occurs and most ATP is generated: ______________________

   (d) Active in membrane synthesis and synthesis of secretory proteins: ______________________

   (e) Active in lipid and hormone synthesis and secretion: ______________________

   (f) Small organelle responsible for the destruction of toxic substances: ______________________

3. Explain the importance of membrane systems and organelles in providing compartments within the cell:

   ____

   ____

4. (a) Explain why non-polar (lipid-soluble) molecules diffuse more rapidly through membranes than polar molecules:

   ____

   ____

   (b) Explain the implications of this to the transport of substances into the cell through the plasma membrane:

   ____

   ____

5. Identify three substances that need to be transported **into** all kinds of human cells, in order for them to survive:

   (a) ______________ (b) ______________ (c) ______________

6. Identify two substances that need to be transported **out** of all kinds of human cells, in order for them to survive:

   (a) ______________ (b) ______________

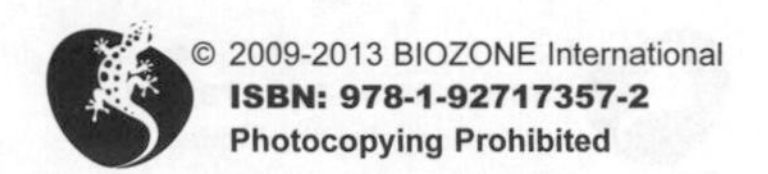

ISBN: 978-1-92717357-2

Cell Structures and Organelles

This activity requires you to summarize information about the components of a typical eukaryotic cell. Complete the table using the list provided and by referring to other pages in this chapter. Fill in the final column with either 'YES' or 'NO'. The first has been completed for you as a guide and the log scale of measurements (next page) illustrates the relative sizes of some cells and cell structures. **List of components**: nucleus, ribosome, centrioles, mitochondrion, lysosome and vacuole (given), endoplasmic reticulum, Golgi apparatus, plasma membrane (given), cell cytoskeleton, flagella or cilia (given), cellular junctions (given).

Cell Component	Details	Visible under light microscope
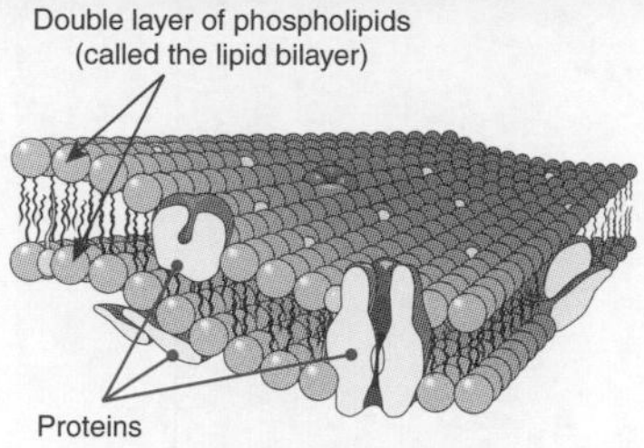 (a) Double layer of phospholipids (called the lipid bilayer); Proteins	Name: Plasma (cell surface) membrane Location: Surrounding the cell Function: Gives the cell shape and protection. It also regulates the movement of substances into and out of the cell.	YES *(but not at the level of detail shown in the diagram)*
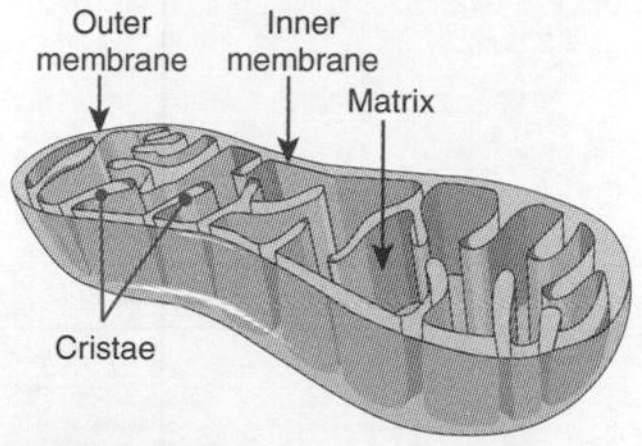 (b) Outer membrane; Inner membrane; Matrix; Cristae	Name: Location: Function:	
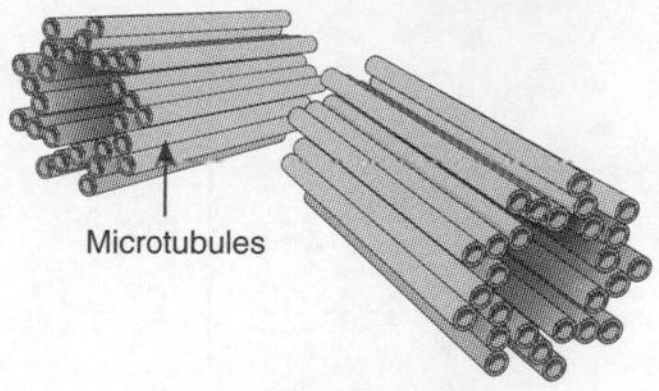 (c) Microtubules	Name: Location: Function:	
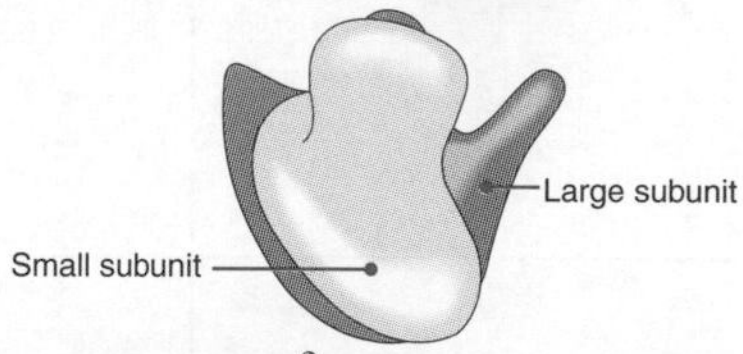 (d) Large subunit; Small subunit	Name: Location: Function:	
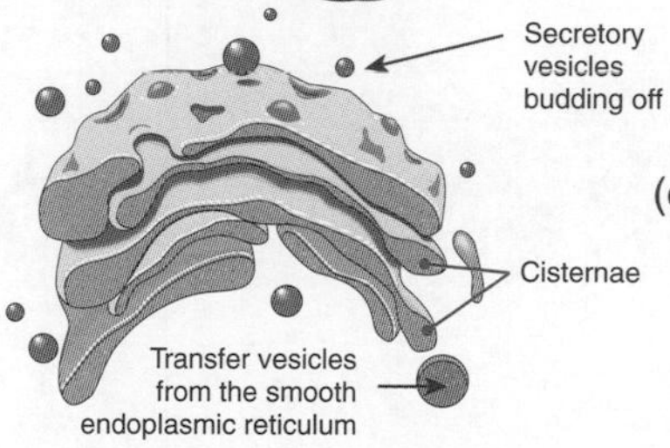 (e) Secretory vesicles budding off; Cisternae; Transfer vesicles from the smooth endoplasmic reticulum	Name: Location: Function:	
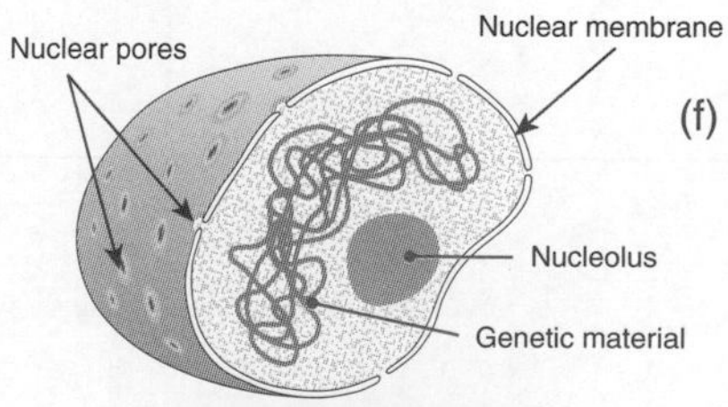 (f) Nuclear pores; Nuclear membrane; Nucleolus; Genetic material	Name: Location: Function:	
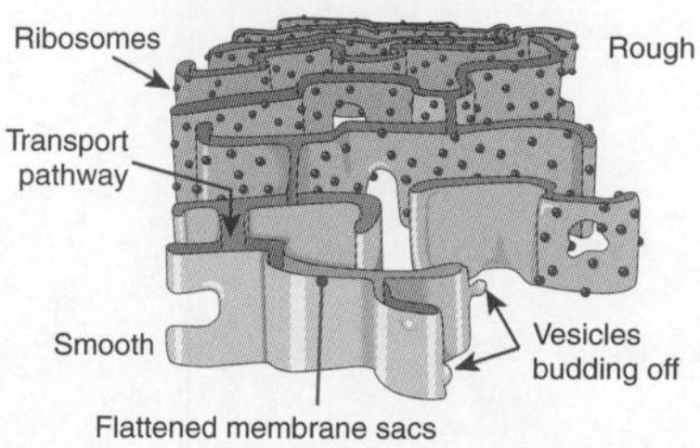 (g) Ribosomes; Rough; Transport pathway; Smooth; Vesicles budding off; Flattened membrane sacs	Name: Location: Function:	

Cells and Tissues

ISBN: 978-1-92717357-2

Related activities: *Basic Cell Structure*
Weblinks: *Eukaryotic Cells Interactive Animation, Animal Cell Junctions*

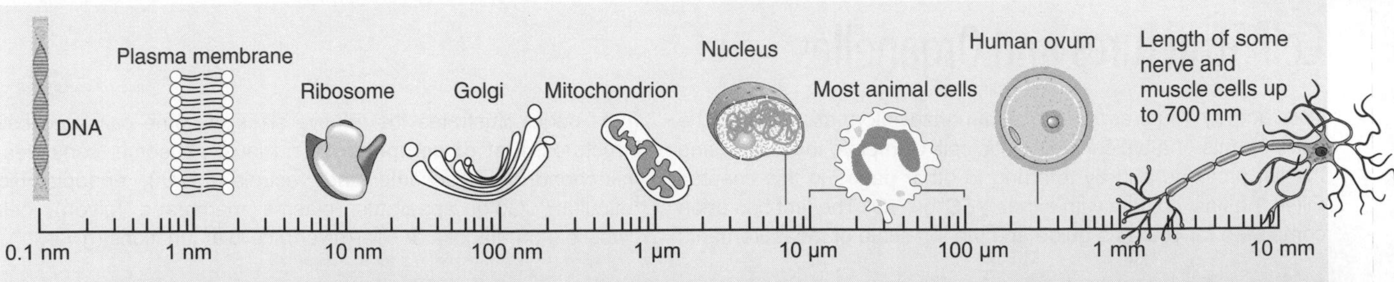

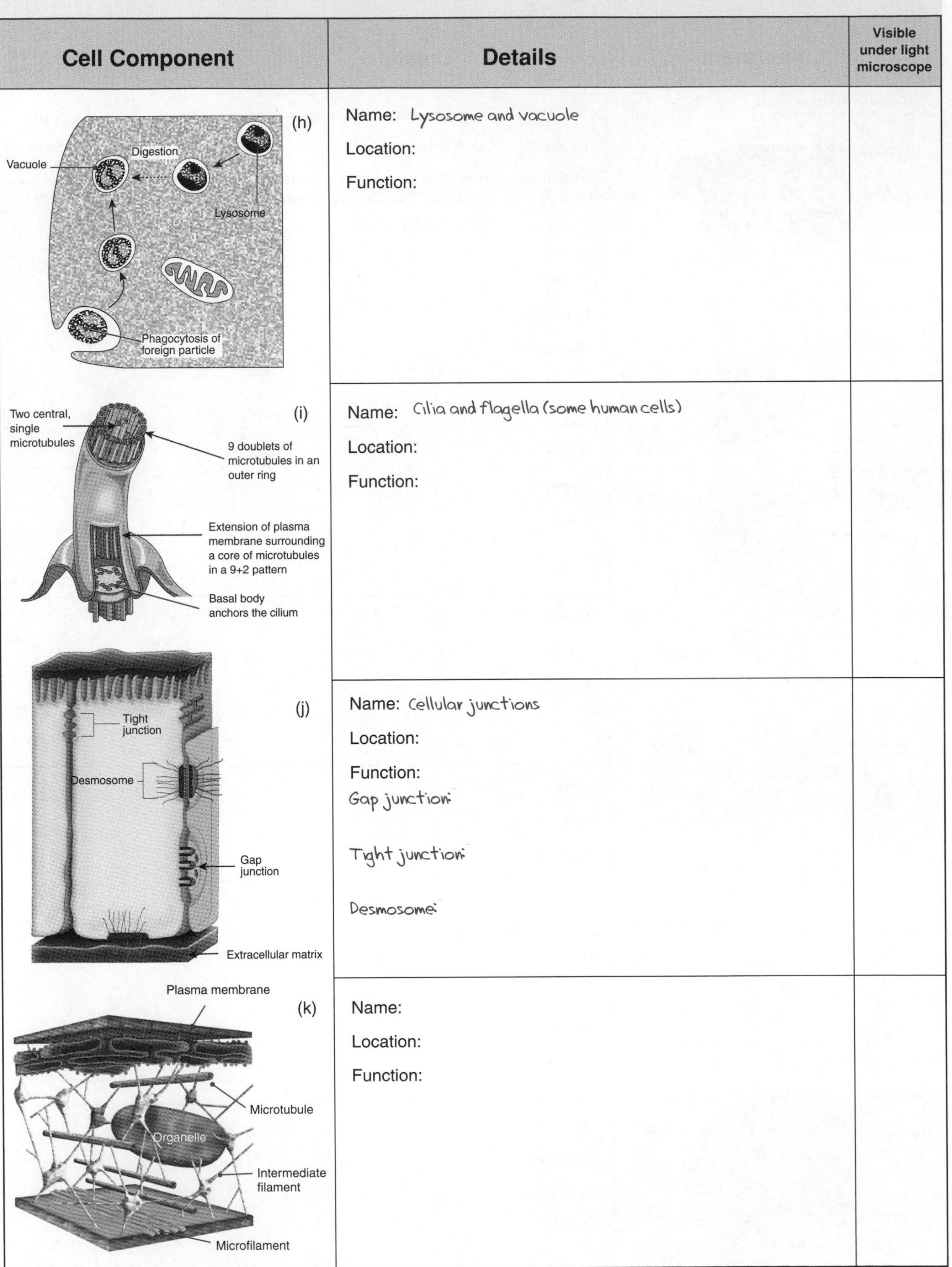

Cell Component	Details	Visible under light microscope
(h)	Name: Lysosome and vacuole Location: Function:	
(i)	Name: Cilia and flagella (some human cells) Location: Function:	
(j)	Name: Cellular junctions Location: Function: Gap junction: Tight junction: Desmosome:	
(k)	Name: Location: Function:	

ISBN: 978-1-92717357-2

The Cell's Cytoskeleton

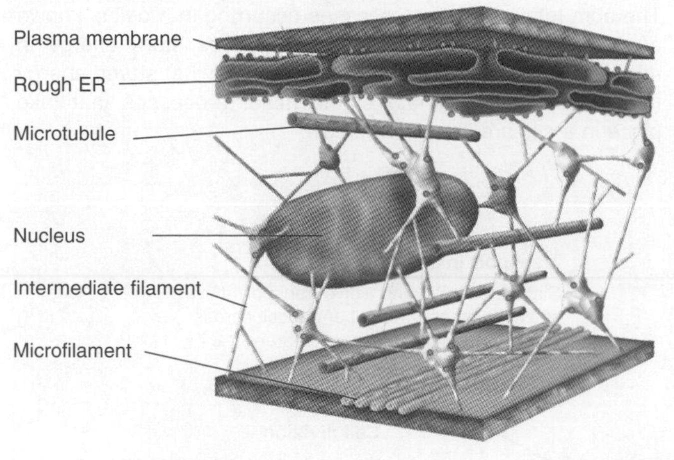

The cell's cytoplasm is not a fluid filled space; it contains a complex network of fibers called the **cytoskeleton**. The cytoskeleton provides tension and so provides structural support to maintain the cell's shape. The cytoskeleton is made up of three proteinaceous elements: microfilaments, intermediate filaments, and microtubules. Each has a distinct size structure and protein composition, and a specific role in cytoskeletal function. Cilia and flagella are made up of microtubules and for this reason they are considered to be part of the cytoskeleton.

The elements of the cytoskeleton are dynamic; they move and change to alter the cell's shape, move materials within the cell, and move the cell itself. Movement of materials is achieved through the action of motor proteins, which transport material by 'walking' along cytoskeletal 'tracks', hydrolyzing ATP at each step.

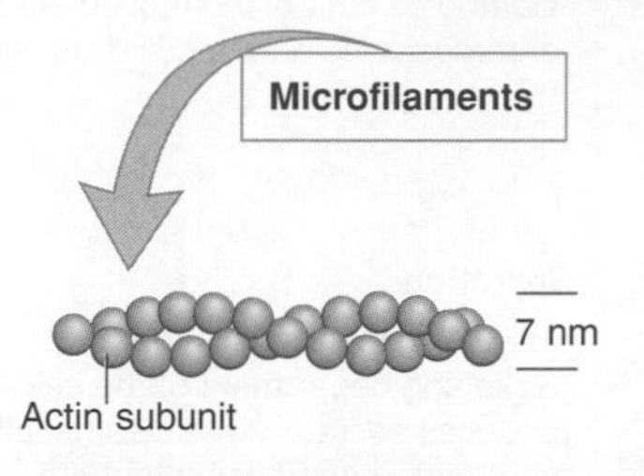

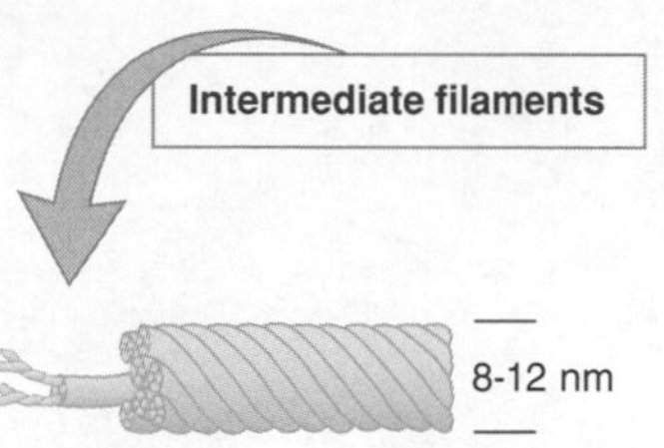

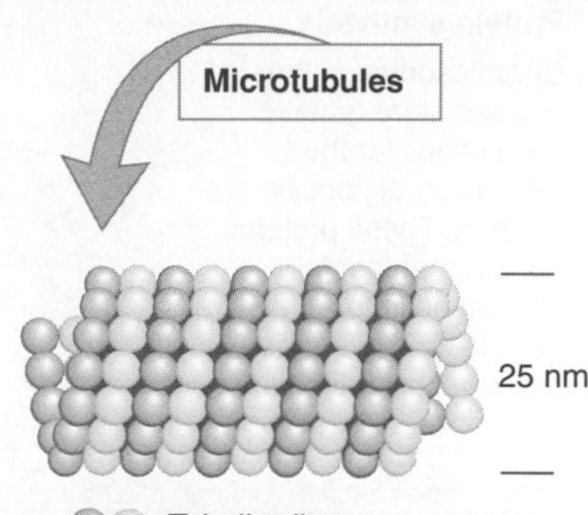

	Microfilaments	Intermediate filaments	Microtubules
Protein subunits	Actin	Fibrous proteins, e.g. keratin	α and β tubulin dimers
Structure	Two intertwined strands	Fibers wound into thicker cables	Hollow tubes
Functions	• Maintain cell shape • Motility (pseudopodia) • Contraction (muscle) • Cytokinesis of cell division	• Maintain cell shape • Anchor nucleus and organelles	• Maintain cell shape • Motility (cilia and flagella) • Move chromosomes (spindle) • Move organelles

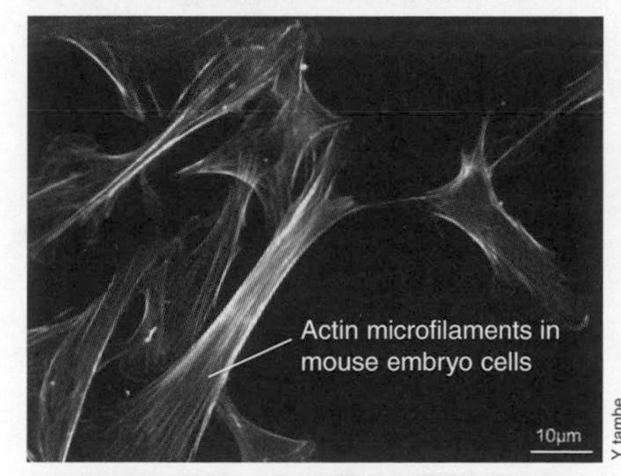

Microfilaments are long polymers of the protein actin. Microfilaments can grow and shrink as actin subunits are added or taken away from either end. Networks of microfilaments form a matrix that helps to define the cell's shape. Actin microfilaments are also involved in cell division (during cytokinesis) and in muscle contraction.

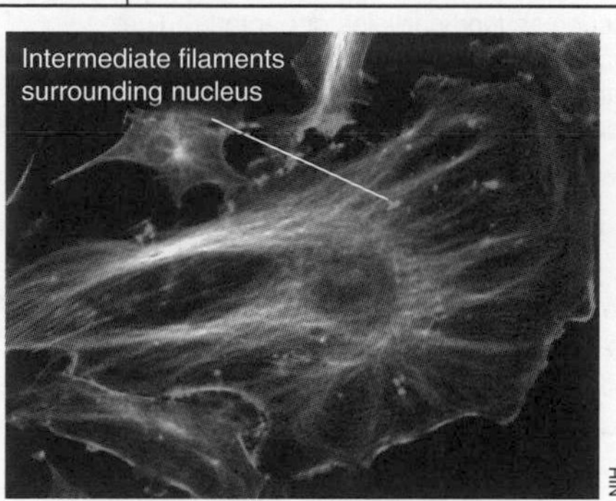

Intermediate filaments can be composed of a number of different fibrous proteins and are defined by their size rather than composition. The protein subunits are wound into cables around 10 nm in diameter. Intermediate filaments form a dense network within and projecting from the nucleus, helping to anchor it in place.

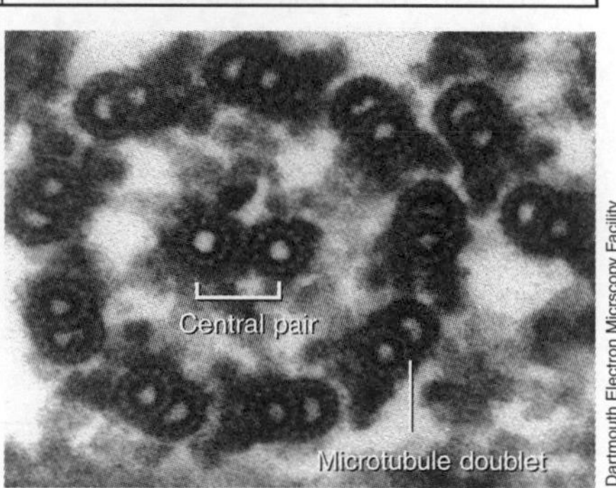

Microtubules are the largest cytoskeletal components and grow or shrink in length as tubulin subunits are added or subtracted from one end. The are involved in movement of material within the cell and in moving the cell itself. This EM shows a cilia from *Chlamydomonas*, with the 9+2 arrangement of microtubular doublets.

1. Describe the role that all components of the cytoskeleton have in common: ______

2. Explain the importance of the cytoskeleton being a dynamic structure: ______

3. Explain how the presence of a cytoskeleton could aid in directing the movement of materials within the cell: ______

ISBN: 978-1-92717357-2

Cell Processes

All of the organelles and other structures in the cell have specific functions. The cell can be compared to a factory with an assembly line. Organelles in the cell provide the equivalent of the power supply, assembly line, packaging department, repair and maintenance, transport system, and the control center. The sum total of all the processes occurring in a cell is known as **metabolism**. Some of these processes store energy in molecules (**anabolism**) while others release that stored energy (**catabolism**). A summary of the major processes that take place in a cell are illustrated below.

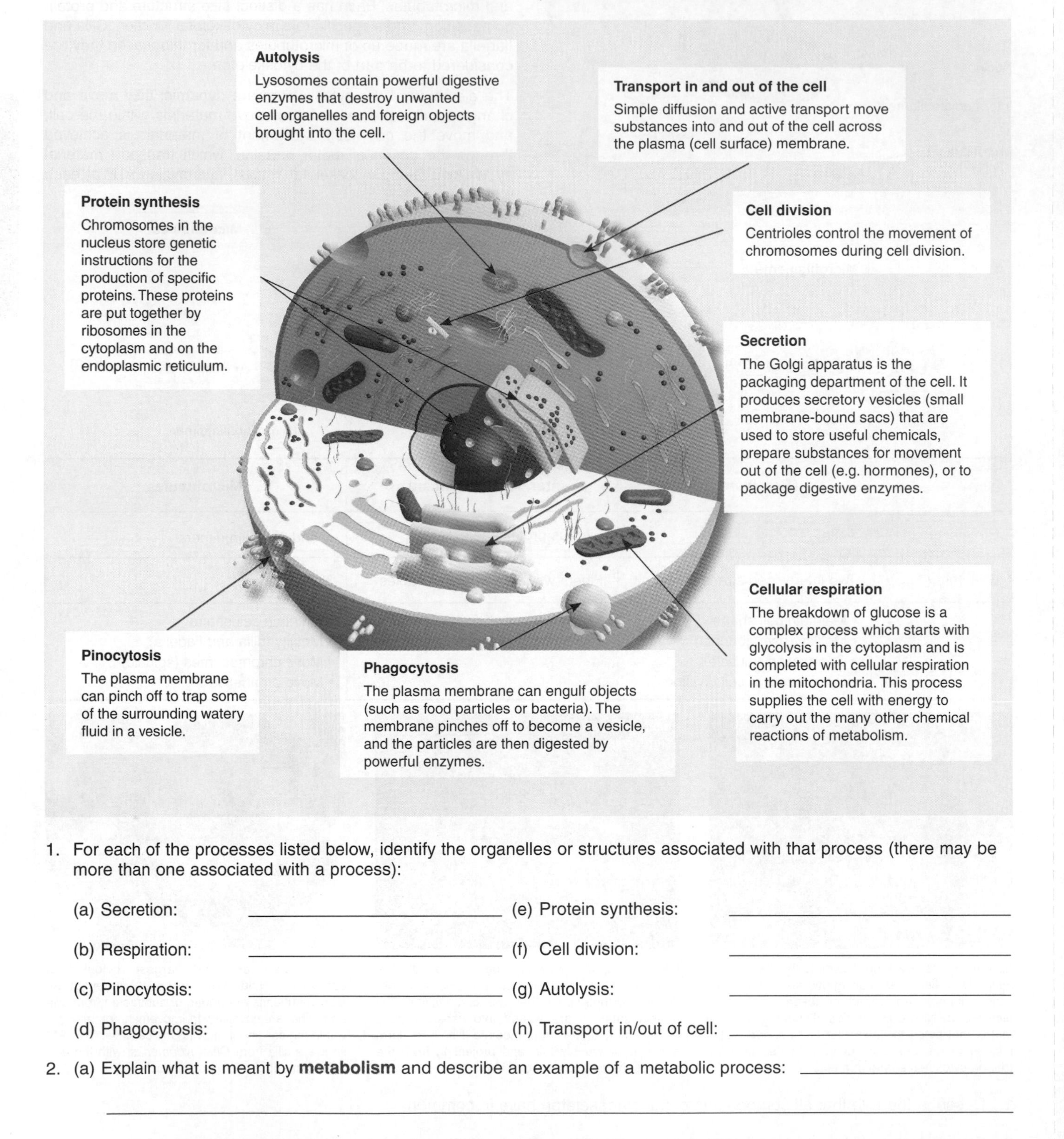

1. For each of the processes listed below, identify the organelles or structures associated with that process (there may be more than one associated with a process):

   (a) Secretion: ______________________ (e) Protein synthesis: ______________________

   (b) Respiration: ______________________ (f) Cell division: ______________________

   (c) Pinocytosis: ______________________ (g) Autolysis: ______________________

   (d) Phagocytosis: ______________________ (h) Transport in/out of cell: ______________________

2. (a) Explain what is meant by **metabolism** and describe an example of a metabolic process: ______________________

   (b) Identify one catabolic process in the diagram above and explain your choice: ______________________

   (c) Identify one anabolic process in the diagram above and explain your choice: ______________________

ISBN: 978-1-92717357-2

Related activities: Basic Cell Structure

Passive Transport Processes

The molecules that make up substances are constantly moving in a random way. This random motion causes molecules to disperse from areas of high to low concentration (i.e. down a **concentration gradient**). This process is called **diffusion**. All types of diffusion, including **osmosis**, are **passive transport** processes, in that they require no expenditure of energy. Diffusion occurs freely across membranes, as long as the membrane is permeable to that molecule (**partially permeable membranes** allow the passage of some molecules but not others). Simple diffusion may occur directly across the lipid bilayer, whereas facilitated diffusion uses trans-membrane proteins to assist the diffusion of specific molecules. **Filtration** is also a passive process, in which fluid pressure is used to push substances through a membrane or capillary wall. Filtration obeys the same rules of movement as diffusion, but the gradient involved is a pressure gradient rather than a concentration gradient.

Diffusion of Molecules Down Concentration Gradients

Diffusion is the movement of molecules (and ions) from regions of high to low concentration, with the end result being that the molecules become evenly distributed. In biological systems, diffusion often occurs across **partially permeable membranes**. Each type of diffusing molecule (gas, solvent, solute) moves down its own **concentration gradient**. Various factors (right) determine the rate at which this occurs.

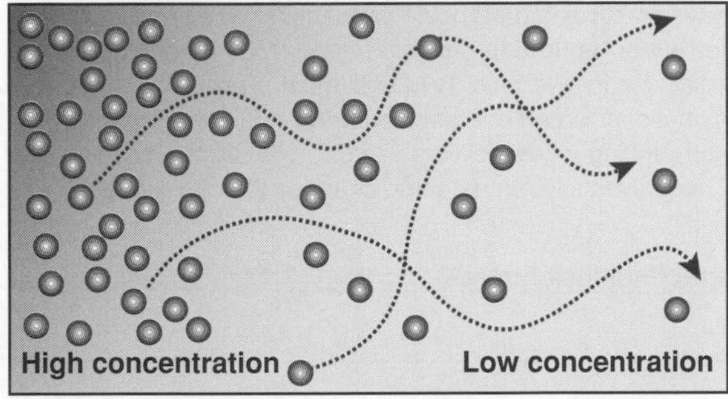

Concentration gradient

If molecules can move freely, they move from high to low concentration (down a concentration gradient) until evenly dispersed.

Factors affecting rates of diffusion for any given diffusing molecule

- **Concentration gradient**: The greater the concentration gradient, the higher the diffusion rate.
- **Surface area:** The larger the area across which diffusion occurs, the greater the rate of diffusion.
- **Barriers to diffusion**: Thicker barriers slow diffusion rate. Pores in a barrier enhance diffusion.
- **Temperature**: Diffusion rates are higher at higher temperatures (within the body this is a negligible factor).

These factors are expressed in **Fick's law**, which governs the rate of diffusion of substances across membranes. It is described by:

Fick's law

$$\frac{\text{Surface area of membrane} \times \text{Difference in concentration across the membrane}}{\text{Length of the diffusion path (thickness of the membrane)}}$$

Diffusion Through Membranes

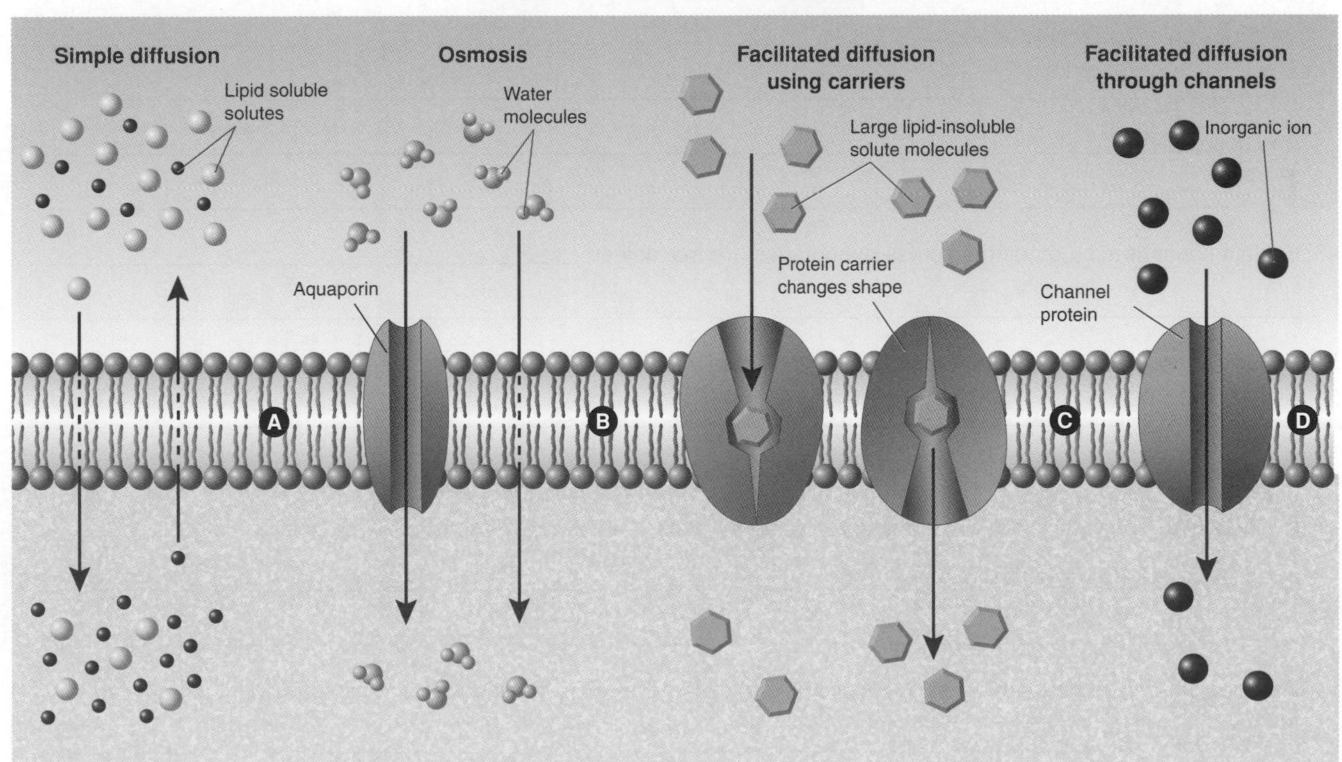

A: Some molecules (e.g. gases and lipid soluble molecules) diffuse directly across the plasma membrane. Two-way diffusion is common in biological systems, e.g. at the alveolar surface of the lung, CO_2 diffuses out and oxygen diffuses into the blood.

B: Osmosis describes the diffusion of water across a partially permeable membrane (in this case, the plasma membrane). Some water can diffuse directly through the lipid bilayer, but movement is also aided by specific protein channels called **aquaporins**.

C: In **carrier-mediated facilitated diffusion**, a lipid-insoluble molecule is aided across the membrane by a transmembrane carrier protein specific to the molecule being transported (e.g. glucose transport into red blood cells).

D: Small polar molecules and ions diffuse rapidly across the membrane by **channel-mediated facilitated diffusion**. Protein channels create hydrophilic pores that allow some solutes, usually inorganic ions, to pass through.

ISBN: 978-1-92717357-2

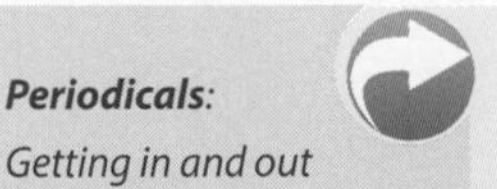

Cellular Tonicity and Osmotic Pressure

In physiology, it is important to understand the consequences of changes to the solute concentrations of cellular environments. The tendency of a solution to 'pull' water into it is called the **osmotic pressure** and it is directly related to the concentration of solutes in the solution. The higher the solute concentration, the greater the osmotic pressure and the greater the tendency of water to move into the solution. In biology, **relative tonicity** (isotonic, hypotonic, or hypertonic) is used describe the difference in osmotic pressure between solutions. Only solutes that cannot cross the plasma membrane affect tonicity.

Tonicity of solution relative to the cytosol	Extracellular environment (solution)	Intracellular environment (cytosol)	Consequence to a cell in the solution
Isotonic	Equal osmotic environment		Normal shape and form
Hypotonic	Lower solute concentration	Higher solute concentration	Water enters cell causing the cell to burst (cell **lysis**)
Hypertonic	Higher solute concentration	Lower solute concentration	Water leaves cell causing shrinkage (**crenation**)

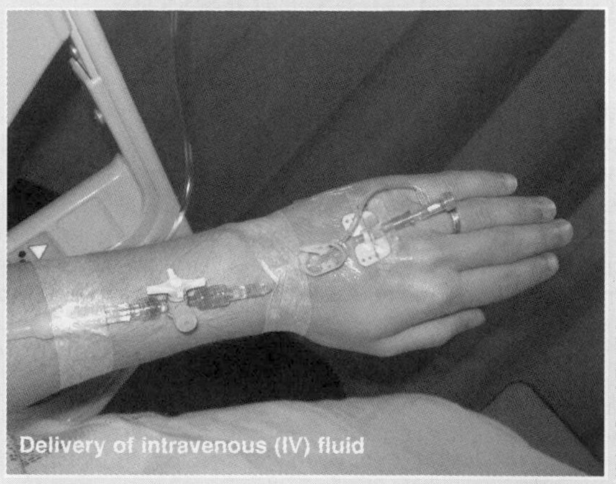
Delivery of intravenous (IV) fluid

Michael Berry (Wikipedia)

The relative tonicity of cells can be used to predict the consequences of changes in solute concentration either side of a partially permeable membrane (e.g. the plasma membrane around each body cell). Such predictions have practical importance. For example, when delivering **intravenous fluid** to patients (*intravenous means within vein*), the intravenous (IV) fluids must present the same osmotic environment as the blood cells they will be surrounding when delivered (e.g. 0.9% saline solution). This prevents life-threatening changes to cell volumes.

1. Describe two properties of an exchange surface that would facilitate rapid diffusion rates:

   (a) ______________________ (b) ______________________

2. Describe two biologically important features of diffusion:

   (a) ______________________

   (b) ______________________

3. Describe how facilitated diffusion is achieved for:

   (a) Small polar molecules and ions: ______________________

   (b) Glucose: ______________________

4. Explain concentration gradients across membranes are maintained: ______________________

5. Explain the role of aquaporins in the rapid movement of water through some cells: ______________________

6. Fluid replacements are usually provided for heavily perspiring athletes after endurance events.

   (a) Identify the preferable tonicity of these replacement drinks (isotonic, hypertonic, or hypotonic): ______________________

   (b) Give a reason for your answer: ______________________

7. Describe what would happen to a patient's red blood cells if they were treated with an intravenous drip containing:

   (a) Pure water: ______________________

   (b) A hypertonic solution: ______________________

   (c) A hypotonic solution: ______________________

8. The malarial parasite lives in human blood. Relative to the tonicity of the blood, the parasite's cell contents would be hypotonic / isotonic / hypertonic (circle the correct answer).

ISBN: 978-1-92717357-2

Ion Pumps

Diffusion alone cannot supply all the cell's needs in all situations. Sometimes molecules must be moved in a particular direction to meet the requirements of the cell. The movement of molecules against their concentration gradient requires **energy expenditure** and is achieved by **active transport** (ion pumps and cytosis). **Ion pumps** are specific trans-membrane carrier proteins that harness the energy of ATP to move molecules from a low to a high concentration. When ATP transfers a phosphate group to the carrier protein, the protein changes its shape, moving the bound molecule across the membrane. Three types of membrane pump are shown below. The sodium-potassium pump (center) is almost universal in animal cells. The concentration gradient created by ion pumps such as this and the proton pump (left) is frequently coupled to the transport of other larger molecules, as shown below right. Note that glucose enters most cells by facilitated diffusion (i.e. passively) but moves by active transport into the intestinal epithelial cells. In this way, uptake of glucose from food continues despite highly fluctuating glucose levels in the gut.

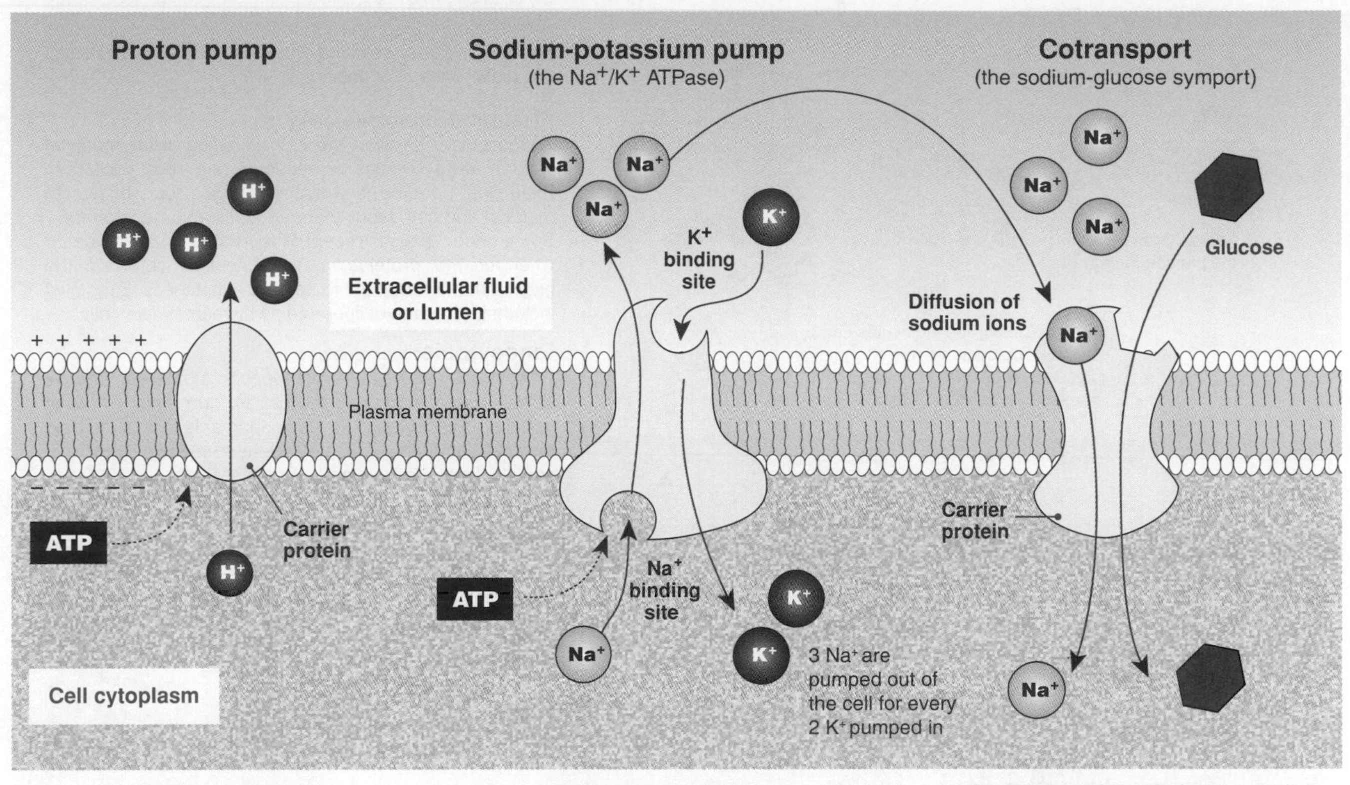

Proton pumps

Proton pumps use energy to move H^+ from the inside of a cell to the outside. This creates a large potential difference across the membrane, which can be coupled to the transport of other molecules. In cellular respiration, the energy for moving the H^+ comes from electrons, and the flow of H^+ back across the membrane is used to drive the synthesis of ATP by ATP synthase.

Sodium-potassium pump

The sodium-potassium pump is a specific protein in the membrane that uses energy in the form of ATP to exchange sodium ions (Na^+) for potassium ions (K^+) across the membrane. The unequal balance of Na^+ and K^+ across the membrane creates large concentration gradients that can be used to drive transport of other substances (e.g. cotransport of glucose).

Cotransport (coupled transport)

A gradient in sodium ions drives the active transport of **glucose** in intestinal epithelial cells. The specific transport protein couples the return of Na^+ down its concentration gradient to the transport of glucose into the intestinal epithelial cell. A low intracellular concentration of Na^+ (and therefore the concentration gradient) is maintained by a sodium-potassium pump.

1. Explain why the ATP is required for membrane pump systems to operate: ______________________

2. (a) Explain what is meant by cotransport: ______________________

   (b) Explain how cotransport is used to move glucose into the intestinal epithelial cells: ______________________

   (c) Explain what happens to the glucose that is transported into the intestinal epithelial cells: ______________________

3. Describe two consequences of the extracellular accumulation of sodium ions: ______________________

ISBN: 978-1-92717357-2

Periodicals: *How biological membranes achieve selective transport*

Related activities: *Absorption & Transport*
Weblinks: *Cellular Transport, Symport*

Endocytosis and Exocytosis

Most cells carry out **cytosis**: a form of **active transport** involving infolding or outfolding of the plasma membrane. Cells are able to do this because of the flexibility of the plasma membrane. Cytosis results in the bulk transport of materials into or out of the cell and is achieved through the localized activity of microfilaments and microtubules in the cell cytoskeleton. Engulfment of material is termed **endocytosis.** Endocytosis typically occurs in certain white blood cells of the human defense system (e.g. neutrophils, macrophages). **Exocytosis** is the reverse of endocytosis and involves the release of material from vesicles or vacuoles that have fused with the plasma membrane. Exocytosis is typical of cells that export material (secretory cells).

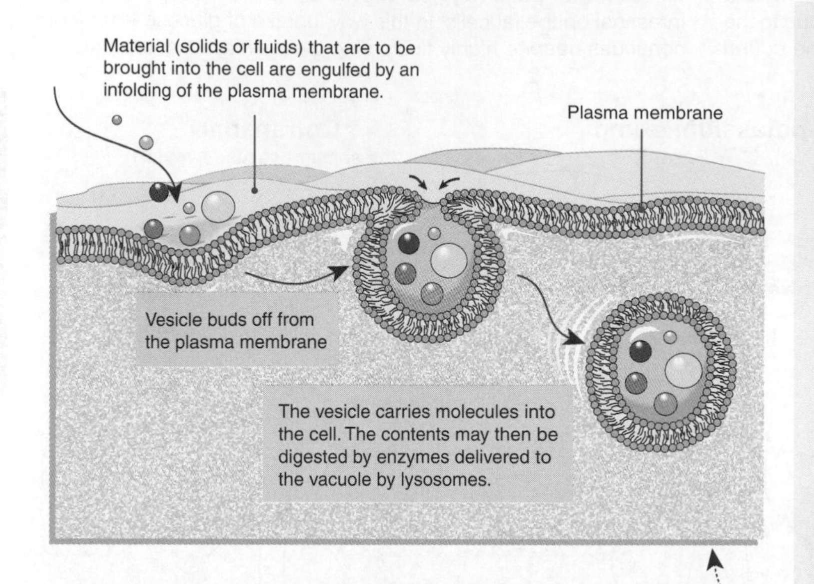

Endocytosis

Endocytosis (left) occurs by invagination (infolding) of the plasma membrane, which then forms vesicles or vacuoles that become detached and enter the cytoplasm. There are two main types of endocytosis:

Phagocytosis: 'cell-eating'
Phagocytosis involves the cell engulfing **solid material** to form large vesicles or vacuoles (e.g. food vacuoles). Examples: Feeding in *Amoeba*, phagocytosis of foreign material and cell debris by neutrophils and macrophages. Some endocytosis is **receptor mediated** and is triggered when receptor proteins on the extracellular surface of the plasma membrane bind to specific substances. Examples include the uptake of lipoproteins by mammalian cells.

Pinocytosis: 'cell-drinking'
Pinocytosis involves the non-specific uptake of **liquids** or fine suspensions into the cell to form small pinocytic vesicles. Pinocytosis is used primarily for absorbing extracellular fluid. Examples: Uptake in many protozoa, some cells of the liver, and some plant cells.

Both endocytosis and exocytosis require energy in the form of ATP.

Areas of enlargement

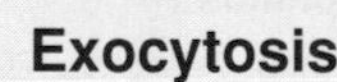

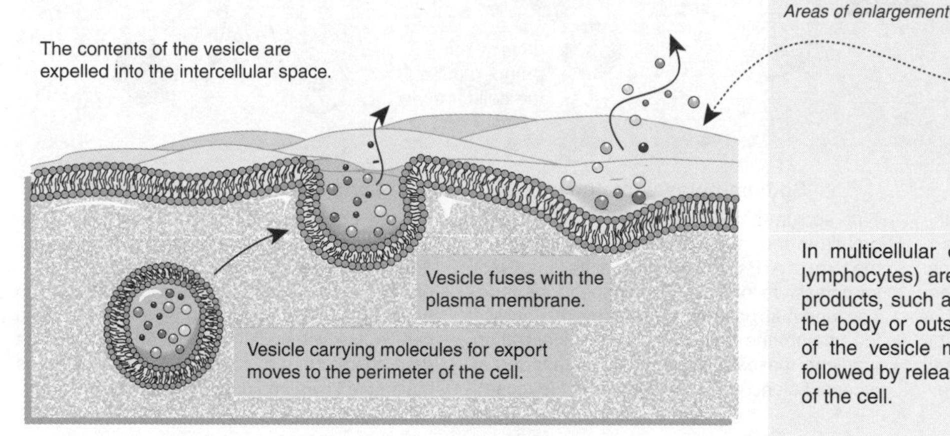

Exocytosis

In multicellular organisms, several types of cells (e.g. lymphocytes) are specialized to manufacture and export products, such as proteins, from the cell to elsewhere in the body or outside it. Exocytosis (left) occurs by fusion of the vesicle membrane and the plasma membrane, followed by release of the vesicle's contents to the outside of the cell.

1. Distinguish between **phagocytosis** and **pinocytosis**: ______________________

2. Describe an example of phagocytosis and identify the cell type involved: ______________________

3. Describe an example of exocytosis and identify the cell type involved: ______________________

4. Explain why cytosis is affected by changes in oxygen level, whereas diffusion is not: ______________________

5. Identify the processes by which the following substances enter a living macrophage:

   (a) Oxygen: ______________ (c) Water: ______________

   (b) Cellular debris: ______________ (d) Glucose: ______________

Related activities: *The Role of Membranes in Cells, The Action of Phagocytes, Passive Transport Processes*

Periodicals: *What is endocytosis?*

ISBN: 978-1-92717357-2

Mitosis and the Cell Cycle

Mitosis is part of the **cell cycle** in which an existing cell (the parent cell) divides into two (the **daughter cells**). In humans and other multicellular organisms, mitosis has a role in growth and development, and in repairing damaged cells and tissues. Although mitosis is part of a continuous cell cycle, it is divided into stages (below). The example below illustrates the cell cycle in a generic animal cell. Unlike meiosis, which is involved in gamete production, mitosis does not reduce the chromosome number in the daughter cells, which are identical to the parent cell. The two types of cell division are compared on the following page. Note that in animal cells, **cytokinesis** (with formation of a cleavage furrow) is usually well underway by the end of telophase.

The Cell Cycle and Stages of Mitosis

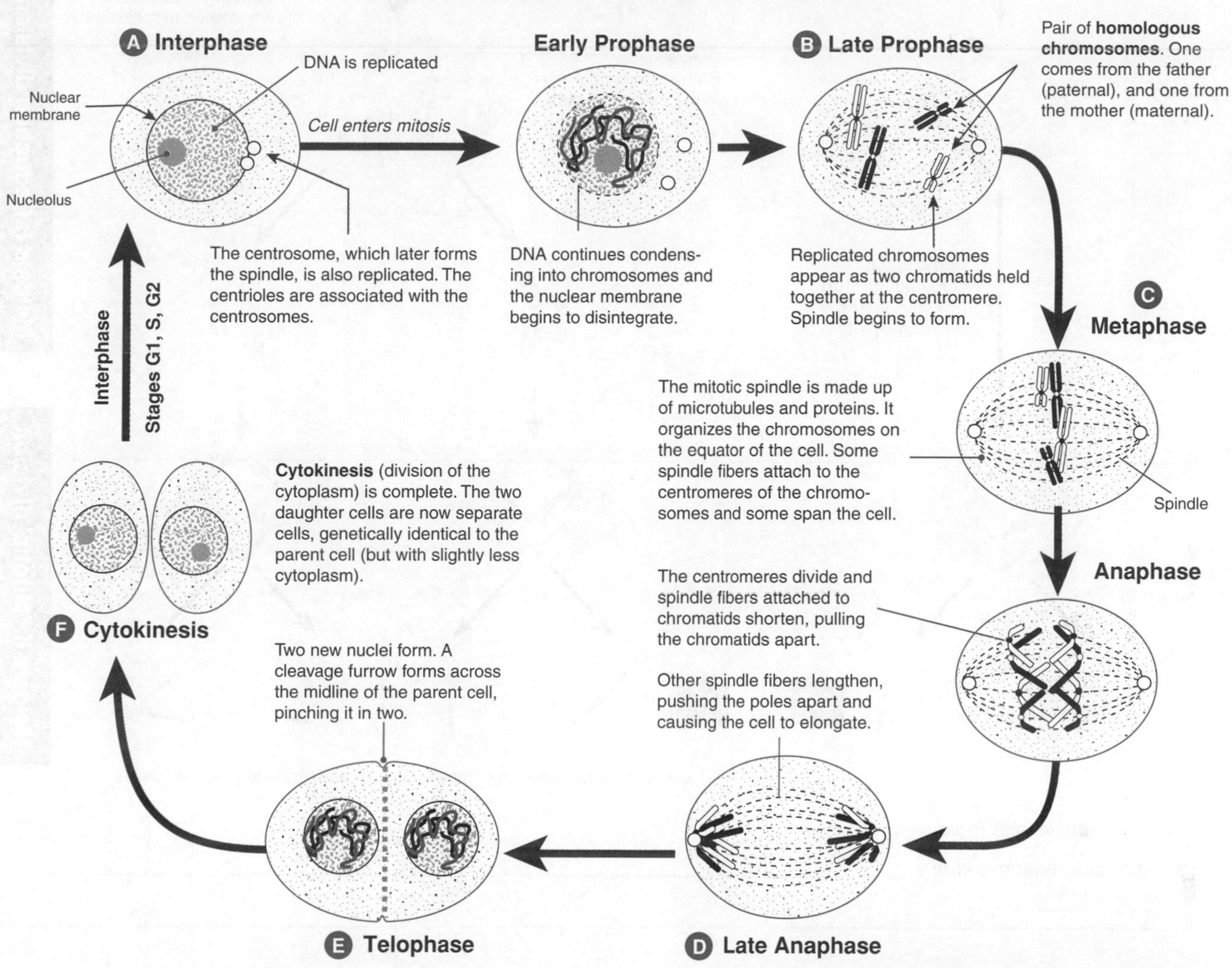

The Cell Cycle Overview

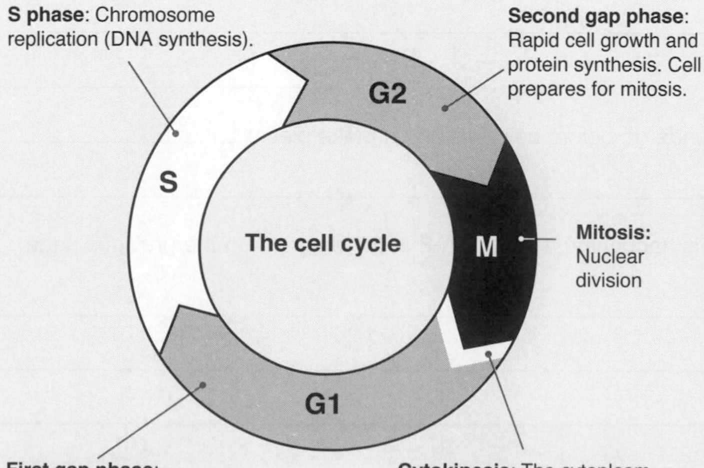

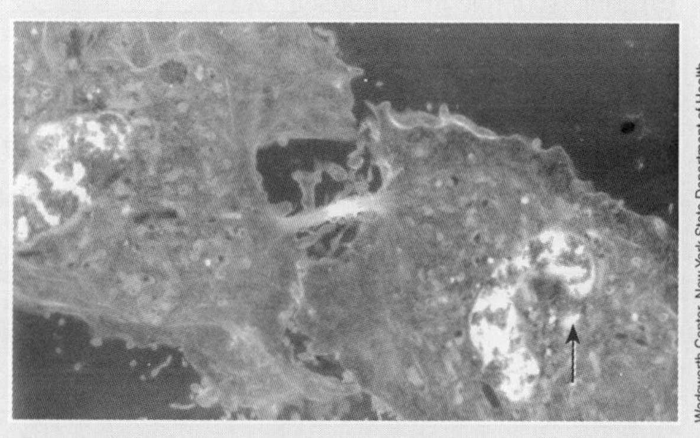

Animal cell cytokinesis (above) begins shortly after the sister chromatids have separated in anaphase of mitosis. A contractile ring of microtubular elements assembles in the middle of the cell, next to the plasma membrane, constricting it to form a **cleavage furrow**. In an energy-using process, the cleavage furrow moves inwards, forming a region of abscission where the two cells will separate. In the photograph above, an arrow points to a centrosome, which is still visible near the nucleus.

ISBN: 978-1-92717357-2

Periodicals: *The cell cycle and mitosis*

Related activities: *Apoptosis and Development*
Weblinks: *Mitosis in an Animal Cell*

Mitosis
(growth and repair)

Meiosis
(gamete formation)

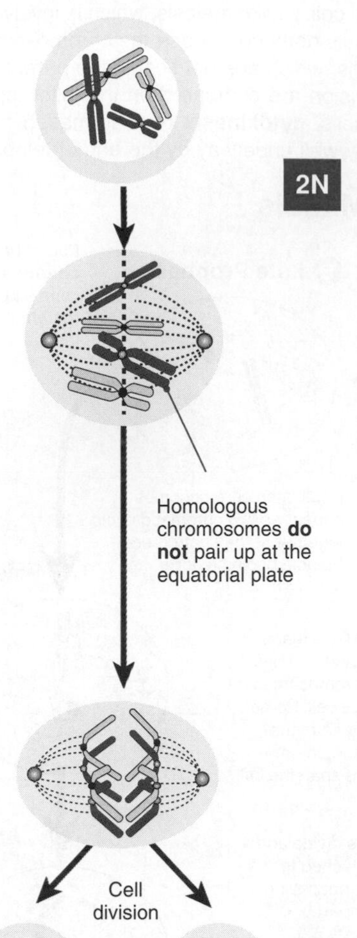

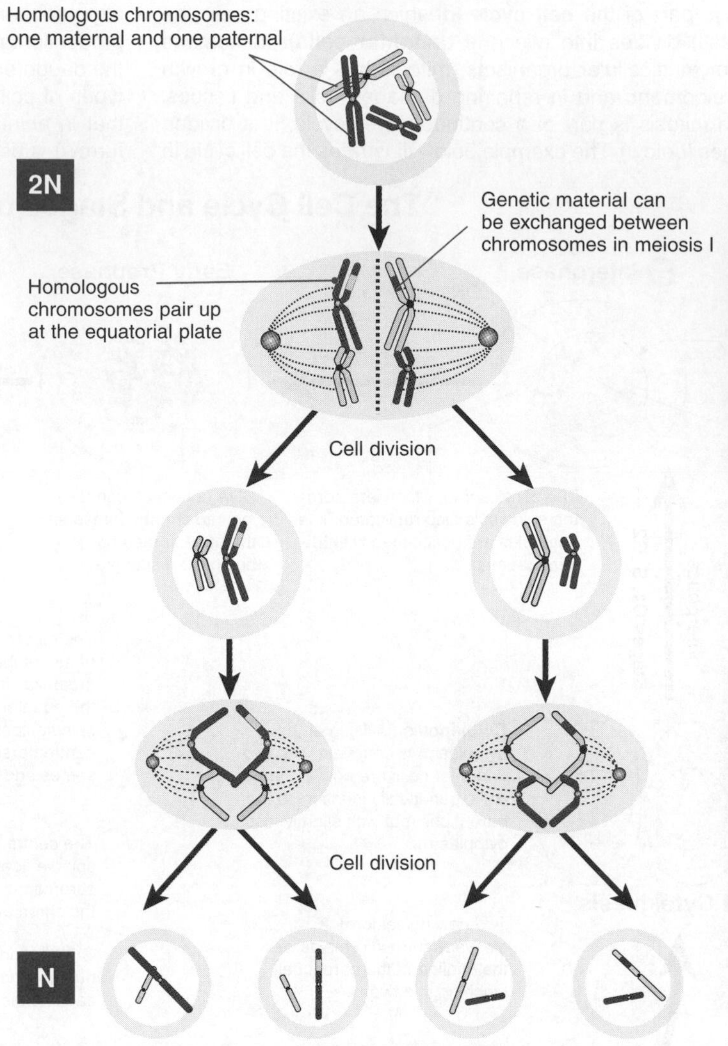

1. Contrast mitosis with meiosis in terms of:
   (a) Final chromosome status: ______

   (b) Biological role in humans: ______

2. State two important changes that chromosomes must undergo before cell division can take place: ______

3. Summarize stages of the cell cycle by describing what is happening at points **A-F** in the diagram on the previous page:

   A. ______
   B. ______
   C. ______
   D. ______
   E. ______
   F. ______

ISBN: 978-1-92717357-2

Control of the Cell Cycle

The events of mitosis are virtually the same for all eukaryotes. However, aspects of the cell cycle can vary enormously between species and even between cells of the same organism. For example, the length of the cell cycle varies between cells such as intestinal and liver cells. Intestinal cells divide around twice a day, while cells in the liver divide once a year. However, if these tissues are damaged, cell division increases rapidly until the damage is repaired. Variation in the length of the cell cycle can be explained by the existence of a regulatory mechanism that slows down or speeds up the cell cycle in response to changing conditions.

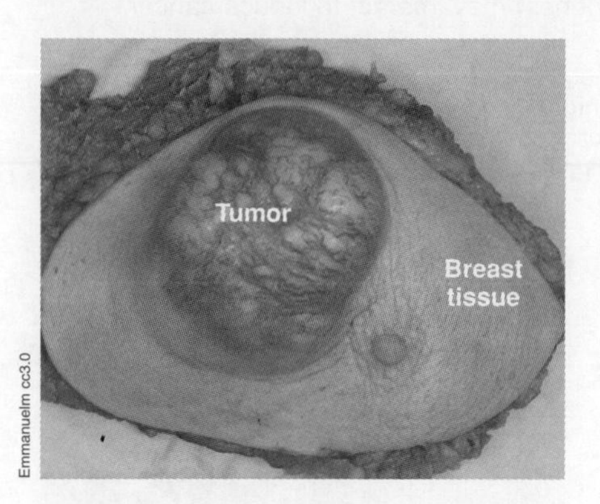

Regulation of the cell cycle is important in detecting and repairing of genetic damage, and preventing uncontrolled cell division. Tumors and cancers, such as this breast cancer (above) are the result of uncontrolled cell division.

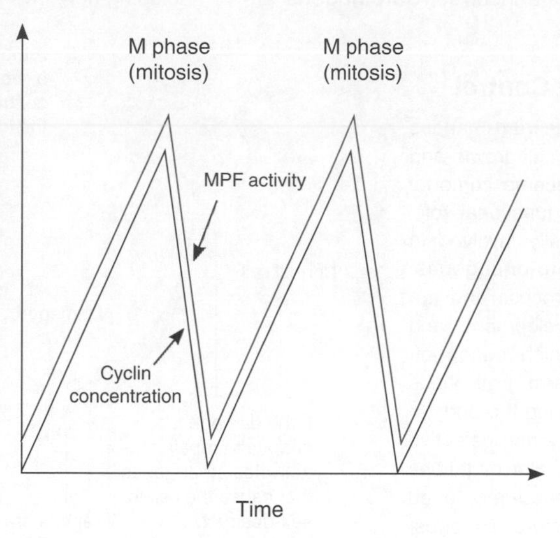

A substance called an M-phase promoting factor (MPF) controls cell regulation. MPF is made up of two regulatory molecules, **cyclins** and **cyclin-dependent kinases** (CDKs).

Cyclins are proteins that control the progression of cells through the cell cycle by activating CDKs (which are enzymes).

CDKs phosphorylate other proteins to signal a cell is ready to proceed to the next stage in the cell cycle. Without cyclin, CDK has little kinase activity; only the cyclin-CDK complex is active. CDK is constantly present in the cell, cyclin is not.

Checkpoints During the Cell Cycle

There are three **checkpoints** during the cell cycle. A checkpoint is a critical regulatory point in the cell cycle. At each checkpoint, a set of conditions determines whether or not the cell will continue into the next phase. For example, cell size is important in regulating whether or not the cell can pass through the G_1 checkpoint.

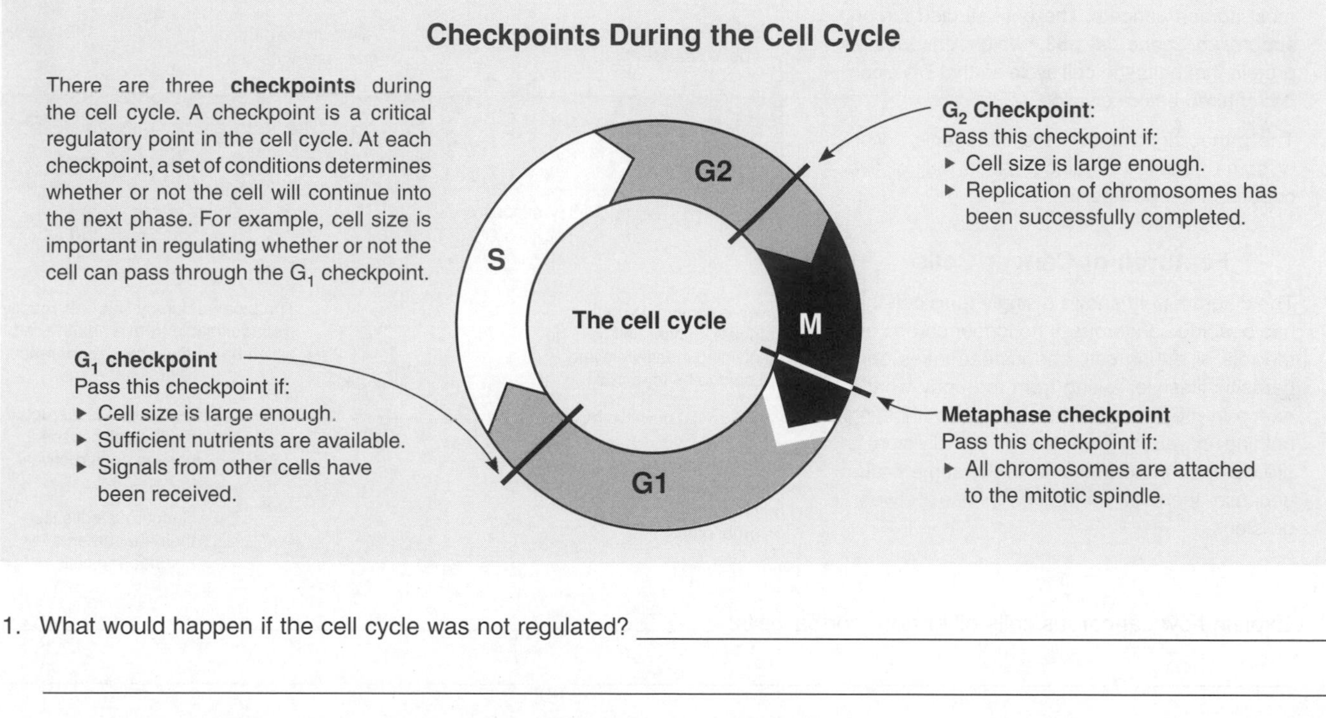

1. What would happen if the cell cycle was not regulated? ______________________________

2. (a) Suggest why the cell cycle is shorter in epithelial cells (such as intestinal cells) than in liver cells:

____

(b) Describe another situation in which the cell cycle shortens to allow for a temporary rapid rate of cell division:

____

ISBN: 978-1-92717357-2

Related activities: Mitosis and the Cell Cycle, Cancer: Cells Out of Control
Weblinks: Checkpoints and Cell Cycle Control

Cancer: Cells Out of Control

Normal cells do not live forever. Under certain circumstances, cells are programed to die, particularly during development. Cells that become damaged beyond repair will normally undergo this programed cell death (called **apoptosis** or **cell suicide**). Cancer cells evade this control and become immortal, continuing to divide regardless of any damage incurred. **Carcinogens** are agents capable of causing cancer. Roughly 90% of carcinogens are also mutagens, i.e. they damage DNA. Chronic exposure to carcinogens accelerates the rate at which dividing cells make errors. Susceptibility to cancer is also influenced by genetic make-up. Any one or a number of cancer-causing factors (including defective genes) may interact to induce cancer.

Cancer: Cells out of Control

Cancerous transformation results from changes in the genes controlling normal cell growth and division. The resulting cells become immortal and no longer carry out their functional role. Two types of gene are normally involved in controlling the cell cycle: **proto-oncogenes**, which start the cell division process and are essential for normal cell development, and **tumor-suppressor genes**, which switch off cell division. In their normal form, both kinds of genes work as a team, enabling the body to perform vital tasks such as repairing defective cells and replacing dead ones. But mutations in these genes can disrupt these finely tuned checks and balances. Proto-oncogenes, through mutation, can give rise to **oncogenes**; genes that lead to uncontrollable cell division. Mutations to tumor-suppressor genes initiate most human cancers. The best studied tumor-suppressor gene is **p53**, which encodes a protein that halts the cell cycle so that DNA can be repaired before division.

The panel, right, shows the mutagenic action of some selected carcinogens on four of five codons of the **p53 gene**.

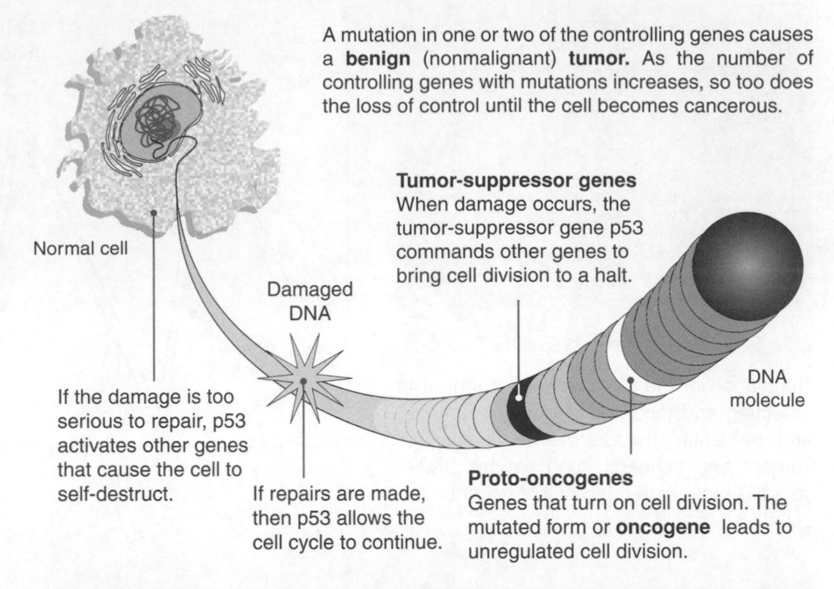

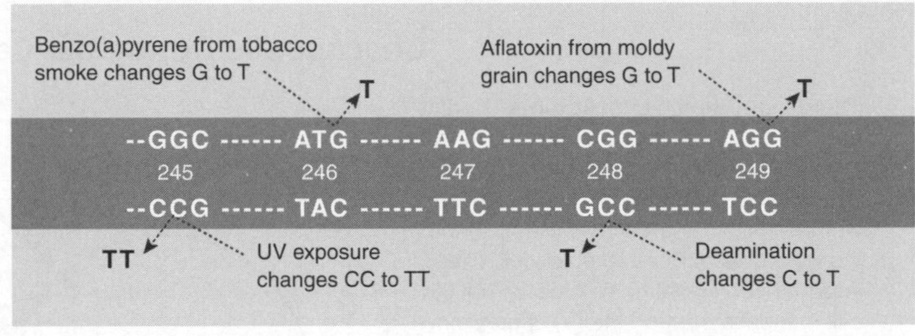

Features of Cancer Cells

The diagram right shows a single **lung cell** that has become cancerous. It no longer carries out the role of a lung cell, and instead takes on a parasitic lifestyle, taking from the body what it needs in the way of nutrients and contributing nothing in return. The rate of cell division is greater than in normal cells in the same tissue because there is no *resting phase* between divisions.

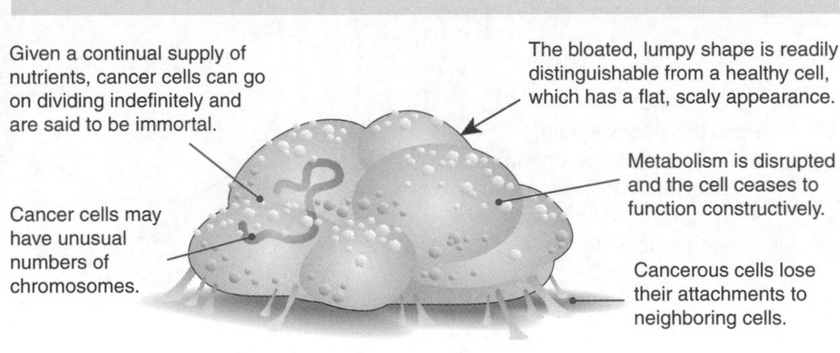

1. Explain how cancerous cells differ from normal cells:

2. Explain how the cell cycle is normally controlled, including reference to the role of **tumor-suppressor genes**:

3. With reference to the role of **oncogenes**, explain how the normal controls over the cell cycle can be lost:

Related activities: *Diseases of the Reproductive System*
Weblinks: *p53*

Periodicals:
Living with the enemy

ISBN: 978-1-92717357-2

Levels of Organization

Organization and the emergence of novel properties in complex systems are two of the defining features of living organisms. Organisms are organized according to a hierarchy of structural levels (below), each level building on the one below it. At each level, novel properties emerge that were not present at the simpler level. Hierarchical organization allows specialized cells to group together into tissues and organs to perform a particular function. This improves efficiency of function in the organism.

In the spaces provided below, assign each of the examples listed to one of the levels of organization as indicated.

1. Examples: *blood, bone, brain, cardiac muscle, cartilage, epinephrine (adrenaline), collagen, DNA, heart, leukocyte, lysosome, pancreas, mast cell, nervous system, phospholipid, reproductive system, ribosomes, neuron, Schwann cell, spleen, squamous epithelium, astrocyte, respiratory system, muscular system, peroxisome, ATP, collagen, testis, liver.*

   (a) Chemical level: ______________

   (b) Organelles: ______________

   (c) Cells: ______________

   (d) Tissues: ______________

   (e) Organs: ______________

   (f) Organ system: ______________

2. State the name given to the microscopic study of cells and tissues:

   ____

CHEMICAL LEVEL

Atoms and molecules form the most basic level of organization. This level includes all the chemicals essential for maintaining life, e.g. water, ions, fats, carbohydrates, amino acids, proteins, and nucleic acids.

ORGANELLE LEVEL

Many diverse molecules may associate together to form complex, highly specialized structures within cells called **cellular organelles**, e.g. mitochondria, Golgi apparatus, endoplasmic reticulum, chloroplasts.

CELLULAR LEVEL

Cells are the basic structural and functional units of an organism. Each cell type has a different structure and function (the result of cellular differentiation during development).

***Human examples** include: epithelial cells, osteoblasts, muscle fibers.*

TISSUE LEVEL

Tissues are composed of groups of cells of similar structure that perform a particular, related function.

***Human examples** include: epithelial tissue, bone, muscle.*

ORGAN LEVEL

Organs are structures of definite form and structure, comprising two or more tissues with related functions.

***Human examples** include: heart, lungs, brain, stomach, kidney.*

ORGAN SYSTEM LEVEL

In animals, organs form parts of even larger units known as **organ systems**. An organ system is an association of organs with a common function, e.g. digestive system, cardiovascular system, and the urinogenital system. In all, 11 organ systems make up the body, or **organism**.

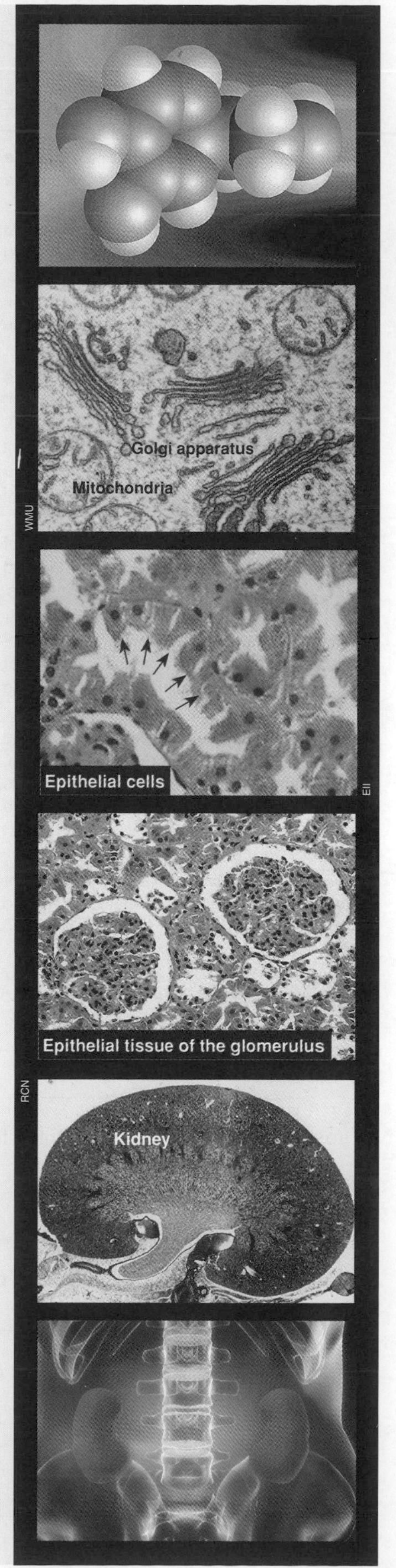

ISBN: 978-1-92717357-2

***Related activities:** Epithelial Tissues, Muscle Tissue, Connective Tissue, Nervous Tissue, Human Organ Systems*

Epithelial Tissues

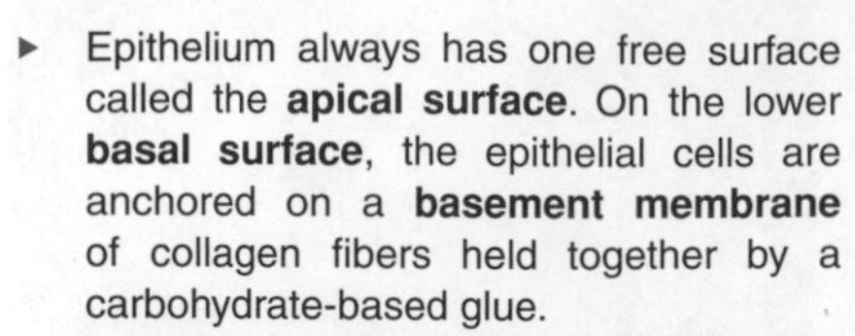

The microscopic study of tissues is called **histology**. The cells of a tissue, and their associated extracellular substances, e.g. collagen, are grouped together to perform particular functions. Tissues improve the efficiency of operation because they enable tasks to be shared amongst various specialized cells. **Epithelial tissues** make up one of the four broad groups of tissues found in humans and other animals. Epithelial tissues line internal and external surfaces (e.g. blood vessels, ducts, gut lining) and protect the underlying structures from wear and tear, infection, and pressure. They are found associated with other tissue types (e.g. muscle and connective tissues) in every organ system of the body.

- Epithelium always has one free surface called the **apical surface**. On the lower **basal surface**, the epithelial cells are anchored on a **basement membrane** of collagen fibers held together by a carbohydrate-based glue.
- Except for glandular epithelium, epithelial cells form fitted continuous sheets, held in place by desmosomes and tight junctions.
- Epithelial tissues are **avascular**, i.e. they have no blood supply and rely on diffusion from underlying capillaries.
- Epithelia are classified as **simple** (single layered) or **stratified** (two or more layers), and the cells may be **squamous** (flat), **cuboidal**, or **columnar** (rectangular). Thus at least two adjectives describe any particular epithelium (e.g. stratified cuboidal).
- **Pseudostratified epithelium** is a type of simple epithelium that appears layered because the cells are of different heights. All cells rest on the basement membrane.
- **Transitional epithelium** is a type of stratified epithelium which is capable of considerable stretching. It lines organs such as the urinary bladder.
- Epithelia may be modified, e.g. ciliated or specialized for secretion, absorption, or filtration. Glandular epithelium is specialized for secretion. Epithelia may be also be ciliated e.g. in the respiratory tract or specialized for absorption or filtration.
- Glandular epithelium is specialized for secretion. Epithelia may be also be ciliated e.g. in the respiratory tract or specialized for absorption or filtration.

Features of Epithelial Tissue

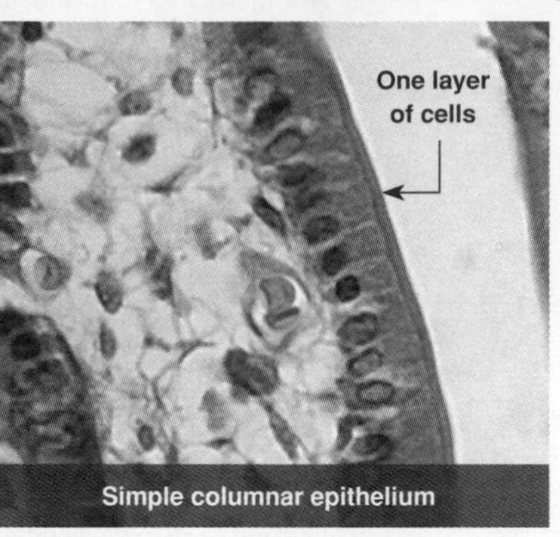

The simple epithelium of the gastrointestinal tract is easily recognized by the regular column-like cells. it is specialized for secretion and absorption.

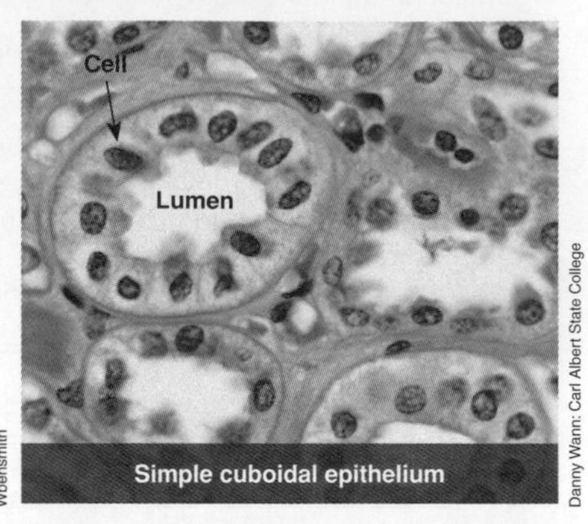

Simple cuboidal epithelium is common in glands and their ducts and also lines the kidney tubules (above) and the surface of the ovaries.

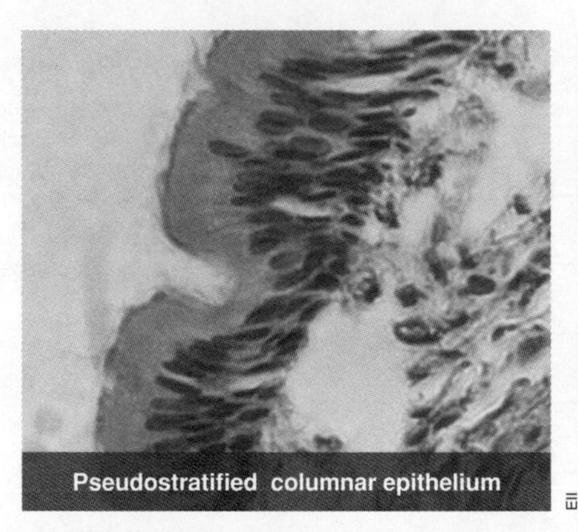

This epithelium lines much of the respiratory tract (above). Mucus produced by goblet cells in the epithelium traps dust particles.

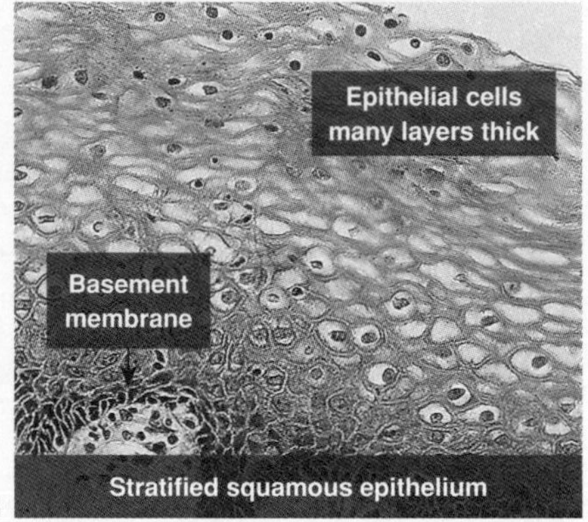

Stratified epithelium is more durable than simple epithelium because it has several layers. It has a protective role, e.g. in the vagina above.

1. (a) Describe the basic components of a tissue: ______________________

   (b) Explain how the development of tissues improves functional efficiency: ______________________

2. Describe the general functional role of epithelial tissue: ______________________

3. Describe the particular features that contribute to the functional role of each of the following types of epithelial tissue:

   (a) Transitional epithelium: ______________________

   (b) Stratified epithelium: ______________________

ISBN: 978-1-92717357-2

Muscle Tissue

The muscle tissue of the body is responsible for producing movement. This includes movement of the body as in locomotion, and also internal movements such as heartbeat, intestinal peristalsis, blood vessel constriction and dilation, and contraction and expansion of the iris of the eye. Muscle tissue is composed of specialized elongated cells called fibers, held together by connective tissue. The contractile protein filaments within these fibers gives the muscle cells their ability to contract. Muscle is classed as skeletal, cardiac, or smooth according to its structure, function, and location in the body. Each type is described below.

Features of Muscle Tissue

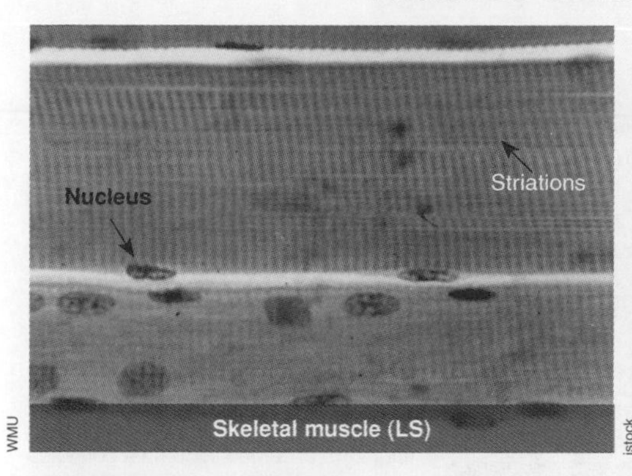

Skeletal muscle (LS)

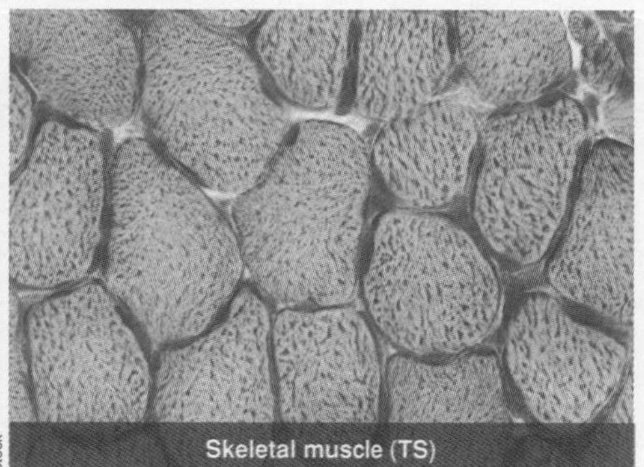

Skeletal muscle (TS)

Skeletal muscle is the major muscle type in the body. It brings about **voluntary movement** of the skeleton as well as the facial skin, tongue, and eyeball. The contraction and relaxation of skeletal muscle is under conscious control (hence voluntary). The fibers are large with many peripheral nuclei, and the regular arrangement of the contractile elements gives them a striated appearance. Skeletal fibers are innervated by motor neurons. If they lose their nerve supply, they lose function and waste away.

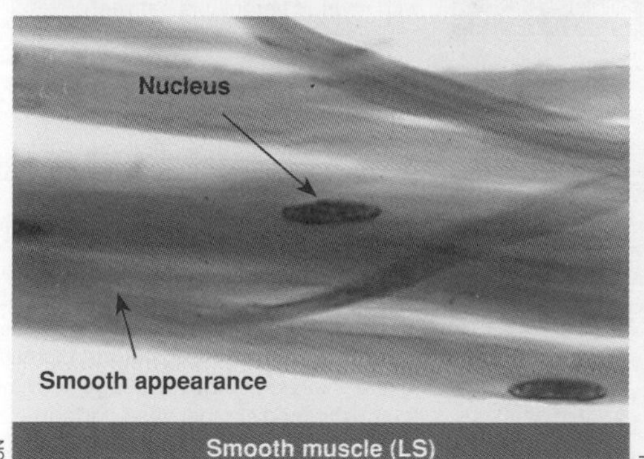

Smooth muscle (LS)

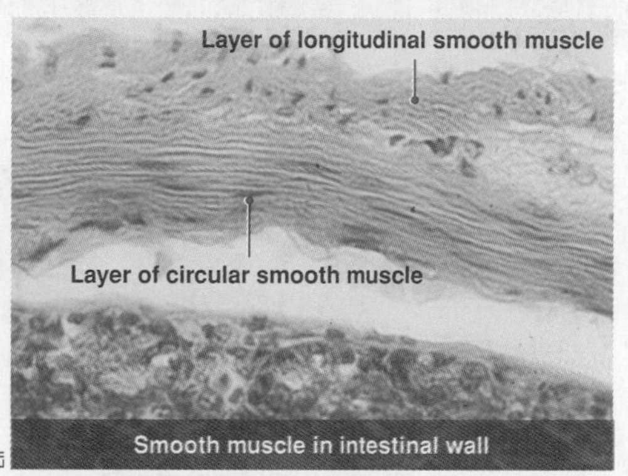

Smooth muscle in intestinal wall

The spindle-shaped cells of **smooth muscle** have only one nucleus per cell. The contractile elements are not regularly arranged so the tissue appears smooth. Smooth muscle is responsible for involuntary movements (e.g. peristalsis in the gut wall) and is found predominantly lining the visceral organs and blood vessels. Smooth muscle cells make contact with each other at specialized regions called gap junctions. They also receive input from the neurons of the autonomic nervous system.

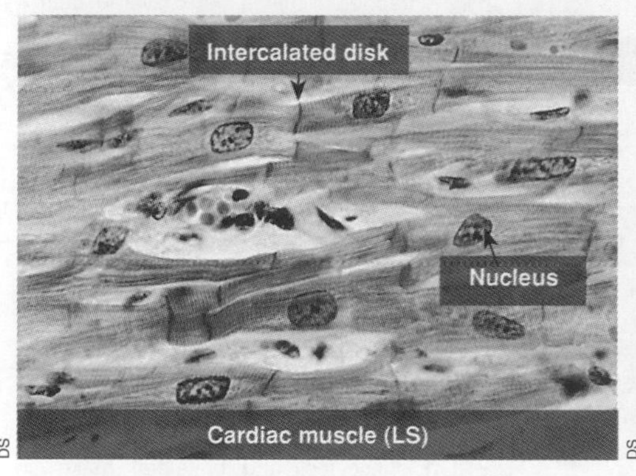

Cardiac muscle (LS)

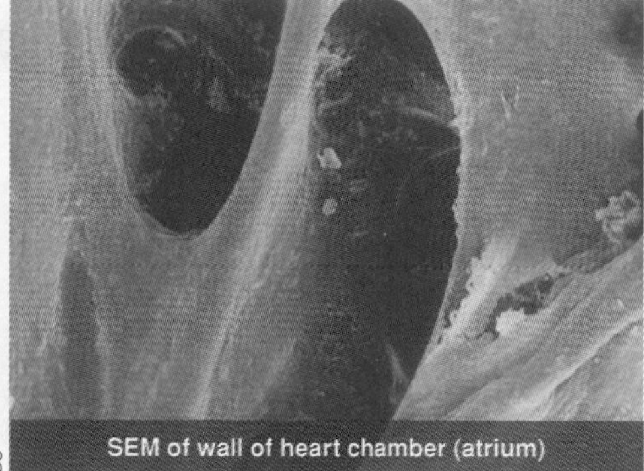

SEM of wall of heart chamber (atrium)

Cardiac muscle is found only in the heart. It has striations, like skeletal muscle, but the fibers are short and branched and usually have only one centrally located nucleus. The fibers are joined at specialized **intercalated disks** with gap junctions to allow rapid communication between the cells. Contraction of cardiac muscle is involuntary - it contracts spontaneously and rhythmically throughout life, although the rate of beating is influenced by nervous input and hormones.

1. Summarize features of each type of muscle tissue in the following table:

	Skeletal muscle	Smooth muscle	Cardiac muscle
Appearance of cells	Large, striated		
Nuclei (number, location)		One, central	
Control			Involuntary

2. Compare and contrast the functional role of each type of muscle tissue, relating structure to function in each case:

ISBN: 978-1-92717357-2

Related activities: *Muscles of the Human Body*
Weblinks: *Animal Tissues*

Connective Tissues

Connective tissue (CT) is the major supporting tissue of the body. Connective tissues bind other structures together and provide support and protection against damage, infection, or heat loss. Most CTs have a plentiful blood supply, although tendons and ligaments are poorly vascularized and cartilage is avascular. Connective tissues range from very hard to fluid and include so-called ordinary CTs (dense and loose CT) and special CTs (e.g. bone, cartilage, fat, and blood). In most ordinary CTs, collagen is the predominant fiber type. All CTs have three common elements: fibers, cells, and non-cellular matrix material. The most common cells in all CTs are fibroblasts, which synthesize the collagen and extracellular matrix.

Ordinary Connective Tissues

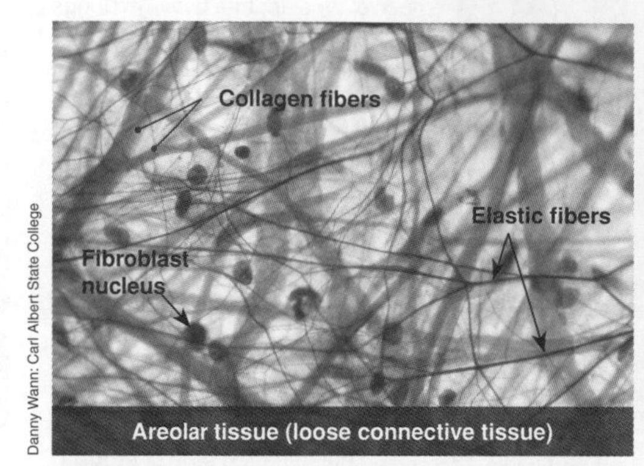

Areolar tissue (loose connective tissue)

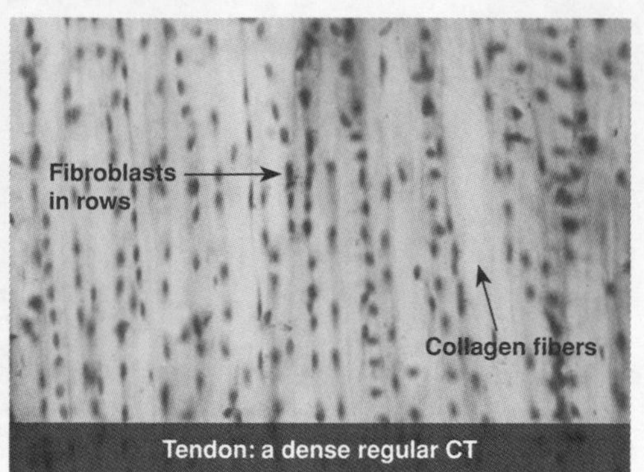

Tendon: a dense regular CT

Danny Wann: Carl Albert State College

The so-called **ordinary connective tissues** are categorized as either loose or dense depending on the relative abundance of cells, fibers, and ground substance. **Loose connective tissues** have more cells and fewer fibers than denser connective tissues. The fibers are loosely organized as the name suggests. A typical example is areolar tissue, which helps to hold internal organs in position. **Dense regular CT** is found where strength is required , such as in ligaments and tendons. The collagen fibers are arranged in compact bundles with rows of fibroblasts between.

Special Connective Tissues

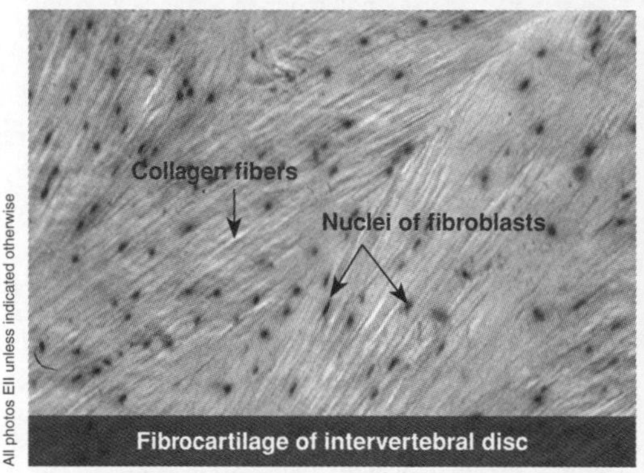

Fibrocartilage of intervertebral disc

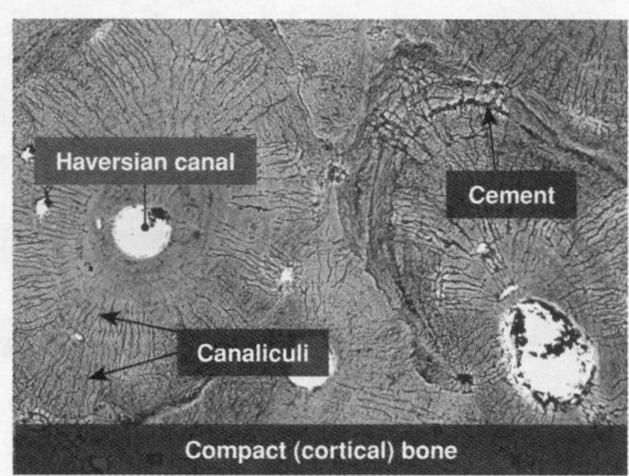

Compact (cortical) bone

All photos EII unless indicated otherwise

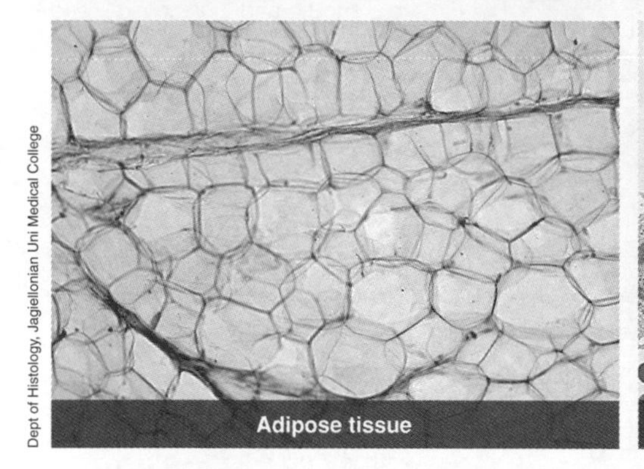

Adipose tissue

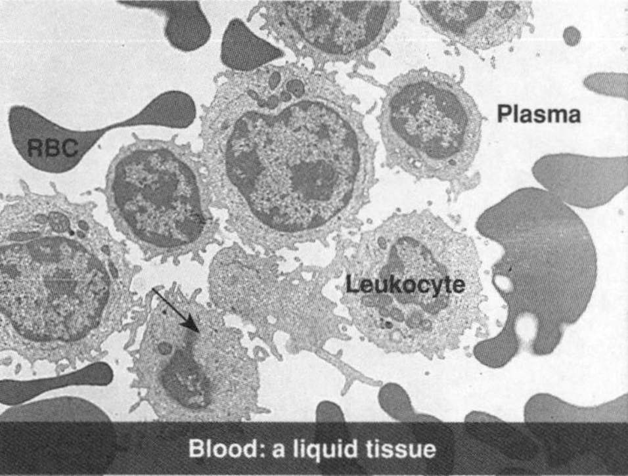

Blood: a liquid tissue

Dept of Histology, Jagiellonian Uni Medical College

Special connective tissues differ from proper CTs in being specialized for particular functions (e.g. elasticity or compression resistance). They are also more restricted in their distribution than the 'ordinary' CTs described above. **Cartilage** is more flexible than bone and forms supporting structures in the skeleton. **Fibrocartilage** forms the cushiony disks between vertebrae. **Bone** is the hardest connective tissue. In **compact bone**, the bone cells are trapped within a hard layered matrix. Canals (canaliculi) penetrate the matrix allowing the cells to make exchanges with the blood vessels supplying the bone tissue.

Adipose (fat) tissue is made up of large cells, each of which contains a single large droplet of fat surrounded by a thin ring containing the cellular cytoplasm and nucleus. The adipose cells are surrounded by a fine supporting network of collagen fibers.

Blood is a liquid tissue, composed of cells floating in a liquid matrix, which includes soluble fibers. The cells (red blood cells, white blood cells or leukocytes, and platelets) make up the formed elements of the blood. The liquid matrix or plasma makes up ~55% of the blood volume.

1. (a) Identify the most common cell type in connective tissue: ______________________

   (b) What is the role of these cells? ______________________

   (c) Identify a common fiber type in connective tissue: ______________________

2. Giving examples, discuss the roles of different connective tissues in the body: ______________________

____

____

____

____

____

ISBN: 978-1-92717357-2

Nervous Tissue

Nervous tissue makes up the structures of the nervous system: the brain, spinal cord and all the peripheral nervous tissue, including sensory structures and sense organs. Nervous tissue contains densely packed **neurons**, which are specialized for receiving and transmitting electrochemical impulses. Neurons are usually associated with supporting cells (neuroglia) and connective tissue containing blood vessels. A **nerve** (or nerve fiber) is a cable-like structure in the peripheral nervous system containing the axons of many neurons and their associated insulating sheaths. Nerves are enclosed and protected by connective tissue. The equivalent structure in the central nervous system is called a **tract**.

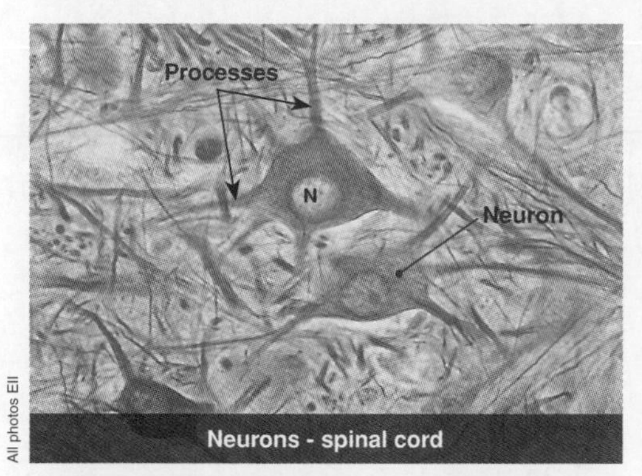

Neurons - spinal cord

All photos EII

The large multipolar (=many-processes) motor neurones of the central nervous system (left) have a large central nucleus (N), several radiating cell processes called dendrites, and a long thin axon. The long processes allow nerve impulses to be transmitted over long distances. Note how the cell processes form a dense network, together with their supporting glial cells.

Astrocytes (right) are a type of glial cell. They have a small cell body and numerous processes. They provide physical and metabolic support to the neurons of the CNS and help maintain the composition of the extracellular fluid.

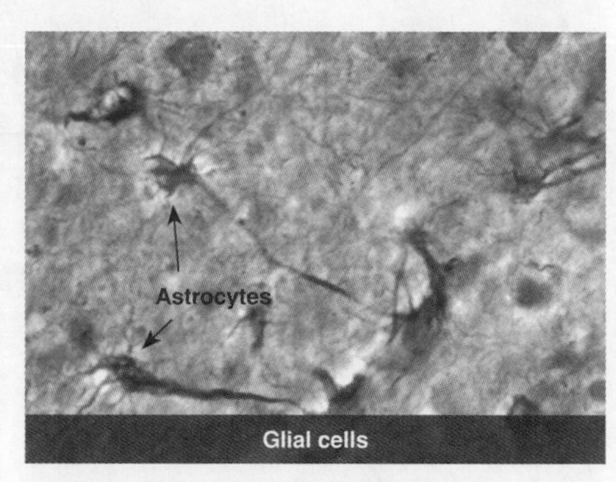

Glial cells

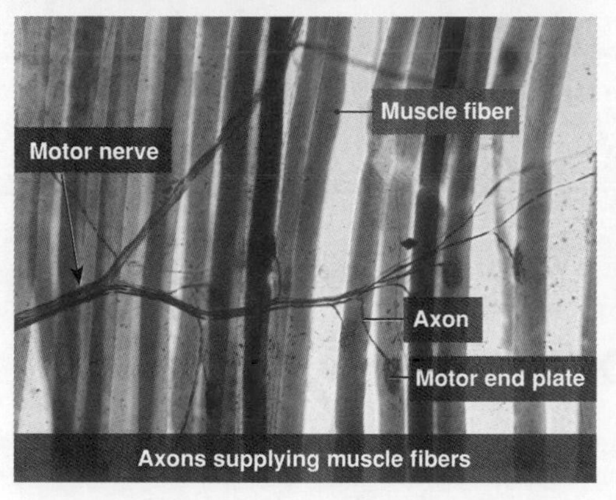

Axons supplying muscle fibers

The photograph (left) shows the terminating axons of a motor nerve. The nerve distributes its axons to the individual muscle fibers. Each axon ends in a specialized junction called the motor end plate.

In the TS of the spinal cord (right), staining reveals outer white matter (the myelinated axons of neurons) and inner gray matter (mostly the cell bodies of neurons). The gray matter is connected across the midline where the central canal is located. The central canal contains nutritive cerebrospinal fluid and is continuous with the ventricles of the brain. Like the brain, it is surrounded by connective tissue meninges.

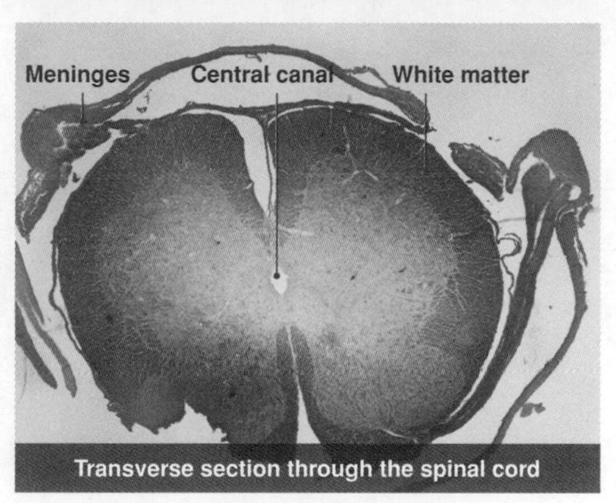

Transverse section through the spinal cord

1. (a) Describe the characteristic features of nervous tissue: ______

   (b) What is the role of glial cells in nervous tissue? ______

   (c) What is the role of the cerebrospinal fluid and how is it supplied to the tissue of the central nervous system?

   ______

2. Discuss the role of neural tissue in the body: ______

ISBN: 978-1-92717357-2

Related activities: *Neuron Structure and Function, Neuroglia*
Weblinks: *Animal Tissues*

Human Organ Systems

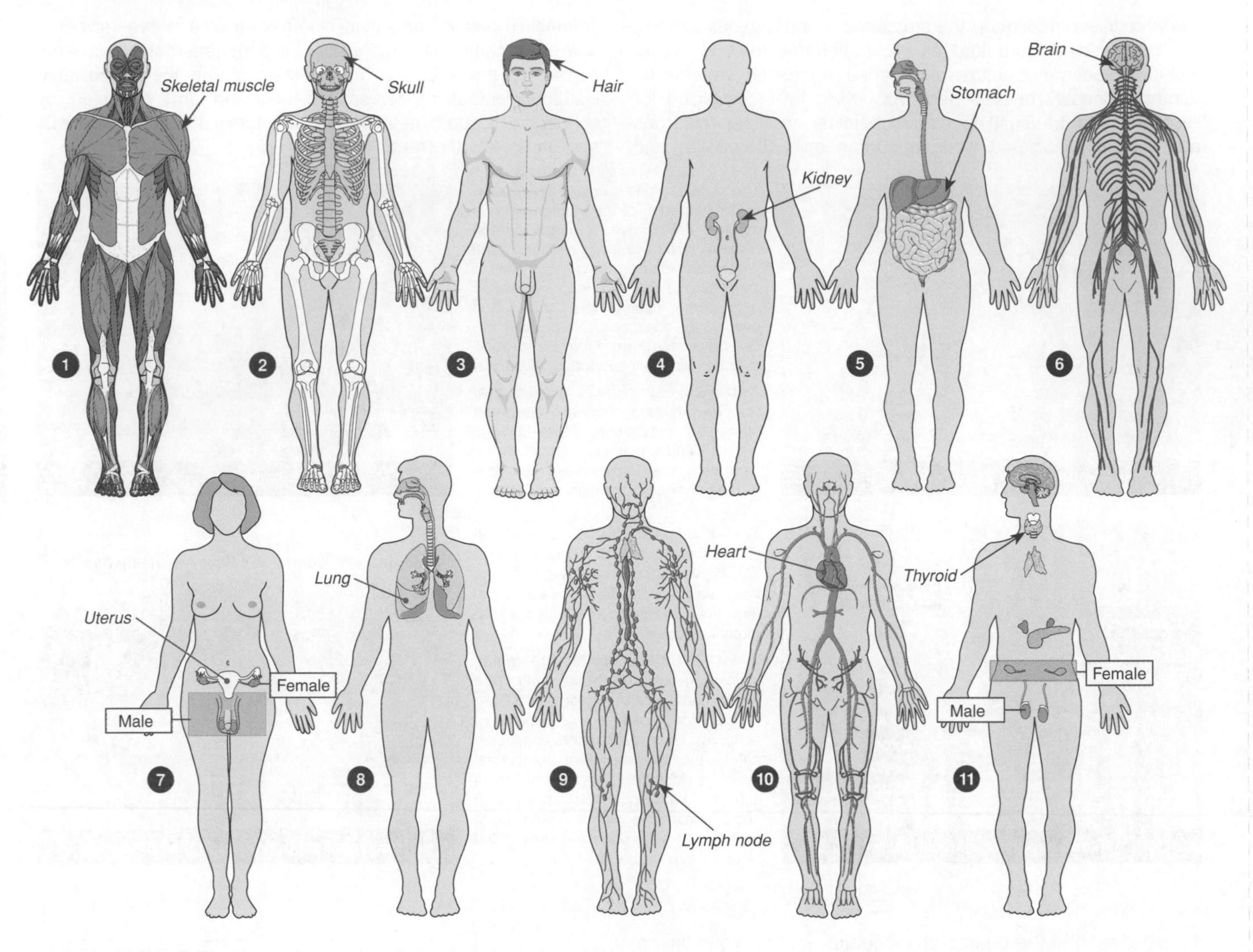

There are 11 organ systems in the human body, each comprising a number of components with specific functions. Identify each organ system (1-11) pictured and state its basic functional role in a few words. Some labels are given as clues and you can use the Weblinks, your textbook or other chapters in this workbook to help you. In addition, identify the main components of organ system #11, briefly state the primary role of each organ, and identify which organs are specific to each gender:

1. ____________________

2. ____________________

3. ____________________

4. ____________________

5. ____________________

6. ____________________

7. ____________________

8. ____________________

9. ____________________

10. ____________________

11. ____________________

Weblinks: *Human Anatomy Online Interactive*

ISBN: 978-1-92717357-2

KEY TERMS: Mix and Match

INSTRUCTIONS: Test your vocabulary by matching each term to its definition, as identified by its preceding letter code.

active transport ....

cell ....

cellular differentiation ....

chromosome ....

connective tissue ....

cytokinesis ....

diffusion ....

endocytosis ....

epithelial tissue ....

exocytosis ....

hypertonic ....

hypotonic ....

interphase ....

ion pump ....

isotonic ....

metaphase ....

mitosis ....

muscle tissue ....

nervous tissue ....

organ ....

organ system ....

organelle ....

osmosis ....

partially permeable membrane ....

passive transport ....

phospholipid ....

prophase ....

telophase ....

tissue ....

A A common term in animal biology for a solution with a lower total solute concentration relative to another solution (across a membrane).

B The division of the cytoplasm of a parent eukaryotic cell into two daughter cells during the late stages of mitosis.

C Active transport process by which cells take in molecules from outside the cell by engulfing it with their plasma membrane.

D A group of organs that work together to perform a specific task.

E The energy-requiring movement of substances across a biological membrane against a concentration gradient.

F A class of amphipathic lipids having both a polar and non-polar region, they are a major component of all cell membranes as they can form lipid bilayers.

G A structural and functional part of the cell usually bound within its own membrane. An example is the mitochondria.

H Tissue that makes up the structures of the nervous system.

I A stage of mitosis in a eukaryotic cell in which the chromatin condenses into chromosomes and becomes visible.

J Tissue that lines internal and external surfaces.

K A membrane that acts selectively to allow some substances, but not others, to pass.

L Process by which a less specialized cell becomes more specialized in order to perform a specific function.

M The phase of the cell cycle prior to cell division in which all the DNA in the nucleus is copied.

N Structures comprising two or more tissues with related functions.

O A stage of mitosis in which condensed chromosomes align in the middle of the cell.

P Tissue responsible for producing both external and internal movement. Types include skeletal, cardiac, and smooth.

Q Movement of substances across a biological membrane without energy expenditure.

R A transmembrane protein that moves ions across a plasma membrane against their concentration gradient.

S The phase of a cell cycle resulting in nuclear division.

T The passive movement of molecules from high to low concentration.

U Support tissue that binds other structures together and provides support and protection against damage, infection, or heat loss.

V Active transport process by which membrane-bound secretory vesicles fuse with the plasma membrane and release the vesicle contents into the external environment.

W A common term in animal biology for a solution with a higher total solute concentration relative to another solution (across a membrane).

X A stage of mitosis in which condensed chromosomes align in the middle of the cell.

Y Passive movement of water molecules across a partially permeable membrane down a concentration gradient.

Z The mitosis stage where the daughter chromosomes reach the opposite poles of the cell and reform the nuclei.

AA An organized structure of DNA and protein found in cells.

BB In animal biology, solutions of equal solute concentration are often termed this.

CC A collection of cells from the same origin, which together carry out a specific function.

DD The basic structural, functional, and biological unit of any living organism.

Cells and Tissues

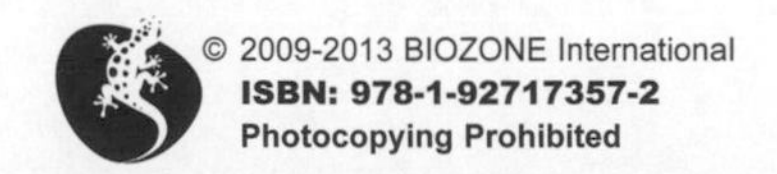

ISBN: 978-1-92717357-2

The Integument and **Homeostasis**

Key concepts

- Homeostasis is maintained using hormonal and nervous mechanisms via negative feedback.
- Thermoregulation enables maintenance of an optimum body temperature for metabolism.
- The integument plays an important role in thermoregulation and other homeostatic processes.
- Modern medical technology enables the diagnosis of homeostatic imbalances.

Key terms

abdominal cavity
anterior/ventral
dermis
distal
dorsal body cavity
epidermis
epithelial membranes
homeostasis
hypodermis
inferior
lateral
medial
mucous membranes (mucosa)
negative feedback
pelvic cavity
positive feedback
posterior/dorsal
proximal
serous membranes (serosa)
superficial
superior
synovial membranes
thoracic cavity
ventral body cavity

Learning Objectives

☐ 1. Use the **KEY TERMS** to compile a glossary for this topic.

Body Cavities and Membranes pages 7, 37-38

☐ 2. Review terms used in relation to the position of anatomical parts. Correctly refer to the location of body parts, using: **anterior** (or ventral), **posterior** (or dorsal), **superior**, **inferior**, **lateral**, **distal**, **proximal**, **medial**, **superficial**.

☐ 3. Describe the location of the **dorsal body cavity** and the **ventral body cavity**. Identify the organs associated with each of these cavities.

☐ 4. Describe the features, location, and functional role of the body's membranes:
(a) **Epithelial membranes**: **cutaneous membranes** (the skin), **mucous membranes**, and **serous membranes**
(b) Connective tissue membranes: **synovial membranes**.

Homeostasis pages 35-36, 39-40, 45-50

☐ 5. Explain the need for **homeostasis** and identify the role of nerves and hormones in achieving a steady state. Appreciate that the ability to maintain homeostasis independently is a developmental milestone.

☐ 6. Explain how **negative feedback** mechanisms maintain dynamic homeostasis by stabilizing systems against excessive change. Use examples (e.g. thermoregulation) to explain how negative feedback operates.

☐ 7. Describe examples of how **positive feedback** operates in biological systems (e.g. in lactation, labor (parturition), and blood clotting). Appreciate that positive feedback is unstable because it amplifies disturbance, but it has certain physiological roles.

☐ 8. Describe how body temperature is regulated in humans, including the role of the **hypothalamus**, the **autonomic nervous system**, the circulatory system (including the blood), the skin and its associated sensory receptors.

The Integumentary System pages 43-44, 109

☐ 9. Describe the functions of the integument and identify structural features associated with these functions.

☐ 10. Describe the structure of the skin including features of its two tissues, the **epidermis** and **dermis**. Appreciate that the **hypodermis** (or subcutaneous tissue) is not considered part of the skin but is closely associated with it.

Medical Imaging and Diagnosis pages 41-42

☐ 11. Discuss the technologies available for medical diagnosis and identify their advantages and disadvantages. You could include reference to any of the following: x-ray imaging and radionuclide scanning, computer imaging techniques (e.g. MRI, CT, and PET), endoscopy, electrocardiography, and biosensors. In each case, describe the basic principle of the technique and situations appropriate to its use.

Periodicals:
Listings for this chapter are on page 279

Weblinks:
www.thebiozone.com/weblink/AnaPhy-3572.html

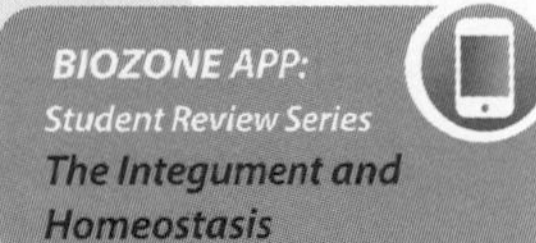

Principles of Homeostasis

Homeostasis the relative physiological constancy of the body, despite external fluctuations. Homeostasis of the internal environment is an essential feature of complex animals and it is the job of the body's **organ systems** to maintain it, even as they make necessary exchanges with the environment. Homeostatic control systems have three functional components: a receptor to detect change, a control centre, and an effector to direct an appropriate response. In **negative feedback** systems, movement away from a steady state triggers a mechanism to counteract further change in that direction. Using negative feedback systems, the body counteracts disturbances and restores the steady state. **Positive feedback** is also used in physiological systems, but to a lesser extent since positive feedback leads to the response escalating in the same direction.

Organ systems maintain a constant internal environment that provides for the needs of all the body's cells, making it possible for animals to move through different and often highly variable external environments. This representation shows how organ systems permit exchanges with the environment. The exchange surfaces are usually internal, but may be connected to the environment via openings on the body surface.

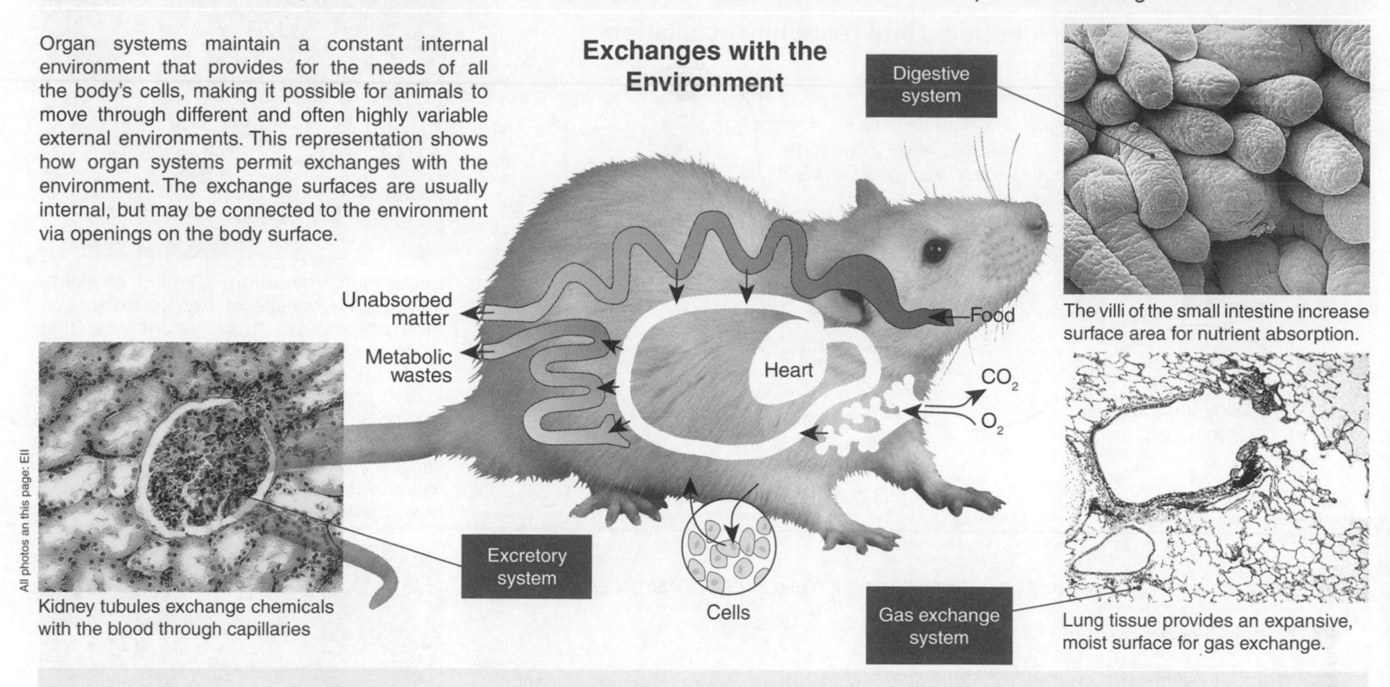

Kidney tubules exchange chemicals with the blood through capillaries

The villi of the small intestine increase surface area for nutrient absorption.

Lung tissue provides an expansive, moist surface for gas exchange.

Negative feedback and control systems

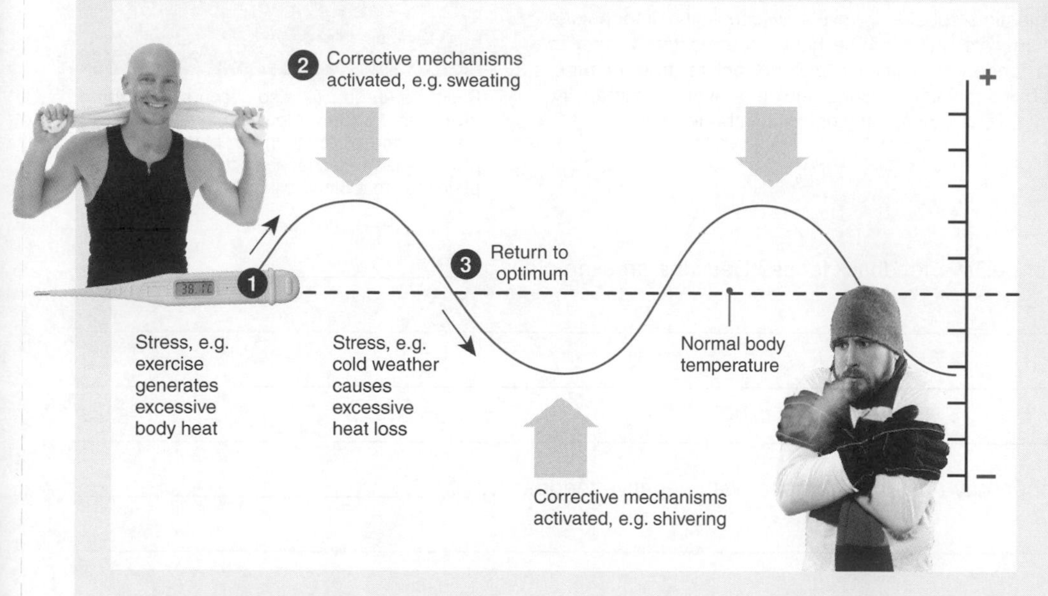

1. A stress or disturbance, e.g. exercise, takes the internal environment away from optimum.
2. Stress is detected by receptors and corrective mechanisms (e.g. sweating) are activated.
3. The corrective mechanisms act to restore optimum.

Negative feedback acts to counteract any departures from a steady physiological state. The diagram shows how a stress (disturbance) is counteracted by corrective mechanisms in the case of body temperature.

In contrast to negative feedback, positive feedback will push physiological levels out of the normal range. While it is inherently unstable, it has a useful role at certain times, e.g. during childbirth.

1. Describe the three main components of a regulatory control system in the human body: ______________________

2. Explain how negative feedback mechanisms maintain homeostasis in a variable environment: ______________________

Positive Feedback

Positive feedback mechanisms amplify a physiological response in order to achieve a particular result. Labor, lactation, fever, and blood clotting all involve positive feedback mechanisms. Normally, a positive feedback loop is ended when a natural resolution is reached (e.g. baby is born, pathogen is destroyed, blood clot forms). Very few physiological processes involve positive feedback because such mechanisms are unstable. If left unchecked, they can be dangerous or even fatal.

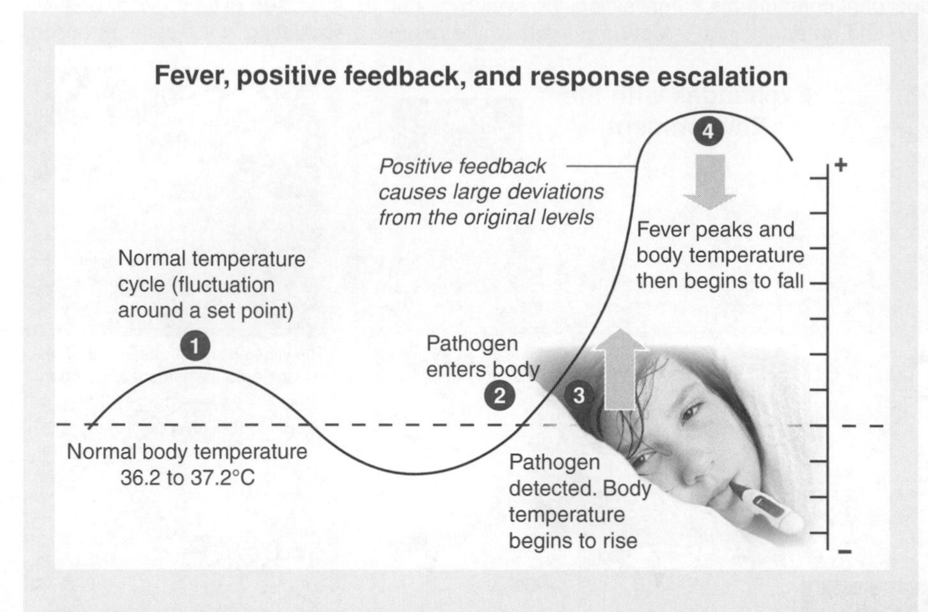

1. Body temperature fluctuates on a normal, regular basis around a narrow set point.
2. Pathogen enters the body.
3. The body detects the pathogen and macrophages attack it. Macrophages release interleukins which stimulate the hypothalamus to increase prostaglandin production and reset the body's thermostat to a higher 'fever' level by shivering (the chill phase).
4. The fever breaks when the infection subsides. Levels of circulating interleukins (and other fever-associated chemicals) fall, and the body's thermostat is reset to normal. This ends the positive feedback escalation and normal controls resume. If the infection persists, the escalation may continue, and the fever may intensify. Body temperatures in excess of 43°C are often fatal or result in brain damage.

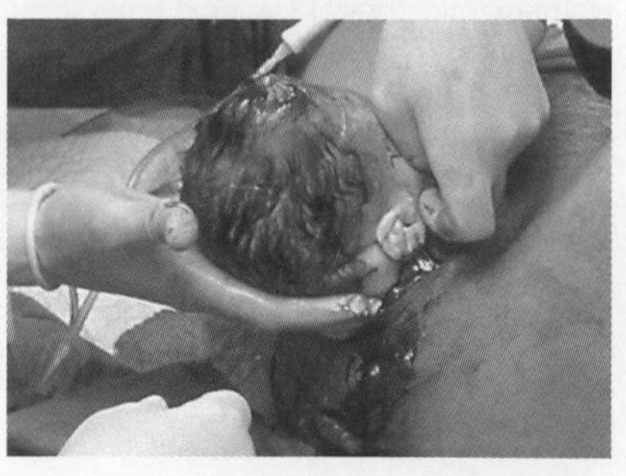

Labor and lactation: During childbirth (above), the release of oxytocin intensifies the contractions of the uterus so that labor proceeds to its conclusion. The birth itself restores the system by removing the initiating stimulus. After birth, levels of the milk-production hormone prolactin increase. Suckling maintains prolactin secretion and causes the release of oxytocin, resulting in milk release. The more an infant suckles, the more these hormones are produced.

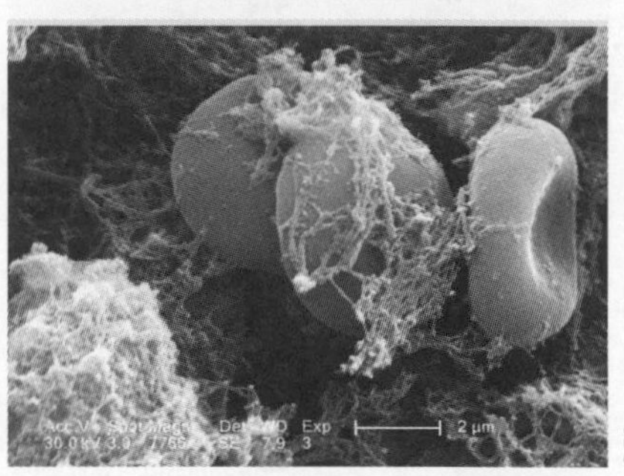

CDC

Positive feedback also occurs in **blood clotting**. A wound releases signal chemicals that activate platelets in the blood. Activated platelets release chemicals that activate more platelets, so a blood clot is rapidly formed.

1. (a) What is the biological role of positive feedback loops? Describe an example: ______________________

(b) Why is positive feedback inherently unstable (compare with negative feedback)? ______________________

(c) How is a positive feedback loop normally stopped? ______________________

(d) Describe a situation in which this might not happen. What would be the result? ______________________

Related activities: *Principles of Homeostasis, Maintaining Homeostasis*
Weblinks: *Control, Regulation and Feedback*

ISBN: 978-1-92717357-2

Body Membranes and Cavities

The study of anatomy and physiology requires a basic understanding of anatomical terms, including the directional (e.g. distal) and regional (e.g. pelvic) terms used to describe the position of body parts, the location of the body's **cavities** (dorsal and ventral), and the way in which the body's **membranes** line those cavities and protect the organs within. A membrane is a thin layer of tissue that covers a structure or lines a cavity. The body's membranes fall into two broad categories: **epithelial membranes** (the skin, mucosa, and serosa), and **synovial membranes**, which lack epithelium. Membranes line and cover the internal and external surfaces of the body, protecting and, in some cases, lubricating them. Epithelial membranes are formed from epithelium and the connective tissue on which it rests. Whereas the skin (**cutaneous membrane**) is exposed to air and is a dry membrane, **mucous membranes** (mucosa) and **serous membranes** (serosa) are moist and bathed in secretions.

Body Cavities

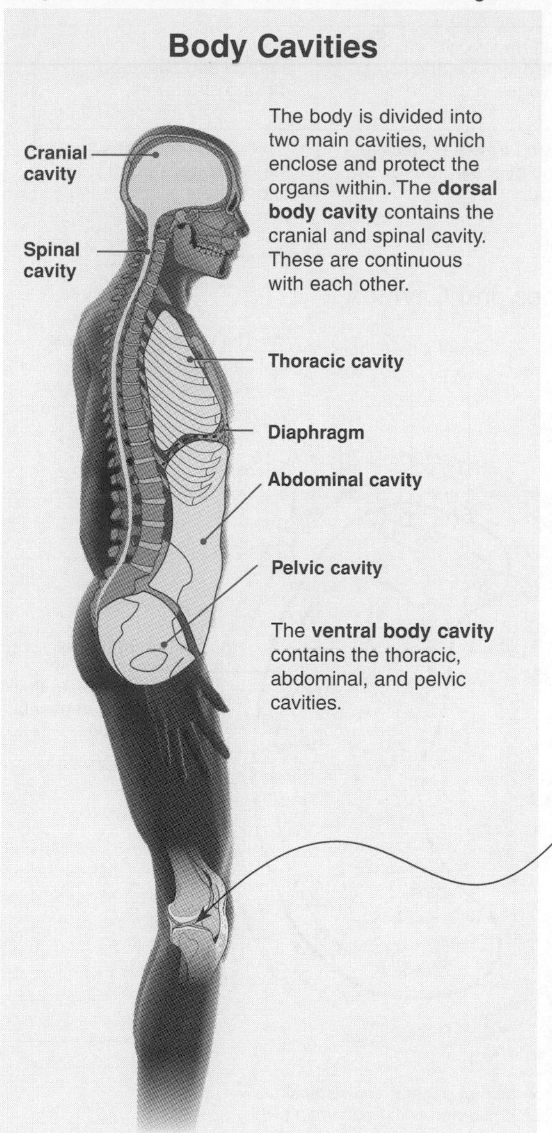

Above: Location of dorsal and ventral body cavities in the human body. A knee joint shows a typical location of connective tissue (synovial) membranes.

Location of the Body's Membranes

Cutaneous membrane (the **skin**) forms a protective covering over the surface of the body.

Mucous membranes (muscosa) lines all body cavities that open to the exterior, i.e. the hollow organs of the respiratory, digestive, urinary, and reproductive tracts.

Serous membranes (serosa) line internal body cavities that are closed to the exterior. Serosa occur in pairs: the **parietal layer** lines the body wall and the **visceral layer** lines the organ within that cavity. The membranes are separated by a thin film of serous fluid.

They are named according to their location in the body:
- **peritoneum** (abdomen)
- **pleura** (lungs)
- **pericardium** (heart).

Synovial membranes line the capsules around joints and secrete a lubricating fluid.

A B C D

Right: Location of the mucous and serous membranes in the thorax.

Cavity		Main contents	Membranous lining
Dorsal body cavity	Cranial cavity	Brain	Meninges
	Vertebral canal	Spinal cord	Meninges
Ventral body cavity	Thoracic cavity	Lungs Heart	Pleural cavities Pericardium
	Abdominal cavity	Digestive organs, spleen, kidneys	Peritoneum
	Pelvic cavity	Bladder, reproductive organs	Peritoneum

1. Use the information given above to name the **serous membranes** labeled A-D above right:

(a) A: ______________________ (c) C: ______________________

(b) B: ______________________ (d) D: ______________________

2. (a) Describe the general role of epithelial membranes in the body: ______________________

(b) Explain how epithelial membranes differ from synovial membranes: ______________________

ISBN: 978-1-92717357-2

Related activities: Terms in Anatomy and Physiology, Joints

Weblinks: The Body's Cavities, Quizlet: Body Cavities and Membranes

Features of the Body's Membranes

Epithelial Membranes			
Serous membranes	Mucous membranes	Cutaneous membrane	Synovial membranes
• Moist - bathed in secretions	• Moist - bathed in secretions	• Dry, open to the air	• Moist - bathed in secretions
• Made of a thin layer of squamous epithelium resting on a thin layer of loose connective tissue.	• Composed of some type of simple epithelium (e.g. columnar or squamous) resting on loose connective tissue.	• It is made up of an epidermis of stratified squamous epithelium and an underlying dermis of connective tissue.	• They are composed of connective tissue and contain no epithelial cells.
• The parietal and visceral membranes are separated by a thin film of **serous fluid**.	• The epithelium of mucosae is often absorptive or secretory. Many of them, but not all, produce mucus.	• The outermost cells of skin are protected by a keratin layer, which varies in thickness.	• They provide a smooth surface and cushion moving structures.
• Line internal body cavities that are closed to the exterior. Occur in pairs and named according to their location, e.g. visceral pleura (see previous page).	• Lines all body cavities that open to the exterior, i.e. the hollow organs of the respiratory, digestive, urinary, and reproductive tracts.	• The **skin** forms a protective covering over the surface of the body.	• Synovial membranes line the capsules around joints and secrete a lubricating **synovial fluid**.

The Relationship Between Membranes and Cavities

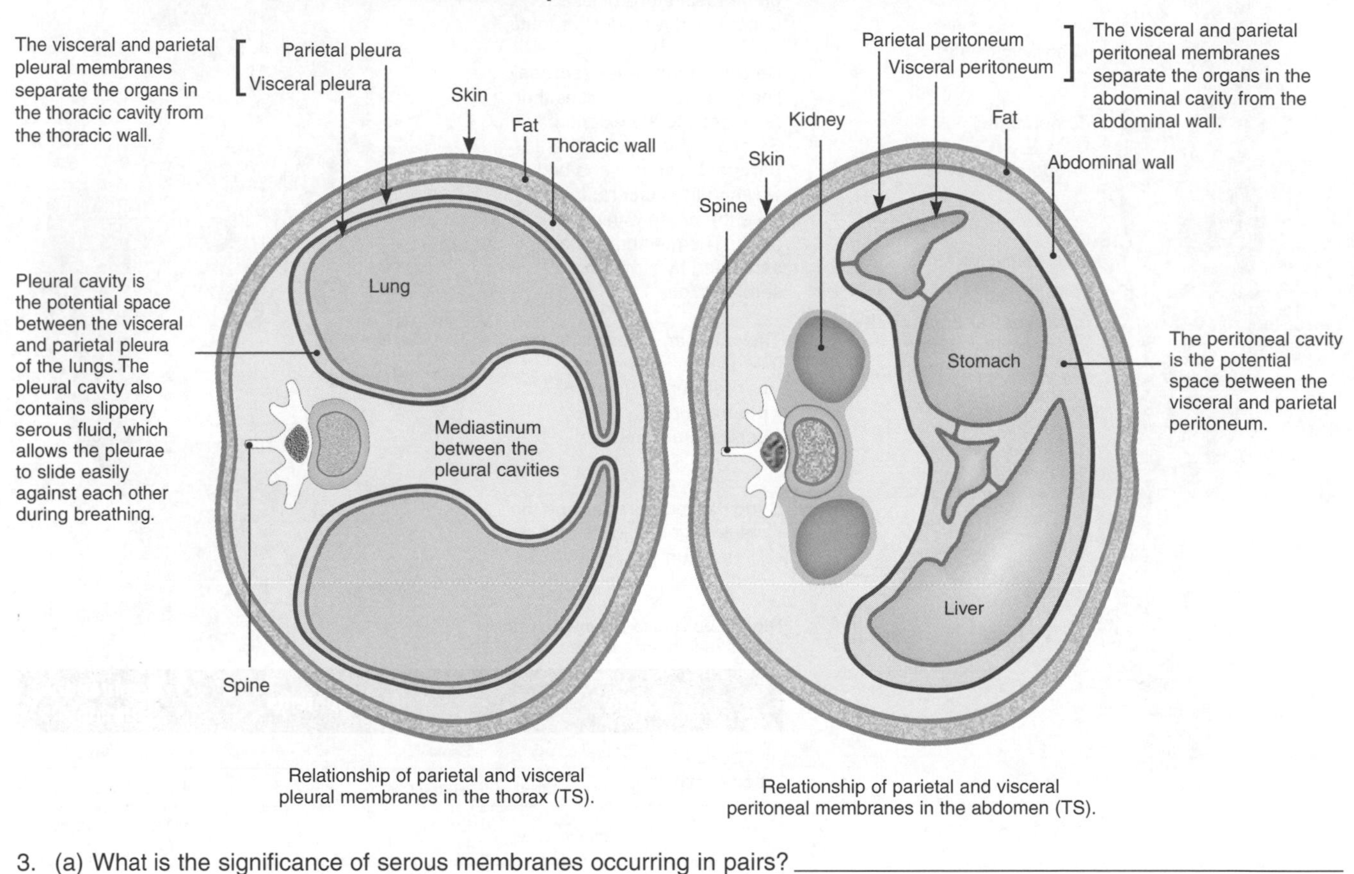

Relationship of parietal and visceral pleural membranes in the thorax (TS).

Relationship of parietal and visceral peritoneal membranes in the abdomen (TS).

3. (a) What is the significance of serous membranes occurring in pairs? ____________________

____

____

(b) Describe the role of the serous fluid secreted by these membranes: ____________________

____

____

4. (a) The stomach (labeled above, right) is covered in visceral peritoneum. What lines its inside surface (the lumen)?

____

(b) What is the importance of the cutaneous membrane being a dry membrane? ____________________

____

____

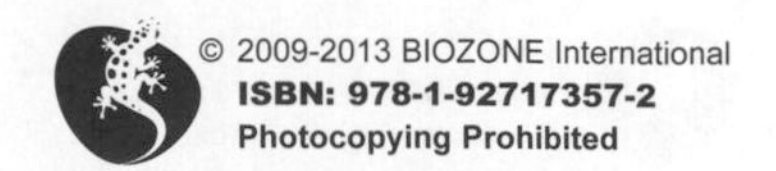

ISBN: 978-1-92717357-2

Maintaining Homeostasis

The various organ systems of the body act to maintain homeostasis through a combination of hormonal and nervous mechanisms. In everyday life, the body must regulate respiratory gases, protect itself against agents of disease (pathogens), maintain fluid and salt balance, regulate energy and nutrient supply, and maintain a constant body temperature. All these must be coordinated and appropriate responses made to incoming stimuli. In addition, the body must be able to repair itself when injured and be capable of reproducing (leaving offspring).

Regulating Respiratory Gases

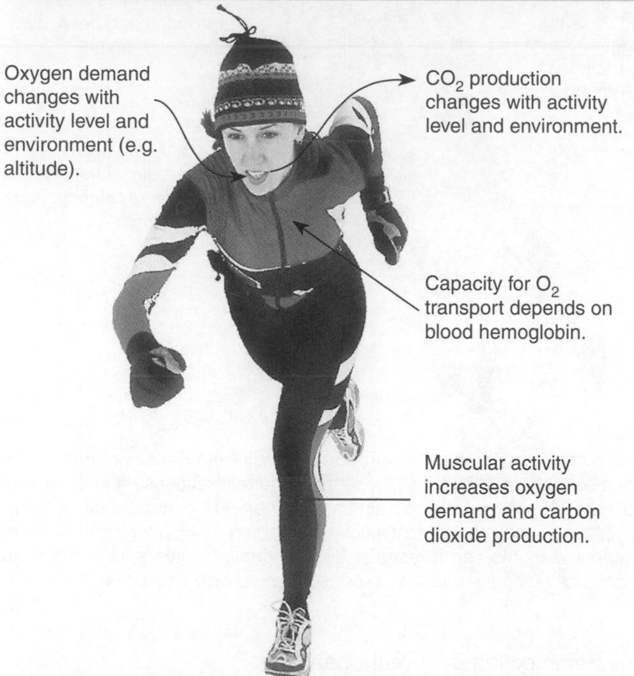

Oxygen must be delivered to all cells and carbon dioxide (a waste product of cellular respiration) must be removed. **Breathing** brings in oxygen and expels CO_2, and the cardiovascular and lymphatic systems circulate these respiratory gases (the oxygen mostly bound to hemoglobin). The rate of breathing is varied according to oxygen demands (as detected by CO_2 levels in the blood).

Coping with Pathogens

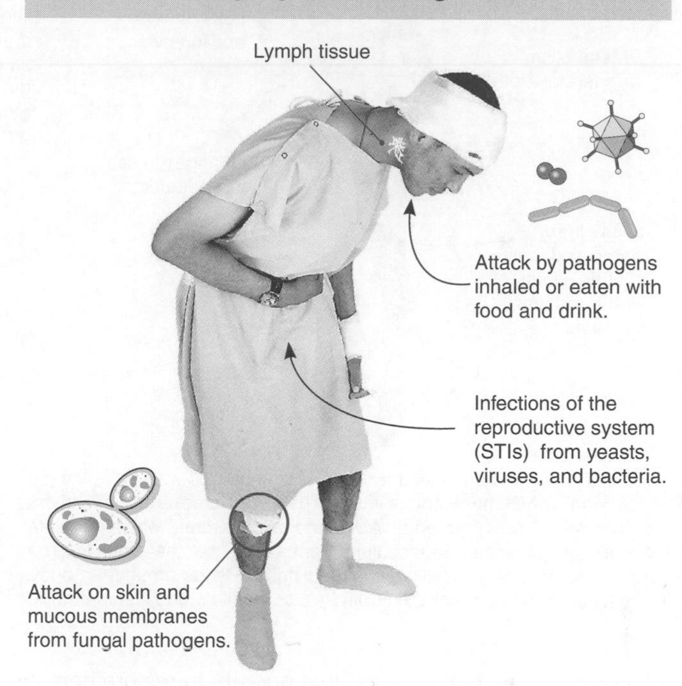

All of us are under constant attack from pathogens (disease causing organisms). The body has a number of mechanisms that help to prevent the entry of pathogens and limit the damage they cause if they do enter the body. The skin, the digestive system, and the immune system are all involved in the body's defense, while the cardiovascular and lymphatic systems circulate the cells and antimicrobial substances involved.

Maintaining Nutrients and Removing Wastes

Repairing Injuries

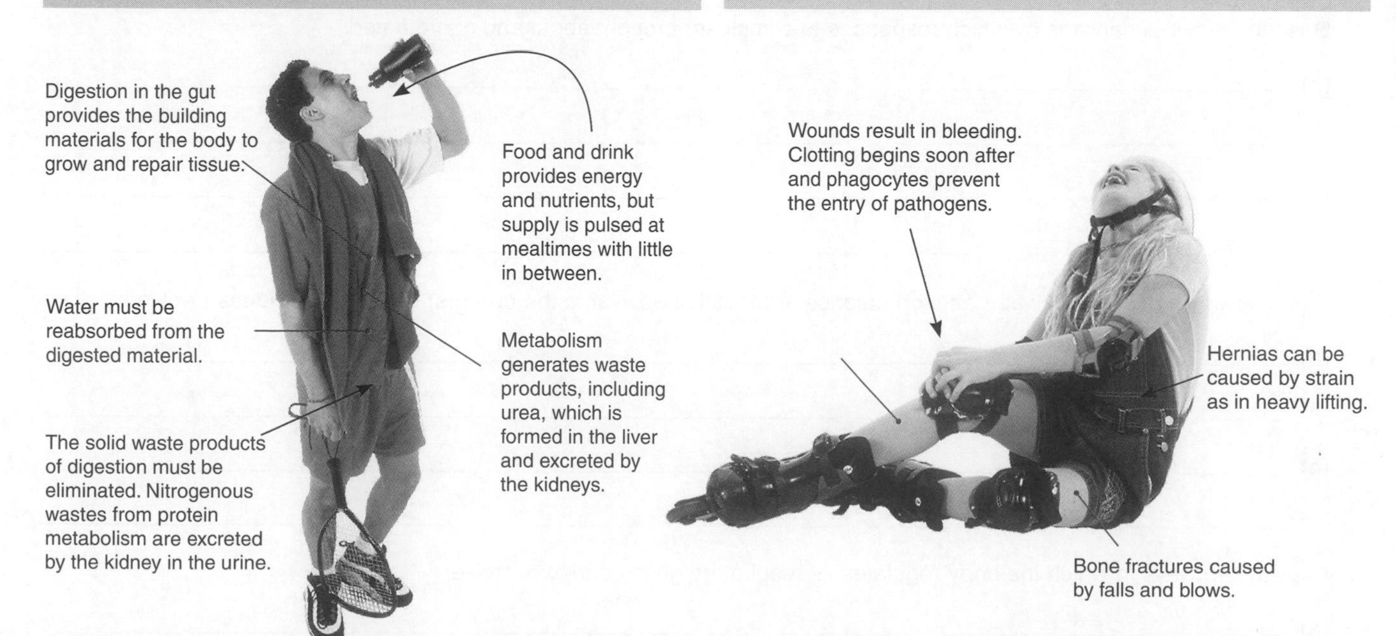

Food and drink is taken in to maintain energy supplies. The digestive system makes these nutrients available, and the cardiovascular system distributes them throughout the body. Food intake is regulated largely through nervous mechanisms, while hormones control the cellular uptake of glucose. The liver metabolizes proteins to form urea, which is excreted by the kidneys.

Damage to body tissues triggers the **inflammatory response** and white blood cells move to the injury site. The inflammatory response is started (and ended) by chemical signals (e.g. from histamine and prostaglandins) released when tissue is damaged. The cardiovascular and lymphatic systems distribute the cells and molecules involved.

ISBN: 978-1-92717357-2

Regulating Temperature, Fluid and Electrolytes

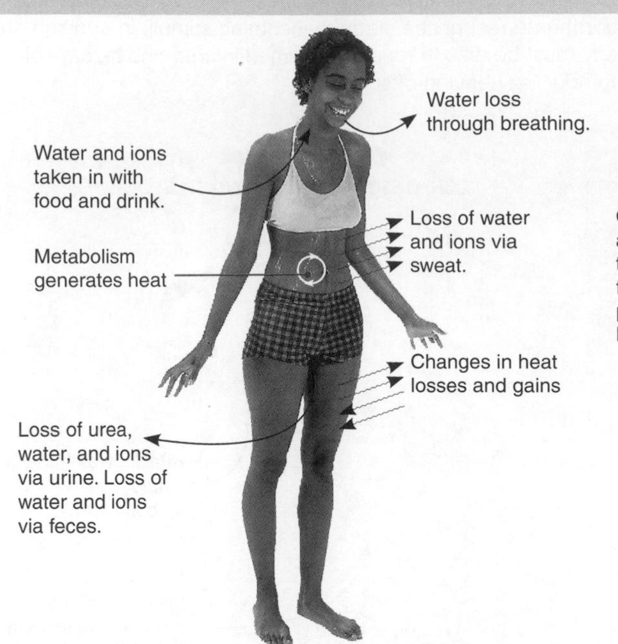

The balance of fluid and electrolytes (and excretion of wastes) is the job of the kidneys. Osmoreceptors monitor blood volume and bring about the release of the hormones ADH and aldosterone, which regulate reabsorption of water and sodium from blood via the kidneys. The cardiovascular and lymphatic systems distribute fluids around the body. The circulatory system and skin both help to maintain body temperature.

Coordinating Responses

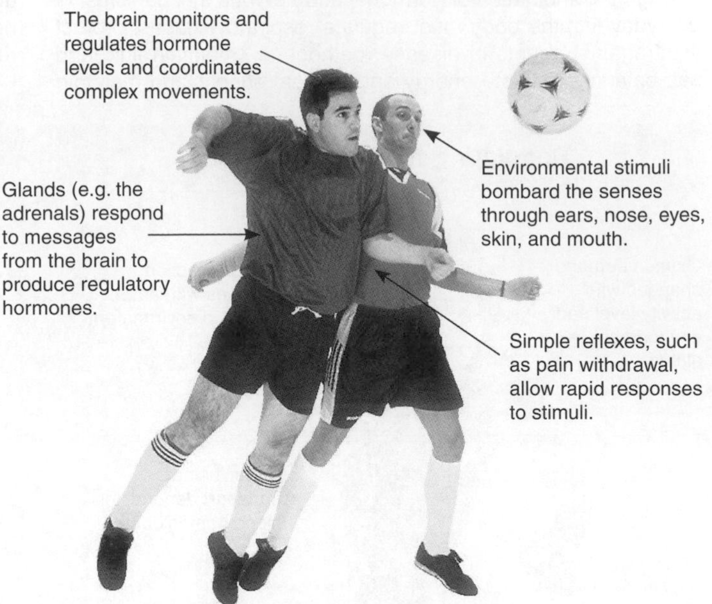

The body is constantly bombarded by stimuli from the environment. The brain sorts these stimuli into those that require a response and those that do not. Responses are coordinated via nervous or hormonal controls. Simple nervous responses (reflexes) act quickly. Hormones, which are distributed by the cardiovascular and lymphatic systems, take longer to produce a response and the response is more prolonged.

1. Describe two mechanisms that operate to restore homeostasis after infection by a pathogen:

   (a) ______________________________

   (b) ______________________________

2. Describe two mechanisms by which responses to stimuli are brought about and coordinated:

   (a) ______________________________

   (b) ______________________________

3. Explain two ways in which water and ion balance are maintained. Name the organ(s) and any hormones involved:

   (a) ______________________________

   (b) ______________________________

4. Explain two ways in which the body regulates its respiratory gases during exercise:

   (a) ______________________________

   (b) ______________________________

ISBN: 978-1-92717357-2

Diagnostic Medicine

Sophisticated **medical imaging** has provided the means to look in detail at the tissues and organs of the body, making it possible to accurately and rapidly diagnose disorders and therefore treat people more effectively for medical problems. As well as imaging techniques, simpler methods, such as blood tests, are also widely used for diagnostic purposes.

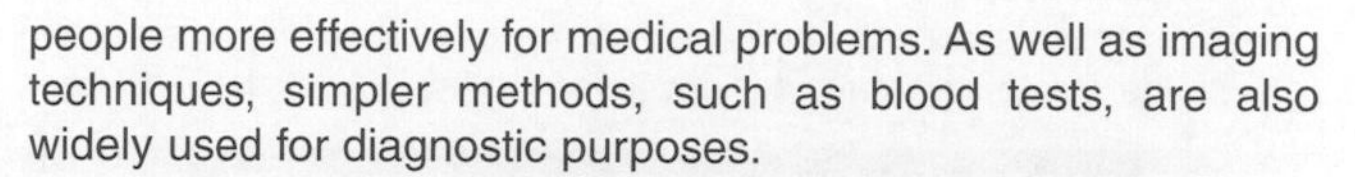

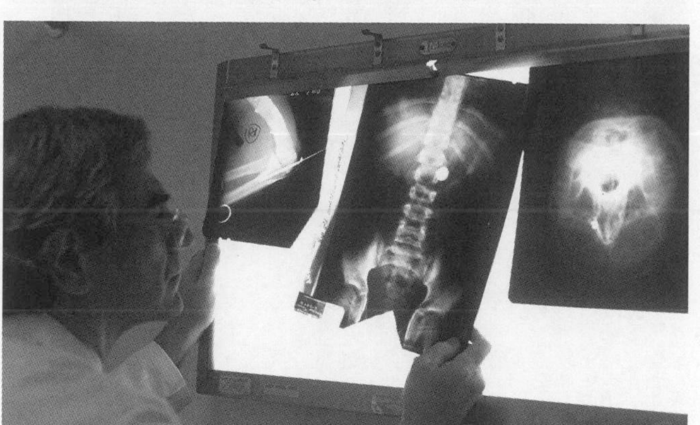

X-ray imaging

X-rays are a form of electromagnetic radiation that can pass through tissues and expose photographic film. The X-rays are absorbed by dense body tissues (e.g. bone) which appear as white areas, but they pass easily through less dense tissues (e.g. muscle), which appear dark. X-rays are used to identify fractures or abnormalities in bone. X-ray technology is also used in conjunction with computer imaging techniques (see below).

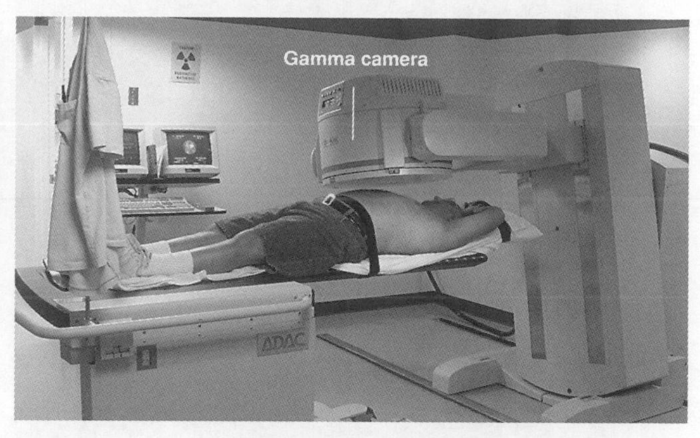

Radionuclide scanning

Radionuclide scanning involves introducing a radioactive substance (the radionuclide) into the body, where it is taken up in different amounts by different tissues (e.g. radioactive iodine is taken up preferentially by the thyroid). The radiation emitted by the tissues that take up the radionuclide is detected by a gamma camera. Radionuclide scanning provides better detail of function than other techniques, but gives less anatomical detail.

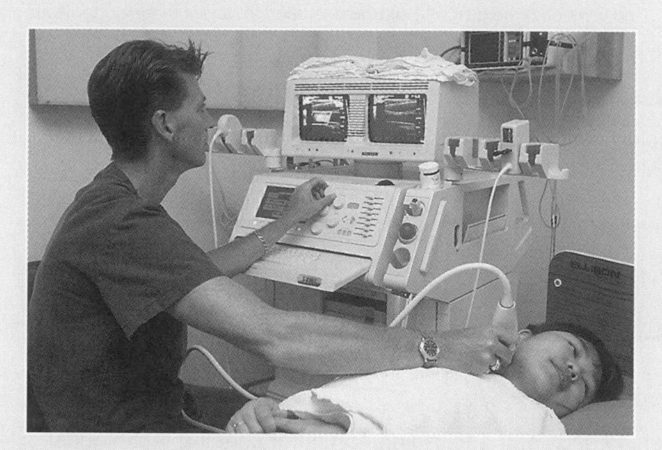

Diagnostic uses of ultrasound

Ultrasound is a diagnostic tool used to visualize internal structures without surgery or X-rays. Ultrasound imaging is based on the fact that tissues of different densities reflect sound waves differently. Sound waves are directed towards a structure (e.g. uterus, heart, kidney, liver) and the reflected sound waves are recorded. An image of the internal structures is analyzed by computer and displayed on a screen.

Echocardiography uses ultrasound to investigate heart disorders such as congenital heart disease and valve disorders. The liver and other abdominal organs can also be viewed with ultrasound for diagnosis of disorders such as cirrhosis, cysts, blockages, and tumors. Ultrasound scans of the uterus are commonly used during pregnancy to indicate placental position and aspects of foetal growth and development. This information aids better pregnancy management.

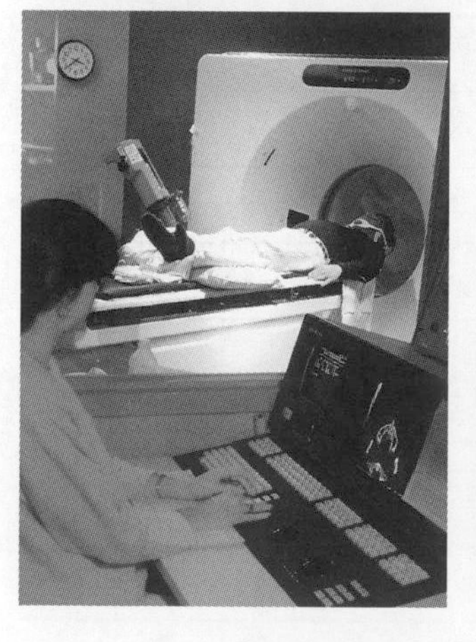

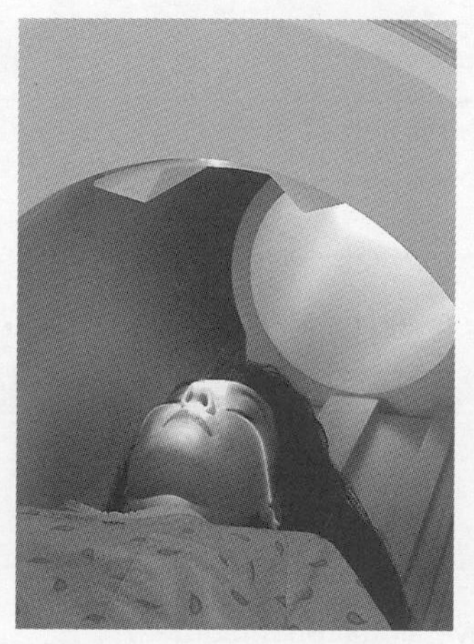

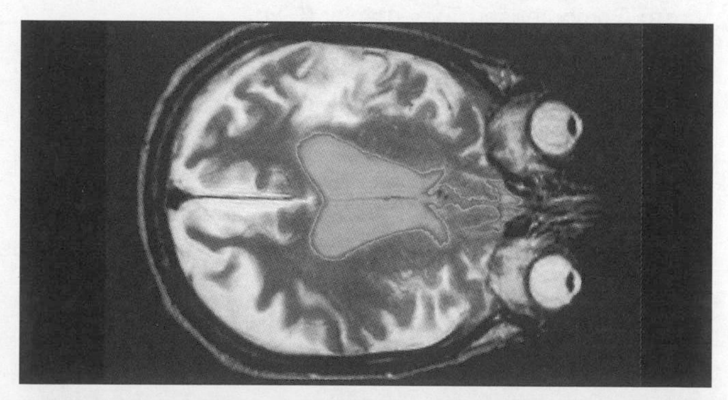

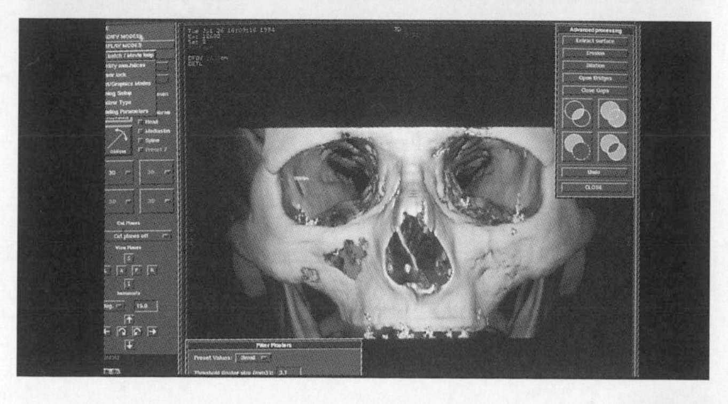

Computer imaging techniques

Computers are used extensively to examine the soft tissues of the body for diagnostic purposes. The photos directly above show **magnetic resonance imaging** (MRI), which uses computer analysis of high frequency radio waves to map out variations in tissue density, especially of the central nervous system (above, far right). In **computerized tomography** (CT) scans, a series of X-rays is made through an organ and the picture from each X-ray slice is reconstructed (using computer software) into a 3-D image (e.g. the skull, right). Such images can be used to detect abnormalities such as tumors.

ISBN: 978-1-92717357-2

Related activities: *The Skeleton, Aging, and Disease, Monoclonal Antibodies, Type 1 Diabetes Mellitus, Type 2 Diabetes Mellitus*

Endoscopy

An **endoscope** is an illuminated tube comprising fiber-optic cables with lenses attached. Endoscopy can be used for a visual inspection of the inside of organs (or any body cavity) to look for blockages or damage. Endoscopes can also be fitted with devices to remove foreign objects, temporarily stop bleeding, remove tissue samples (biopsy), and remove polyps or growths.

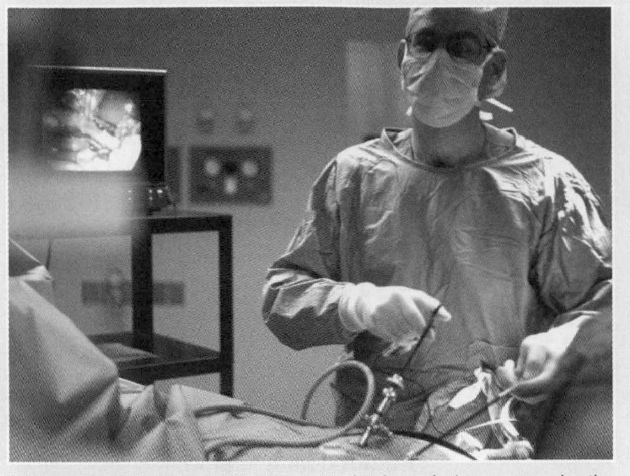

Laparoscopy is the endoscopic examination of the organs in the abdominal cavity, and is used during simple surgical operations (e.g. tubal ligation). Endoscopic examination of the stomach is called gastroscopy.

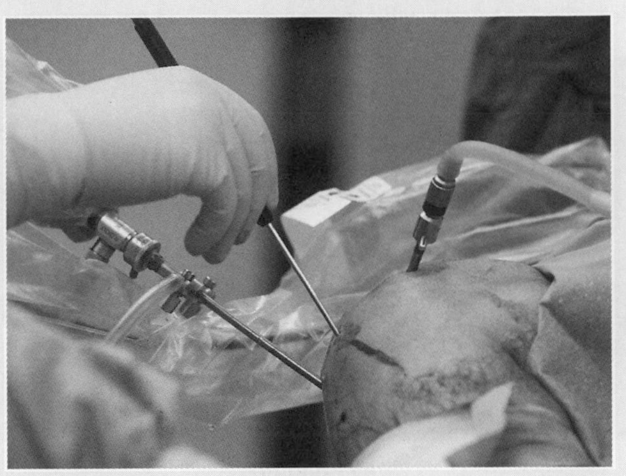

Arthroscopy is used for inspecting joints, usually knee joints (above), while the patient is under a general anesthetic. Using very small incisions, damaged cartilage can be removed from the joint using other instruments.

Biosensors

Biosensors are electronic monitoring devices that use biological material to detect the presence or concentration of a particular substance. Because of their specificity and sensitivity, enzymes are ideally suited for use in biosensors. The example below illustrates how the enzyme **glucose oxidase** is used to detect blood sugar level in diabetics.

The enzyme, *glucose oxidase* is immobilized in a semi-conducting silicon chip. It catalyses the conversion of glucose (from the blood sample) to gluconic acid.

Hydrogen ions from the gluconic acid cause a movement of electrons in the silicon which is detected by a transducer. The strength of the electric current is directly proportional to the blood glucose concentration.

Results are shown on a liquid crystal display.

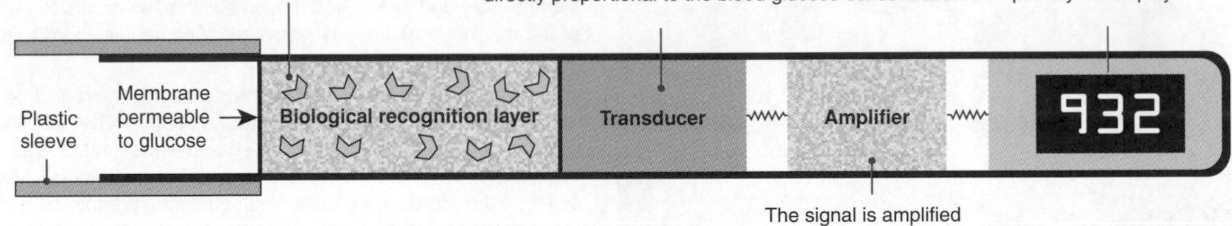

The signal is amplified

1. Describe the basic principle of the scanning technology behind each of the following computer imaging techniques:

   (a) Computerised Tomography (CT): ______________________________

   (b) Magnetic Resonance Imaging (MRI): ______________________________

2. Describe the benefits of using computer imaging techniques such as MRI or CT: ______________________________

3. Explain how radionuclide scanning differs from X-rays: ______________________________

4. Describe the benefits of surgery using endoscopy over conventional open surgery: ______________________________

5. Describe the basic principle of a biosensor: ______________________________

ISBN: 978-1-92717357-2

The Integumentary System

The skin, or **cutaneous membrane**, and its associated structures (hair, sweat glands, nails) collectively make the **integumentary system**. The skin is the body's largest organ. Unlike other epithelial membranes, it is a dry membrane, made up of an outer **epidermis** and underlying **dermis**. The subcutaneous tissue beneath the dermis, which is largely fat (a loose connective tissue) is called the **hypodermis**. It is not part of the skin, but it does anchor the skin to underlying organs, thereby insulating and protecting them. The homeostatic interactions of the skin with other body systems are described below (highlighted panels).

Endocrine system

- Estrogens help to maintain skin hydration.
- Androgens activate the sebaceous glands and help to regulate the growth of hair.
- Changes in skin pigmentation are associated with hormonal changes during pregnancy and puberty.

Respiratory system

- Provides oxygen to the cells of the skin via gas exchange with the blood.
- Removes carbon dioxide (gaseous metabolic waste) from the cells of the skin via gas exchange with the blood.

Cardiovascular system

- Blood vessels transport O_2 and nutrients to the skin and remove wastes (via the blood).
- The skin prevents fluid loss and acts as a reservoir for blood.
- Dilation and constriction of blood vessels is important thermoregulatory mechanism.
- The blood supplies substances required for functioning of the skin's glands.

Digestive system

- Skin synthesizes vitamin D, which is required for absorption of calcium from the gut.
- Digestive system provides nutrients for growth, repair, and maintenance of the skin (delivered via the cardiovascular system).

Skeletal system

- Skin absorbs ultraviolet light and produces a vitamin D precursor. Vitamin D is involved in calcium and phosphorus metabolism, and is needed for normal calcium absorption and deposition of calcium salts in bone.

Nervous system

- Many sensory organs (e.g. Pacinian corpuscle) and simple receptors (e.g. thermoreceptors), are located in the skin.
- The nervous system regulates blood vessel dilation and sweat gland secretion.
- CNS interprets sensory information from the skin's sensory receptors.
- Nervous stimulation of arrector pili muscles causes erection of hair (thermoregulatory response).

Lymphatic system and immunity

- Tissue fluid bathes and nourishes skin cells. Lymphatic vessels prevent edema by collecting and returning tissue fluid to the general circulation.

Urinary system

- Vitamin D synthesis begins in the skin but final activation of vitamin D occurs in the kidneys.
- Urination controlled by a voluntary sphincter in the urethra.

Muscular system

- Muscular activity generates heat, which is dissipated via an increase in blood flow to the skin's surface.
- Muscular activity and increased blood flow increases secretion from the skin's glands (e.g. sweating).

Reproductive system

- Mammary glands, which are modified sweat glands, nourish the infant in lactating women.
- Skin stretches during pregnancy to accommodate enlargement of the uterus.
- Changes in skin pigmentation are associated with pregnancy and puberty.

Fingernail

Fingernails (inset, right) and **toenails** are modifications of the epidermis that protect the ends of the digits. Nails are formed from the horny layer of the epidermis (stratum corneum) that contains hard keratin, and they grow from the basal layer of cells in an area called the nail matrix.

Hairy skin

Hair is found over almost all of the body. Each hair has a shaft, which protrudes above the skin's surface, and a root and hair bulb beneath. The root and shaft are made of dead, keratinized epithelial cells in three layers (a central medulla, thick inner cortex, and outer cuticle). As with skin, melanin is responsible for the color of the hair.

General Functions and Effects on all Systems

The skin is the body's largest organ and the integumentary system covers and provides physical and chemical protection for most parts of all other body systems. It has a critical role in thermoregulation and in the absorption of sunlight and synthesis of a vitamin D precursor.

ISBN: 978-1-92717357-2

Related activities: *Human Organ Systems, Skin Senses*
Weblinks: *The Integumentary System, Skin Structure and Function*

The Structure of the Skin

The skin is made up of two kinds of tissue resting on a underlying layer of fat.

1. The upper **epidermis** of **stratified squamous epithelium**. There are up to five layers of cells in the epidermis, with the regenerative **basal layer** furthest from the surface. The basal layer houses the melanocytes containing the pigment **melanin**, which gives skin a dark color. Cells divide at the basal layer and migrate towards the surface, becoming flatter and more **keratinized**. Like all epithelial tissue, the skin's epidermis lacks blood vessels.

2. The lower **dermis** of dense connective tissue contains collagen (strength), elastic fibers (flexibility) and reticular fibers (support). The dermis also contains the skin's sensory receptors. The upper papillary region is uneven with small projections into the epidermis. The lower reticular layer contains blood vessels, sweat and oil glands, and deep dermal sensory receptors.

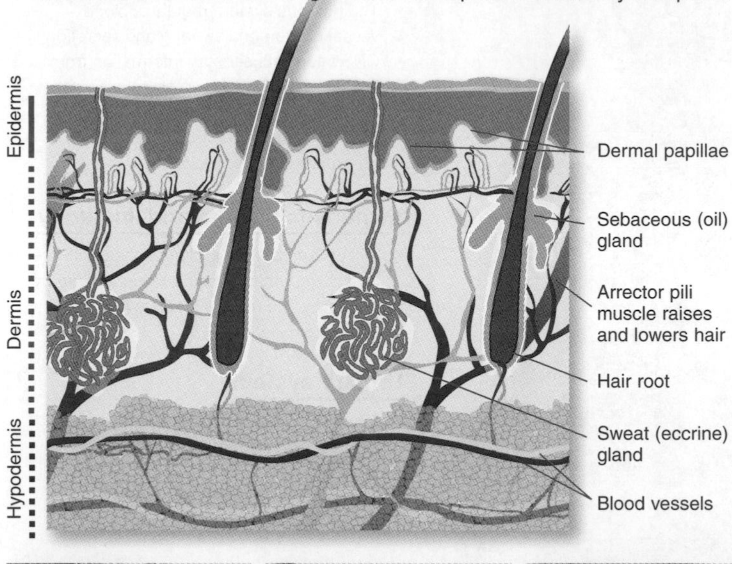

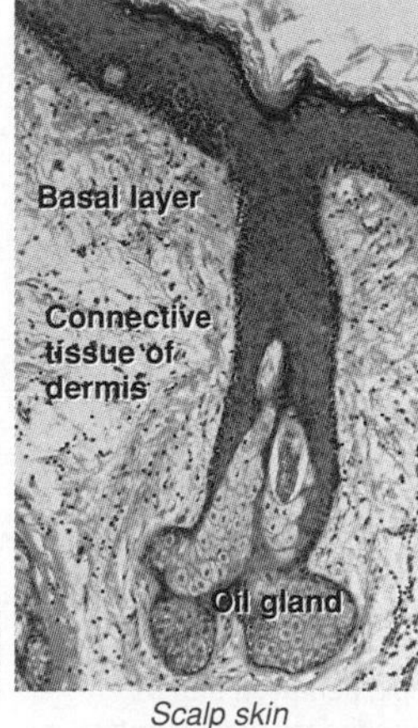

Scalp skin

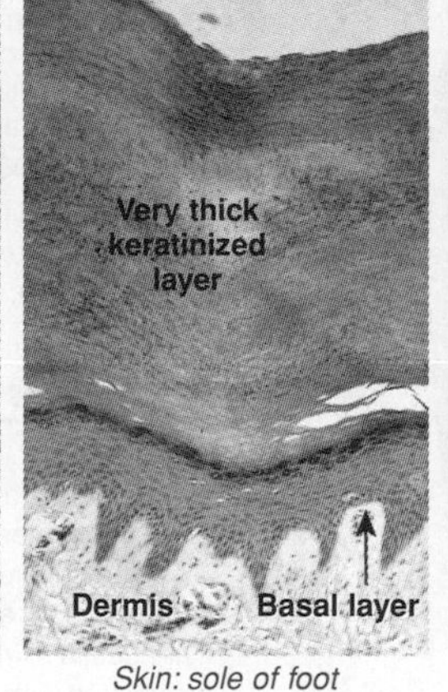

Skin: sole of foot

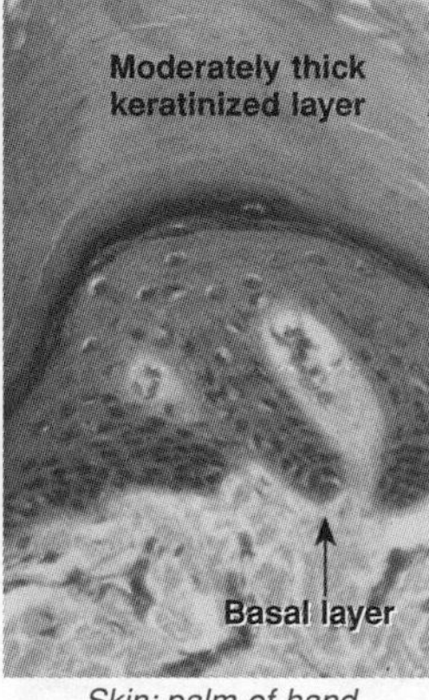

Skin: palm of hand

Photos: EII

There are up to five layers of cells in the epidermis. The thickness of the layers, particularly the outermost heavily keratinized layer, varies depending on where the skin is. Keratin protects the deeper cell layers, and skin subjected to regular wear and tear is heavily keratinized.

Changes in Skin

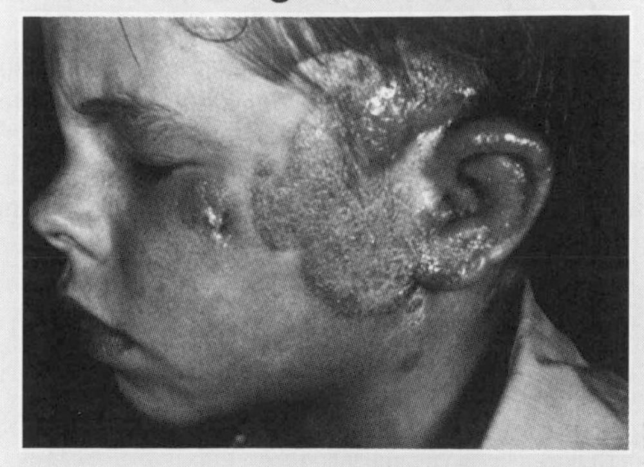
CDC

Ringworm is a fungal infection of the skin. It can affect any area of the body surface and causes raised ring-like scaly lesions, which may be itchy and inflamed.

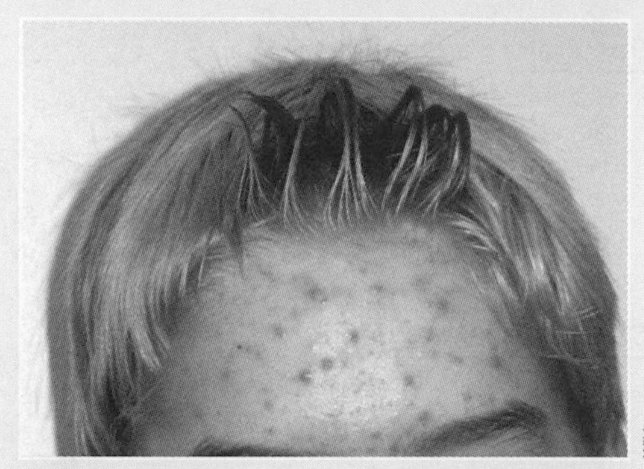
Wiki

Acne is a common skin condition in adolescents in which the sebaceous glands become infected and **pimples** appear on the skin. Acne can be extremely severe, causing deep abcesses and scarring.

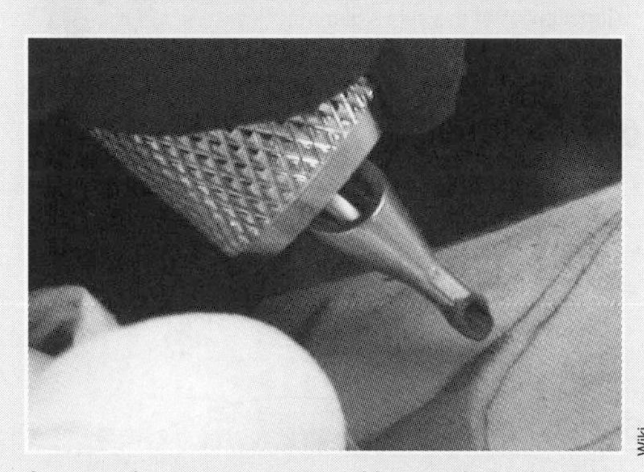
Wiki

A **tattoo** is a permanent marking made by inserting ink into the layers of skin. The pigment remains stable in the upper layer of the dermis, trapped within fibroblasts but, in the long term, it tends to migrate deeper into the dermis and the tattoo degrades.

1. Describe two homeostatic functions of the skin:

   (a) ______________________________

   (b) ______________________________

2. Explain why a tattoo stays permanently in the skin: ______________________________

   ____

3. Identify the location (dermis or epidermis) of each of the following and identify its role as part of the skin:

   (a) Basal layer (stratum basale): ______________________________

   (b) Outermost keratinized layer (stratum corneum): ______________________________

   (c) Sweat glands: ______________________________

   (d) Collagen fibers: ______________________________

ISBN: 978-1-92717357-2

Thermoregulation

In humans (and other placental mammals), the temperature regulation center of the body is in the **hypothalamus**. In humans, it has a 'set point' temperature of 36.7°C. The hypothalamus responds directly to changes in core temperature and to nerve impulses from receptors in the skin. It then coordinates appropriate nervous and hormonal responses to counteract the changes and restore normal body temperature. Like a thermostat, the hypothalamus detects a return to normal temperature and the corrective mechanisms are switched off (negative feedback). Toxins produced by pathogens, or substances released from some white blood cells, cause the set point to be set to a higher temperature. This results in fever and is an important defense mechanism in the case of infection.

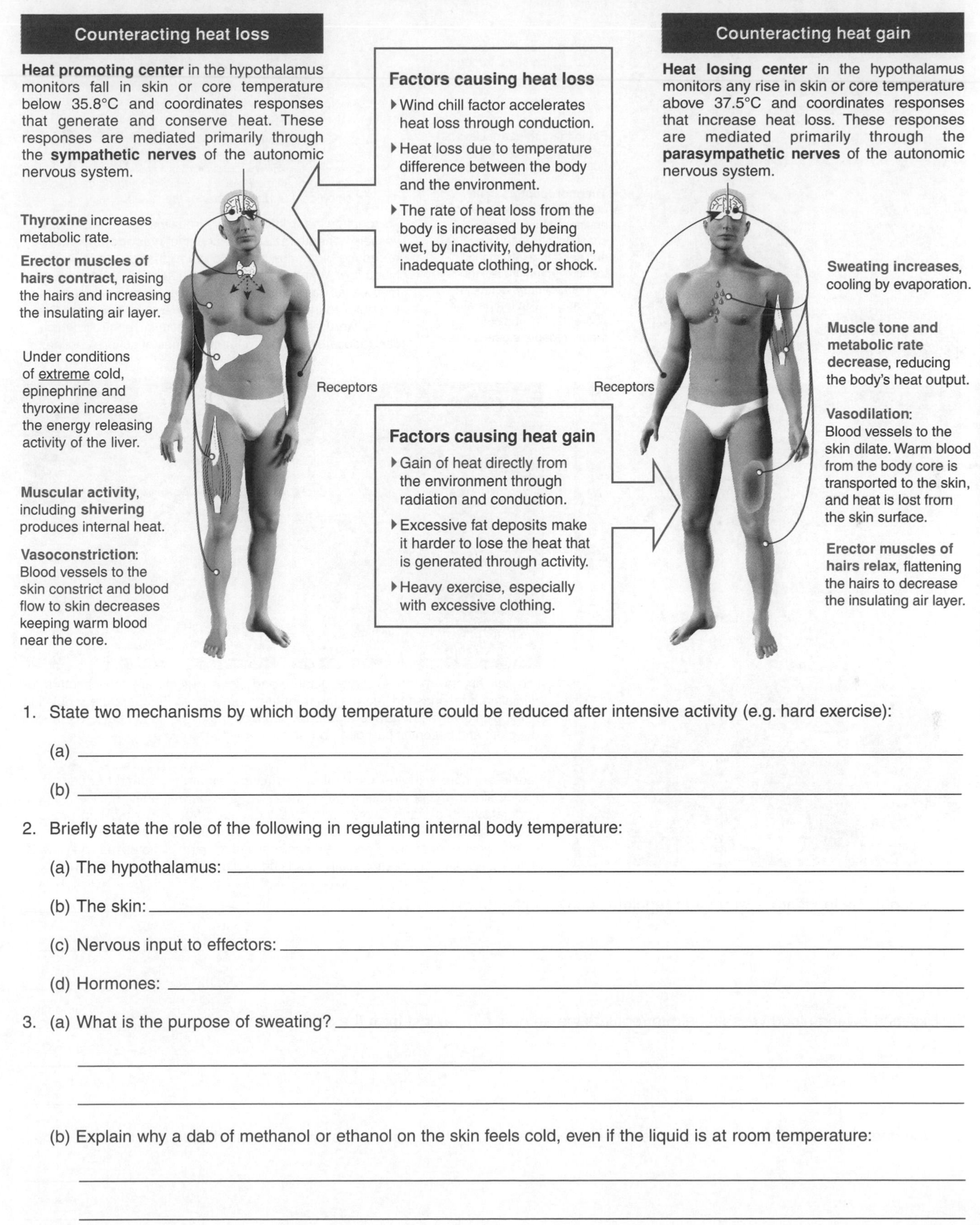

1. State two mechanisms by which body temperature could be reduced after intensive activity (e.g. hard exercise):

   (a) ______________________

   (b) ______________________

2. Briefly state the role of the following in regulating internal body temperature:

   (a) The hypothalamus: ______________________

   (b) The skin: ______________________

   (c) Nervous input to effectors: ______________________

   (d) Hormones: ______________________

3. (a) What is the purpose of sweating? ______________________

   (b) Explain why a dab of methanol or ethanol on the skin feels cold, even if the liquid is at room temperature:

ISBN: 978-1-92717357-2

Skin Section

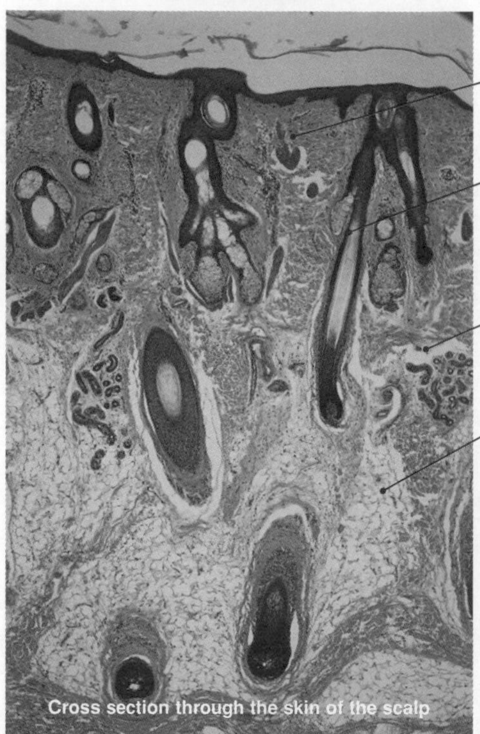
Cross section through the skin of the scalp

EII

Blood vessels in the dermis dilate or constrict to promote or restrict heat loss.

Hairs raised or lowered to increase or decrease the thickness of the insulating air layer between the skin and the environment.

Sweat glands produce sweat, which cools through evaporation.

Fat in the sub-dermal layers insulates the organs against heat loss.

Thermoreceptors in the dermis are free nerve endings, which respond to changes in skin temperature and send that information to the hypothalamus. Hot thermoreceptors detect an increase in skin temperature above 37.5°C and cold thermoreceptors detect a fall below 35.8°C.

Regulating Blood Flow to the Skin

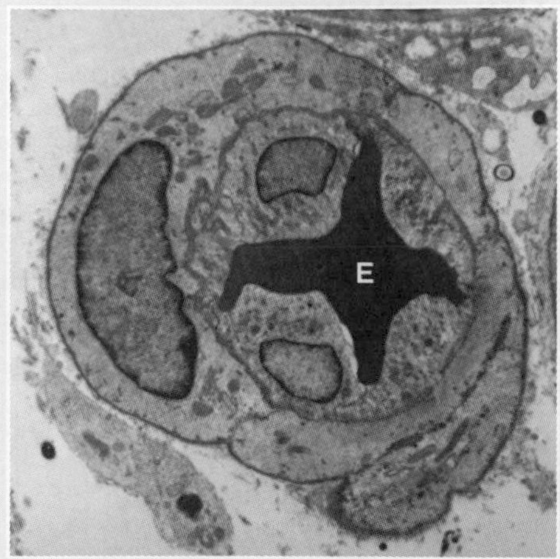

RM Hunt CC 3.0

Constriction of a small blood vessel. An erythrocyte (E) (red blood cell) is in the center of the vessel.

To regulate heat loss or gain from the skin, the blood vessels beneath the surface constrict (**vasoconstriction**) to reduce blood flow or dilate (**vasodilation**) to increase blood flow. When blood vessels are fully constricted there may be as much as a 10°C temperature gradient from the outer to inner layers of the skin. Extremities such the hands and feet have additional vascular controls which can reduce blood flow to them in times of severe cooling.

Graham Crumb CC 2.0

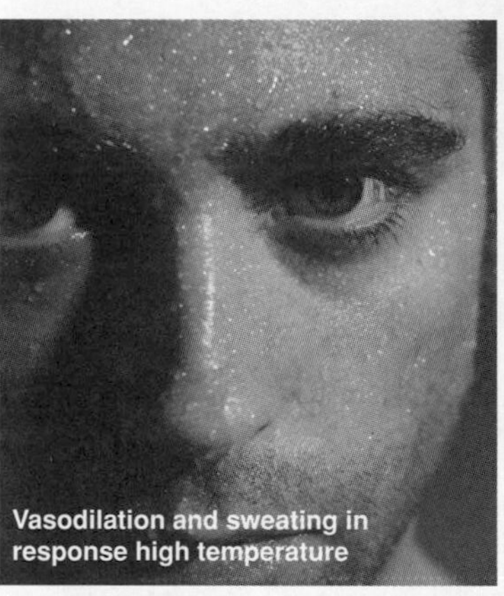
Vasodilation and sweating in response high temperature

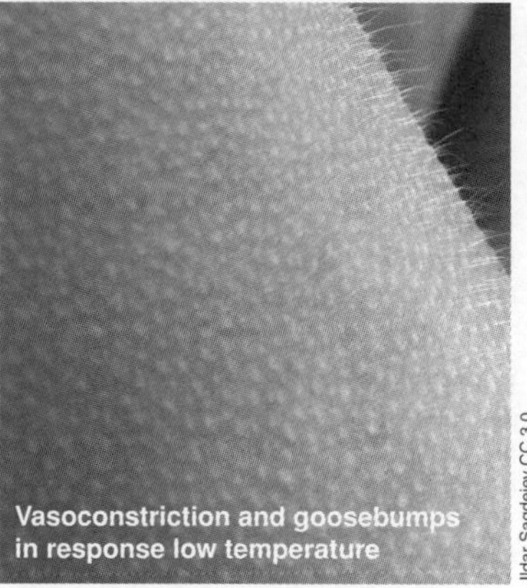
Vasoconstriction and goosebumps in response low temperature

Ildar Sagdejev CC 3.0

The hair erector muscles, sweat glands, and blood vessels are the effectors for mediating a response to information from thermoreceptors. Temperature regulation by the skin involves **negative feedback** because the output is fed back to the skin receptors and becomes part of a new stimulus-response cycle.

People are born with an excess of sweat glands, but if they spend the first years of their life in a cold climate most of these become permanently inactive. People acclimatised to a warm climate (such as the **young Vanuatu boy**, left) produce sweat profusely (up to 3 L h^{-1}) and the sweat is distributed uniformly. This increases the efficiency of heat loss. People not acclimatised to warm climates may only sweat up to 1 L h^{-1} and the sweat usually beads up and drips off the body.

4. Describe the feedback system that regulates body temperature: ______________________

5. Explain how the blood vessels help to regulate the amount of heat lost from the skin and body: ______________________

6. Why does a person from a cool climate have difficulty regulating their body temperature when visiting a hot climate?

ISBN: 978-1-92717357-2

Hypothermia

Hypothermia is a condition experienced when the core body temperature drops below 35°C. Hypothermia is caused by exposure to low temperatures, and results from the body's inability to replace the heat being lost to the environment. The condition ranges from mild to severe depending how low the body temperature has dropped. Severe hypothermia results in severe mental confusion, including inability to speak and amnesia, organ and heart failure, and death.

A body temperature of around 37°C allows for optimal metabolic function. Below 35°C, metabolism slows, causing a loss of coordination, lethargy, and mental fatigue. Hypothermia can result from exposure to very low temperatures for a short time or to moderately low temperatures for a long time. Exposure to cold water produces symptoms of hypothermia far more quickly than exposure to the same temperature of air. This is because water is much more effective than air at conducting heat away from the body.

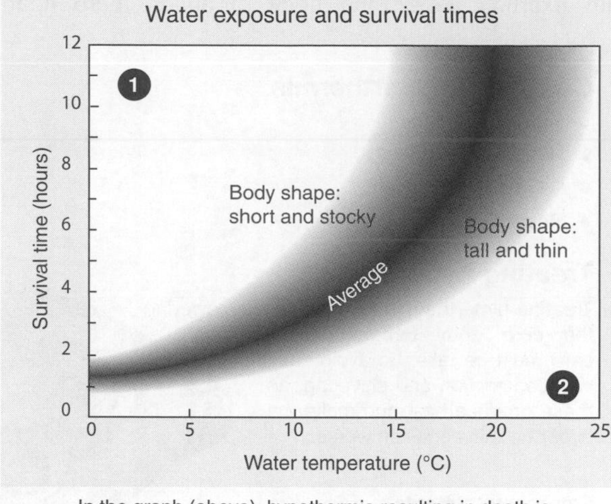

In the graph (above), hypothermia resulting in death is highly likely in region 1 and highly unlikely in region 2.

Normal body temperature → 37°

Mild hypothermia: Shivering. Vasoconstriction reduces blood flow to the extremities. Hypertension and cold diuresis (increased urine production due to the cold).

Moderate hypothermia: Muscle coordination becomes difficult. Movements slow or laboured. Blood vessels in ears, nose, fingers, and toes constrict further resulting in these turning a blue color. Mental confusion sets in.

Severe hypothermia: Speech fails. Mental processes become irrational, victim may enter a stupor. Organs and heart eventually fail resulting in death.

Treating Hypothermia

Hypothermia is treated by rewarming. This must be done carefully, because rewarming the body too quickly or with the wrong method can actually cause the body to attempt to remove the sudden excess of heat and so cause more heat loss and/or death.

Mild hypothermics can be rewarmed by **passive rewarming**, using their own body heat coupled with warm, dry, insulated clothing in a warm dry environment. Moderate hypothermia requires **active external rewarming**. This involves using warming devices such as hot water bottles or warm water baths. Severe hypothermics must be treated with **active internal** or **core warming**. Methods include delivery of warm intravenous fluids, inhaling warm moist air, or warming the blood externally by using a heart-lung machine.

1. What conditions might cause a person to become hypothermic? ______

2. (a) With reference to the graph (above), which body shape has best survival at 15°C? ______

   (b) Explain your choice: ______

3. Describe the methods used to rewarm hypothermics and the importance of using the correct methods.

ISBN: 978-1-92717357-2

Hyperthermia

Hyperthermia is a physiological state in which the core body temperature exceeds 38.5°C without a change to the set-point of the heat control centre in the hypothalamus. The most common cause is heat stroke caused by prolonged exposure to excessive heat or humidity, often associated with exertion. When the body produces more heat than it can dissipate, the heat regulating systems of the body can become overwhelmed and body temperature will rise uncontrollably. Prolonged elevation in body temperature is potentially fatal and is thus regarded as a medical emergency. Hyperthermia is different from a fever, which involves resetting the body's thermostat to a higher level in response to infection.

Causes of hyperthermia:

- Dehydration
- Hot environment
- Exercise
- Response to some drugs

Treating hyperthermia

Treating hyperthermia involves rapidly lowering the core body temperature. However, care must be taken to avoid causing vasoconstriction and shivering, as these produce heat and make the hyperthermic condition worse.

External treatments

Mild hyperthermia, e.g. exercising on a hot day, is treated by drinking water, removing excess clothing, and resting in a cool place. In more serious cases, cooling is achieved by sponging the body with cool water, or using cooling blankets or ice-packs. A person may also be wrapped in wet sheets and have a fan directed on them to increase evaporative cooling (sweating). Placing the patient in a bathtub of cool water removes a significant amount of heat in a short period of time.

Internal treatments

Internal cooling is required when the core temperature exceeds 40°C, or the patient is unconscious or confused. These methods are more aggressive and invasive. Treatments include administering cool intravenous fluids, flushing the stomach or rectum with cold water, or hemodialysis, where a machine cools the blood externally before it is returned to the body.

Phases in heat exhaustion and hyperthermia

Phase 1

Overexertion is usually accompanied with a flushed red face and rapid short breaths. Correction is by seeking shade and drinking plenty of fluid.

Phase 2

Heat exhaustion is a more serious problem. It is indicated by profuse sweating, a dry mouth, cramps, and nausea. The skin will appear red as blood is directed to the skin to reduce core temperature. Physical activity should be stopped immediately and shade should be sought. Cool drinks and ice packs on the skin may be needed.

Phase 3

Heat stroke is the final and most serious stage. The body's core temperature may have risen to 41°C. Thermoregulatory mechanisms fail. Sweat is no longer produced and the skin becomes hot and dry. Disorientation is followed by collapse and unconsciousness. Metabolic processes become uncoupled and enzymes denature. Death follows.

1. (a) What is **hyperthermia**? ______________________________

   (b) How does hyperthermia differ from a fever? ______________________________

2. (a) Why would bringing down a hyperthermia patient's temperature down so fast that they began to shiver be dangerous?

   ____

   (b) Why are internal cooling treatments used on patients with temperatures over 40°C? ______________________________

3. Explain why untreated heat stroke can rapidly lead to death: ______________________________

ISBN: 978-1-92717357-2

Drugs and Thermoregulation

Ecstasy, or **MDMA**, is an illegal stimulant popular in the clubbing or dance party scene for the feelings of euphoria it induces. It has profound psychological and physiological effects, brought about by an increase in the levels of **serotonin** in the brain. These effects include changes to temperature regulation, heart rate, blood pressure, and appetite control. The effects of ecstasy start 20-60 minutes after taking it and last for six hours. It can take days or weeks for serotonin levels to return to normal.

Ecstasy Induced Hyperthermia

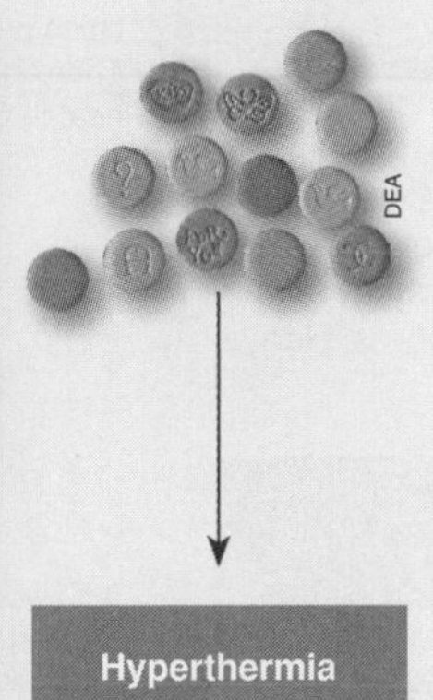

Physiological effect of ecstasy	Consequence
Reduces blood flow to skin	Reduces heat loss
Decreases sweating	Reduces heat loss
Increases metabolic rate	Increases heat generation
Decreases thirst recognition	Increases dehydration

Other contributing factors:
Hot environment
Dancing (= heat generation)

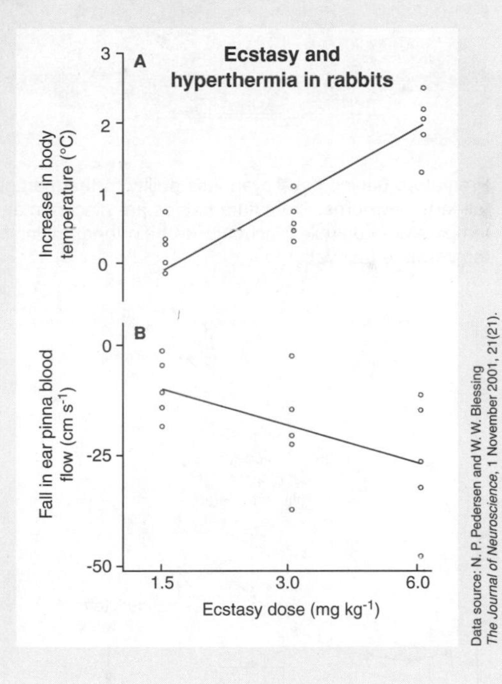

Data source: N. P. Pedersen and W. W. Blessing *The Journal of Neuroscience*, 1 November 2001, 21(21).

Hyperthermia is one of the major side effects of ecstasy use and can result in organ failure and death if it is not treated. It has been implicated in the deaths of people attending dance-parties.

To understand how humans may be affected by ecstasy, researchers investigated how it causes hyperthermia in rabbits. The data (right) shows a clear positive relationship between the ecstasy dose and body temperature (figure A). The same research also showed that blood flow to the skin decreases (due to vasoconstriction) after ecstasy is taken (figure B). One of the body's main cooling mechanisms is to increase blood flow to the skin, so that heat can be lost. Ecstasy shuts down this mechanism, so it becomes more difficult to lose heat. In humans, the problem is exacerbated because ecstasy use is associated with dance parties or clubs, where the environment is hot and crowded and the physical effort of dancing generates heat. The combination of these behavioral and physiological factors results in abnormal increases in body temperature, heat exhaustion, and heat stroke.

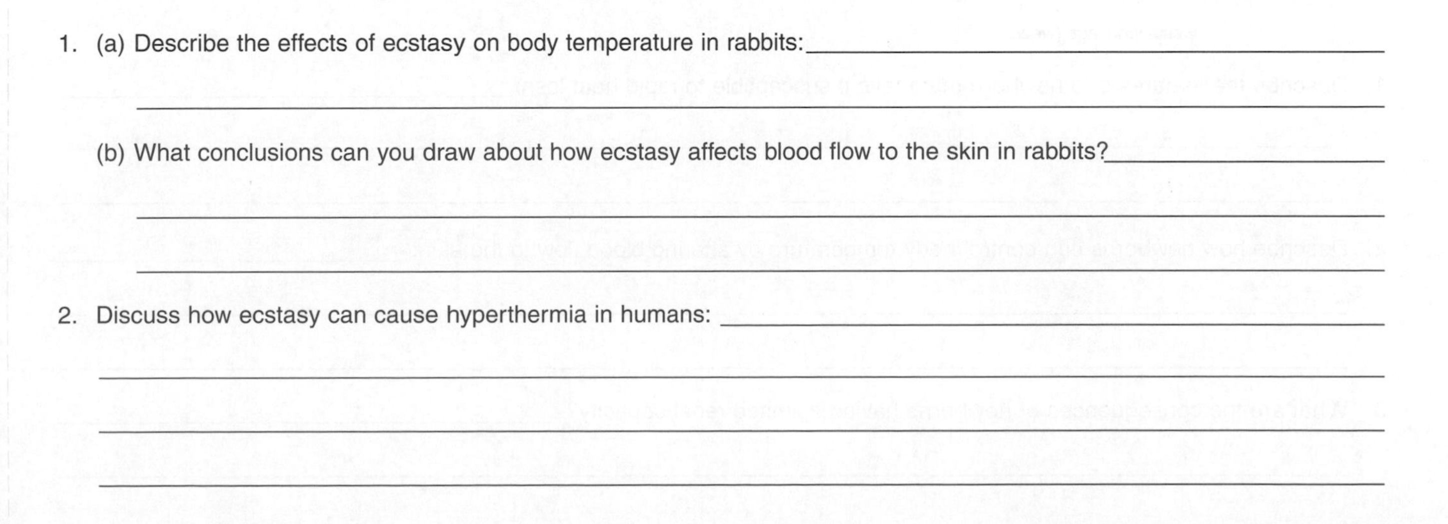

1. (a) Describe the effects of ecstasy on body temperature in rabbits: ________

   (b) What conclusions can you draw about how ecstasy affects blood flow to the skin in rabbits? ________

2. Discuss how ecstasy can cause hyperthermia in humans: ________

ISBN: 978-1-92717357-2

Homeostasis in Newborns

While in the uterus, many of the physiological processes of a fetus are supported by the mother. Once delivered, the newborn must function independently of a placental connection. This requires a number of significant physiological changes associated with temperature regulation, blood glucose homeostasis, and fluid and electrolyte balance. Newborns require considerable help to maintain homeostasis (e.g. dressing them appropriately so they do not become too hot or too cold).

Thermoregulation in Newborns

Newborn babies cannot fully regulate their body temperature until six months of age. As a result, they can become **hypothermic** or **hyperthermic** very quickly.

Newborns cannot shiver to produce heat, and have limited capacity to generate heat from large body movements (because their ability to move is limited). Much of an infant's heat production comes from the metabolic activity of **brown fat**, a mitochondrial rich organ abundant in newborns. Heat is also generated by metabolic activity.

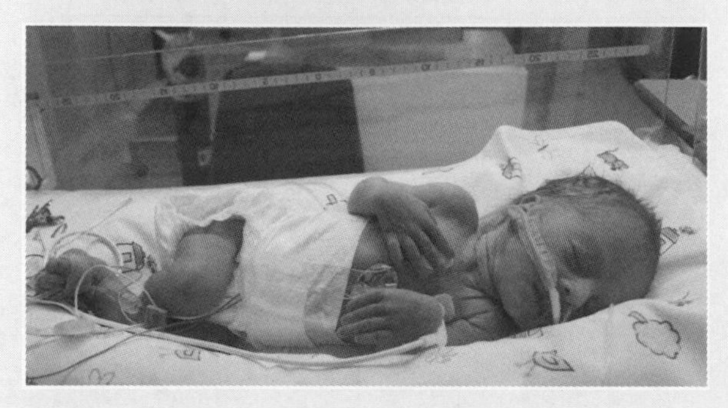

Premature babies have even less ability to thermoregulate than full-term newborns. Premature babies are placed in an enclosed temperature controlled incubator to help them maintain a stable temperature (above).

A baby's body surface is three times greater than an adult's. There is greater surface area for heat to be lost from.

The head is very large compared to the rest of the body. The skull is about 1/4 of the total body length. Heat losses from the head are high.

Newborns have thin skin, and blood vessels that run close to the skin, these features allow heat to be lost easily.

Newborns have very little subcutaneous (white) fat to insulate them against heat loss. The smaller the baby, the less white fat they have.

Newborns minimize heat loss by reducing the blood supply to the periphery (skin, hands, and feet). This helps to maintain the core body temperature. Increased brown fat activity and general metabolic activity generate heat. Newborns are often dressed in a hat to reduce heat loss from the head, and tightly wrapped to trap heat next to their bodies.

Newborns lower their temperature by increasing peripheral blood flow. This allows heat to be lost, cooling the core temperature. Newborns can also reduce their body temperature by sweating, although their sweat glands are not fully functional until four weeks after birth.

Glomerular filtration rate (ml/min)
6
5
4
3
2
1
28 30 32 34 36 38 40 42
Gestational age (weeks)
Kidney development is fully completed
Full term gestation
Data source: Al-Dahhan *et al.* Archives of Disease in Childhood, 1983, 58.

Renal Homeostasis

Newborn babies have inefficient kidneys. Ten days after birth, the filtration rate of the kidneys has increased significantly, but is still half that of an adult. It takes one to two years for renal function to reach adult levels. The data (left) shows how kidney glomerular filtration rate (a measure of kidney efficiency) increases through gestation and after birth.

Low renal efficiency affects the ability of the newborn to maintain fluid and electrolyte balances. A newborn's kidneys are unable to concentrate urine very well, and absorption of some electrolytes (sodium and bicarbonate) is poor. Newborn babies have a greater risk of becoming dehydrated. This is partially because of their limited ability to concentrate urine, but also because their large surface area increases the rate of water loss through sweating and breathing.

1. Describe the features of a newborn that make it susceptible to rapid heat loss: ______

2. Describe how newborns can control body temperature by altering blood flow to the skin: ______

3. What are the consequences of newborns having a limited renal capacity? ______

Related activities: *Thermoregulation, Hypothermia, Hyperthermia*

ISBN: 978-1-92717357-2

KEY TERMS: Mix and Match

INSTRUCTIONS: Test your vocabulary by matching each term to its definition, as identified by its preceding letter code.

abdominal cavity

anterior/ventral

dermis

distal

dorsal body cavity

epidermis

epithelial membranes

homeostasis

hypodermis

inferior

lateral

medial

mucous membranes (mucosa)

negative feedback

pelvic cavity

positive feedback

posterior/dorsal

proximal

serous membranes (serosa)

superficial

superior

synovial membranes

thoracic cavity

ventral body cavity

A A term meaning the back of an organism, organ, or body part.

B The body cavity comprising the cranial and spinal cavities.

C Regulation of the internal environment to maintain a stable, constant condition.

D To the side. A position away from the midline of the body.

E A term meaning farthest from the point of attachment of a limb or origin of a structure.

F Paired epithelial membranes lining enclosed internal body cavities.

G A term referring to the lower part of a structure, or a location below or under another structure or organ.

H A connective tissue membrane that lines joint capsules and secretes a lubricating fluid.

I The outermost layer of skin, which covers and protects the dermis.

J The body cavity comprising the thoracic, abdominal, and pelvic cavities.

K Epithelial membranes that line all body cavities that open to the exterior (e.g. the hollow organs of the respiratory tract).

L A destabilizing mechanism in which the output of the system causes an escalation in the initial response.

M The connective tissue layer underlying and anchoring the skin. Also called subcutaneous tissue.

N A term meaning close to the body surface.

O Linings of organs and cavities formed from epithelium and connective tissue. They include the skin, mucosa, and serosa.

P The connective tissue layer of the skin beneath the epidermis.

Q Located near the head or upper body region.

R A term meaning the front of an organism, organ, or body part.

S A mechanism in which the output of a system acts to oppose changes to the input of the system. The net effect is to stabilize the system and dampen fluctuations.

T The ventral body cavity containing the lungs and heart.

U Located towards the midline of the body.

V A ventral body cavity containing a number of organs including the digestive organs, spleen and kidneys.

W A ventral body cavity containing the reproductive organs and bladder.

X A term meaning towards the attached end of a limb or the origin of a structure.

ISBN: 978-1-92717357-2

Interacting Systems

Muscular and Skeletal Systems

Endocrine system

- The skeleton protects the endocrine organs especially in the pelvis, chest, and brain.
- Bone growth, remodeling, and repair occurs in response to hormones.
- Androgens and growth hormone promote muscle strength and increase in mass.

Respiratory system

- Skeleton encloses and protects lungs
- Flexible ribcage enables ventilation of the lungs for exchange of gases (O_2/CO_2).
- Diaphragm and intercostals produce volume changes in breathing.

Cardiovascular system

- Heart and blood vessels transport O_2, nutrients, and waste products to all the body.
- Bone marrow produces red blood cells
- Bone matrix stores calcium, which is required for contraction of muscle in the heart and blood vessels.

Digestive system

- Skeleton provides some protection and support for the abdominal organs.
- Digestive system provides nutrients for growth, repair, and maintenance of muscle and connective tissues.

Skeletal system

- Muscular activity maintains bone strength and helps determine bone shape.
- Muscles pull on bones to create movement.

Integumentary system

- Skin absorbs and produces precursor of vitamin D, which is involved in calcium and phosphorus metabolism.
- Skin covers and protects the muscle tissue.

Nervous system

- The skeleton protects the CNS.
- Bone acts as a store of calcium ions required for nerve function.
- Innervation of bone and joint capsules provides sensation and positional awareness.
- Muscular activity is dependent on innervation.

Lymphatic system and immunity

- Stem cells in the bone marrow give rise to the lymphocytes involved in the immune response.

Urinary system

- The skeleton protects the pelvic organs.
- Final activation of vitamin D, which is involved in calcium and phosphorus metabolism, occurs in the kidneys.
- Urination controlled by a voluntary sphincter in the urethra.

Reproductive system

- The skeleton protects the reproductive organs.
- Reproductive (sex) hormones influence skeletal development.

Muscular system

- Skeleton acts as a system of levers for muscular activity.
- Bone provides a store of calcium for muscle contraction.

General Functions and Effects on all Systems

The skeletal system provides bony protection for the internal organs, especially the brain and spinal cord, and the lungs, heart, and pelvic organs. The muscular system acts with the skeletal system to generate voluntary movements. Smooth and cardiac muscle provide motility for involuntary activity.

Disease

Symptoms of disease
- Pain (moderate to severe)
- Inflammation
- Limitations in function

Disorders of the bones and joints
- Growth disorders
- Trauma (fractures and sprains)
- Infection
- Tumors
- Degenerative diseases

Diseases of the skeletal muscles
- Inherited diseases
- Fibrosis (scarring)
- Strains, tears, and cramps
- Denervation and atrophy

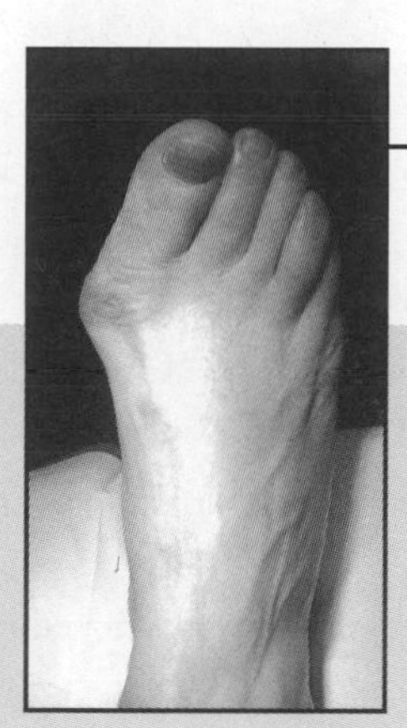

- Osteomalacia
- Osteoarthritis
- Osteoporosis
- Sarcomas
- Muscular dystrophy

Medicine & Technology

Diagnosis of disorders
- Blood tests
- Bone scans
- Medical imaging techniques
- Athroscopy

Treatment of injury
- Surgery
- Physical and drug therapies
- Prosthetics and orthotics

Treatment of inherited disorders
- Surgery
- Radiotherapy (for cancers)
- Physical and drug therapies
- Prosthetics and orthotics
- Gene therapy

- Joint replacement
- Grafts
- Genetic counselling
- X-rays
- MRI

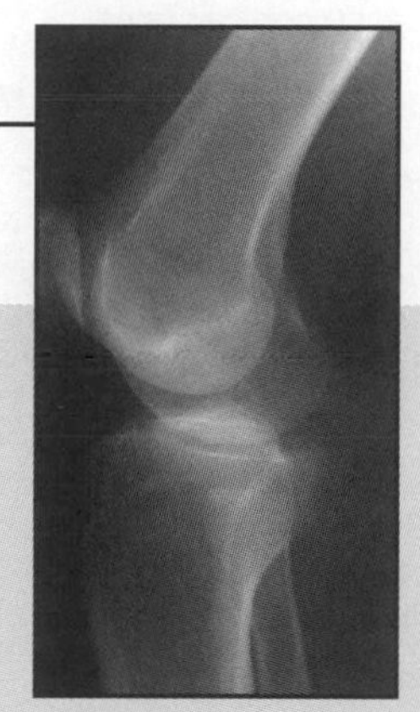

Support & Movement

The Musculoskeletal System

The musculoskeletal system can be affected by disease and undergoes changes associated with aging.

Medical technologies and exercise can be used to diagnose, treat, and delay the onset of musculoskeletal disorders.

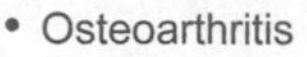

- Osteoarthritis
- Osteoporosis
- Muscular atrophy

Aging and the bones, joints, and muscles
- Bone loss
- Loss of muscle mass
- Accumulated trauma
- Increased incidence of cancers

The Effects of Aging

- Muscle fatigue
- Fast vs slow twitch
- Aerobic training
- Anaerobic training

Effects of exercise on bones, joints, and muscles
- Increased bone density
- Increased lean muscle mass
- Changes in flexibility & joint mobility
- Changes in fiber type & recruitment
- Changes in oxidative capacity

Exercise

The Skeletal System

Key terms

appendicular skeleton
articular cartilage
axial skeleton
bone (tissue)
calcitonin
cartilage
cartilaginous joint
compact (=cortical) bone
diaphysis
epiphyses
fibrous joint
flat bones
growth hormone
Haversian canals
irregular bones
joint (articulation)
lacunae
ligament
long bones
matrix (of bone)
ossification
osteoarthritis
osteoblasts
osteocytes
osteroporosis
parathyroid hormone
pectoral girdle
pelvic girdle
periosteum
reticular connective tissue
sex hormones
short bones
spongy (=cancellous) bone
synovial fluid
synovial joint
tendon

Periodicals:
Listings for this chapter are on page 279

Weblinks:
www.thebiozone.com/weblink/AnaPhy-3572.html

BIOZONE APP:
Student Review Series
The Skeletal & Muscular Systems

Key concepts

- The skeleton is the internal supporting structure of the body, composed of mineralized connective tissue.
- The skeleton, together and the body's system of muscles, enables movement of the body.
- Movement occurs articulations in the skeleton called joints. The amount of movement permitted depends on the joint type.
- Bone is a dynamic tissue, undergoing growth, remodeling, and repair. Aging is associated with degenerative changes in the skeleton.

Learning Objectives

☐ 1. Use the **KEY TERMS** to compile a glossary for this topic.

Bone and Cartilage

pages 30, 59-61

☐ 2. Recall the characteristics of connective tissues (CTs). Describe the CTs that contribute to the components of the skeleton: **bone**, **cartilage**, **tendons** and **ligaments**, and reticular connective tissue.

☐ 3. Describe the functions of bone, including its role in homeostasis.

☐ 4. Describe the basic composition of **bone**. Compare and contrast **compact** (cortical) **bone** and **spongy** (cancellous) **bone**.

☐ 5. Using examples, classify bones according to their size and shape. Recognize **long bones**, **short bones**, **flat bones**, and **irregular bones**.

☐ 6. Use a diagram to describe the gross structure of a long bone, including the features conferring strength and shock absorption. Indicate the **diaphysis** (shaft), **periosteum**, and the **epiphyses** and associated **articular cartilage**. Describe the location of yellow and red bone marrow and explain their functions.

☐ 7. Describe the ultrastructure of compact bone, identifying the **periosteum**, **osteoblasts**, **osteocytes**, **matrix**, **lacunae**, and **Haversian canals**.

☐ 8. Describe **ossification** (bone formation), explaining the role of the osteoblasts and the process by which hyaline cartilage is replaced with hard bone. Describe how bone is remodeled during growth and repaired in response to injury.

☐ 9. Explain how bone growth and repair is regulated by hormones, including:
- **parathyroid hormone** and **calcitonin** (calcium metabolism and remodeling)
- growth hormone and sex hormones (growth of long bones).

The Skeleton

pages 30, 55-58, 62-66

☐ 10. Identify the components of the **skeleton** and describe its two functional regions:
- the **axial skeleton** (skull, spine, ribcage, sternum)
- the **appendicular skeleton** (limbs and pectoral and pelvic girdles).

☐ 11. Describe the role of **joints** in the skeleton. Classify joints structurally (e.g. **synovial**) and functionally (based on the amount of movement permitted).

☐ 12. Describe the structure and function of a typical **synovial joint** (e.g. the elbow or knee joint). Include reference to the role and properties of **synovial fluid**.

☐ 13. Describe the degenerative changes in the skeleton that occur with increasing age, including a reduction in the rate of bone remodeling, accelerated rates of bone loss, **osteroporosis**, and **osteoarthritis**.

The Human Skeleton

The human skeleton consists of two main divisions: the **axial skeleton** (made up of the **skull**, **rib cage**, and **spine**) and the **appendicular skeleton** (made up of the limbs and the shoulder and pelvic girdles). Bones are identified by their location and described by their shape (e.g. irregular, flat, long, or short), which is related to their functional position in the skeleton. Most of the bones of the upper and lower limbs, for example, are long bones.

Bones also have features such as processes, holes (foramina, *sing.* **foramen**), and depressions (**fossae**), associated with nerves, blood vessels, ligaments, and muscles. Understanding the basic organization of the skeleton, the particular features associated with its component bones, and the nature of skeletal articulations (**joints**) is essential to understanding how the movement of body parts is achieved.

WORD LIST:
phalanges, humerus, patella, scapula, tibia, clavicle, sternum, lumbar vertebra, femur, phalanges, cranium, sacrum, metacarpals, rib, ilium of hip bone, fibula, carpals, tarsals, metatarsals, facial bones

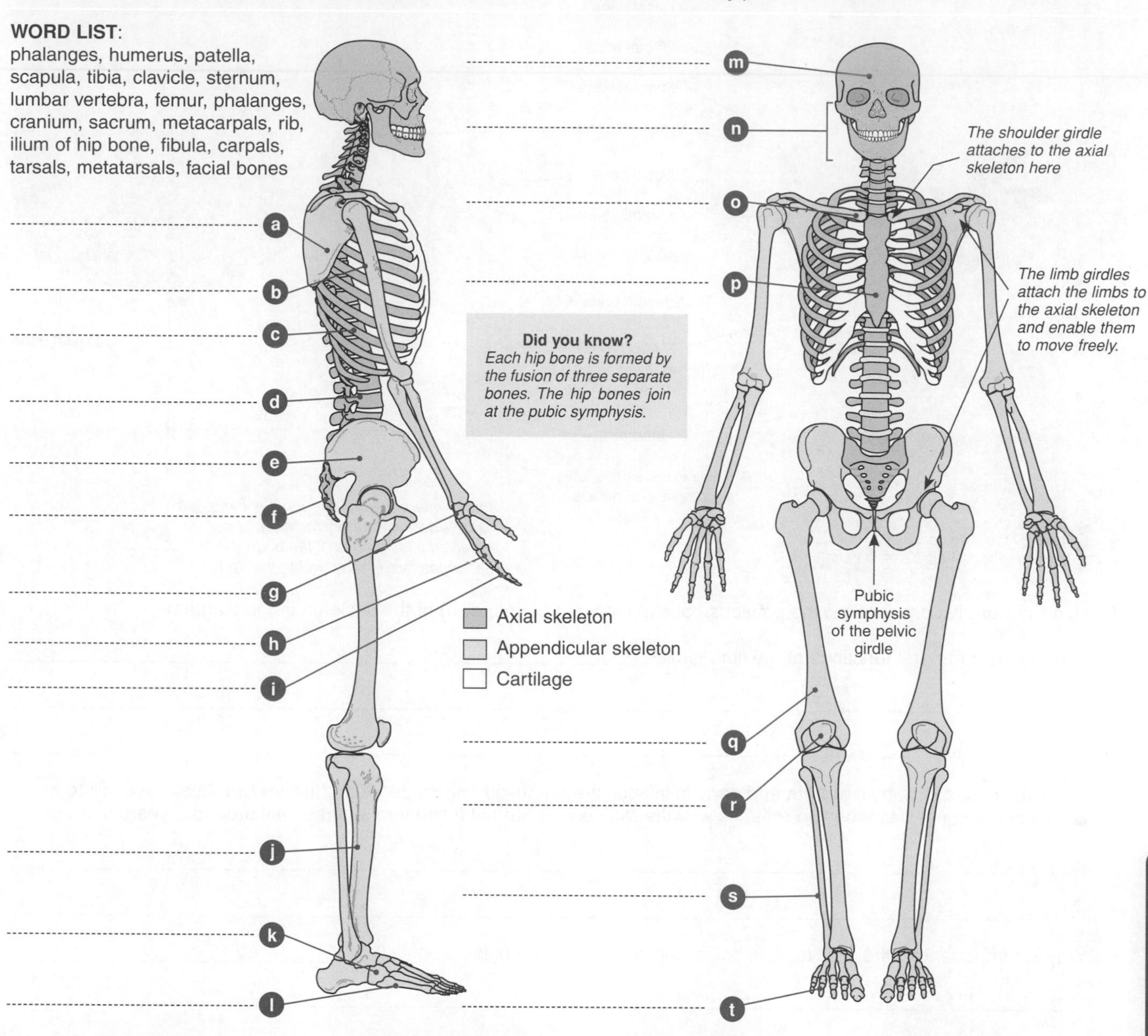

Did you know?
Each hip bone is formed by the fusion of three separate bones. The hip bones join at the pubic symphysis.

Bone Shapes

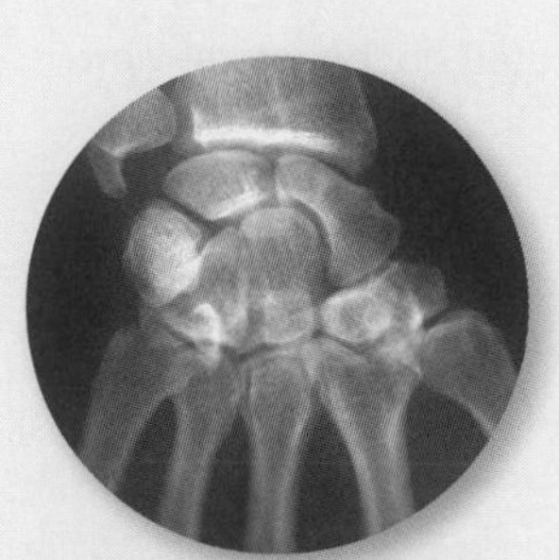

Short bones are roughly cube shaped and contain mostly spongy bone:
- carpals (above)
- tarsals
- patella

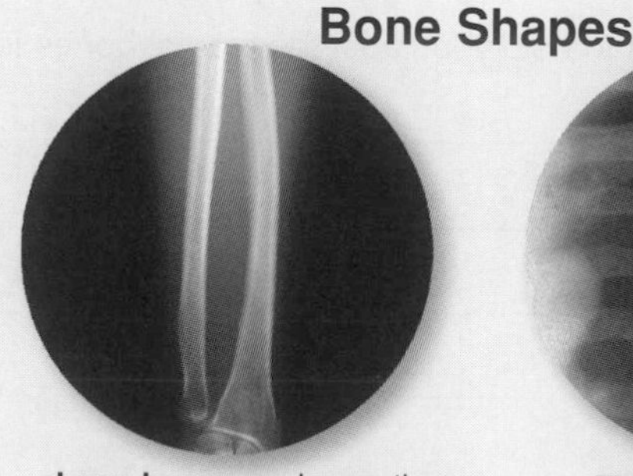

Long bones are longer than they are wide:
- most bones of the upper limbs. e.g. ulna, radius
- most bones of the lower limbs, e.g. femur, tibia

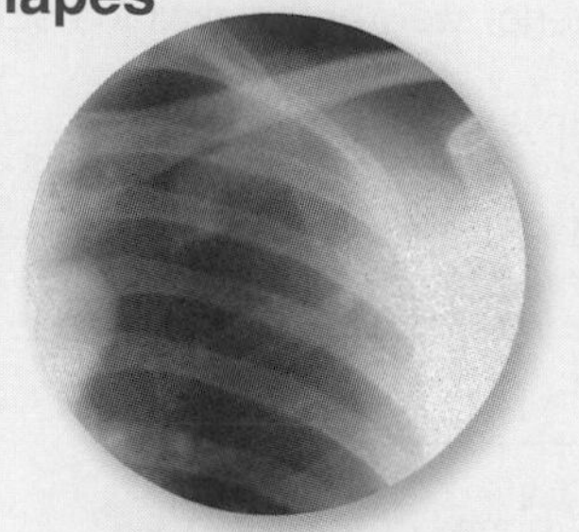

Flat bones have a thin flattened shape:
- ribs (above)
- sternum
- scapulae
- some skull bones

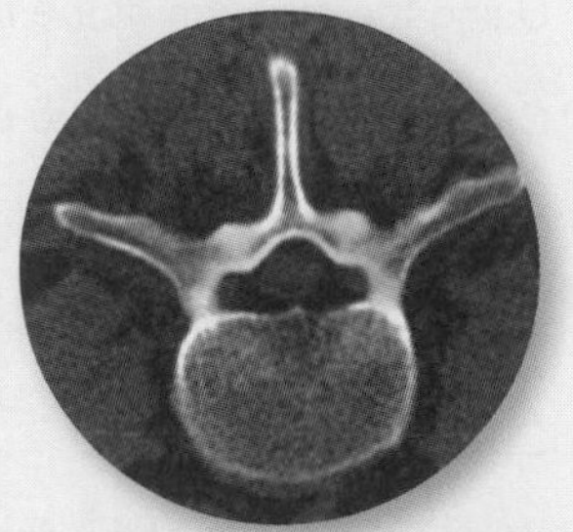

Irregular bones have an irregular shape and do not fit into the other groups:
- vertebrae (above)
- hip bones
- facial bones

ISBN: 978-1-92717357-2

Related activities: *The Bones of the Spine, The Limb Girdles, Joints*
Weblinks: *The Axial Skeleton, The Appendicular Skeleton*

Bones of the Skull

The skull is formed from the cranial bones and the facial bones. The cranium is composed of eight large flat bones, and forms a protective dome enclosing the brain. The parietal and temporal bones are paired, but the rest of the cranial bones are single. The fourteen facial bones hold the eyes in position and enable attachment of the facial muscles. Twelve are paired and only the mandible and the small vomer bone in the nasal cavity are single. Of all the skull bones, only the mandible is freely movable. The rest are joined by **sutures** (immovable joints).

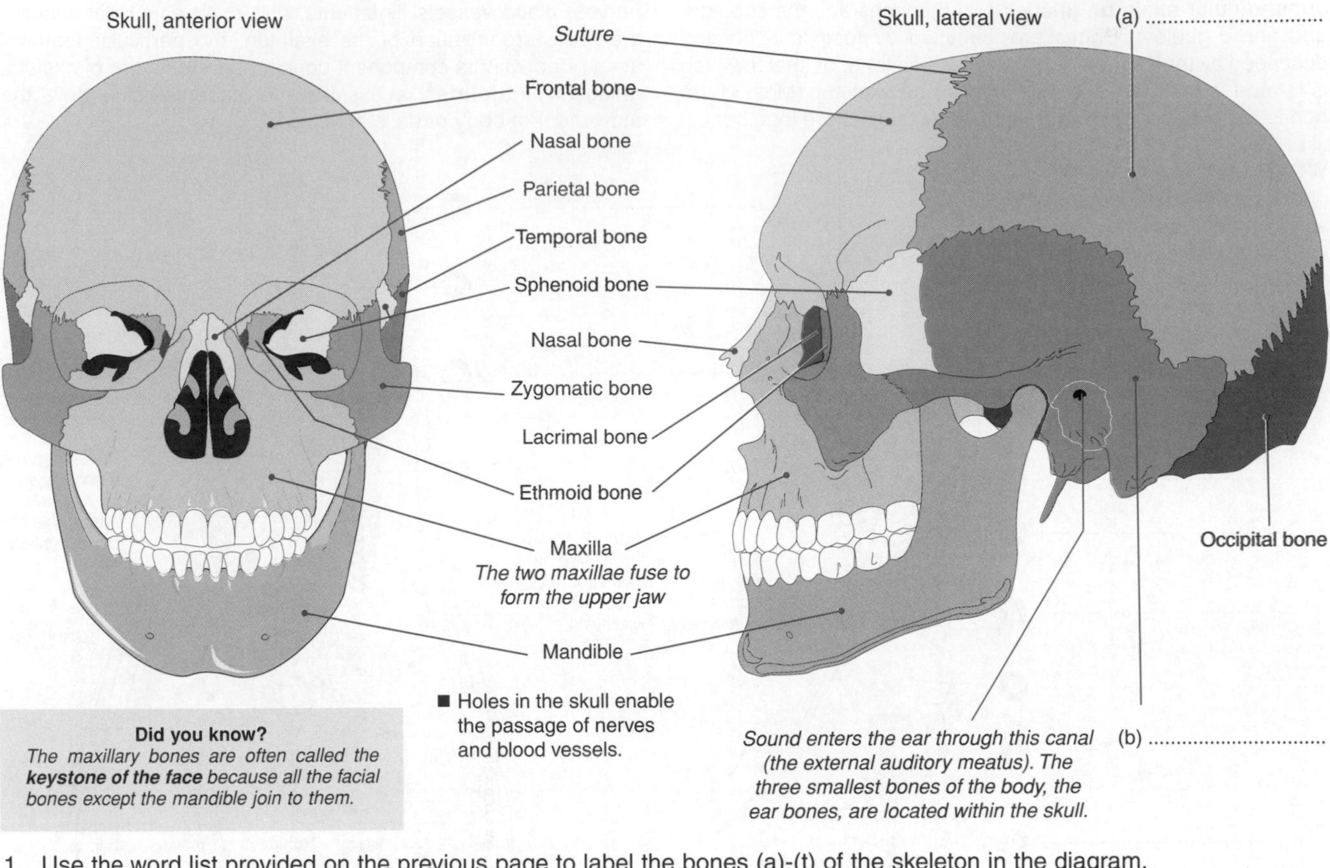

- Holes in the skull enable the passage of nerves and blood vessels.

Did you know?
The maxillary bones are often called the ***keystone of the face*** *because all the facial bones except the mandible join to them.*

1. Use the word list provided on the previous page to label the bones (a)-(t) of the skeleton in the diagram.

2. Describe two general functions of the limb girdles: ______________________________

3. The skull bones of babies at birth and early in infancy are not fused and some areas (the **fontanelles**) have still to be converted to bone. Describe two reasons why the skull bones are not fused into sutures until around 2 years of age:

   (a) ______________________________

   (b) ______________________________

4. Why is it important for the skull to have holes (called foramina) through the bones? ______________________________

5. Using the diagram of the anterior view of the skull to help you, label the cranial bones indicated on the lateral view:

6. Classify the shape of the patella: ______________________________

7. Classify the shape of the parietal bone: ______________________________

8. What is the purpose of the domed skull? ______________________________

9. What is purpose of the facial bones? ______________________________

10. If someone is rapidly moving the only freely movable bone in the skull, what might they be doing? ______________________________

ISBN: 978-1-92717357-2

The Bones of the Spine

The spine supports the skull and shoulder girdle and transmits the weight of the upper body to the lower limbs. It also forms a protective tube for the spinal cord. The spine is formed from 26 bones, separated and connected by discs of cartilage called the **intervertebral discs**. Together the vertebrae form an S-shaped bend which brings the center of mass to the mid-line of the body.

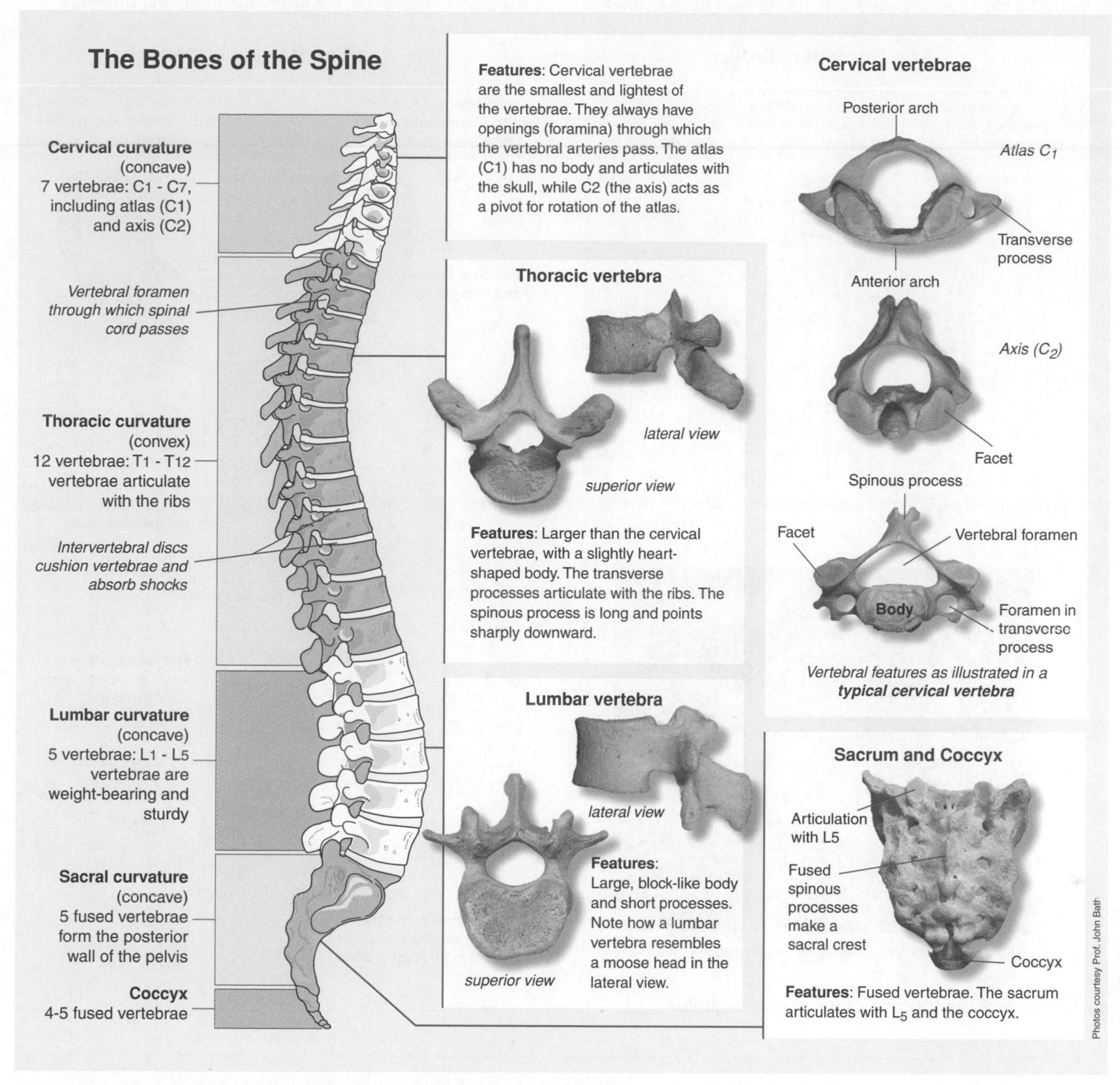

1. Identify the vertebrae associated with each of the following features:

   (a) Functional role in bearing much of the spinal load: ______________________

   (b) Articulate with the ribs. Vertebral body is heart shaped (highlight this on the diagram): ______________________

   (c) Articulates with the skull and lacks a vertebral body: ______________________

   (d) Typically has a small body and foramina (openings) in the transverse processes: ______________________

   (e) Forms the posterior wall of the bony pelvis: ______________________

2. Suggest a function of the S-shape of the spine: ______________________

____

3. At birth, the spine consists of 33 bones, 9 more than an adult. What happens to these extra bones? ______________________

____

ISBN: 978-1-92717357-2

Related activities: *The Human Skeleton*
Weblinks: *The Vertebral Column*

The Limb Girdles

The pectoral (shoulder) and pelvic girdles attach the limbs to the axial skeleton and allow for the free and wide-ranging movement of the arms and legs. The shoulder girdle consists of two **scapulae** (shoulder blades) and two **clavicles** (collar bones). The clavicles articulate with the sternum (breastbone) so that the girdle forms an incomplete ring around the thorax. The pelvic (hip) girdle is formed of two hip bones (also called pelvic, innominate, or coxal bones) connected anteriorly at the pubic symphysis and posteriorly by the sacrum. Each hip bone arises by fusion of three bones: the **ilium**, **ischium**, and **pubis**.

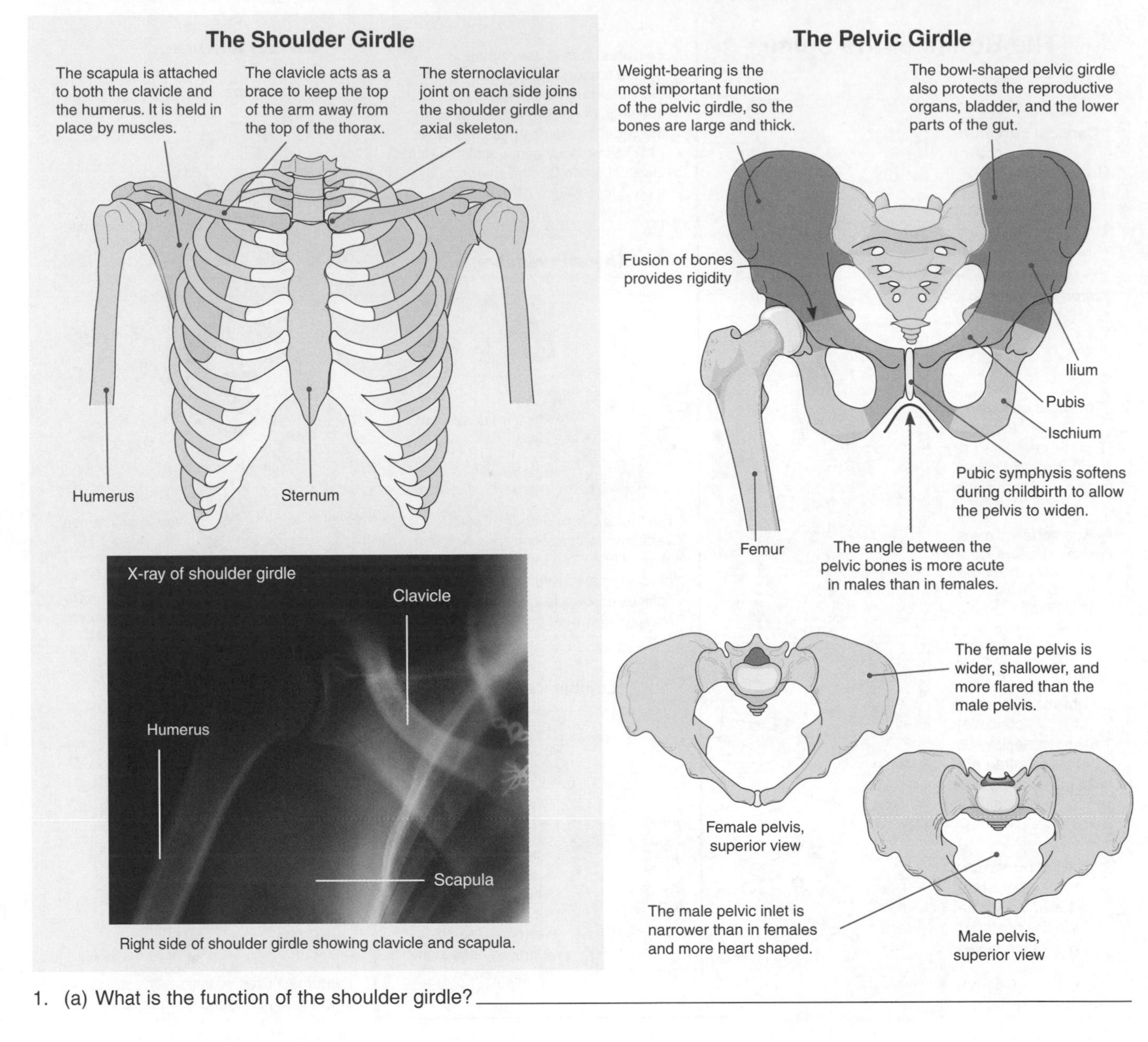

Right side of shoulder girdle showing clavicle and scapula.

1. (a) What is the function of the shoulder girdle? ______________________________

   ____

   (b) Identify the single point of attachment of shoulder girdle to the axial skeleton: ______________

2. Relate the particular features of the pelvic girdle to its functional roles:

   ____

   ____

   ____

3. Explain how and why the male and female pelves (*sing.* pelvis) differ:

   ____

   ____

   ____

4. On the X-ray (right), label the femur, ilium, ischium, pubis, and pelvic inlet:

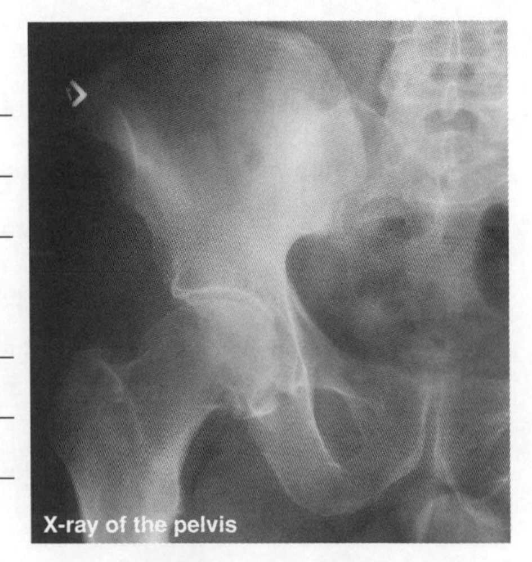

ISBN: 978-1-92717357-2

Related activities: The Human Skeleton

Bone

The skeleton is formed from two stiffened connective tissues: **bone** and **cartilage**. Although bone is hard, it is dynamic and is continually **remodeled** and repaired according to needs and in response to blood calcium levels and the pull of gravity and muscles. Hormones from the thyroid, parathyroids, and gonads, as well as growth hormone, are involved in this activity. Most bones of the skeleton are formed from hyaline cartilage by a process of **ossification** (bone formation) and they grow by **bone remodeling**. Bone remodeling is also important in bone repair. Bones have a simple gross structure, as illustrated by a long bone such as the humerus (below). The hard (dense) bone surrounds spongy (cancellous) bone filled with red bone marrow.

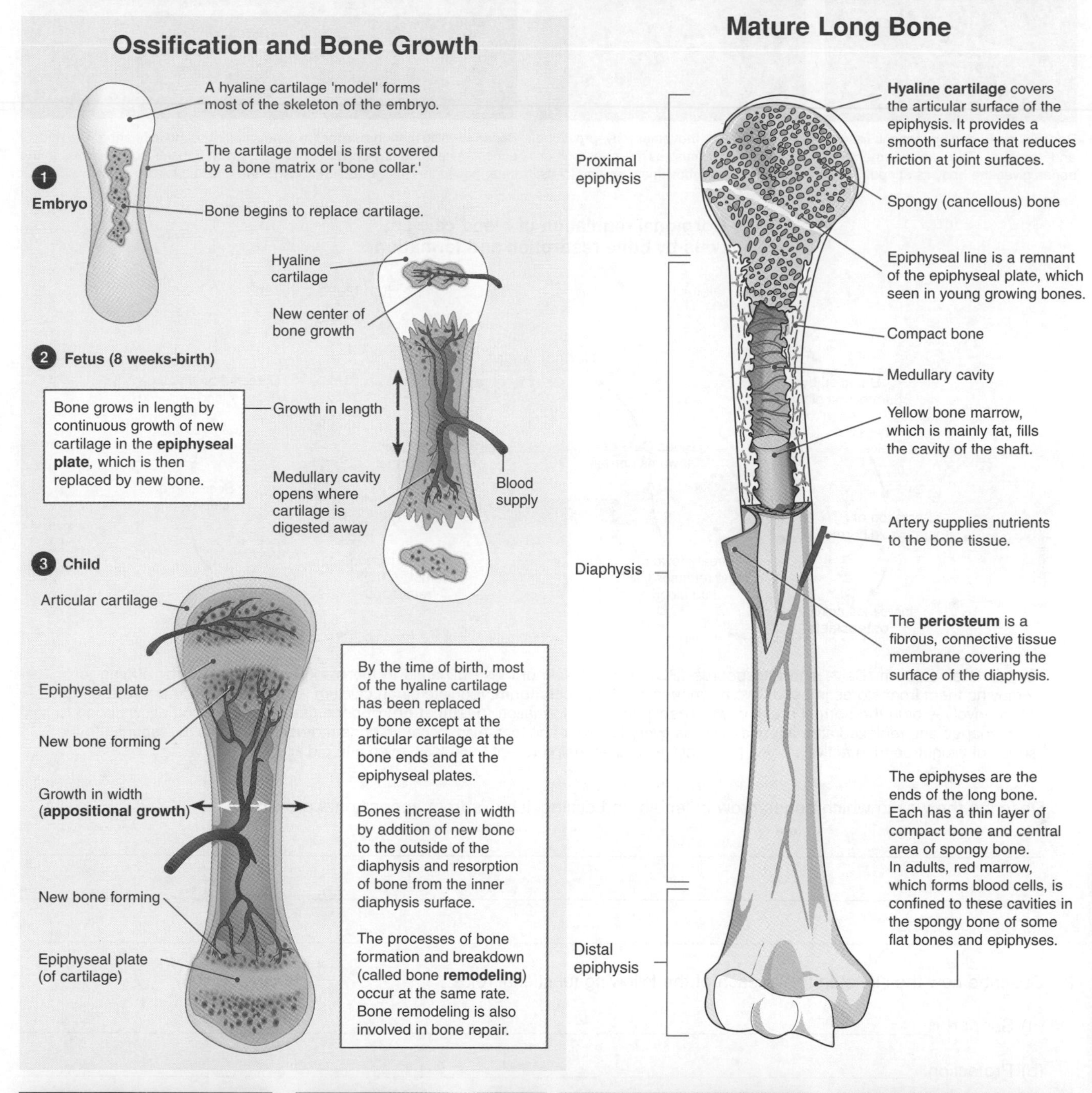

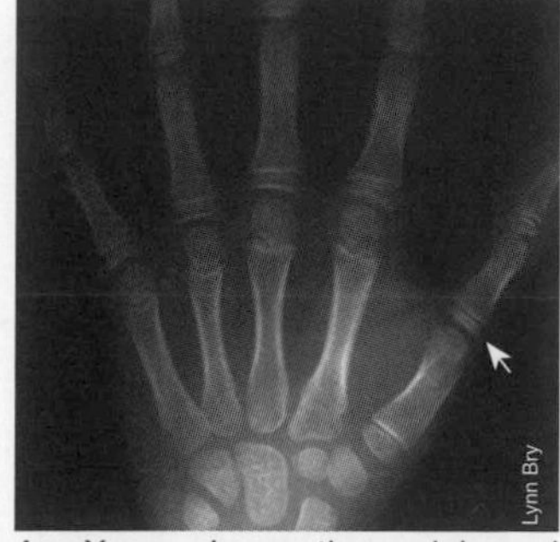

An X-ray shows the epiphyseal plates (growth plates) of a child's hand, seen as separate from the longer bones.

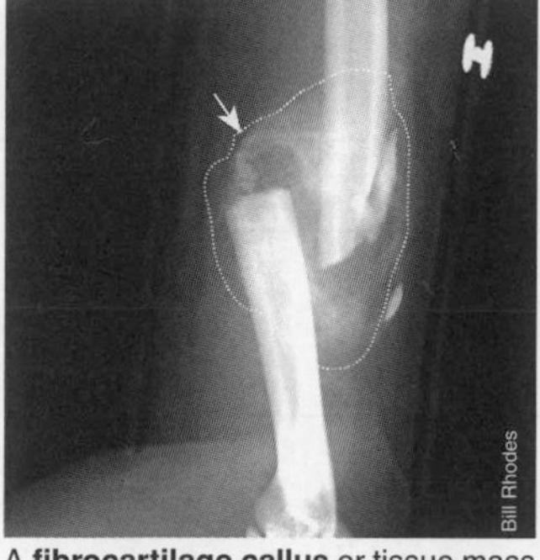

A **fibrocartilage callus** or tissue mass (indicated) begins the repair process on a fractured humerus. Cigarette smoking slows bone healing markedly.

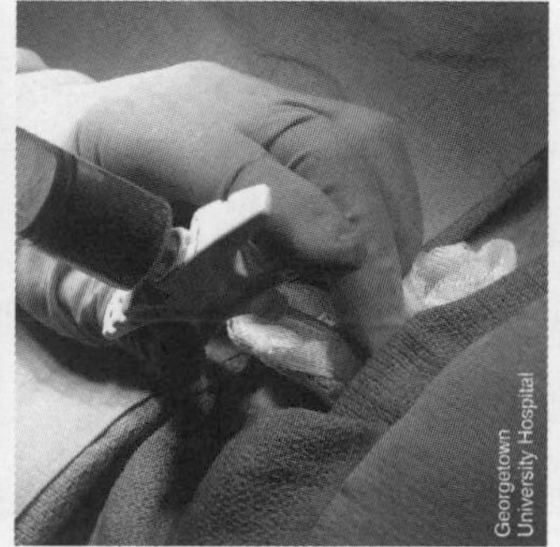

Red bone marrow is stored in the cavities of spongy bone. Here it is being extracted for transplant. Bone marrow is a source of stem cells.

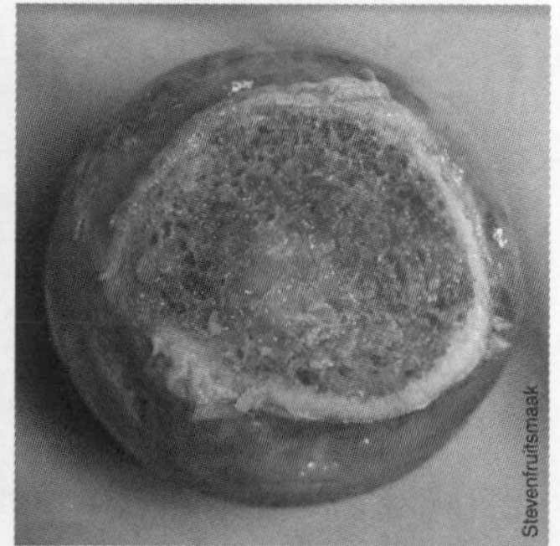

A section of a femur head shows the compact bone surrounding inner spongy bone and marrow. Blood cells are formed in the red marrow.

The Skeletal System

ISBN: 978-1-92717357-2

Related activities: *Connective Tissue, Hematopoiesis, Growth and Development*
Weblinks: *Bone Growth, How Bone Grows, Hormonal Regulation of Calcium*

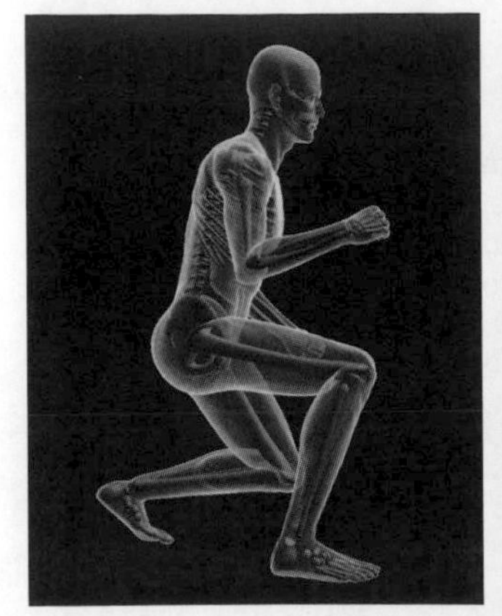

Bones support and protect soft tissues and organs. The arrangement of bones gives the body its shape.

Bones enable movement by providing attachment for muscles. Muscles pull on bones to create movement around joints.

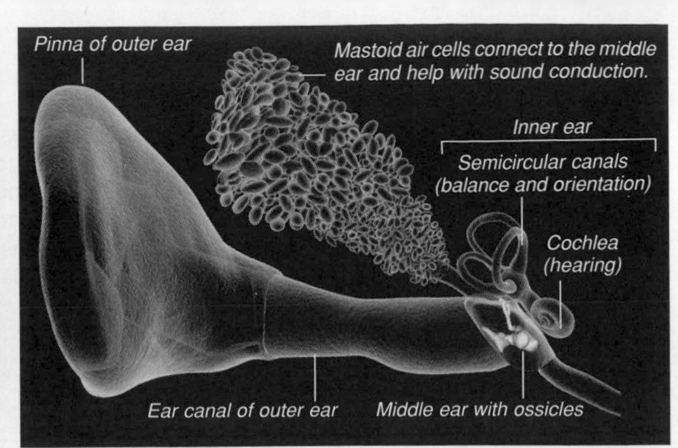

Bones are also responsible for the conduction of sound in hearing. The middle ear contains three tiny bones or ossicles, which transmit sound waves to the inner ear. In this image (anterior view), the mastoid is also shown.

Hormonal regulation of blood calcium levels by bone resorption and formation

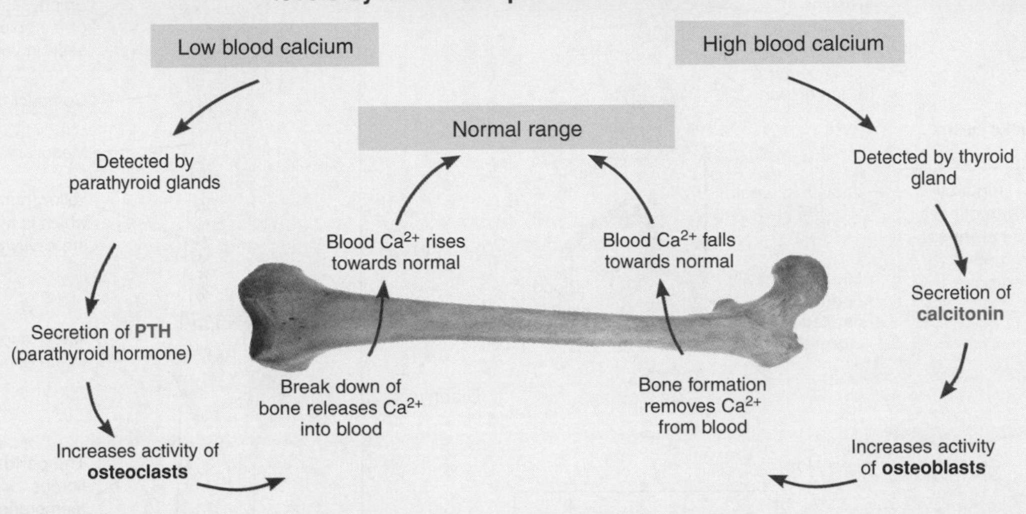

Bone stores calcium (Ca^{2+}) and phosphorus (PO_4^{3-}). The levels of these minerals in the blood are maintained by adding to or removing them from stores in bone. Two hormones, **PTH** and **calcitonin**, regulate blood calcium levels through **bone remodeling**, which involves both the normal break down (**resorption**) and formation (**ossification**) of bone tissue. Remodeling allows bone to be reshaped and replaced after injury as well as everyday wear and tear. Also, because bone remodeling occurs in response to the stress of weight-bearing activity, being physically active can help prevent bone loss, even into old age.

1. Describe the way in which bones grow in length and distinguish this from appositional growth:

2. Describe how the skeleton fulfills each of the following functional roles:

   (a) Support:

   (b) Protection:

   (c) Movement:

   (d) Blood cell production:

   (e) Mineral storage:

3. Identify the feature described by each of the following definitions:

   (a) A feature of bones that are still increasing in length:

   (b) The long shaft of a mature bone:

   (c) Fibrous, connective tissue membrane covering the surface of the bone shaft:

   (d) The end of a long bone, covered in articular hyaline cartilage:

ISBN: 978-1-92717357-2

The Ultrastructure of Bone

The cells that produce bone are called **osteoblasts**. They secrete the matrix of **calcium phosphate** and **collagen fibers** that forms the rigid bone. When they are mature, the bone cells are called **osteocytes**. They are trapped within the matrix but have many thin cytoplasmic extensions, which lie within small channels (called **canaliculi**). **Dense bone** has a very regular structure, composed of repeating units called **osteons** or Haversian systems (after British physician Clopton Havers). Spongy bone is found inside dense bone. It has a much looser structure with irregular spaces filled with red bone marrow.

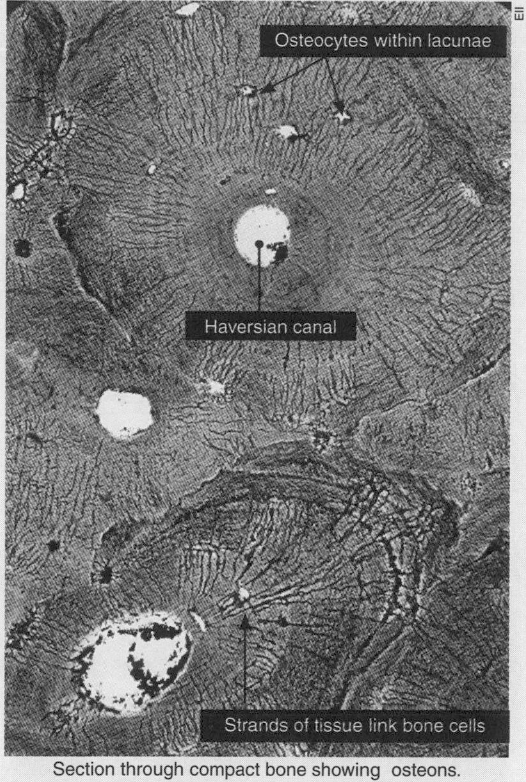

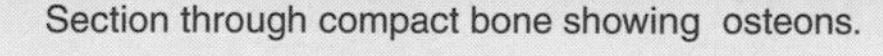
Section through compact bone showing osteons.

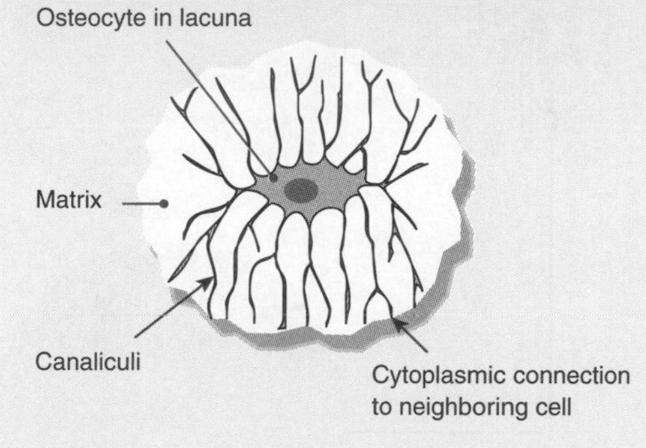

Osteocyte (mature bone cell) embedded in a lacuna within the matrix. Osteocytes maintain the bone tissue (as opposed to osteoblasts, which form the bone).

The Structure of Dense Bone

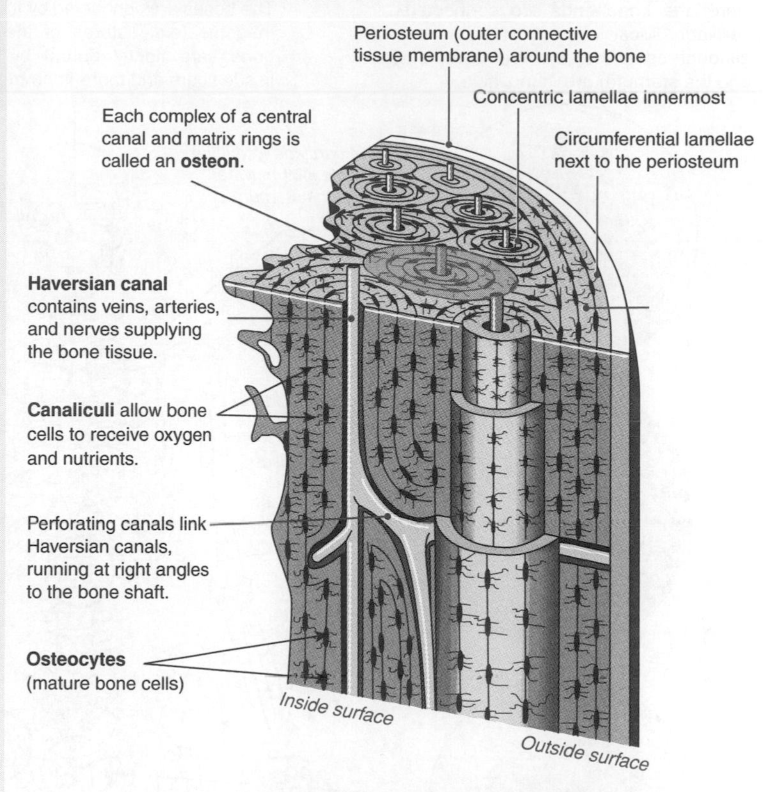

Synonyms for Bone Types

You may come across several terms for the same thing:

Bone type	Also called	Features
Dense bone	Hard bone Cortical bone Compact bone	• Haversian canals surrounded by a regular arrangement of osteocytes. • A single Haversian system is called an osteon.
Spongy bone	Cancellate bone Cancellous bone Trabecular bone	• Less dense structure and higher surface area than dense bone. • Large irregular spaces containing red bone marrow.

1. Distinguish between the function of **osteocytes** and **osteoblasts**: ______

2. Draw lines on the photograph above to mark the boundary of the two most obvious osteons.

3. What is the function of the Haversian canals in dense bone tissue? ______

4. (a) Outline the differences between dense and spongy bone: ______

   (b) Suggest why spongy bone is more susceptible to becoming brittle in old age: ______

The Skeletal System

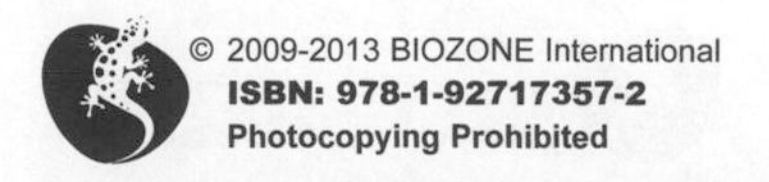

ISBN: 978-1-92717357-2

Related activities: *Bone, The Human Skeleton, Aging and Diseases of the Skeleton*

Joints

Bones are too rigid to bend without damage. To allow movement, the skeletal system consists of many bones held together at **joints** by flexible connective tissues called **ligaments**. All movements of the skeleton occur at joints: points of contact between bones, or between cartilage and bones. Joints may be classified structurally as fibrous, cartilaginous, or synovial based on whether fibrous tissue, cartilage, or a joint cavity separates the bones of the joint. Each of these joint types allows a certain degree of movement. Bones move about a joint by the force of muscles acting upon them.

Cartilaginous Joints

Here, the bone ends are connected by cartilage. Most allow limited movement although some (e.g. between the first ribs and the sternum) are immovable.

Immovable Fibrous Joints

The bones are connected by fibrous tissue. In some (e.g. sutures of the skull), the bones are tightly bound by connective tissue fibers and there is no movement.

Synovial Joints

These allow free movement in one or more planes. The articulating bone ends are separated by a joint cavity containing lubricating synovial fluid (see overleaf).

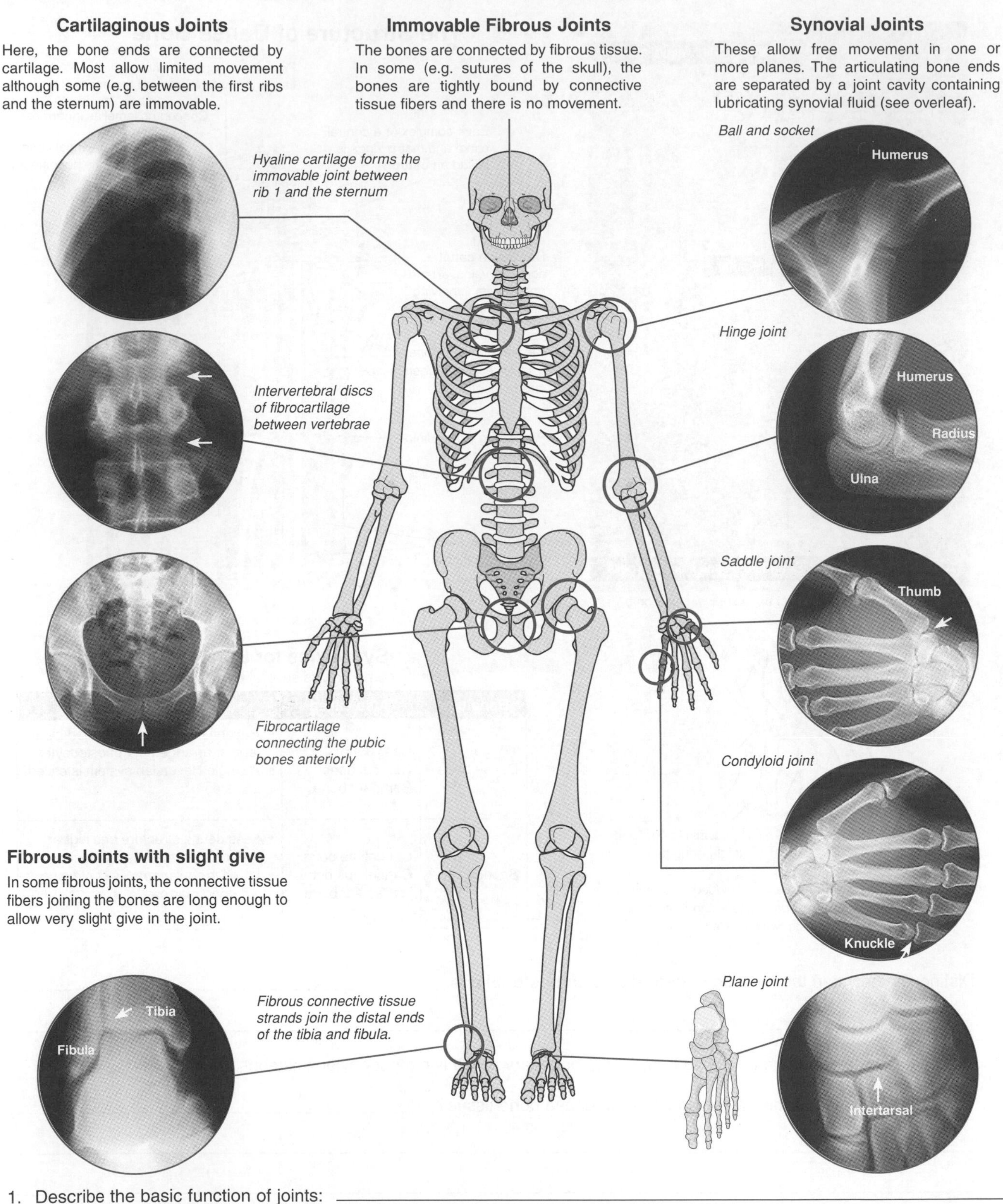

Fibrous Joints with slight give

In some fibrous joints, the connective tissue fibers joining the bones are long enough to allow very slight give in the joint.

1. Describe the basic function of joints: ______

2. How are bones held together at joints? ______

ISBN: 978-1-92717357-2

Related activities: Body Membranes and Cavities, The Mechanics of Movement
Weblinks: Skeleton and Joints

Synovial Joints

Synovial joints are distinguished from other joint types by the presence of a fluid-filled joint capsule surrounding the articulating surfaces of the bones. Synovial joints are the most common and most movable joints in the body. They allow free movement of body parts in varying directions (one, two or three planes). The most freely movable synovial joints are also the least stable and the most prone to injury. Restricting the amount of movement gives less freedom, but also makes the joint more stable.

Structure of an Elbow Joint

The elbow joint is a hinge joint and typical of a synovial joint. Like most synovial joints, it is reinforced by ligaments (not all shown). In the diagram, the brachialis muscle, which inserts into the ulna and is the prime mover for flexion of the elbow, has been omitted to show the joint structure. Muscles are labeled blue and bones bold.

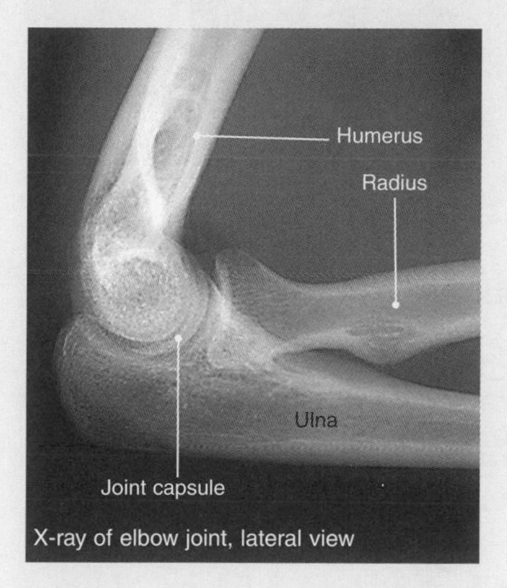

X-ray of elbow joint, lateral view

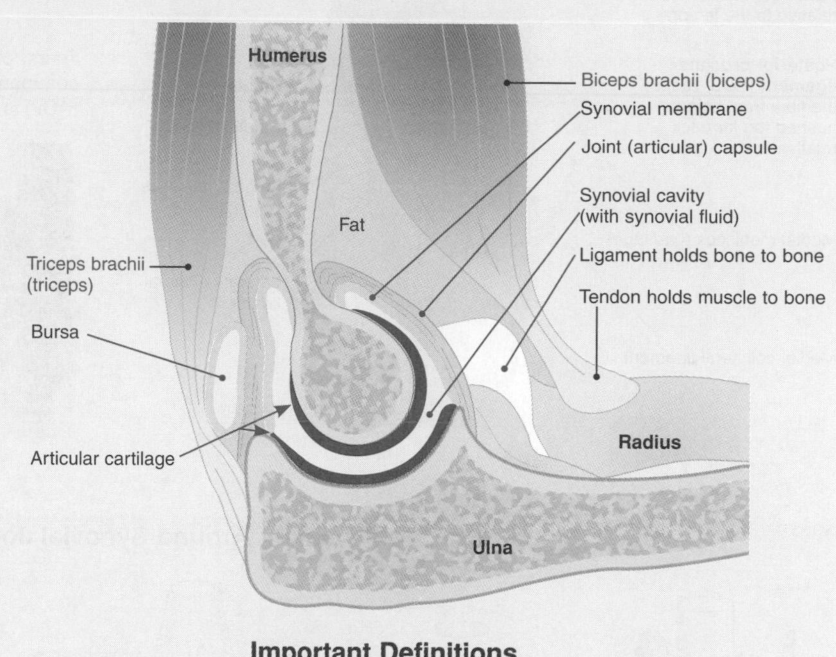

Important Definitions

A **bursa** is a fluid filled cavity lined with synovial membrane. It acts as a cushion, e.g. between tendon and bone, or between bones.

Cartilage is a flexible connective tissue. It protects a joint surface against wear but has no blood supply (it is **avascular**).

Structure of a Knee Joint

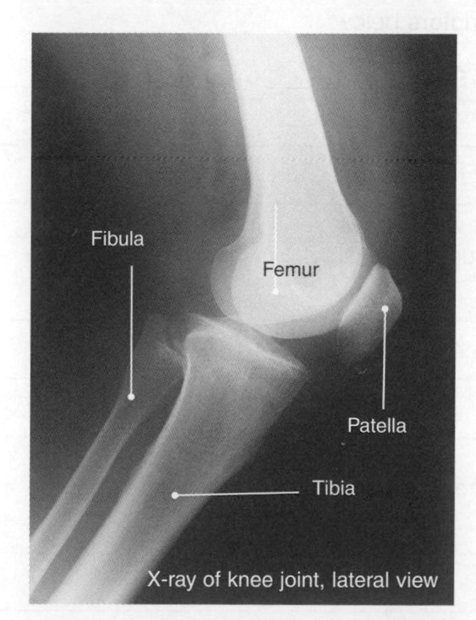

X-ray of knee joint, lateral view

Structurally the knee joint is a **condyloid joint**, where a convex surface fits a concave surface. However, **it functions as a hinge joint** because the cartilages and the ligaments of the joint prevent any lateral (sideways) movement. Note that the patella ligament is the only ligament shown in the diagram (right), but other ligaments stabilize the joint (see next page).

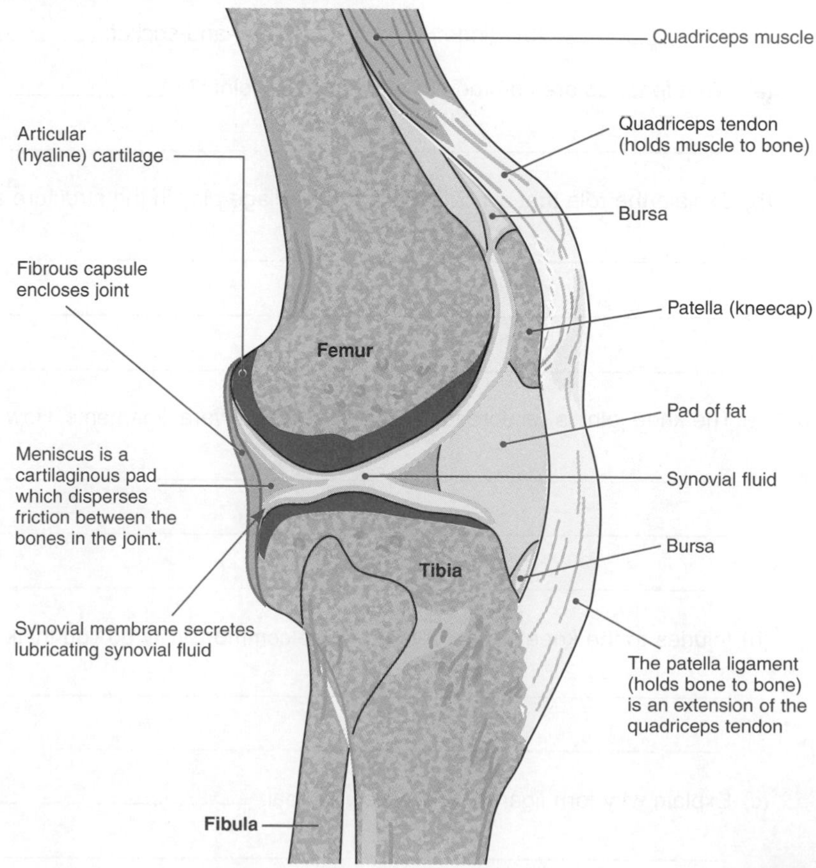

ISBN: 978-1-92717357-2

Related activities: *Joints, Connective Tissue*
Weblinks: *Synovial Joint Movements, Types of Synovial Joints*

About Ligaments

Ligaments are made of connective tissue and hold bones together. They are strong and flexible but have a poor blood supply.

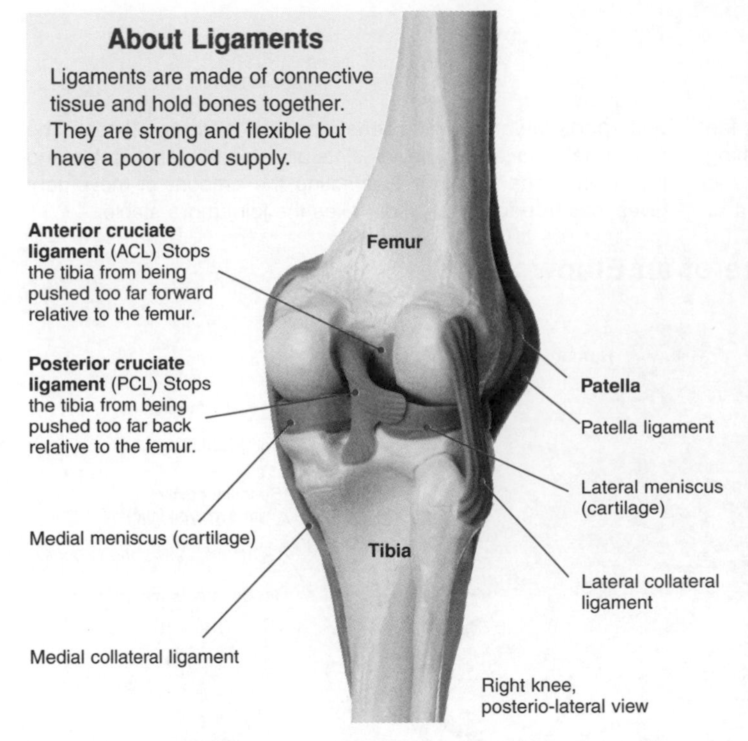

Right knee, posterio-lateral view

Ligaments of the Knee Joint

The knee is the largest joint in the human body and bears most of the weight of the body during locomotion. The femur articulates with the tibia, but also with the patella. The joint allows flexion and extension and a small amount of medial and lateral rotation. Excessive rotation and lateral movements are prevented by two sets of ligaments, which hold the joint in place and stabilize it. The diagram (left) shows the positions of these ligaments. The cruciate ligaments form an X shape through the center between the articulating surfaces.

Ligament Injuries

Tears to the ligaments of the knee are among the most common sporting injuries. The anterior cruciate ligament (ACL) is often injured by twisting and must be repaired.

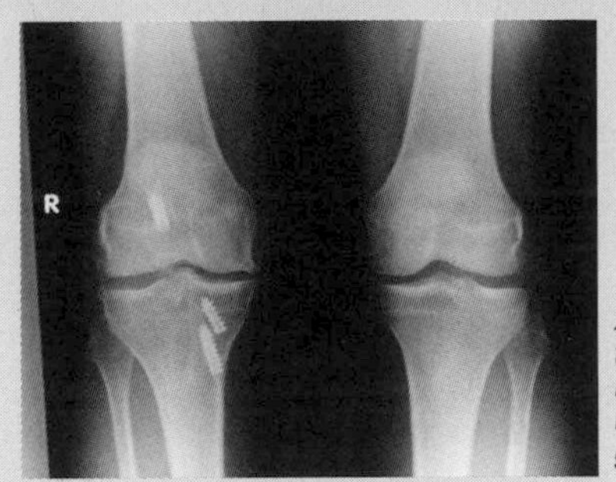

X-ray of repair to the ACL of the right knee. Screws hold the ligament graft in place.

Movement Around Synovial Joints

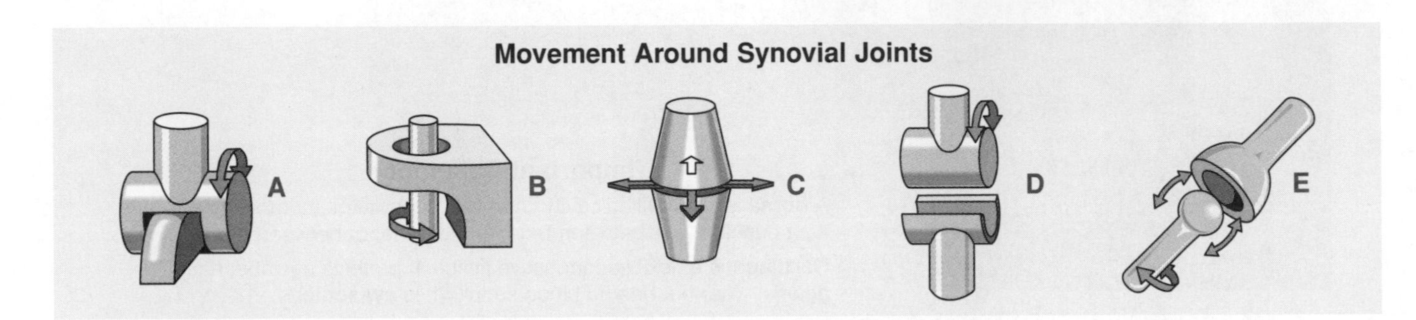

1. Classify each of the synovial joint models (**A-E**) above, according to the descriptors below:

   (a) Pivot: ________ (b) Hinge: ________ (c) Ball-and-socket: ________ (d) Saddle: ________ (e) Gliding: ________

2. (a) What features are common to most synovial joints? ________

   (b) Explain the role that synovial fluid and cartilage play in the structure and function of a synovial joint:

3. (a) The knee joint is reinforced and stabilized by several ligaments. How do these ligaments assist the joint function?

   (b) Injuries to the knee's ligaments are quite common. Why do you think this is the case? ________

   (c) Explain why torn ligaments are slow to heal: ________

ISBN: 978-1-92717357-2

Aging and Diseases of the Skeleton

Aging refers to the degenerative changes that occur in the body as a result of cell renewal rates slowing. Aging affects all tissues, including bone and the other connective tissues that make up the skeleton. As people age, the rate of bone remodelling slows and bone resorption rates begin to exceed rates of deposition. As a result, the skeleton loses strength and there is an increased tendency for bones to fracture. The joints also tend to become less flexible. Some specific diseases of the skeleton (such as osteoporosis) are also more common in older people, although they are not exclusively associated with aging.

Osteoporosis

Loss of height

Hunching of spine

Osteoporosis affecting the spine

RA

Osteoporosis is an age-related disorder where bone mass decreases, and there is a loss of height and an increased tendency for bones to break (**fracture**). Women are at greater risk of developing the disease than men because their skeletons are lighter and their estrogen levels fall after menopause (estrogen provides some protection against bone loss). Younger women with low hormone levels and/or low body weight are also affected. Osteoporosis affects the whole skeleton, but especially the spine, hips, and legs.

Osteoarthritis

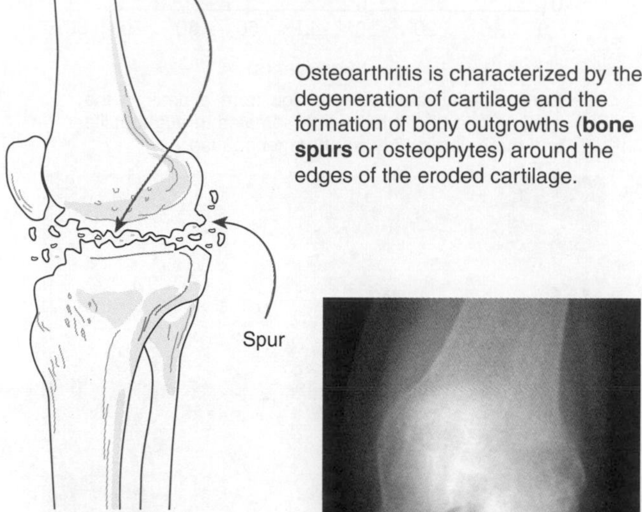

Osteoarthritis is characterized by the degeneration of cartilage and the formation of bony outgrowths (**bone spurs** or osteophytes) around the edges of the eroded cartilage.

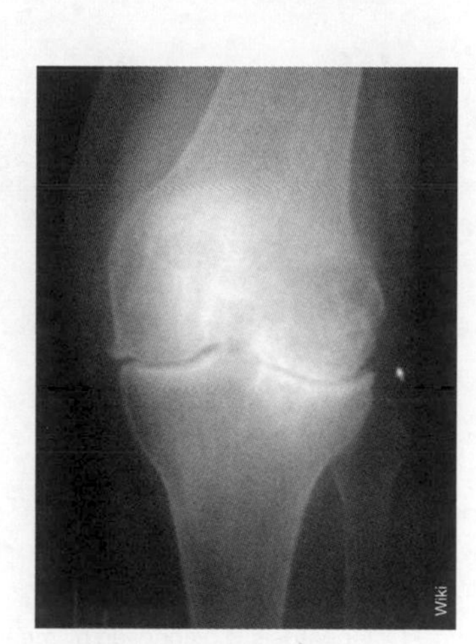

In severe osteoarthritis, the cartilage can become so thin that it no longer covers the bone. The bone ends touch, rub against each other, and start to wear away, as shown in this X-ray of an osteoarthritic knee (right).

Osteoarthritis (OA) a chronic degenerative disease aggravated by **mechanical stress** on bone joints. This leads to pain, stiffness, inflammation, and full or partial loss of joint function. OA occurs in almost all people over the age of 60 and affects three times as many women as men. Weight bearing joints such as those in the knee, foot, hips, and spine are the most commonly affected. Although there is no cure, the symptoms can be greatly relieved by painkillers, anti-inflammatory drugs, and exercises to maintain joint mobility.

1. Describe the main reason for age-related loss of bone mass: __________

2. How do the structural changes in an osteoathritic joint relate to the following symptoms of the disease:

   (a) Loss of function: __________

   (b) Pain: __________

3. Suggest why the weight-bearing joints of the body are most commonly affected by OA? __________

ISBN: 978-1-92717357-2

Bone Density vs Age

Peak bone mass

Bone mass (% total max): 0, 50, 100

Age (years): 0, 10, 20, 30, 40, 50, 60, 70, 80

Puberty

Menopause

Male

Female

Bone density declines with age from a peak in the early to mid 20s. It is lower in females throughout life and more particularly so after menopause.

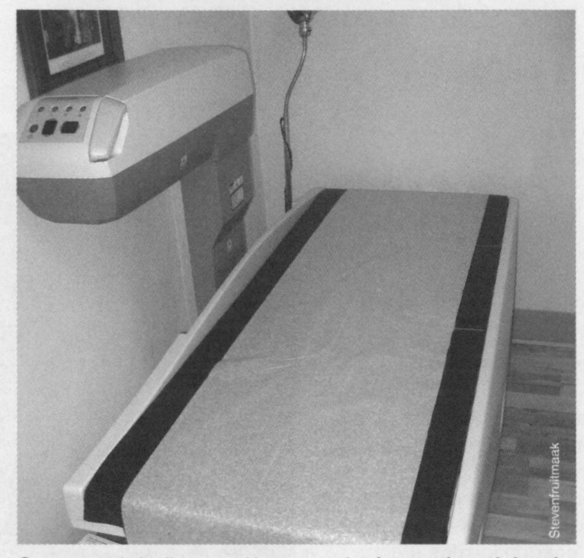

Osteoporosis is diagnosed by measuring **bone mineral density** (BMD). The most widely-used method is dual energy X-ray absorptiometry or **DXA** (above). Two X-ray beams with different energy levels are aimed at the bone. BMD is determined from the bone's absorption of each beam.

Diagnosis of Osteoarthritis

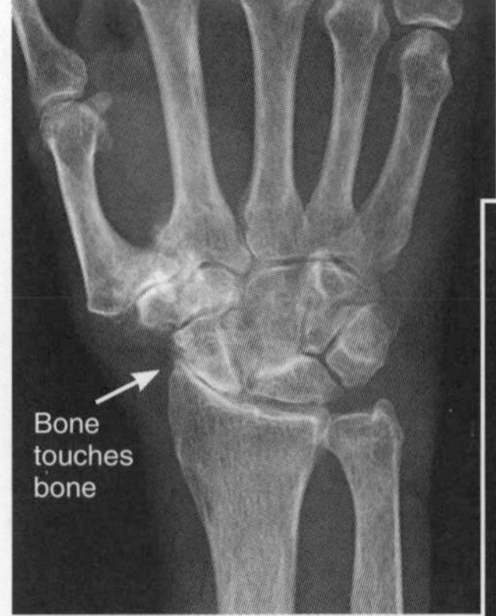

X rays are one of the commonly used diagnostic method for osteoarthritis as the features associated with the disease often show up clearly.

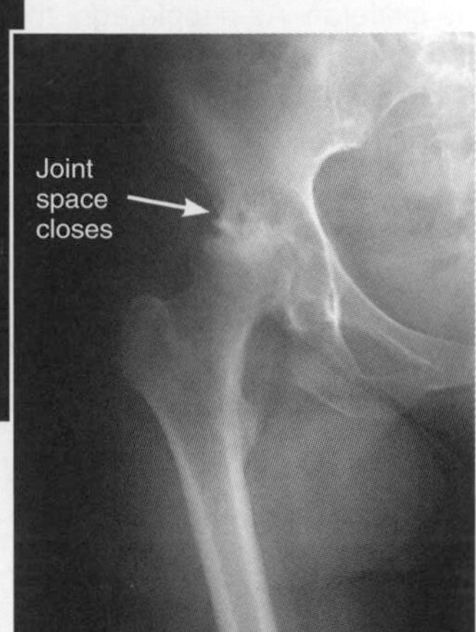

The following features may be seen on X-rays:

- Narrowing of joint spaces
- Presence of bone spurs
- Hardening of bone on the articular surfaces
- Partial dislocation of the bone

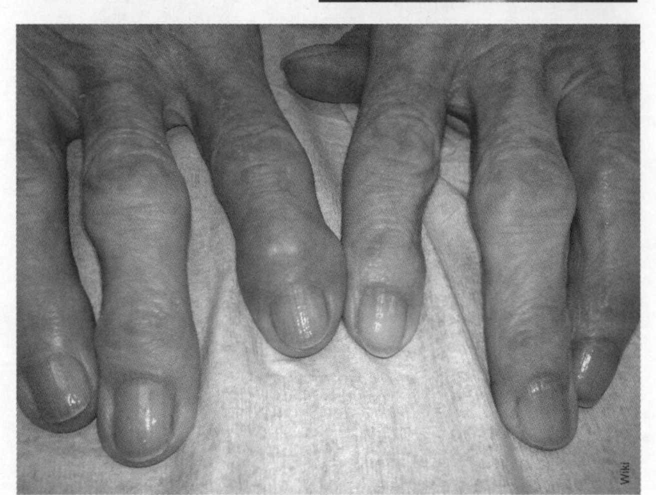

The loss of cartilage, the wearing of bone, and the growth of spurs all combine to change the shape of joints affected by osteoarthritis. This forces the bones out of their normal positions and causes deformity, as seen in the fingers of this elderly patient. Finger joint changes in particular are more common in women and may be hereditary.

4. Explain why joint injury and jobs involving repetitive actions (such as typing) can increase the risk of developing OA:

5. Explain why loss of bone mineral density is associated with increased risk of bone fracture:

6. Approximately how much of peak bone mass will a be lost by age 70 in:

   (a) A male: (b) A female:

7. Adequate dietary intake of calcium is very important for people in their first 20 years or so. Explain why:

8. Lack of physical activity increases the risk of osteoporosis. From what you know of bone remodeling, suggest why:

ISBN: 978-1-92717357-2

KEY TERMS: Mix and Match

INSTRUCTIONS: Test your vocabulary by matching each term to its definition, as identified by its preceding letter code.

appendicular skeleton ....

axial skeleton ....

bone (tissue) ....

cartilage ....

cartilaginous joint ....

compact (=cortical) bone ....

fibrous joint ....

flat bones ....

Haversian canals ....

irregular bones ....

joint (articulation) ....

ligament ....

long bones ....

matrix (of bone) ....

ossification ....

osteoarthritis ....

osteoblasts ....

osteocyte ....

osteroporosis ....

parathyroid hormone ....

pectoral girdle ....

pelvic girdle ....

periosteum ....

synovial fluid ....

synovial joint ....

tendon ....

A Canals running through the bone. They contain blood vessels and nerves to supply oxygen and nutrients to the bone cells, and remove waste materials.

B The junction where two or more bones meet.

C The descriptor for a bone which is thin and flattened. An example is the sternum.

D Name given to bones which are longer than they are wide (e.g. femur).

E A fibrous, connective tissue membrane covering the shaft of a long bone.

F The set of bones connecting the upper limb to the axial skeleton on each side. It consists of the clavicle and scapula. Also called the shoulder girdle.

G The bones of the skull, rib cage, and spine.

H Fluid contained with the synovial capsule. Its function is to reduce friction between the cartilaginous ends of the bones in a synovial joint.

I Cells that secrete the matrix for bone formation.

J Mineralized connective tissue composed mainly of calcium and phosphorus that forms the skeleton.

K Fibrous connective tissue that connects muscles to bones.

L A disease characterized by loss of bone mass and increased likelihood of bone fracture. Often (but not always) associated with old age.

M The bones making up the limbs and the pectoral and pelvic girdles.

N A joint that allows free movement of body parts in varying directions, on one, two or three planes. A typical example is the knee joint.

O A type of joint where the bones are joined by fibrous tissue. This type of joint has little or no movement, so is sometimes called a fixed or immovable joint.

P Name given to bones which have a peculiar or complex shape (e.g. vertebrae).

Q Joints that are held together by cartilage. No joint cavity is present.

R A disease characterized by the degeneration of cartilage and the formation of bony outgrowths around the edges of the eroded cartilage.

S The weight-bearing girdle of the lower body, consists of the coxal (hip) bones, and sacrum.

T A dense type of bone which makes up 80% of the human skeleton and provides strength and rigidity.

U The intercellular substance of bone, consisting of collagenous fibers, ground substance, and inorganic salts such as calcium and phosphate.

V A bone cell.

W A connective tissue structure that connects two or more bones.

X This hormone stimulates break down of bone and is important in bone remodeling.

Y A strong but flexible connective tissue that covers the ends of bones in synovial joints (and some other areas) and provides a smooth surface for joint movement.

Z The formation of bone tissue.

ISBN: 978-1-92717357-2

The Muscular System

Key terms

actin (thin filament)
antagonistic pair
blood lactate
cardiac muscle
cross bridge
fast twitch
filament (=myofilament)
muscle fatigue
muscle fiber
myofibril
myosin (thick filament)
neuromuscular junction
oxygen debt
prime mover
sarcomere
sarcoplasmic reticulum
skeletal (=striated) muscle
sliding filament hypothesis
slow twitch
smooth muscle
tropomyosin
troponin

Key concepts

- The muscular system is organized into discrete muscles, which work as antagonistic pairs.
- Skeletal muscle tissue acts with the bones of the skeleton to produce movements.
- In muscle, contraction results from the movement of actin filaments against myosin filaments.
- ATP is required for muscle contraction.

Learning Objectives

☐ 1. Use the **KEY TERMS** to compile a glossary for this topic.

Muscle Structure pages 29, 69-72

☐ 2. Recall the distinguishing features and roles of **cardiac muscle**, **smooth muscle**, and **skeletal muscle** tissue.

☐ 3. Describe the ultrastructure of skeletal muscle, including:
- The organization of the muscle into bundles of **fibers**.
- The organization of the fibers (cells) into bundles of smaller **myofibrils**.
- The composition and arrangement of the **filaments** within each myofibril.

☐ 4. Describe the organization of the body's musculature into functional groups. Recognize major muscle groups and their functions. Explain the functional significance of muscles as **antagonistic pairs**.

The Physiology of Muscle Contraction pages 71-73

☐ 5. Recognize a **sarcomere** as a complete contractile unit. Relate the structure of the sarcomere to the distribution of **actin** and **myosin** within the myofibril.

☐ 6. Describe the **sliding-filament hypothesis** of muscle contraction. Identify the role of calcium ions (Ca^{2+}), the troponin-tropomyosin complex, and ATP. Recognize that contraction of a muscle fiber is an all-or-nothing event,.

☐ 7. Describe the structure and function of the **neuromuscular junction**, including the role of acetylcholine (Ach), the T-tubules, and the **sarcoplasmic reticulum**.

☐ 8. Describe the structural differences between **fast twitch** and **slow twitch** muscle fibers. Explain the physiological basis of these differences.

The Functioning Muscle pages 74-80

☐ 9 . Explain how joints, with muscles, bring about specific movements of the skeleton. With reference to muscles and their activity, define the terms **origin** and **insertion**. Distinguish between **isotonic** and **isometric contractions** and explain how muscle tone is achieved and maintained.

☐ 10. Describe movements of the skeleton, including **extension**, **flexion**, **rotation**, **abduction**, **adduction**, and **circumduction**.

☐ 11. Explain how muscles as a whole produce graded responses by (1) changing the frequency of stimulation (to tetany) or (2) by recruitment of motor units.

☐ 12. Identify sources of energy for muscle contraction. Compare aerobic and anaerobic pathways as sources of ATP for muscle contraction.

☐ 13. Explain **muscle fatigue** and relate it to the increase in **blood lactate**, depletion of carbohydrate supplies, and decreased pH. Explain how these changes provide the stimulus for increased breathing (and heart) rates.

☐ 14. Define the term **oxygen debt** and explain how it is repaid after intense exercise.

☐ 15. Discuss the effects of exercise, inactivity, and aging on muscle.

Periodicals:
Listings for this chapter are on page 279

Weblinks:
www.thebiozone.com/weblink/AnaPhy-3572.html

BIOZONE APP:
Student Review Series
The Skeletal & Muscular Systems

Muscles of the Human Body

The muscles of the human body occur as groups which work together to achieve an outcome. For example, the raising of the forearm is achieved by the contraction of the biceps brachii and the brachialis. This muscle group is sometimes referred to simply as the biceps. Similarly, the abdominals is used to refer to the muscle layers covering the body's anterior midsection. Muscle groups are divided between the head, trunk, upper and lower arms, thorax and midsection, and upper and lower legs, each with anterior and posterior muscles. Some common muscle groupings are illustrated below.

Muscle Groups

Word list:
Facial muscles, pectorals, obliques (abdominal group), rectus abdominis (abdominal group), trapezius, latissimus dorsi, deltoid, biceps, triceps, gluteals, quadriceps, hamstrings, gastrocnemius

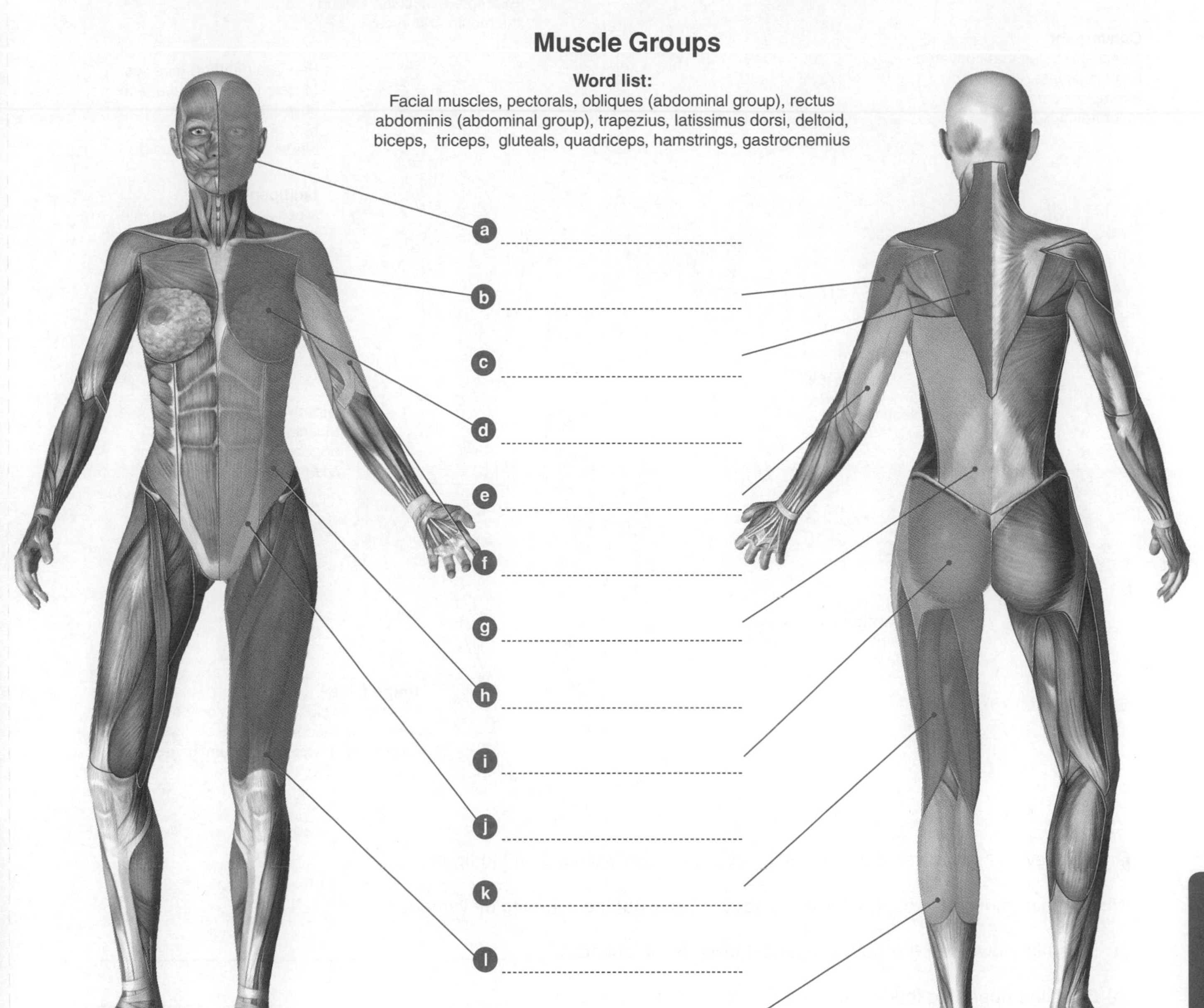

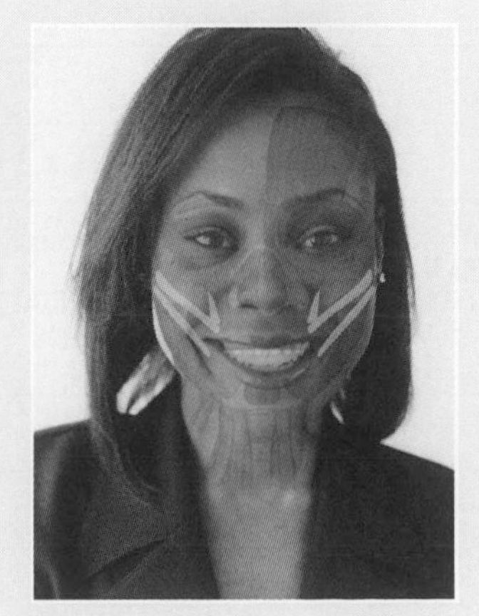

Head Muscles

Head muscles are divided into the **facial muscles**, which make expressions, and the **chewing muscles**. Facial muscles are inserted into soft tissues (e.g. skin) and enable a range of facial expressions.

Smiling involves about 12 muscles. Major muscles involved include:
- Zygomaticus major raises the corners of the mouth and produces the cheek dimples
- Zygomaticus minor raises the upper edges of the lips
- Levator anguli oris raises the upper lip to show the canine teeth.

Frowning involves about 11 muscles. Muscles involved include:
- Procerus pulls the skin between the eyebrows down towards the nose producing the "fighters fold"
- Depressor anguli oris pulls the corners of the mouth down to form the lips into an inverted U.

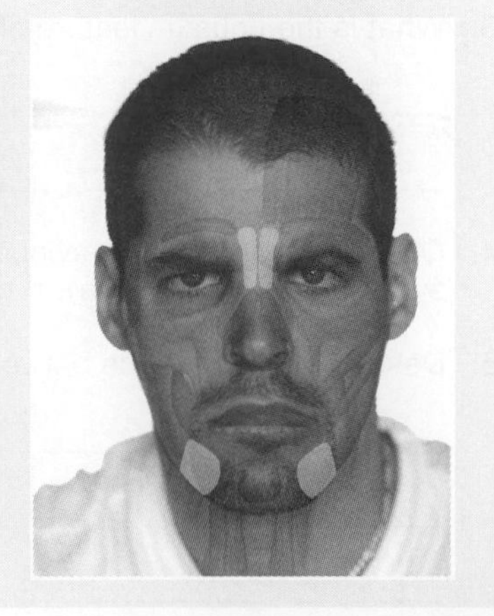

ISBN: 978-1-92717357-2

Muscle Fascicles and Muscle Structure

Skeletal muscles consist of **fascicles** (bundles of muscles fibers surrounded by connective tissue layer). The arrangement of fascicles varies, producing a variety of muscle structures.

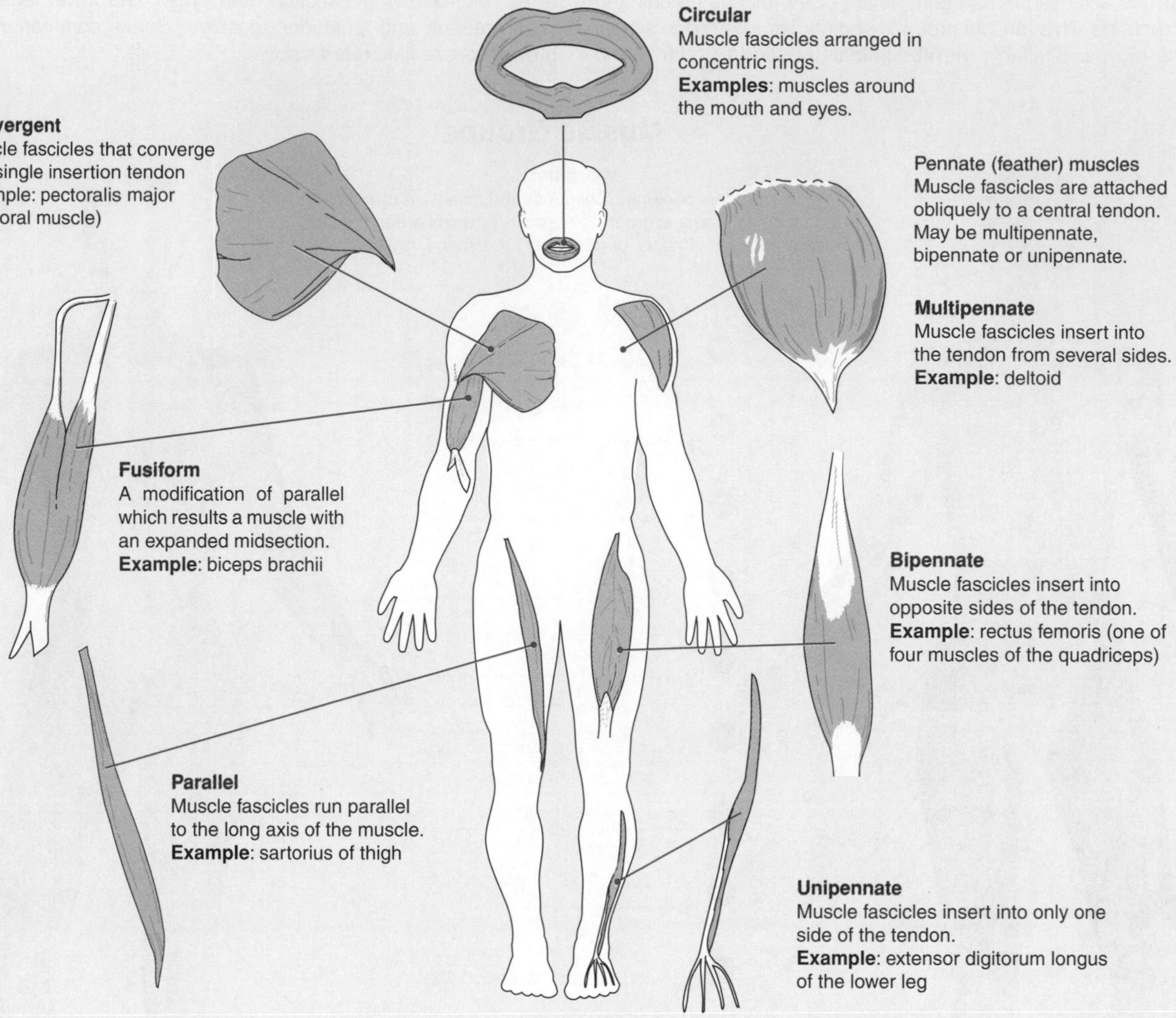

Circular
Muscle fascicles arranged in concentric rings.
Examples: muscles around the mouth and eyes.

Convergent
Muscle fascicles that converge to a single insertion tendon
Example: pectoralis major (pectoral muscle)

Pennate (feather) muscles
Muscle fascicles are attached obliquely to a central tendon. May be multipennate, bipennate or unipennate.

Multipennate
Muscle fascicles insert into the tendon from several sides.
Example: deltoid

Fusiform
A modification of parallel which results a muscle with an expanded midsection.
Example: biceps brachii

Bipennate
Muscle fascicles insert into opposite sides of the tendon.
Example: rectus femoris (one of four muscles of the quadriceps)

Parallel
Muscle fascicles run parallel to the long axis of the muscle.
Example: sartorius of thigh

Unipennate
Muscle fascicles insert into only one side of the tendon.
Example: extensor digitorum longus of the lower leg

1. On the previous page, use the word list to label the muscle groups on the figure:
2. Which major muscles group(s) would be used to carry out the following movements:
   (a) Raise the lower leg (i.e. tibia and fibula) towards the buttocks: ________
   (b) Bring the upper leg forward (i.e. the femur) as in taking a step: ________
   (c) Rotate the wrist: ________
   (d) Raise the arm from the side of the body up over the head: ________
3. What is the unusual feature of facial muscles? ________
4. On the photos on the previous page identify and label the facial muscles mentioned using the following shorthand: zygomaticus major (ZMa), zygomaticus minor (ZMi), levator anguli oris (LAO), procerus (P), depressor anguli oris (DAO)
5. Describe the difference between parallel and fusiform muscle structure: ________

ISBN: 978-1-92717357-2

Skeletal Muscle Structure and Function

Skeletal muscle (also called striated or voluntary muscle) is organized into bundles of muscle cells or fibers. Each fiber is a single cell with many nuclei and each fiber is itself a bundle of smaller **myofibrils** arranged lengthwise. Each myofibril is in turn composed of two kinds of **myofilaments** (thick and thin), which overlap to form light and dark bands. The alternation of these light and dark bands gives skeletal muscle its striated or striped appearance. The sarcomere, bounded by the dark Z lines, forms one complete contractile unit. Muscle fibers are innervated by the branches of motor neurons, each of which terminates in a specialized cholinergic synapse called the **neuromuscular junction** (or motor end plate). A motor neuron and all the fibers it innervates (which may be a few or several hundred) are called a **motor unit**. Graded responses in the muscle as a whole are achieved by varying the number of motor units active at any one time (recruitment of motor units).

Structure of Muscle

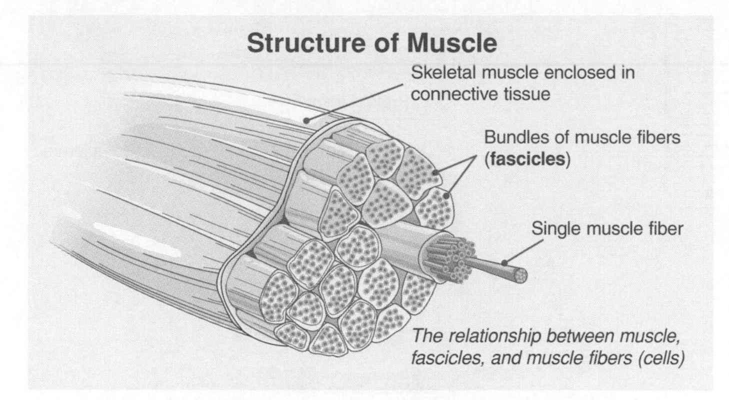

The relationship between muscle, fascicles, and muscle fibers (cells)

Structure of a Muscle Fiber

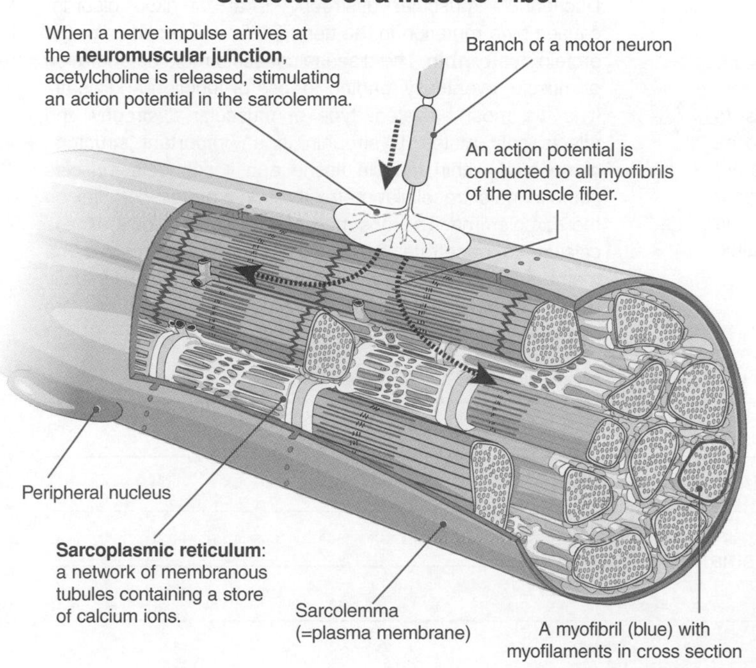

Longitudinal Section of a Sarcomere

I band (light) | A band (dark) | I band (light)

Z line

One sarcomere

WMU

H zone

Thin filament made of **actin**

Thick and thin filaments slide past each other

Thick filament made of **myosin**

Cross section through a region of overlap between thick and thin filaments.

Thick filament

Thin filament

The photograph of a sarcomere (above) shows the banding pattern arising as a result of the highly organized arrangement of thin and thick filaments. It is represented schematically in longitudinal section and cross section.

The Neuromuscular Junction

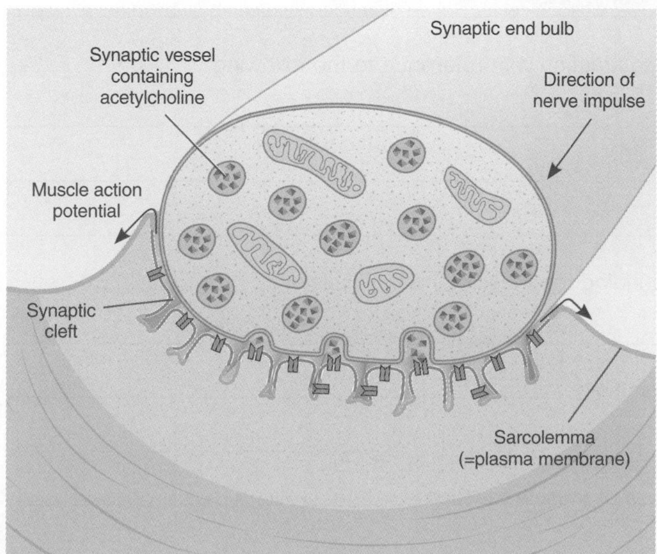

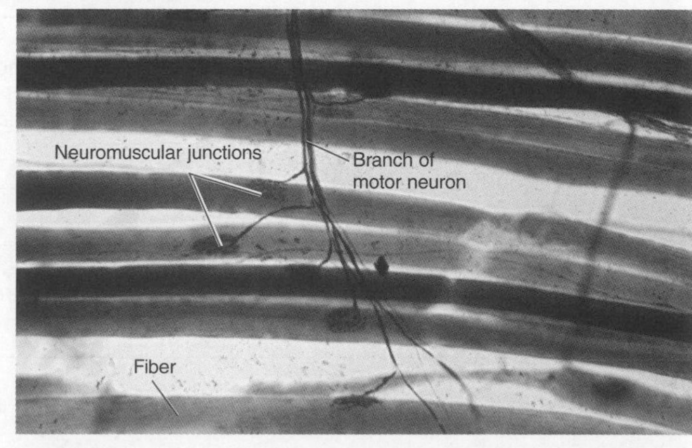

Above: Axon terminals of a motor neuron supplying a muscle. The branches of the axon terminate on the sarcolemma of a fiber at regions called the neuromuscular junction (or motor end plate). Each fiber receives a branch of an axon, but one axon may supply many muscle fibers.

Left: Diagrammatic representation of part of a neuromuscular junction.

The Skeletal System

ISBN: 978-1-92717357-2

Related activities: *The Sliding Filament Theory, Chemical Synapses, Integration at Synapses* ***Weblinks:*** *Muscle Structure and Contraction*

The Banding Pattern of Myofibrils

Within a myofibril, the thin filaments, held together by the **Z lines**, project in both directions. The arrival of an action potential sets in motion a series of events that cause the thick and thin filaments to slide past each other. This is called **contraction** and it results in shortening of the muscle fiber and is accompanied by a visible change in the appearance of the myofibril: the I band and the sarcomere shorten and H zone shortens or disappears (below).

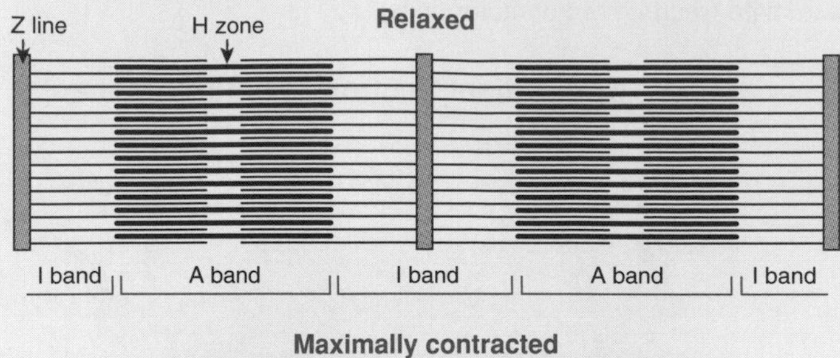

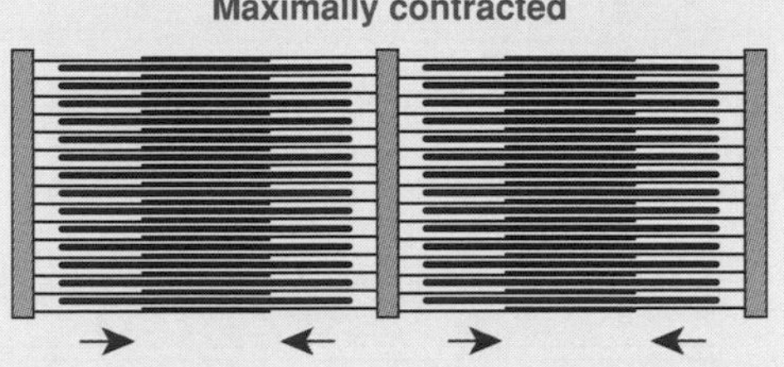

The response of a single muscle fiber to stimulation is to contract maximally or not at all; its response is referred to as the **all-or-none law** of muscle contraction. If the stimulus is not strong enough to produce an action potential, the muscle fiber will not respond. However skeletal muscles as a whole are able to produce varying levels of contractile force. These are called **graded responses**.

When Things Go Wrong

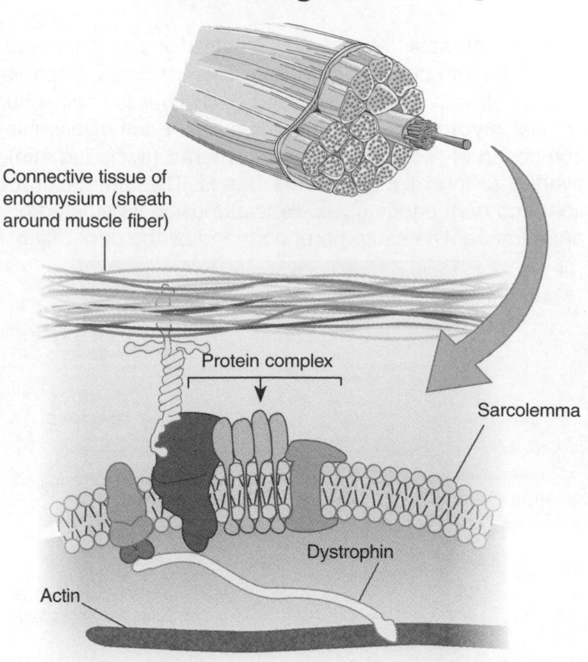

Duchenne's muscular dystrophy is an X-linked disorder caused by a mutation in the gene DMD, which codes for the protein **dystrophin**. The disease causes a rapid deterioration of muscle, eventually leading to loss of function and death. It is the most prevalent type of muscular dystrophy and affects only males. Dystrophin is an important structural component within muscle tissue and it connects muscles fibers to the extracellular matrix through a protein complex on the sarcolemma. The absence of dystrophin allows excess calcium to penetrate the sarcolemma (the fiber's plasma membrane). This damages the sarcolemma, and eventually results in the death of the cell. Muscle fibers die and are replaced with adipose and connective tissue.

1. Describe the **neuromuscular junction**: ______________________________

____

____

2. (a) Explain the cause of the banding pattern visible in striated muscle: ______________________________

____

____

____

(b) Explain the change in appearance of a **myofibril** during contraction with reference to the following:

The I band: ______________________________

The H zone: ______________________________

The sarcomere: ______________________________

3. Describe the purpose of the connective tissue sheaths surrounding the muscle and its fascicles: ______________________________

____

4. Explain what is meant by the all-or-none response of a muscle fiber: ______________________________

____

5. Explain why the inability to produce **dystrophin** leads to a loss of muscle function: ______________________________

____

ISBN: 978-1-92717357-2

The Sliding Filament Theory

The previous activity described how muscle contraction is achieved by the thick and thin muscle filaments sliding past one another. This sliding is possible because of the structure and arrangement of the thick and thin filaments. The ends of the thick myosin filaments are studded with heads or **cross bridges** that can link to the thin filaments next to them. The thin filaments contain the protein actin, but also a regulatory protein complex. When the cross bridges of the thick filaments connect to the thin filaments, a shape change moves one filament past the other. Two things are necessary for cross bridge formation: calcium ions, which are released from the **sarcoplasmic reticulum** when the muscle receives an action potential, and ATP, which is hydrolyzed by ATPase enzymes on the myosin. When cross bridges attach and detach in sarcomeres throughout the muscle cell, the cell shortens. Although a muscle fiber responds to an action potential by contracting maximally, skeletal muscles as a whole can produce varying levels of contractile force. These **graded responses** are achieved by changing the frequency of stimulation (**frequency summation**) and by changing the number and size of motor units recruited (**multiple fiber summation**). Maximal contractions of a muscle are achieved when nerve impulses arrive at the muscle at a rapid rate and a large number of motor units are active at once.

The Sliding Filament Theory

Muscle contraction requires calcium ions (Ca^{2+}) and energy (in the form of ATP) in order for the thick and thin filaments to slide past each other. The steps are:

1. The binding sites on the **actin** molecule (to which myosin 'heads' will locate) are blocked by a complex of two protein molecules: tropomyosin and troponin.
2. Prior to muscle contraction, ATP binds to the heads of the myosin molecules, priming them in an erect high energy state. Arrival of an action potential causes a release of Ca^{2+} from the sarcoplasmic reticulum. The Ca^{2+} binds to the troponin and causes the blocking complex to move so that the myosin binding sites on the actin filament become exposed.
3. The heads of the cross-bridging myosin molecules attach to the binding sites on the actin filament. Release of energy from the hydrolysis of ATP accompanies the cross bridge formation.
4. The energy released from ATP hydrolysis causes a change in shape of the myosin **cross bridge**, resulting in a bending action (*the power stroke*). This causes the actin filaments to slide past the myosin filaments towards the centre of the sarcomere.
5. (Not illustrated). Fresh ATP attaches to the myosin molecules, releasing them from the binding sites and repriming them for a repeat movement. They become attached further along the actin chain as long as ATP and Ca^{2+} are available.

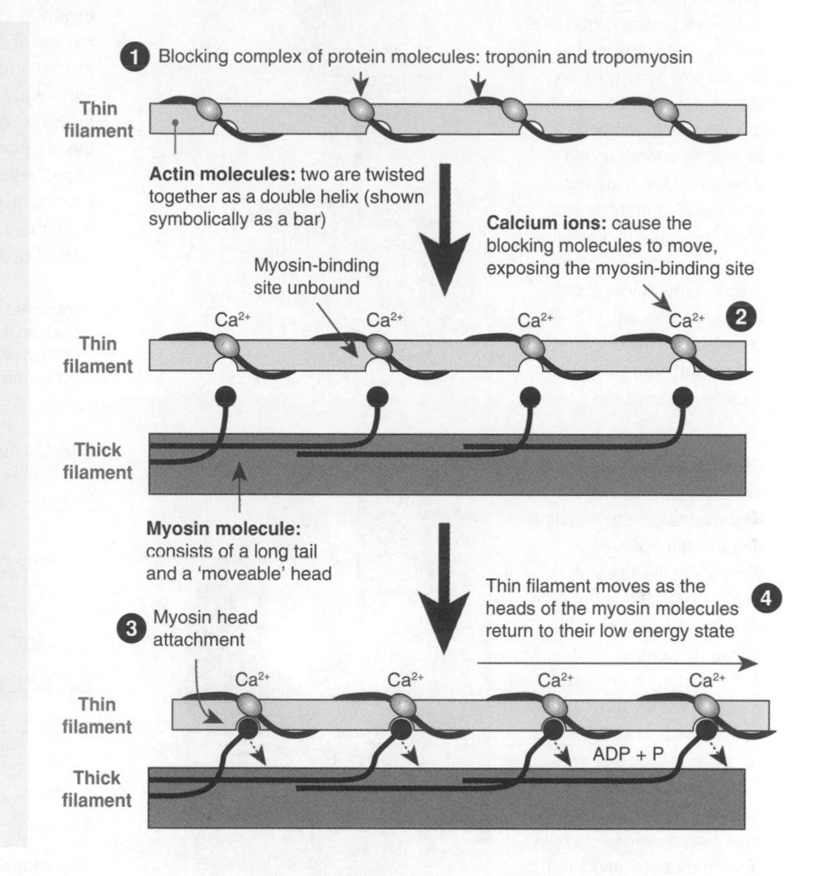

1. Match the following chemicals with their functional role in muscle movement (draw a line between matching pairs):

(a) Myosin	• Bind to the actin molecule in a way that prevents myosin head from forming a cross bridge
(b) Actin	• Supplies energy for the flexing of the myosin 'head' (power stroke)
(c) Calcium ions	• Has a moveable head that provides a power stroke when activated
(d) Troponin-tropomyosin	• Two protein molecules twisted in a helix shape that form the thin filament of a myofibril
(e) ATP	• Bind to the blocking molecules, causing them to move and expose the myosin binding site

2. Describe the two ways in which a muscle as a whole can produce contractions of varying force:

(a) ______________________________

____

(b) ______________________________

____

3. (a) Identify the two things necessary for cross bridge formation: ______________________________

(b) Explain where each of these comes from: ______________________________

____

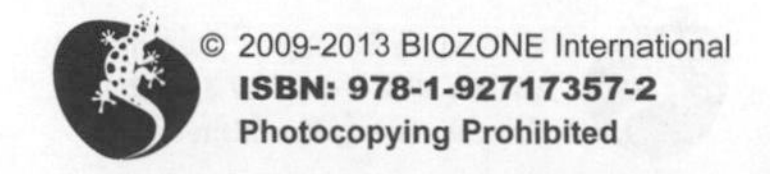

ISBN: 978-1-92717357-2

Periodicals:
How skeletal muscles work

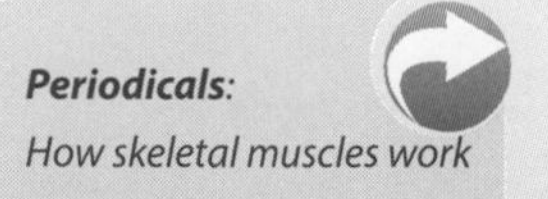

Related activities: *Muscle Structure and Function*
Weblinks: *Muscle Cell Contraction*

Muscle Tone and Posture

Even when we consciously relax a muscle, a few of its fibers at any one time will be involuntarily active. This continuous and passive partial contraction of the muscles is responsible for **muscle tone** and is important in maintaining **posture**. The contractions are not visible but they are responsible for the healthy, firm appearance of muscle. The amount of muscle contraction is monitored by sensory receptors in the muscle called **muscle spindle organs**. These provide the sensory information necessary to adjust movement as required. Abnormally low muscle tone (**hypotonia**) can arise as a result of traumatic or degenerative nerve damage, so that the muscle no longer receives the innervation it needs to contract. The principal treatment for these disorders is physical therapy to help the person compensate for the neuromuscular disability.

We are usually not aware of the skeletal muscles that maintain posture, although they work almost continuously making fine adjustments to maintain body position. Both posture and functional movements of the body are highly dependent on the strength of the body's core (the muscles in the pelvic floor, belly, and mid and lower back). The core muscles stabilize the thorax and the pelvis and lack of core strength is a major contributor to postural problems and muscle imbalances.

Physical therapy is a branch of health care concerned with maintaining or restoring functional movement throughout life. Loss of muscle tone and strength can develop as a result of aging, disease, or trauma. As a result of not being used, muscles will **atrophy**, losing both mass and strength. Although the type of physical therapy depends on the problem, it usually includes therapeutic exercise to help restore mobility and strength, and prevent or slow down the loss of muscle tissue.

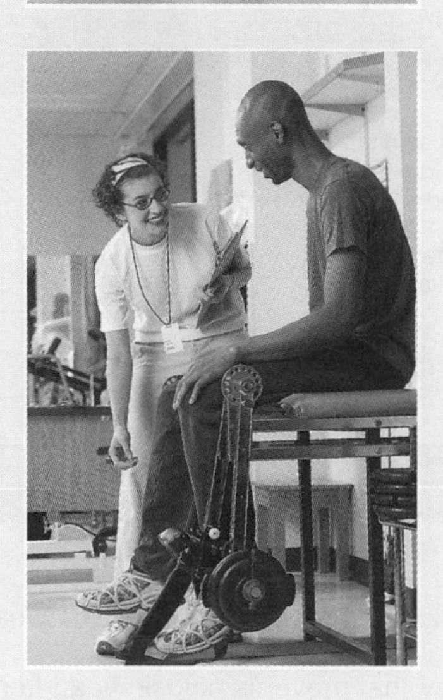

The Role of the Muscle Spindle

Changes in length of a muscle are monitored by the **muscle spindle organ**, a stretch receptor located within skeletal muscle, parallel to the muscle fibers themselves. The muscle spindle is stimulated in response to sustained or sudden stretch on the central region of its specialized intrafusal fibers. Sensory information from the muscle spindle is relayed to the spinal cord. The motor response brings about adjustments to the degree of stretch in the muscle. These adjustments help in the coordination and efficiency of muscle contraction. Muscle spindles are important in the maintenance of muscle tone, postural reflexes, and movement control, and are concentrated in muscles that exert fine control over movement.

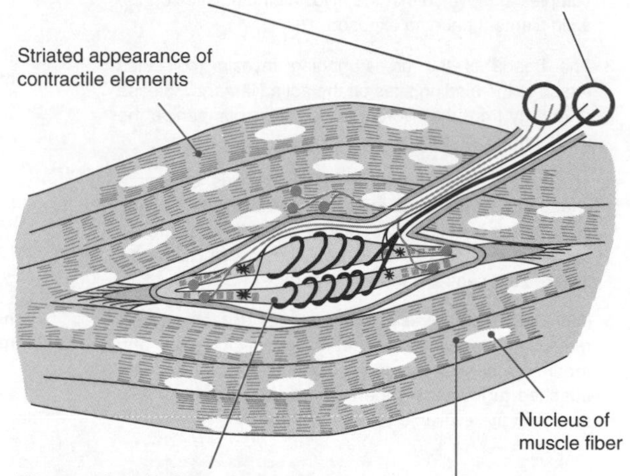

1. (a) Explain what is meant by muscle tone: ______

   (b) Explain how this is achieved: ______

2. (a) Explain the role of the muscle spindle organ: ______

   (b) With reference to the following, describe how the structure of the muscle spindle organ is related to its function:

   Intrafusal fibers lie parallel to the extrafusal fibers: ______

   Sensory neurons are located in the non-contractile region of the organ: ______

   Motor neurons synapse in the extrafusal fibers and the contractile region of the intrafusal fibers: ______

__Related activities:__ Detecting Changing States, The Basis of Sensory Perception
__Weblinks:__ Muscle Sense

ISBN: 978-1-92717357-2

The Mechanics of Movement

We are familiar with the many different bodily movements achievable through the action of muscles. Contractions in which the length of the muscle shortens in the usual way are called **isotonic contractions**: the muscle shortens and movement occurs. When a muscle contracts against something immovable and does not shorten the contraction is called **isometric**. Skeletal muscles are attached to bones by tough connective tissue structures called **tendons**. They always have at least two attachments: the **origin** and the **insertion**. They create movement of body parts when they contract across **joints**. The type and degree of movement achieved depends on how much movement the joint allows and where the muscle is located in relation to the joint. Some common types of body movements are described below (left panel). Because muscles can only pull and not push, most body movements are achieved through the action of opposing sets of muscles (below, right panel).

The Action of Antagonistic Muscles

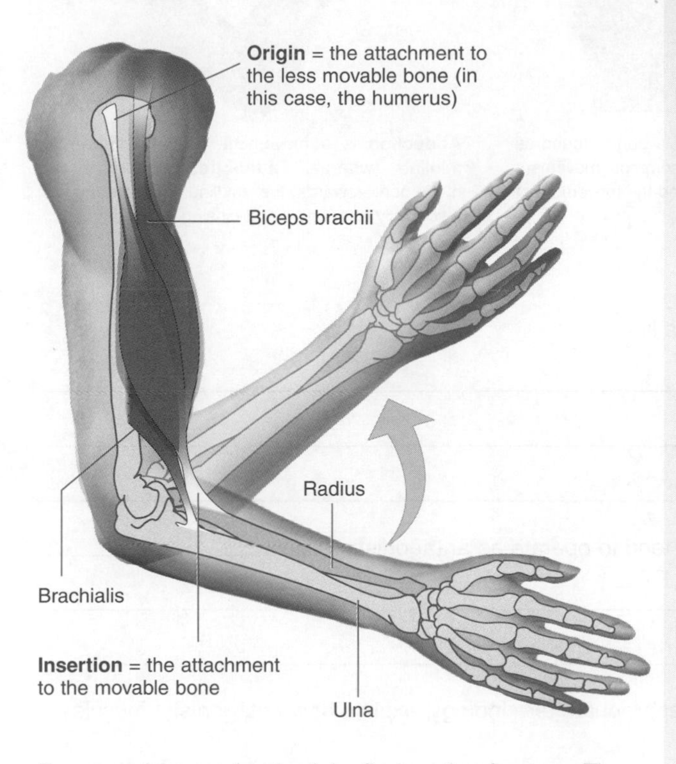

Two muscles are involved in flexing the forearm. The **brachialis**, which underlies the biceps brachii and has an origin half way up the humerus, is the **prime mover**. The more obvious **biceps brachii**, which is a two headed muscle with two origins and a common insertion near the elbow joint, acts as the synergist. During contraction, the insertion moves towards the origin.

The skeleton works as a system of levers. The joint acts as a **fulcrum** (or pivot), the muscles exert the **force**, and the weight of the bone being moved represents the **load**. The flexion (bending) and extension (unbending) of limbs is caused by the action of **antagonistic muscles**. Antagonistic muscles work in pairs and their actions oppose each other. During movement of a limb, muscles other than those primarily responsible for the movement may be involved to fine tune the movement.

Every coordinated movement in the body requires the application of muscle force. This is accomplished by the action of agonists, antagonists, and synergists. The opposing action of agonists and antagonists (working constantly at a low level) also produces muscle tone. Note that either muscle in an antagonistic pair can act as the agonist or **prime mover**, depending on the particular movement (for example, flexion or extension).

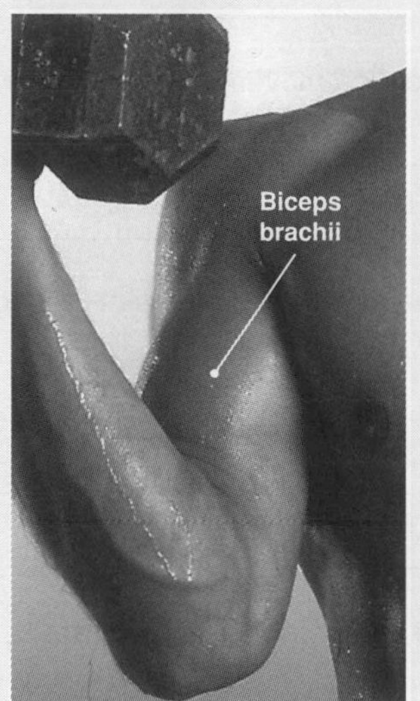

Agonists or prime movers: muscles that are primarily responsible for the movement and produce most of the force required.

Antagonists: muscles that oppose the prime mover. They may also play a protective role by preventing over-stretching of the prime mover.

Synergists: muscles that assist the prime movers and may be involved in fine-tuning the direction of the movement.

During flexion of the forearm (left) the **brachialis** muscle acts as the prime mover and the **biceps brachii** is the synergist. The antagonist, the **triceps brachii** at the back of the arm, is relaxed. During extension, their roles are reversed.

Movement at Joints

The synovial joints of the skeleton allow free movement in one or more planes. The articulating bone ends are separated by a joint cavity containing lubricating synovial fluid. Two types of synovial joint, the shoulder ball and socket joint and the hinge joint of the elbow, are illustrated below.

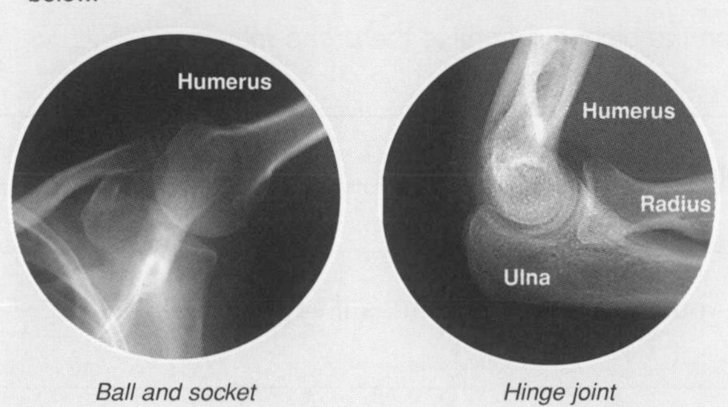

Ball and socket *Hinge joint*

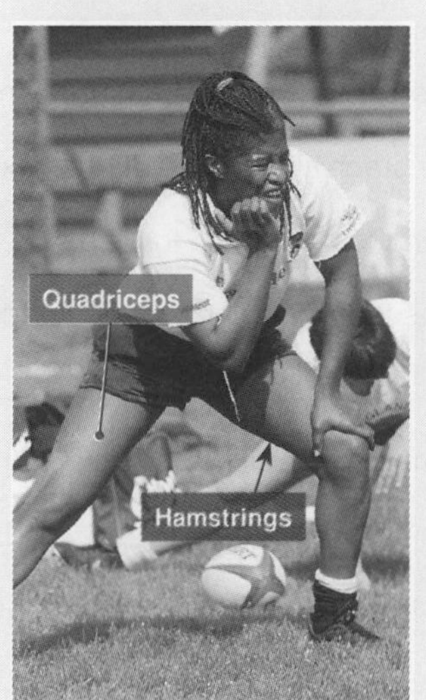

Movement of the upper leg is achieved through the action of several large groups of muscles, collectively called the **quadriceps** and the **hamstrings**.

The hamstrings are actually a collection of three muscles, which act together to flex the leg.

The quadriceps at the front of the thigh (a collection of four large muscles) opposes the motion of the hamstrings and extends the leg.

When the prime mover contracts forcefully, the antagonist also contracts very slightly. This stops over-stretching and allows greater control over thigh movement.

ISBN: 978-1-92717357-2

Types of Body Movement

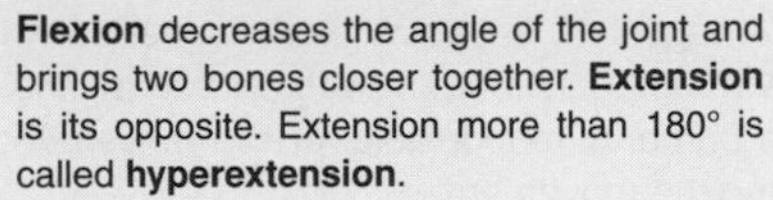

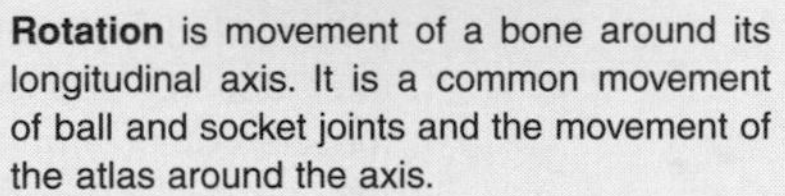

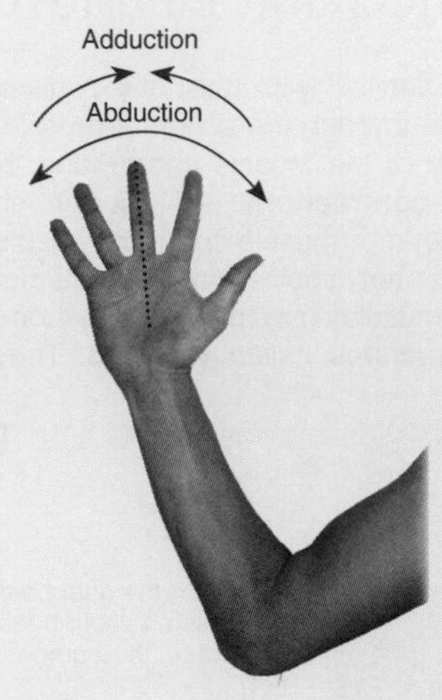

Flexion decreases the angle of the joint and brings two bones closer together. **Extension** is its opposite. Extension more than 180° is called **hyperextension**.

Rotation is movement of a bone around its longitudinal axis. It is a common movement of ball and socket joints and the movement of the atlas around the axis.

Abduction is a movement away from the midline, whereas **adduction** describes movement towards the midline. The terms also apply to opening and closing the fingers.

1. Describe the role of each of the following muscles in moving a limb:

   (a) Prime mover: ______

   (b) Antagonist: ______

   (c) Synergist: ______

2. Explain why the muscles that cause movement of body parts tend to operate as antagonistic pairs: ______

3. Describe the relationship between muscles and joints Using appropriate terminology, explain how antagonistic muscles act together to raise and lower a limb:

4. Explain the role of joints in the movement of body parts: ______

5. (a) Identify the insertion for the biceps brachii during flexion of the forearm: ______

   (b) Identify the insertion of the brachialis muscle during flexion of the forearm: ______

   (c) Identify the antagonist during flexion of the forearm: ______

   (d) Given its insertion, describe the forearm movement during which the biceps brachii is the prime mover:

6. (a) Describe a forearm movement in which the brachialis is the antagonist: ______

   (b) Identify the prime mover in this movement: ______

7. (a) Describe the actions that take place in the neck when you nod your head up and down as if saying "yes":

   (b) Describe the action being performed when a person sticks out their thumb to hitch a ride: ______

ISBN: 978-1-92717357-2

Energy for Muscle Contraction

Exercise places an immediate demand on the body's energy supply systems. During exercise, the metabolic rate of the muscles increases by up to 20 times and the body's systems must respond appropriately to maintain homeostasis. The ability to exercise for any given length of time depends on maintaining adequate supplies of ATP to the muscles. There are three energy systems to do this: the **ATP-CP system**, the **glycolytic system**, and the **oxidative system**. The ultimate sources of energy for ATP generation in muscle via these systems are glucose, and stores of glycogen and triglycerides. Prolonged intense exercise utilizes the oxidative system, and relies on a constant supply of oxygen to the tissues. The **VO_2** is the amount of oxygen (expressed as a volume) used by muscles during a specified interval for cell metabolism and energy production. **VO_2max** is the maximum volume of oxygen that can be delivered and used per minute and therefore represents an individual's upper limit of aerobic metabolism. VO_2max is used as a measure of fitness, and is high in trained athletes. At some percentage of VO_2max (the **anaerobic threshold**) the body is unable to meet its energy demands aerobically and an **oxygen debt** is incurred.

CP provides enough energy to fuel about 10 s of maximum effort (e.g. a 100 m race).

The ATP-CP system

The simplest of the energy systems is the ATP-CP system. CP or **creatine phosphate** is a high energy compound that stores energy sufficient for brief periods of muscular effort. Energy released from the breakdown of CP is not used directly to accomplish cellular work. Instead it rebuilds ATP to maintain a relatively constant supply. This process is anaerobic, occurs very rapidly, and is accomplished without any special structures in the cell.

CP levels decline steadily as it is used to replenish depleted ATP levels. The ATP-CP system maintains energy levels for 3-15 seconds. Beyond this, the muscle must rely on other processes for ATP generation.

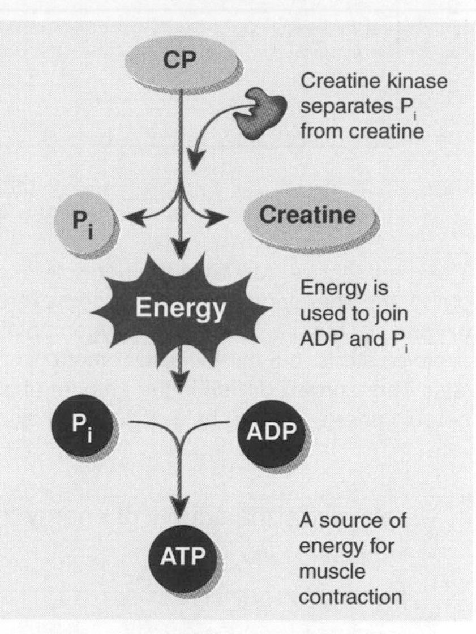

Soccer and other field sports demand brief intense efforts with recovery in-between.

Johnmaxmena

The glycolytic system

ATP can also be provided by **glycolysis**. The ATP yield from glycolysis is low (only net 2 ATP per molecule of glucose), but it produces ATP rapidly and does not require oxygen. The fuel for the glycolytic system is glucose in the blood, or glycogen, which is stored in the muscle or liver and broken down to glucose-6-phosphate. Pyruvate is reduced to lactate, regenerating NAD^+ and allowing further glycolysis.

Glycolysis provides ATP for exercise for just a few minutes. Its main limitation is that it causes an accumulation of H^+ (because protons are not being removed via mitochondrial respiration) and lactate ($C_3H_5O_3$) in the tissues. These changes lead to impairment of muscle function.

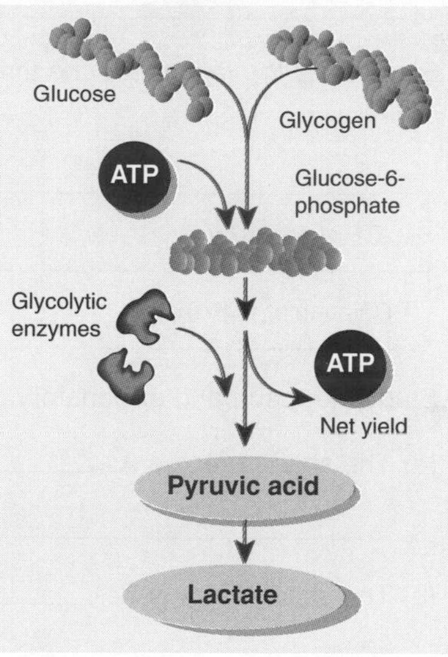

Prolonged aerobic effort (e.g. distance running) requires a sustained ATP supply.

TG

The Oxidative System

In the oxidative system, glucose is completely broken down to yield around 36 molecules of ATP. This process uses oxygen and occurs in the mitochondria. Aerobic metabolism has a high energy yield and is the primary method of energy production during sustained high activity. It relies on a continued supply of oxygen and therefore on the body's ability to deliver oxygen to the muscles. The fuels for aerobic respiration are glucose, stored glycogen, or stored **triglycerides**. Triglycerides provide free fatty acids, which are oxidized in the mitochondria by the successive removal of two-carbon fragments (a process called beta-oxidation). These two carbon units enter the Krebs cycle as acetyl coenzyme A (acetyl CoA).

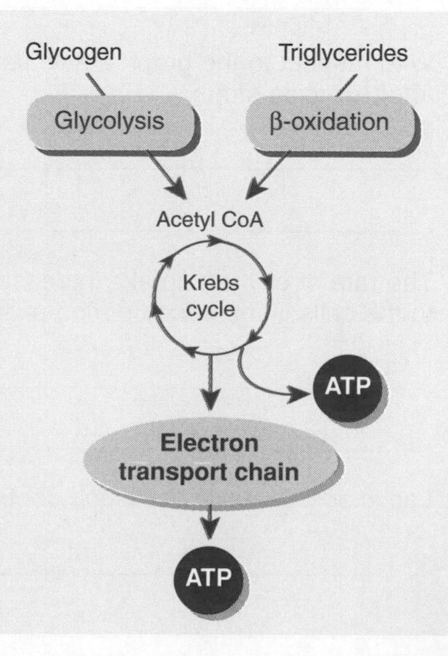

The Skeletal System

ISBN: 978-1-92717357-2

***Related activities**: Muscle Physiology and Performance, Muscle Fatigue*

Oxygen Uptake During Exercise and Recovery

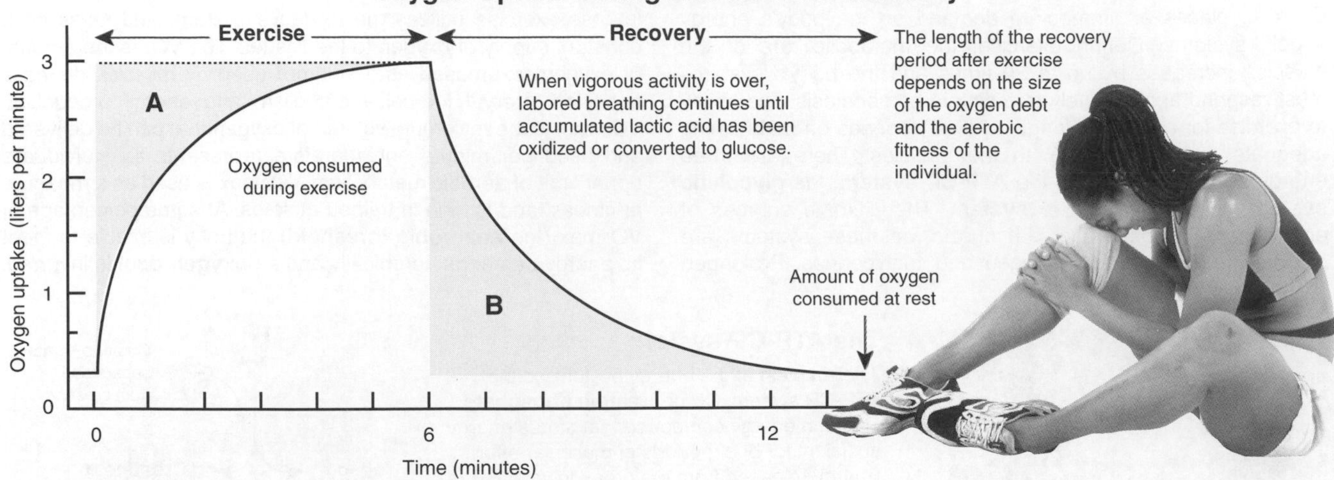

The graph above illustrates the principle of **oxygen debt**. In the graph, the energy demands of aerobic exercise require 3 L of oxygen per minute. The rate of oxygen uptake increases immediately exercise starts, but the full requirement is not met until six minutes later. The **oxygen deficit** is the amount of oxygen needed (for aerobic energy supply) but not supplied by breathing. During the first six minutes, energy is supplied largely from anaerobic pathways: the ATP-CP and glycolytic systems. After exercise, oxygen uptake per minute does not drop immediately to resting level. The extra oxygen that is taken in despite the drop in energy demand is the **oxygen debt**. The oxygen debt is used to replace oxygen reserves, restore creatine phosphate, and oxidise the lactate or convert it to glucose.

1. Explain why the supply of energy through the glycolytic system is limited:

2. Summarize the features of the three energy systems in the table below:

	ATP-CP system	Glycolytic system	Oxidative system
ATP supplied by:			
Duration of ATP supply:			

3. Study the graph and explanatory paragraph above, then identify and describe what is represented by:

   (a) The shaded region **A**:

   (b) The shaded region **B**:

4. With respect to the graph above, explain why the rate of oxygen uptake does not immediately return to its resting level after exercise stops:

5. The rate of oxygen uptake increases immediately exercise starts. Explain how the oxygen supply from outside the body to the cells is increased during exercise:

6. Lactic acid levels in the blood continue to rise for a time after exercise has stopped. Explain why this occurs:

ISBN: 978-1-92717357-2

Muscle Fatigue

Muscle fatigue refers to the decline in a muscle's ability to maintain force in a prolonged or repeated contraction. It is a normal result of vigorous exercise but the reasons for it are complex. Muscles can fatigue because of shortage of fuel or because of the accumulation of metabolites which interfere with the activity of calcium in the muscle. Contrary to older thinking, muscle fatigue is not caused by the toxic effects of lactic acid accumulation in oxygen-starved muscle. In fact, lactate formed during exercise is an important source of fuel (through conversion to glucose) and delays fatigue and metabolic acidosis during moderate activity by acting as a buffer. However, during sustained exhausting exercise, more of the muscle's energy needs must be met by glycolysis, and this leads to the metabolic changes (including accumulation of lactate and phosphate) that contribute to fatigue. Accumulated lactate is metabolized within the muscle itself or transported to the liver and converted back into glucose.

At rest

- Muscles produce a surplus of ATP
- This extra energy is stored in CP (creatine phosphate) and glycogen

During moderate activity

- ATP requirements are met by the aerobic metabolism of glycogen and lipids.
- There is no proton accumulation in the cell

During peak activity

- Effort is limited by ATP. ATP production is ultimately limited by availability of oxygen.
- During short-term, intense activity, more of the muscle's ATP needs must be met by glycolysis. This leads to an increase in H^+.
- Removal of H^+ is slow because mitochondrial respiration is hampered. Lactate may accumulate and is coincident with metabolic acidosis but not the cause of it.
- Muscle contraction is impaired (fatigue).

The complex causes of muscle fatigue

During intense exercise, oxygen is limited and more of the muscle's energy needs must be met through anaerobic metabolism. The effects of this are:

- An increase in H^+ (acidosis) because protons are not being removed via the mitochondrial electron transport system.
- Lactate accumulates faster than it can be oxidized.
- Accumulation of phosphate (Pi) from breakdown of ATP and creatine phosphate.

These metabolic changes lead to a fall in ATP and impaired calcium release from the sarcoplasmic reticulum (SR), both of which contribute to muscle fatigue.

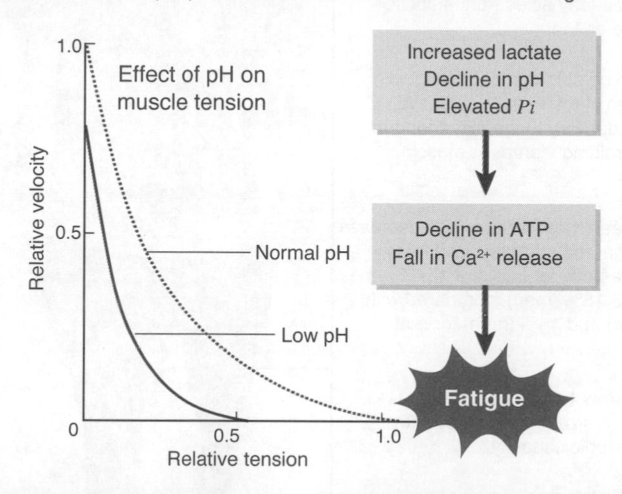

Short term maximal exertion (sprint)

- Lactic acid build-up lowers pH
- Depletion of creatine phosphate
- Buildup of phosphate (P_i) affects the sensitivity of the muscle to Ca^{2+}

Mixed aerobic and anaerobic (5 km race)

- Lactate accumulation in the muscle
- Build-up of ADP and P_i
- Decline in Ca^{2+}release affects the ability of the muscle to contract

Extended sub-maximal effort (marathon)

- Depletion of all energy stores (glycogen, lipids, amino acids) leads to a failure of Ca^{2+}release
- Repetitive overuse damages muscle fibers

The Skeletal System

1. Explain the mechanism by which lactate accumulation is associated with muscle fatigue: ______________________

2. Identify the two physiological changes in the muscle that ultimately result in a decline in muscle performance:

   (a) ______________________ (b) ______________________

3. Suggest why the reasons for fatigue in a long distance race are different to those in a 100 m sprint: ______________________

ISBN: 978-1-92717357-2

Muscle Physiology and Performance

The overall effect of **aerobic training** on muscle is improved oxidative function and better endurance. Regardless of the type of training, some of our ability to perform different types of activity depends on our genetic make-up. This is particularly true of aspects of muscle physiology, such as the relative proportions of different fiber types in the skeletal muscles. Muscle fibers are primarily of two types: **fast twitch** (FT) or **slow twitch** (ST). Fast twitch fibers predominate during anaerobic, explosive activity, whereas slow twitch fibers predominate during endurance activity. In the table below, note the difference in the degree to which the two fiber types show fatigue (a decrease in the capacity to do work). Training can increase fiber size and, to some extent, the makeup of the fiber, but not the proportion of ST to FT, which is genetically determined.

The Effects of Aerobic Training on Muscle Physiology

Improved oxidation of glycogen. Training increases the capacity of skeletal muscle to generate ATP aerobically.

An increased capacity of the muscle to oxidize fats. This allows muscle and liver glycogen to be used at a slower rate. The body also becomes more efficient at mobilizing free fatty acids from adipose tissue for use as fuel.

Increased myoglobin content. Myoglobin stores oxygen in the muscle cells and aids oxygen delivery to the mitochondria. Endurance training increases muscle myoglobin by 75%-80%.

Increase in lean muscle mass and decrease in body fat. Trained endurance athletes typically have body fat levels of 15-19% (women) or 6-18% (men), compared with 26% (women) and 15% (men) for non-athletes.

The size of **slow twitch fibers** increases. This change in size is associated with increased aerobic capacity.

An increase in the size and density of mitochondria in the skeletal muscles and an increase in the activity and concentration of Krebs cycle enzymes.

An increase in the number of capillaries surrounding each muscle fiber. Endurance trained men have 5%-10% more capillaries in their muscles than sedentary men.

Fast vs Slow Twitch Muscle

Feature	Fast twitch	Slow twitch
Color	White	Red
Diameter	Large	Small
Contraction rate	Fast	Slow
ATP production	Fast	Slow
Metabolism	Anaerobic	Aerobic
Rate of fatigue	Fast	Slow
Power	High	Low

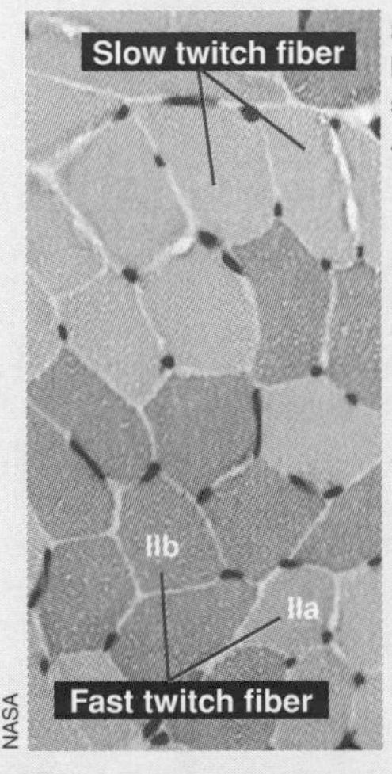

Slow twitch fibers appear light colored when stained with a myofibrillar ATPase stain.

Type II fast twitch fibers are classified further according to their metabolism:
- Type IIa (intermediate) =some oxidative capacity
- Type IIb =fast glycolytic only

There are two basic types of muscle fibers: **slow twitch** (type I) and **fast twitch** (type II) fibers. Both fiber types generally produce the same force per contraction, but fast twitch fibres produce that force at a higher rate. low twitch fibres contain more mitochondria and myoglobin than fast twitch fibers, so they are more efficient at using oxygen to generate ATP without lactate build up. In this way, they can fuel repeated muscle contractions such as those required for endurance events.

1. Explain three ways in which aerobic (endurance) training improves the oxidative function of muscle:

(a) ____________________

(b) ____________________

(c) ____________________

2. Contrast the properties of fast and slow twitch skeletal muscle fibers, identifying how these properties contribute to their performance in different conditions:

____

ISBN: 978-1-92717357-2

Related activities: Exercise and Blood Flow, Muscle Fatigue

KEY TERMS: Mix and Match

INSTRUCTIONS: Test your vocabulary by matching each term to its definition, as identified by its preceding letter code.

actin (thin filament) ....

antagonistic pair ....

blood lactate ....

cardiac muscle ....

cross bridge ....

fast twitch ....

filament (=myofilament) ....

muscle fatigue ....

muscle fiber ....

myofibril ....

myosin (thick filament) ....

neuromuscular junction ....

oxygen debt ....

prime mover ....

sarcomere ....

sarcoplasmic reticulum ....

skeletal (=striated) muscle ....

sliding filament hypothesis ....

slow twitch ....

smooth muscle ....

tropomyosin ....

troponin ....

A Muscle cell containing a bundle of myofibrils.

B Muscle that is primarily responsible for a specific movement and produces most of the force required.

C Name given to a pair of muscles whose actions oppose each other, when one contracts the other relaxes (e.g. the biceps and triceps).

D The muscle fibre that predominates during aerobic, endurance activity. They contain more mitochondria and myoglobin than fast twitch fibre types.

E The decline in a muscle's ability to maintain force in a prolonged or repeated contraction. It is the normal result of vigorous exercise.

F An actin binding protein important in muscle contraction by regulating the binding of myosin.

G A protein found in the thick myofilaments of the sarcomere of muscle.

H The filaments that make up the myofibril, can be thick or thin. Thin filaments consist primarily of the protein actin. Thick filaments consist primarily of the protein myosin

I The amount of lactate in the blood. Its presence is a result of anaerobic metabolism when oxygen delivery to the tissues is insufficient to support metabolic demands (e.g. periods of strenuous exercise).

J A cumulative deficit of oxygen resulting from intense exercise. The oxygen deficit is made up during the recovery (rest) period.

K A contractile protein found in muscle cells.

L Specialized structure of muscle cells. Composed of two kinds of myofilaments (thick and thin). A bundle of these make up a muscle fiber.

M The contractile element of the fiber, it is contained between two Z membranes.

N Specialized smooth endoplasmic reticulum found around myofibrils in skeletal muscle fibres. It stores and releases calcium ions required for muscle contraction.

O The junction between a motor neuron and a skeletal muscle fiber. It is a specialized cholinergic synapse.

P A complex of three proteins that bind to tropomyosin and help regulate muscle contraction by causing tropomyosin to either block or unblock the attachment of myosin to actin.

Q The theory of how thin and thick filaments slide past each other to produce muscles contraction.

R The muscle responsible for automatic movements such as peristalsis. It is not under conscious control. Cells are spindle shaped with one central nucleus.

S The temporary linkage of actin and myosin filaments during muscle contraction.

T Specialized striated muscle that does not fatigue. It is found only in the walls of the heart and is not under conscious control (involuntary muscle).

U The muscle fibre that predominates during anaerobic, explosive activity. It contains less mitochondria and myoglobin than slow twitch fibre types.

V Muscle that is attached to the skeleton and responsible for the movement of bone around joints or movement of some organs, e.g. the eyes.

ISBN: 978-1-92717357-2

Interacting Systems

Nervous and Endocrine Systems

Endocrine system

- The hypothalamus controls the activity of the anterior pituitary gland.
- Sympathetic NS stimulates the release of "fight or flight" hormones from the adrenal medulla.

Cardiovascular system

- ANS regulates heart rate and blood pressure
- Cardiovascular system supplies nervous and endocrine tissues with O_2 and removes wastes
- Hormones influence blood volume and pressure, and heart activity.
- The hormones erythropoietin (EPO) stimulates production of red blood cells.

Respiratory system

- Nervous system regulates breathing
- Epinephrine (adrenaline) dilates bronchioles and increases the rate and depth of breathing
- Respiratory system enables gas exchange, providing O_2 and removing CO_2
- Angiotensin II, which increases blood pressure, is activated in the lung capillaries.

Skeletal system

- Nerves supply bones.
- The skeleton protects the nervous and endocrine systems and provides Ca^{2+} for nerve function.
- Parathyroid hormone regulates blood calcium.
- Growth hormone, and thyroid and sex hormones regulate skeletal growth and development.

Integumentary system

- Sebaceous glands are influenced by sex hormones.
- Sympathetic NS activity regulates sweat glands and blood vessels in the skin (**thermoregulation**).

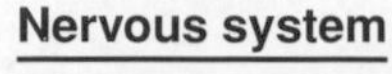

Nervous system

- Hormones (e.g. growth hormone, thyroid hormones, and sex hormones) influence the growth, development, and activity of the nervous system.

Lymphatic system and immunity

- Glucocorticoids (e.g. coritsol) depress the immune response
- Maturation of lymphocytes occurs in response to hormones from the thymus.
- Nervous system innervates and regulates immune system function.

Urinary system

- Final activation of vitamin D occurs in the kidneys (bioactive vitamin D is classed as a steroid hormone).
- Bladder emptying is regulated by both the autonomic NS and voluntary activity (control of the urethral sphincter).
- Autonomic NS regulates renal blood pressure

Digestive system

- Digestion is regulated by the autonomic NS, by local GI hormones, and by adrenal hormones (e.g. epinephrine).
- Bioactive vitamin D is needed to absorb calcium from the diet.

Reproductive system

- Autonomic NS activity regulates erectile tissue in males and females and ejaculation in males.
- Testosterone underlies sexual drive.
- Hormones from the hypothalamus, gonads, and pituitary regulate the development and functioning of the reproductive system.

Muscular system

- Somatic NS controls skeletal muscle activity.
- Muscular activity promotes release of catecholamine hormones from the adrenal medulla.
- Growth hormone is required for normal muscle development.
- Thyroid hormones and catecholamines influence muscle metabolism.

General Functions and Effects on all Systems

The nervous system regulates all the visceral and motor functions of the body, integrating with the endocrine system to provide both short term and longer term responses to stimuli. The endocrine system produces hormones that activate and regulate homeostatic functions, growth, and development.

Disease

Symptoms of disease
- Loss of function or control
- Loss of voluntary control
- Failure to develop normally

Diseases of the nervous system
- Inherited (e.g. Huntington's disease)
- Trauma (e.g. paraplegia)
- Infection (e.g.encephalitis, meningitis)
- Autoimmune (multiple sclerosis)
- Tumors
- Degenerative diseases

Diseases of the endocrine system
- Inherited disease (e.g. congenital adrenal hyperplasia)
- Cancers (e.g. thyroid cancer)
- Autoimmune damage
- Dietary related diseases

Medicine & Technology

Diagnosis of disorders
- Genetic testing and screening
- Genetic counselling
- Medical imaging techniques
- Blood tests

Treatment of injury
- Surgery
- Physical therapies
- Drug therapies

Treatment of inherited disorders
- Dietary management
- Physical and drug therapies
- Radiotherapy (for cancers)
- Stem cell therapy and transplants
- Gene therapy (e.g. for Huntington's)

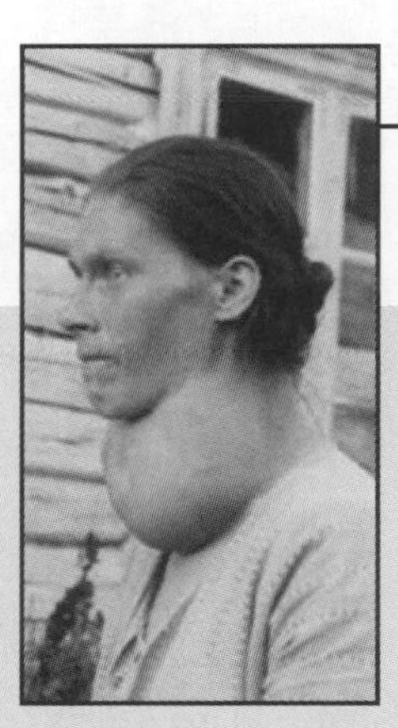

- Multiple sclerosis
- Type 1 & 2 diabetes
- Goiter (left)
- GH deficiency

- Brain scans (right)
- Genetic counselling
- Dietary modification
- Transplants
- Cell therapy

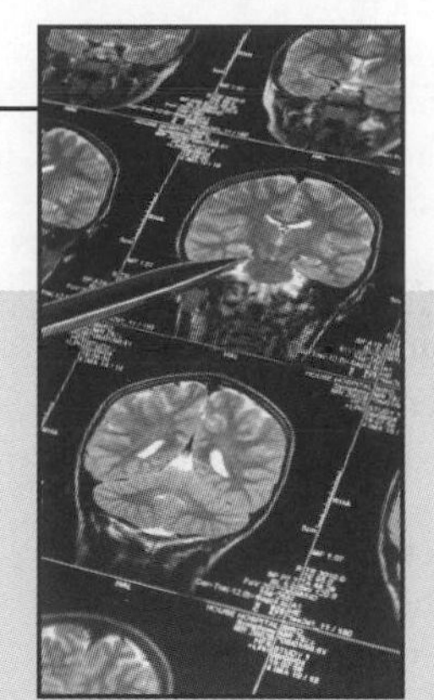

Integration & Control

Nervous & Endocrine Systems

The endocrine and nervous systems are closely linked. They can be affected by disease and undergo marked changes associated with aging.

Nervous and endocrine disorders may respond to exercise and medical treatment.

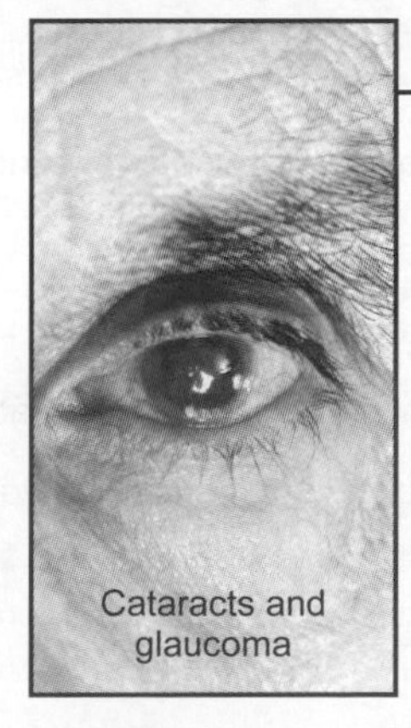

Cataracts and glaucoma

- Alzheimer's disease
- Hearing loss
- Changes in vision
- Menopause

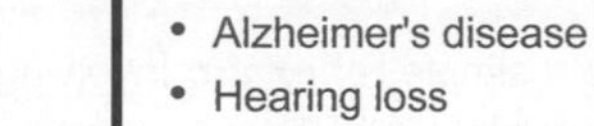

- Control of diabetes
- Improved cognitive function (brain)
- Improved autonomic NS function

Photo: pan Pavel Rycl

Effects of aging on nervous and endocrine function
- Sensory impairment
- Loss of neurons (Alzheimer's)
- Increased risks of falls
- Increased risks of cancers
- Decreased levels of some hormones

The Effects of Aging

Effects of exercise on nervous and endocrine function
- Improved coordination
- Reduced risk of memory loss
- Improved blood glucose management
- Reduced risk of type 2 diabetes

Exercise

The Nervous System

Key terms

action potential
autonomic nervous systems
axon
biological transducer
brain
cell body (soma)
central nervous system (CNS)
cerebellum
cerebrum
chemoreception
dendrites
depolarization
diencephalon
glial cells
integration
mechanoreception
motor neuron
myelinated nerve
nervous system
neurons
nodes of Ranvier
peripheral nervous system (PNS)
photoreception
proprioception
response
saltatory conduction
Schwann cells
sense organ
sensory receptor
somatic nervous system
spinal cord
stimulus
synapse

Key concepts

- The nervous system regulates activity through by sensing and responding to stimuli.
- The central nervous system is the integration center for sensing and responding.
- Neurons are electrically excitable cells capable of transmitting impulses over a considerable distance.
- Action potentials are discrete, all-or-nothing impulses. They transmitted across chemical synapses by diffusion of a neurotransmitter.
- Sensory receptors act as biological transducers.

Learning Objectives

☐ 1. Use the **KEY TERMS** to compile a glossary for this topic.

Nervous System Organization pages 85-90

☐ 2. Review the structure and function of nervous tissue including the general features of **neurons**, **glial cells**, and **Schwann cells**.

☐ 3. Describe the structure and organization of the **nervous system**. Identify the parts of the nervous system associated with monitoring sensory input, integration, and response (via motor output).

☐ 4. Describe the structure and function of a typical **neuron**, e.g. **motor neuron**.

☐ 5. Annotate a diagram of the brain to identify its main regions and their functional roles. Identify **cerebrum**, **cerebellum**, **diencephalon**, **brain stem**.

☐ 6. Describe the organization of the **peripheral nervous system** into the sensory division and motor division (somatic and autonomic divisions).

☐ 7. Describe the generally opposing roles of the sympathetic and parasympathetic divisions of the autonomic nervous system.

Neuron Structure and Function pages 91-101, 112-113

☐ 8. Using a diagram, describe a simple **reflex arc** involving three neurons. Describe examples of reflexes and explain their role.

☐ 9. Describe how an **action potential** is generated and propagated along a **myelinated nerve**. Compare impulse conduction in myelinated and non-myelinated nerves.

☐ 10. Explain how nerve impulses are transmitted across **synapses**. Recall the neuromuscular junction as a specialized cholinergic synapse.

☐ 11. Appreciate the role of synapses in unidirectionality and nervous system **integration**.

☐ 12. Describe the effects of **neurotransmitters** and psychoactive drugs on the nervous system and behavior.

☐ 13. Describe the effects of aging on nervous system function. Describe the degenerative structural and physiological changes associated with Alzheimer's disease (AD).

The Senses pages 74, 102-111

☐ 14. Describe examples of **sensory receptors** and the **stimuli** to which they respond.

☐ 15. Explain how sensory receptors act as **biological transducers** in receiving and responding to stimuli.

☐ 16. Describe the structure and function of some sensory receptors and sense organs:
- Pacinian corpuscle (pressure) and muscle spindle (proprioception)
- Taste buds and olfactory receptors (chemoreception)
- The ear and the structure and function of the cochlea (mechanoreception)
- The eye and the structure and function of the retina (photoreception)

Periodicals:
Listings for this chapter are on page 279

Weblinks:
www.thebiozone.com/weblink/AnaPhy-3572.html

BIOZONE APP:
Student Review Series
The Nervous System

Nervous Regulatory Systems

An essential feature of living organisms is their ability to coordinate their activities. In mammals, such as humans, detecting and responding to environmental change, and regulating the internal environment (**homeostasis**) are brought about by two coordinating systems: the nervous and endocrine systems. Although these two systems are quite different structurally, they frequently interact to coordinate behavior and physiology. The nervous system contains cells called **neurons** (or nerve cells). Neurons are specialized to transmit information in the form of electrochemical impulses (action potentials). The nervous system is a signaling network with branches carrying information directly to and from specific target tissues. Impulses can be transmitted over considerable distances and the response is very precise and rapid. Whilst it is extraordinarily complex, comprising millions of neural connections, its basic plan (below, left) is quite simple, structured around reception of sensory input, integration or processing of the information, and formulation of a response.

Coordination by the Nervous System

The vertebrate nervous system consists of the central nervous system (brain and spinal cord), and the nerves and receptors outside it (peripheral nervous system). Sensory input to receptors comes via stimuli. Information about the effect of a response is provided by feedback mechanisms so that the system can be readjusted. The basic organization of the nervous system can be simplified into a few key components: the sensory receptors, a central nervous system processing point, and the effectors which bring about the response (below):

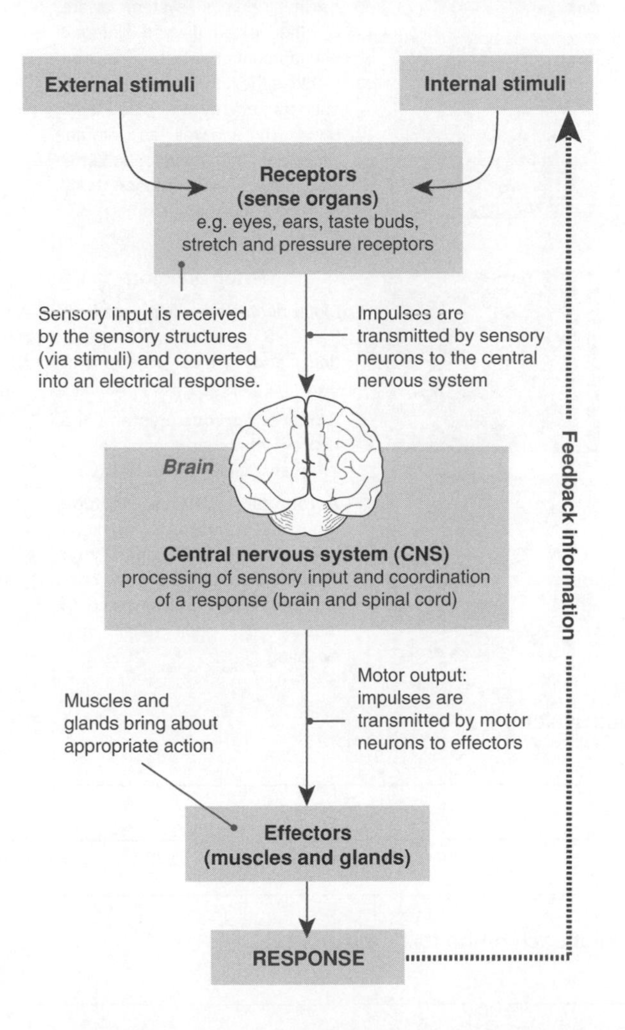

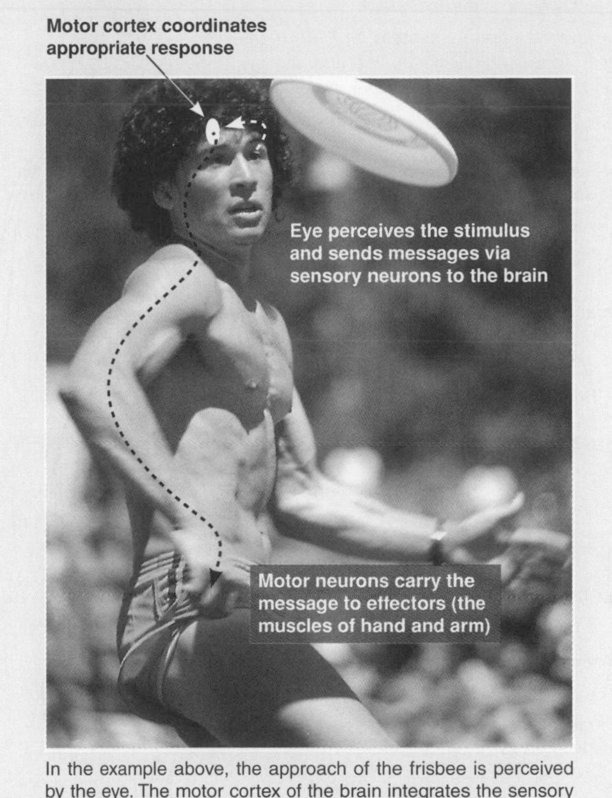

In the example above, the approach of the frisbee is perceived by the eye. The motor cortex of the brain integrates the sensory message. Coordination of hand and body orientation is brought about through motor neurons to the muscles.

Comparison of nervous and hormonal control

	Nervous control	Hormonal control
Communication	Impulses across synapses	Hormones in the blood
Speed	Very rapid (within a few milliseconds)	Relatively slow (over minutes, hours, or longer)
Duration	Short term and reversible	Longer lasting effects
Target pathway	Specific (through nerves) to specific cells	Hormones broadcast to target cells everywhere
Action	Causes glands to secrete or muscles to contract	Causes changes in metabolic activity

1. Identify the three basic components of a nervous system and explain how they function to maintain homeostasis:

   (a) ______

   (b) ______

   (c) ______

2. Describe two differences between nervous control and endocrine (hormonal) control of body systems:

   (a) ______

   (b) ______

ISBN: 978-1-92717357-2

Related activities: The Nervous System, Hormonal Regulation
Weblinks: Nervous System Animation

The Nervous System

The **nervous system** is the body's control and communication center. It has three broad functions: detecting stimuli, interpreting them, and initiating appropriate responses. Its basic structure is outlined below. Further detail is provided in the following pages.

The Human Nervous System

The **central nervous system** (CNS) comprises the brain and spinal cord. The spinal cord is a cylinder of nervous tissue extending from the base of the brain down the back, protected by the spinal column. It transmits messages to and from the brain, and controls spinal reflexes.

The **peripheral nervous system**, or **PNS**, (right, far right) comprises all the nerves and sensory receptors outside the central nervous system.

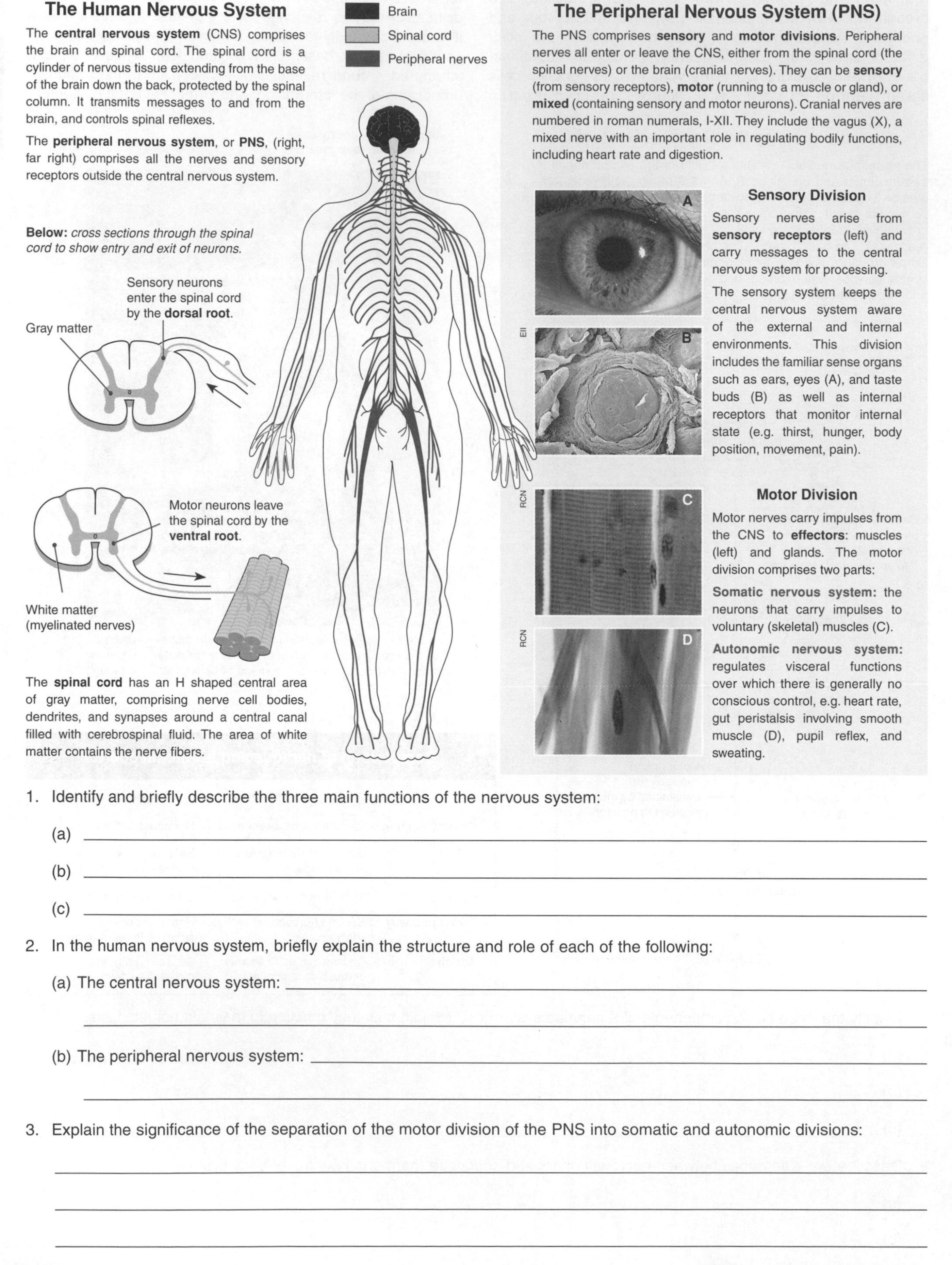

Below: *cross sections through the spinal cord to show entry and exit of neurons.*

The **spinal cord** has an H shaped central area of gray matter, comprising nerve cell bodies, dendrites, and synapses around a central canal filled with cerebrospinal fluid. The area of white matter contains the nerve fibers.

The Peripheral Nervous System (PNS)

The PNS comprises **sensory** and **motor divisions**. Peripheral nerves all enter or leave the CNS, either from the spinal cord (the spinal nerves) or the brain (cranial nerves). They can be **sensory** (from sensory receptors), **motor** (running to a muscle or gland), or **mixed** (containing sensory and motor neurons). Cranial nerves are numbered in roman numerals, I-XII. They include the vagus (X), a mixed nerve with an important role in regulating bodily functions, including heart rate and digestion.

Sensory Division

Sensory nerves arise from **sensory receptors** (left) and carry messages to the central nervous system for processing.

The sensory system keeps the central nervous system aware of the external and internal environments. This division includes the familiar sense organs such as ears, eyes (A), and taste buds (B) as well as internal receptors that monitor internal state (e.g. thirst, hunger, body position, movement, pain).

Motor Division

Motor nerves carry impulses from the CNS to **effectors**: muscles (left) and glands. The motor division comprises two parts:

Somatic nervous system: the neurons that carry impulses to voluntary (skeletal) muscles (C).

Autonomic nervous system: regulates visceral functions over which there is generally no conscious control, e.g. heart rate, gut peristalsis involving smooth muscle (D), pupil reflex, and sweating.

1. Identify and briefly describe the three main functions of the nervous system:

   (a) ______________________________

   (b) ______________________________

   (c) ______________________________

2. In the human nervous system, briefly explain the structure and role of each of the following:

   (a) The central nervous system: ______________________________

   ____

   (b) The peripheral nervous system: ______________________________

   ____

3. Explain the significance of the separation of the motor division of the PNS into somatic and autonomic divisions:

   ____

   ____

   ____

ISBN: 978-1-92717357-2

Related activities: *Nervous Regulatory Systems, The Autonomic Nervous System*

The Autonomic Nervous System

The **autonomic nervous system** (ANS) regulates involuntary visceral functions through **reflexes**. Although most autonomic nervous system activity is beyond our conscious control, voluntary control over some basic reflexes (such as bladder emptying) can be learned. Most visceral effectors have dual innervation, receiving fibers from both branches of the ANS. These two branches, the **parasympathetic** and **sympathetic** divisions, have broadly opposing actions on the organs they control (excitatory or inhibitory). Nerves in the parasympathetic division release acetylcholine. This neurotransmitter is rapidly deactivated at the synapse and its effects are short lived and localized. Most sympathetic postganglionic nerves release norepinephrine (noradrenaline), which enters the bloodstream and is deactivated slowly. Hence, sympathetic stimulation tends to have more widespread and long lasting effects than parasympathetic stimulation. Aspects of ANS structure and function are illustrated below. The arrows indicate nerves to organs or ganglia (concentrations of nerve cell bodies).

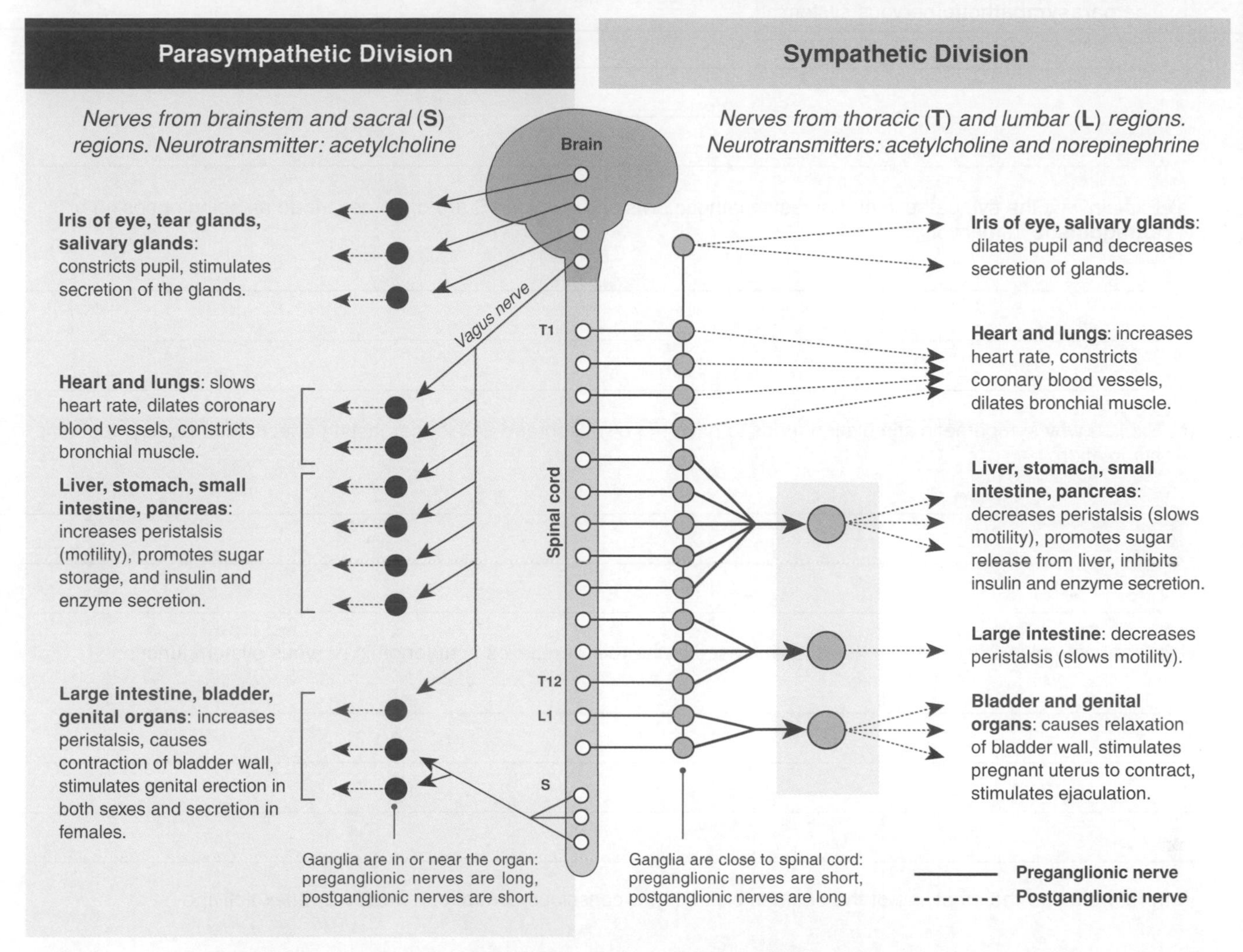

The Effects of the Autonomic Nervous System on the Body

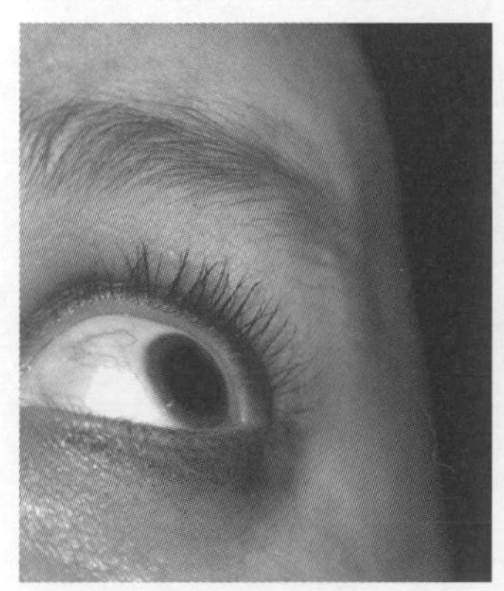

When a person is fearful, their pupils enlarge as a consequence of sympathetic nervous system activity (the fight or flight response). The same response occurs during sexual arousal. Sympathetic stimulation also decreases secretion of the tear (lacrimal) glands.

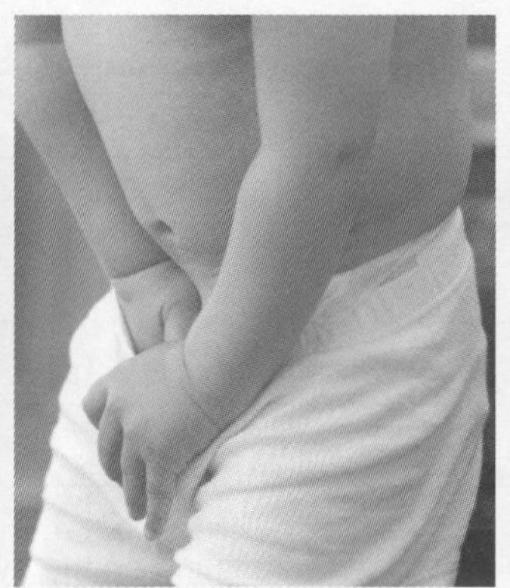

Parasympathetic stimulation is responsible for bladder emptying through contraction of the bladder wall and relaxation of the urethral sphincter. This reflex activity can be inhibited by conscious control, but this ability does not develop until 2-4 years of age.

As part of the fight or flight response, the sympathetic nervous system dilates arteries and increases the rate and force at which the heart contracts. Heart rate is increased when increased blood flow to the heart causes reflex stimulation of the accelerator centre.

The enteric nervous system (ENS) is a recently recognized, interdependent part of the ANS. The ENS regulates itself but is influenced by sympathetic and parasympathetic nerves which are connected to it. The ENS innervates the gut, pancreas and gall bladder.

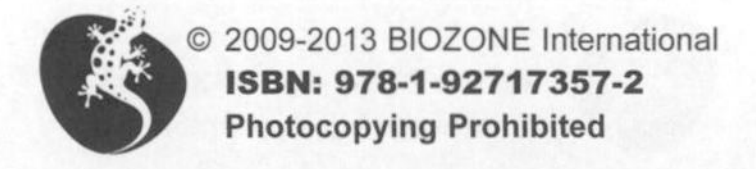

ISBN: 978-1-92717357-2

Periodicals:
The autonomic nervous system

Related activities: The Nervous System, Reflexes

1. Explain the structure and role of each of the following divisions of the autonomic nervous system:

   (a) The **sympathetic** nervous system:

   (b) The **parasympathetic** nervous system:

2. (a) Explain why the sympathetic and parasympathetic divisions of the ANS are often described as being opposing or **antagonistic** in function:

   (b) Explain why sympathetic stimulation tends to have more widespread and longer lasting effects than parasympathetic stimulation:

3. Using an example (e.g. control of heart rate), describe the role of reflexes in autonomic nervous system function:

4. With reference to the emptying of the bladder, explain how conscious control can modify a reflex activity:

5. Predict how the sympathetic reflexes controlling blood vessels would respond to a sudden decrease in blood pressure:

6. Asthma can be treated by inhaling a drug that mimics the action of noradrenaline on the sympathetic NS. Describe how this drug could be used to target the patient's respiratory system:

ISBN: 978-1-92717357-2

The Human Brain

The brain is one the largest organs in the body. It is protected by the skull, the **meninges**, and the **cerebrospinal fluid** (CSF). The brain is the body's control center. It receives a constant flow of sensory information, but responds only to what is important at the time. Some responses are very simple (e.g. cranial reflexes), whilst others require many levels of processing. The human brain is noted for its large, well developed cerebral region, with its prominent folds (**gyri**) and grooves (**sulci**). Each cerebral hemisphere has an outer region of gray matter and an inner region of white matter, and is divided into four lobes by deep sulci or fissures. These lobes: temporal, frontal, occipital, and parietal, correspond to the bones of the skull under which they lie.

Primary Structural Regions of the Brain

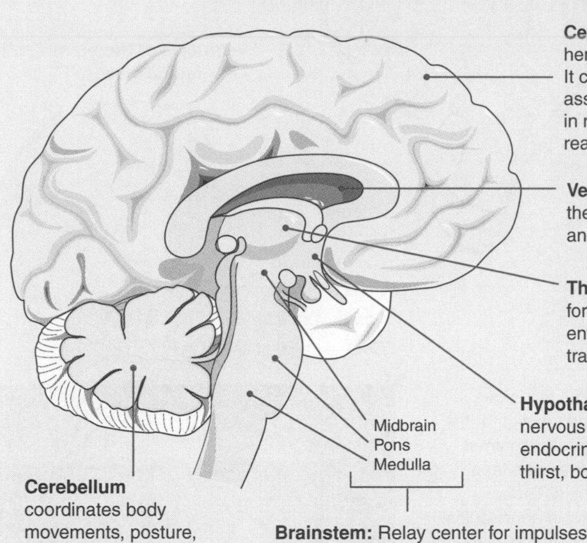

Cerebrum: Divided into two cerebral hemispheres. Many, complex roles. It contains sensory, motor, and association areas, and is involved in memory, emotion, language, reasoning, and sensory processing.

Ventricles: Cavities containing the CSF, which absorbs shocks and delivers nutritive substances.

Thalamus is the main relay center for all sensory messages that enter the brain, before they are transmitted to the cerebrum.

Hypothalamus controls the autonomic nervous system and links nervous and endocrine systems. Regulates appetite, thirst, body temperature, and sleep.

Cerebellum coordinates body movements, posture, and balance.

Brainstem: Relay center for impulses between the rest of the brain and the spinal cord. Controls breathing, heartbeat, and the coughing and vomiting reflexes.

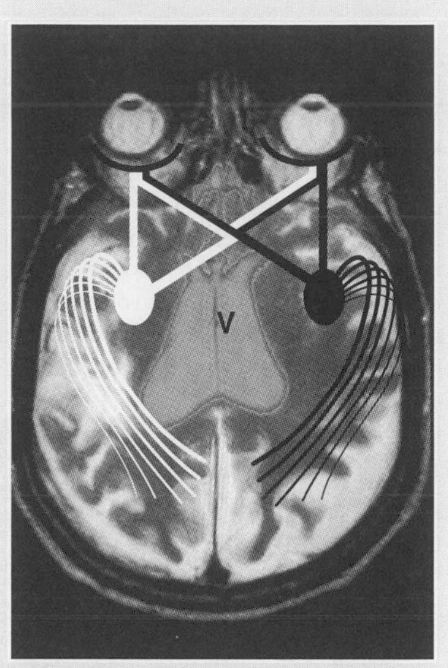

MRI scan of the brain viewed from above. The visual pathway has been superimposed on the image. Note the crossing of some sensory neurons to the opposite hemisphere and the fluid filled ventricles (V) in the center.

Sensory and Motor Regions in the Cerebrum

Frontal lobe

Parietal lobe

Occipital lobe

Temporal lobe

Primary somatic sensory area receives sensations from receptors in the skin, muscles and viscera, allowing recognition of pain, temperature, or touch. Sensory information from receptors on one side of the body crosses to the opposite side of the cerebral cortex where conscious sensations are produced. The size of the sensory region for different body parts depends on the number of receptors in that particular body part.

Visual areas within the occipital lobe receive, interpret, and evaluate visual stimuli. In vision, each eye views both sides of the visual field but the brain receives impulses from left and right visual fields separately (see photo caption above). The visual cortex combines the images into a single impression or **perception** of the image.

Olfactory area

Auditory areas interpret the basic characteristics and meaning of sounds.

Primary motor area controls muscle movement. Stimulation of a point one side of the motor area results in muscular contraction on the opposite side of the body.

Primary gustatory area interprets sensations related to taste.

Sulci (grooves)

Gyri (elevated folds)

Language areas: The motor speech area (Broca's area) is concerned with speech production. The sensory speech area (Wernicke's area) is concerned with speech recognition and coherence.

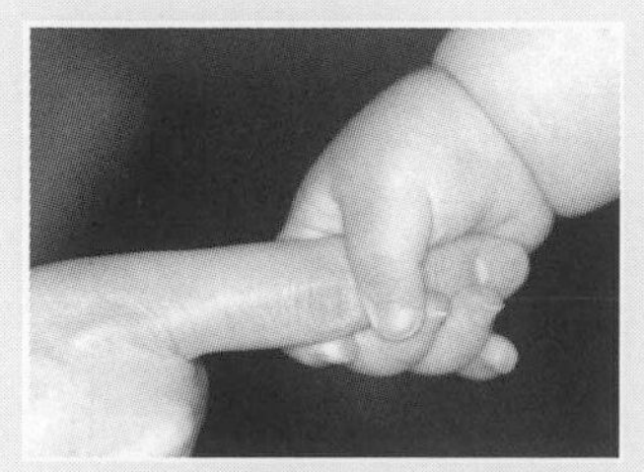

Touch is interpreted in the primary somatic sensory area. The fingertips and the lips have a relatively large amount of area devoted to them.

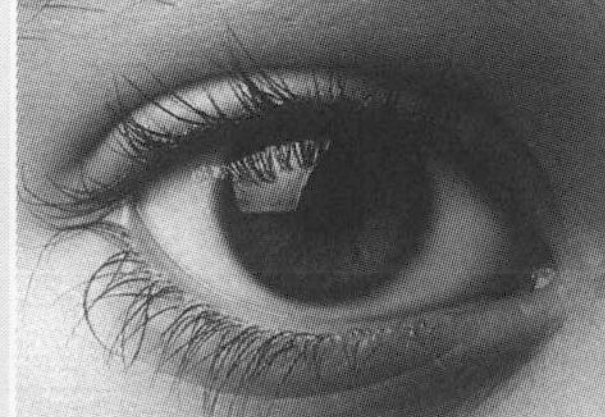

Humans rely heavily on vision. The importance of this **special sense** in humans is indicated by the large occipital region of the brain.

The olfactory tract connects the olfactory bulb with the cerebral hemispheres where olfactory information is interpreted.

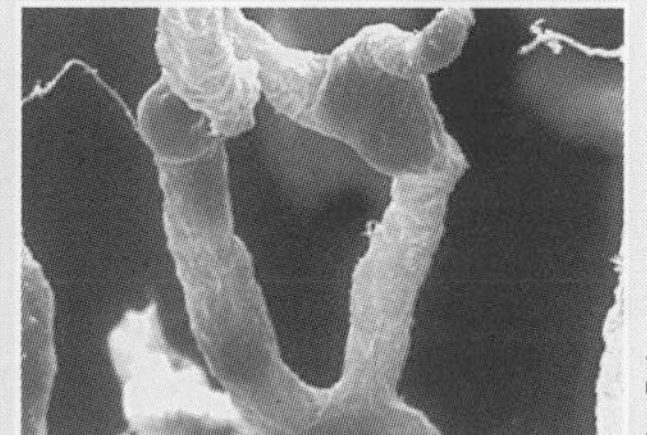

The endothelial tight junctions of the capillaries supplying the brain form a protective **blood-brain barrier** against toxins and infection.

ISBN: 978-1-92717357-2

The Ventricles and CSF

The delicate nervous tissue of the brain and spinal cord is protected against damage by the **bone** of the skull and vertebral column, the membranes overlying the brain (the **meninges**), and the watery but nutritive **cerebrospinal fluid** (CSF), which lies between the inner two of the meningeal layers.

The meninges are collectively three membranes: a tough double-layered outer **dura mater**, a web-like middle **arachnoid mater**, and an inner delicate **pia mater** that adheres to the surface of the brain. The CSF is formed from the blood by clusters of capillaries on the roof of each of the brain's ventricles (choroid plexuses). The CSF is constantly circulated through the ventricles of the brain (and into the spinal cord), returning to the blood via specialized projections of the middle meningeal layer (the arachnoid).

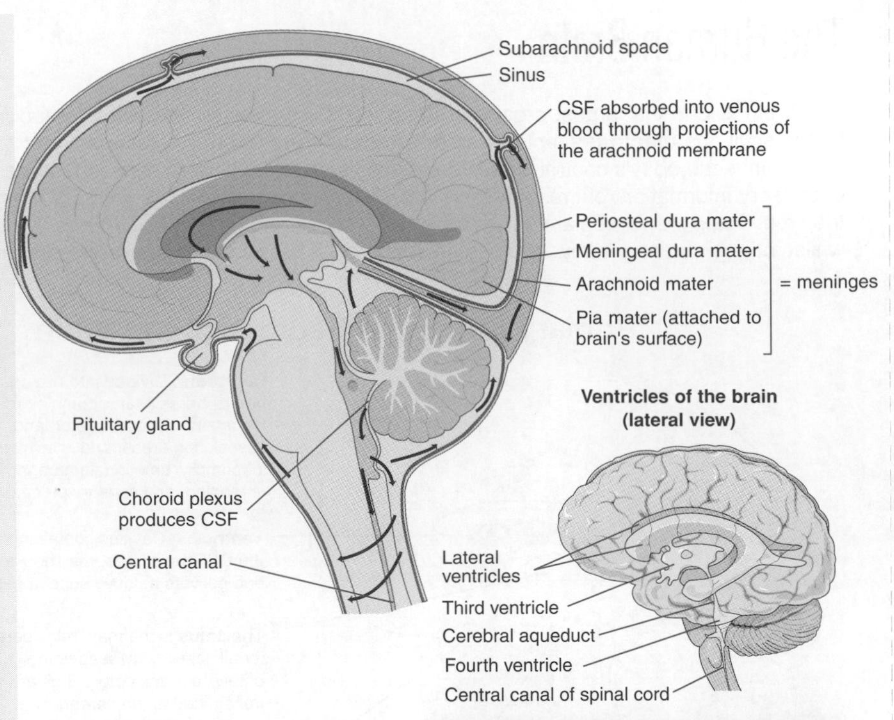

If the passages that normally allow the CSF to exit the brain become blocked, the CSF accumulates within the brain's ventricles causing a condition called hydrocephalus

The accumulated fluid can be seen in this MRI scan.

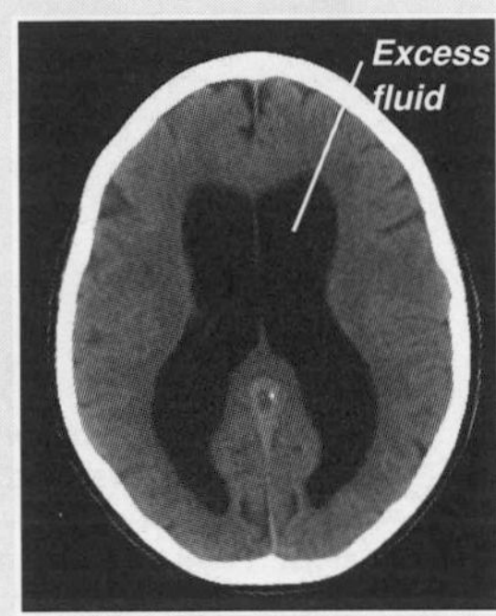

MRI scanning is a powerful technique to visualize the structure and function of the body. It provides much greater contrast between the different soft tissues than computerized tomography (CT) does, making it especially useful in neurological (brain) imaging, especially for indicating the presence of tumors or fluid, and showing up abnormalities in blood supply. In the scan pictured right, the fluid within the lateral and third ventricles is clearly visible.

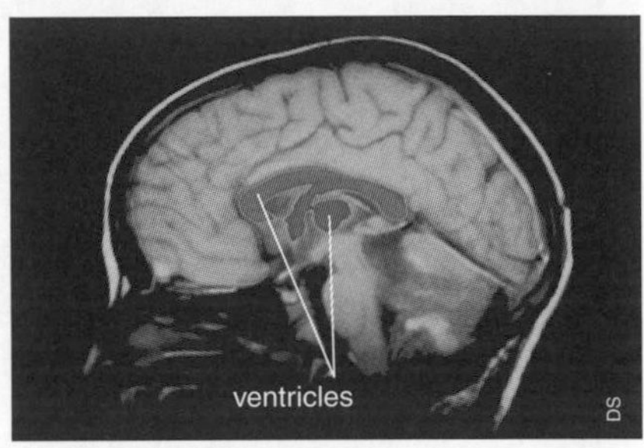

1. For each of the following bodily functions, identify the region(s) of the brain involved in its control:

   (a) Breathing and heartbeat: ______

   (b) Memory and emotion: ______

   (c) Posture and balance: ______

   (d) Autonomic functions: ______

   (e) Visual processing: ______

   (f) Body temperature: ______

   (g) Language: ______

   (h) Muscular movement: ______

2. Explain how the brain is protected against physical damage and infection: ______

3. (a) Describe where CSF is produced and how the CSF returns to the blood: ______

   (b) Explain the consequences of blocking this return flow of CSF: ______

ISBN: 978-1-92717357-2

Neuron Structure

Nervous tissue is made up of two main cell types: **neurons** (nerve cells), which are specialized to transmit nerve impulses, and supporting cells, which are collectively called **neuroglia**. Neurons have a recognizable structure with a cell body (**soma**) and long processes (**dendrites** and **axons**). Most long neurons in the PNS are also supported by a fatty insulating sheath of myelin. Information, in the form of electrochemical impulses, is transmitted along neurons from receptors to effectors. The speed of impulse conduction depends primarily on the axon diameter and whether or not the axon is **myelinated**.

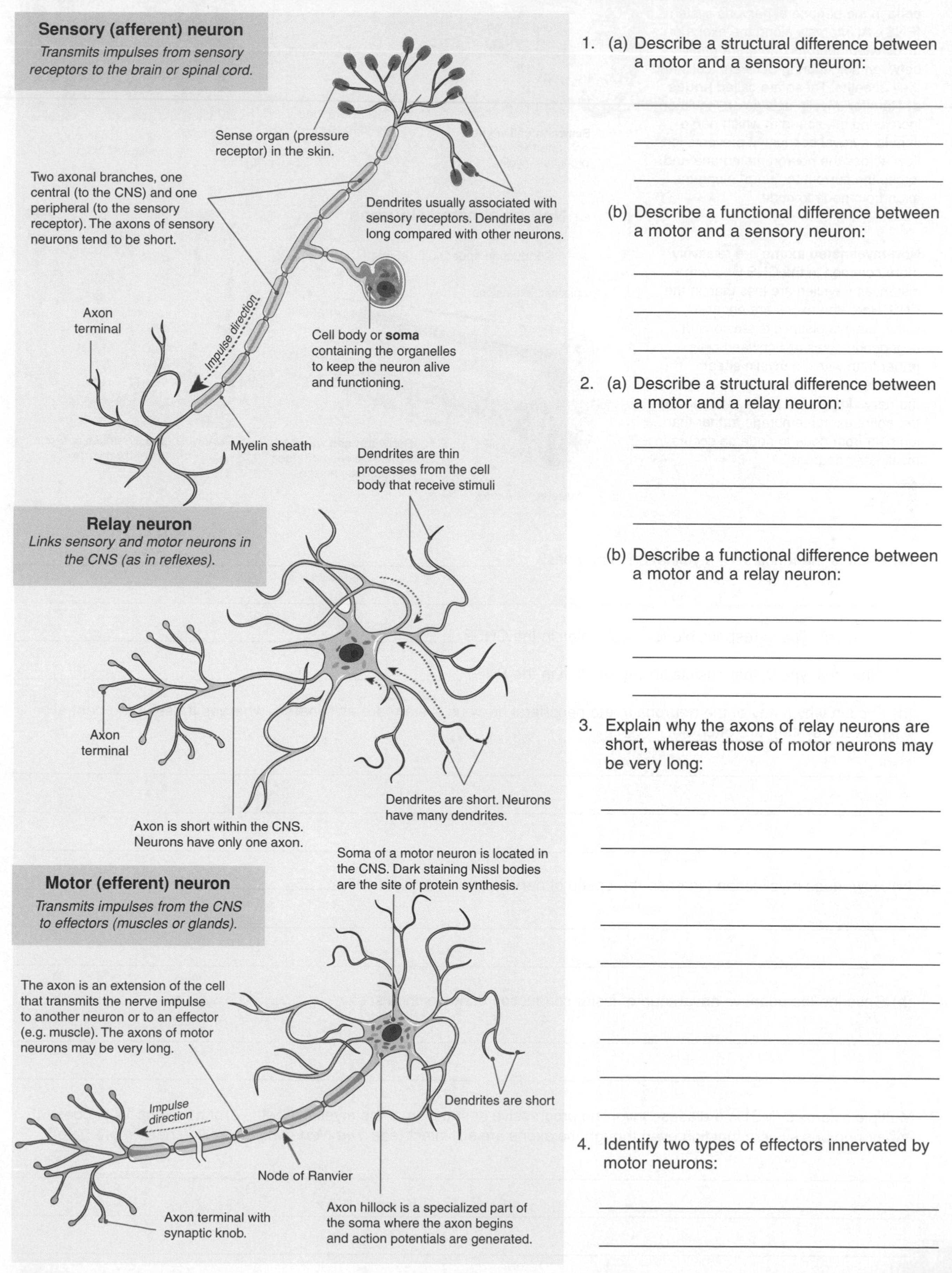

1. (a) Describe a structural difference between a motor and a sensory neuron:

   (b) Describe a functional difference between a motor and a sensory neuron:

2. (a) Describe a structural difference between a motor and a relay neuron:

   (b) Describe a functional difference between a motor and a relay neuron:

3. Explain why the axons of relay neurons are short, whereas those of motor neurons may be very long:

4. Identify two types of effectors innervated by motor neurons:

ISBN: 978-1-92717357-2

Related activities: *Neuroglia, The Nerve Impulse*
Weblinks: *Unipolar and Multipolar Neurons*

Where conduction speed is important, the axons of neurons are sheathed within a lipid and protein rich substance called **myelin**. Myelin is produced by **oligodendrocytes** in the central nervous system (CNS) and by **Schwann cells** in the peripheral nervous system (PNS). At intervals along the axons of myelinated neurons, there are gaps between neighboring Schwann cells and their sheaths. These are called **nodes of Ranvier**. Myelin acts as an insulator, increasing the speed at which nerve impulses travel because it prevents ion flow across the neuron membrane and forces the current to "jump" along the axon from node to node.

Myelinated Neurons

Diameter: 1-25 μm

Conduction speed: 6-120 ms^{-1}

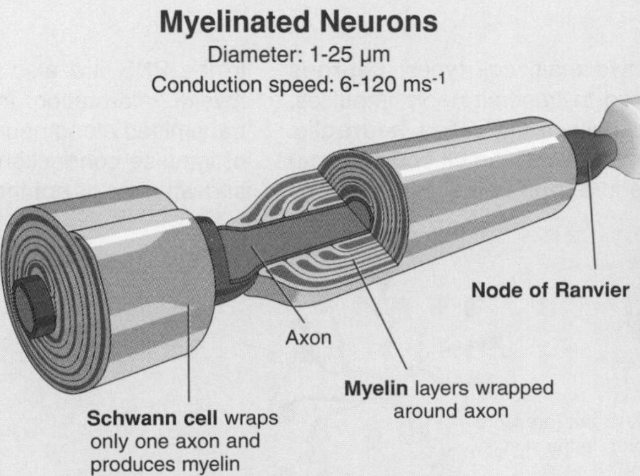

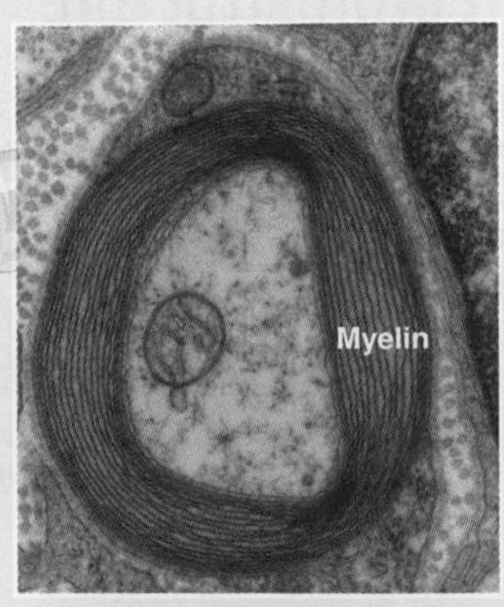

TEM cross section through a myelinated axon

Non-myelinated axons are relatively more common in the CNS where the distances travelled are less than in the PNS. Here, the axons are encased within the cytoplasmic extensions of oligodendrocytes or Schwann cells, rather than within a myelin sheath. **Impulses travel more slowly** because the nerve impulse is propagated along the entire axon membrane, rather than jumping from node to node as occurs in myelinated neurons.

Non-myelinated Neurons

Diameter: <1 μm

Conduction speed: 0.2-0.5 ms^{-1}

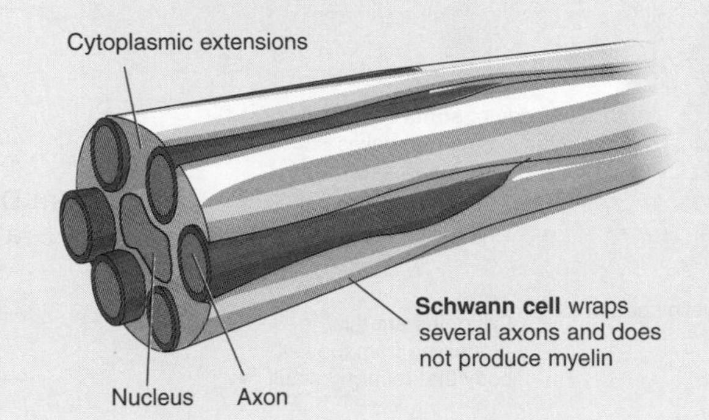

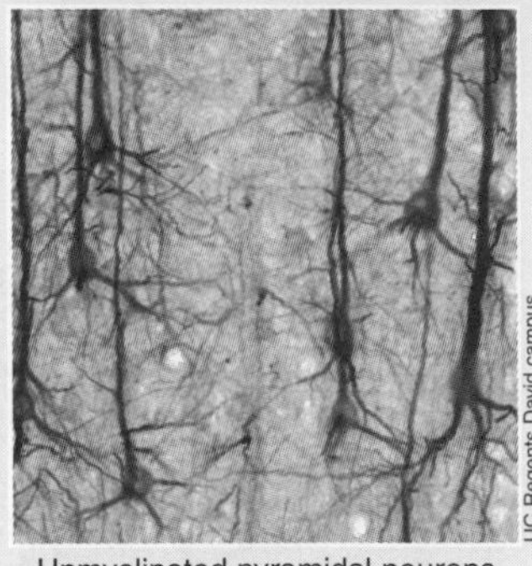

Unmyelinated pyramidal neurons of the cerebral cortex

5. (a) What is the function of myelination in neurons? __________

   (b) What cell type is responsible for myelination in the CNS? __________

   (c) What cell type is responsible for myelination in the PNS? __________

   (d) Explain why many of the neurons in the peripheral nervous system are myelinated, whereas those in the central nervous system are often not:

6. (a) How does myelination increase the speed of nerve impulse conduction? __________

   (b) Describe the adaptive advantage of faster conduction of nerve impulses: __________

7. Multiple sclerosis (MS) is a disease involving progressive destruction of the myelin sheaths around axons. Why does MS impair nervous system function even though the axons are still intact (see *The Nerve Impulse* if you need more help)?

ISBN: 978-1-92717357-2

Neuroglia

Neuroglia, also called glial cells, are the cells that support and protect neurons. Neuroglia have a range of functions. These include holding neurons in place, insulating them, supplying them with nutrients and oxygen, destroying pathogens, and removing dead neurons. There are two main types of neuroglia in the PNS, Schwann cells and satellite cells. The neuroglia of the CNS include astrocytes, microglia, ependymal cells, and oligodendrocytes. Each has a different function in supporting the neurons of the CNS. The structure and function of the CNS neuroglia are described below.

Neuroglia of the CNS

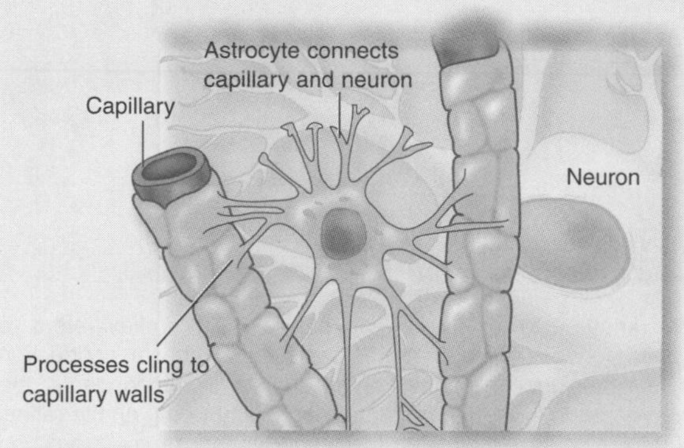

Astrocytes are the most abundant supportive cells in nervous tissue. They anchor neurons to capillaries and support the blood-brain barrier by restricting the passage of certain substances. They are also important in the repair of the brain and spinal cord following injury.

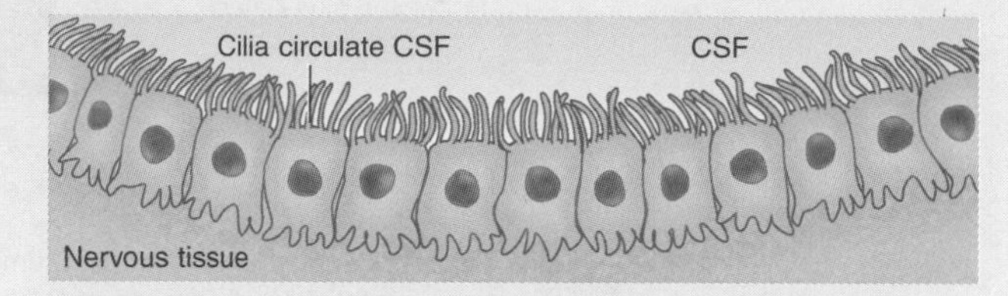

Ependymal cells are epithelial cells lining the ventricles in the brain and the central canal of the spinal cord. The surfaces of these cuboidal cells are covered in cilia and microvilli, which circulate and absorb cerebrospinal fluid (CSF). Specialized ependymal cells and capillaries together form the choroid plexuses, which produce the CSF.

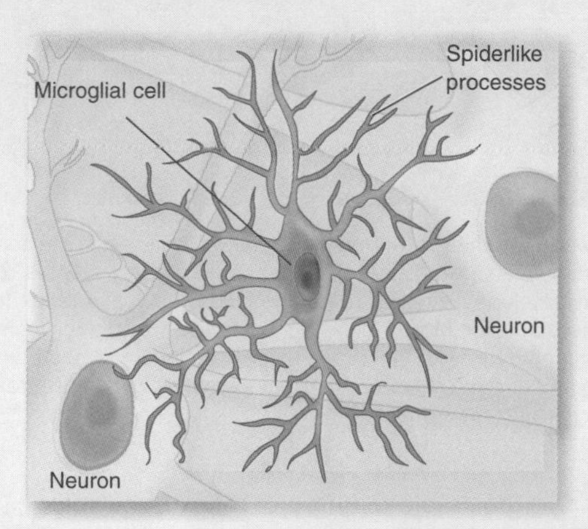

Microglia are the defense cells of nervous tissue. Antibodies are too large to cross the blood-brain barrier, so the phagocytic microglia must be able to recognize and dispose of foreign material and debris.

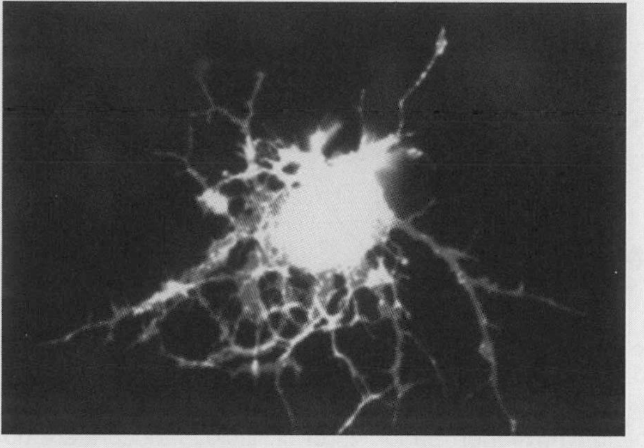
Public domain

Oligodendrocytes produce insulating myelin sheaths around the axons of neurons in the CNS. A single oligodendrocyte can extend to wrap around up to 50 axons. The image above shows an oligodendrocyte genetically altered to fluoresce.

1. What are neuroglia? ___

___

2. Describe the functional role of each of the following neuroglia, with reference to the features associated with that role:

(a) Astrocytes: ___

___

(b) Ependymal cells: ___

___

(c) Microglia: ___

___

(d) Oligodendrocytes: ___

___

ISBN: 978-1-92717357-2

Reflexes

A reflex is an automatic response to a stimulus involving a small number of neurons and a central nervous system (CNS) processing point (usually the spinal cord, but sometimes the brain stem). This type of circuit is called a **reflex arc**. Reflexes permit rapid responses to stimuli. They are classified according to the number of CNS synapses involved; **monosynaptic reflexes** involve only one CNS synapse (e.g. knee jerk reflex), **polysynaptic reflexes** involve two or more (e.g. pain withdrawal reflex). Both are spinal reflexes. The pupil reflex (opening and closure of the pupil) is an example of a cranial reflex.

Pain withdrawal: A polysynaptic reflex arc

Sensory neuron

Stimulus = pin prick

1. Pain receptors in the skin detect stimulus

Spinal cord

Impulse direction

Motor neuron

2. Sensory message is interpreted through a relay neuron. In a monosynaptic reflex arc, the sensory neuron synapses directly with the motor neuron.

3. The impulse reaches the **motor end plate** and causes muscle contraction.

Response = withdraw finger

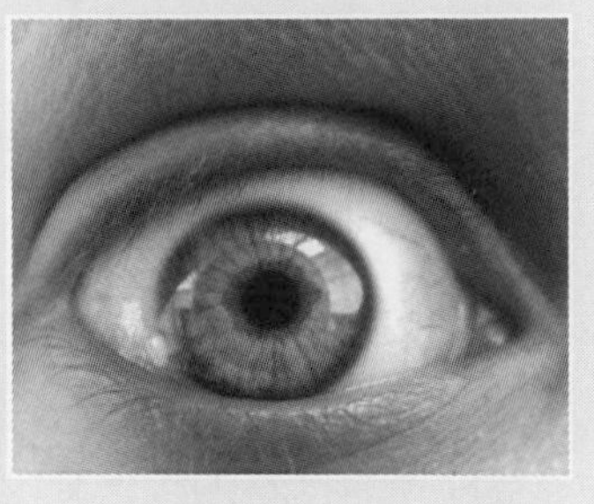

The patella (knee jerk) reflex is a simple deep tendon reflex that is used to test the function of the femoral nerve and spinal cord segments L2-L4. It helps to maintain posture and balance when walking.

The pupillary light reflex refers to the rapid expansion or contraction of the pupils in response to the intensity of light falling on the retina. It is a polysynaptic cranial reflex and can be used to test for brain death.

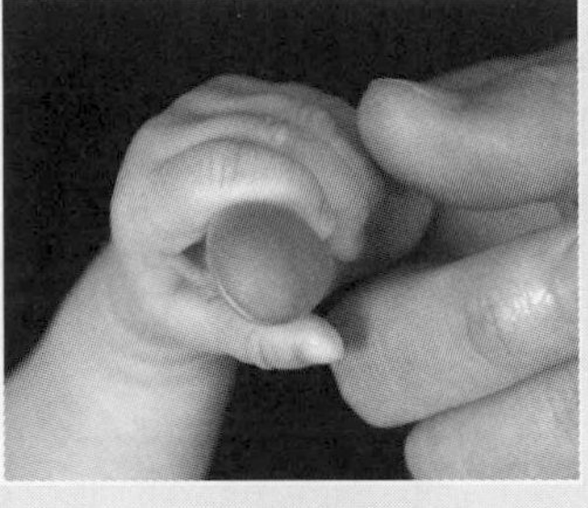

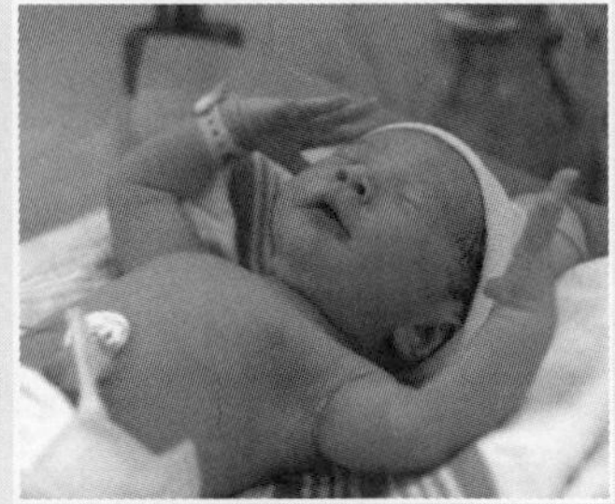

Normal newborns exhibit a number of primitive reflexes in response to particular stimuli. These reflexes disappear within a few months of birth as the child develops. Primitive reflexes include the grasp reflex (above left) and the startle or Moro reflex (right) in which a sudden noise will cause the infant to throw out its arms, extend the legs and head, and cry. The rooting and sucking reflexes are other examples of primitive reflexes.

1. Explain why higher reasoning or conscious thought are not necessary or desirable features of reflex behaviors: ______

2. Distinguish between a spinal reflex and a cranial reflex and give an example of each: ______

3. (a) Distinguish between a monosynaptic and a polysynaptic reflex arc and give an example of each: ______

   (b) Given similar length sensory and motor pathways, identify which would produce the most rapid response and why: ______

4. (a) With reference to specific examples, describe the adaptive value of primitive reflexes in newborns: ______

   (b) Explain why newborns are tested for the presence of these reflexes: ______

ISBN: 978-1-92717357-2

Related activities: The Nervous System, Chemical Synapses
Weblinks: Parasympathetic Eye Response, Knee Jerk Reflex

The Nerve Impulse

The plasma membranes of cells, including neurons, contain **sodium-potassium ion pumps** which actively pump sodium ions (Na^+) out of the cell and potassium ions (K^+) into the cell. The action of these ion pumps in neurons creates a separation of charge (a potential difference or voltage) either side of the membrane and makes the cells **electrically excitable**. It is this property that enables neurons to transmit electrical impulses. The **resting state** of a neuron, with a net negative charge inside, is maintained by the sodium-potassium pumps, which actively move two K^+ into the neuron for every three Na^+ moved out (below left). When a nerve is stimulated, a brief increase in membrane permeability to Na^+ temporarily reverses the membrane polarity (a depolarization). After the nerve impulse passes, the sodium-potassium pump restores the resting potential. The depolarization is propagated along the axon by local current in non-myelinated fibers and by **saltatory conduction** in myelinated fibers. Impulses pass from neuron to neuron by crossing junctions called **synapses**.

The Resting Neuron

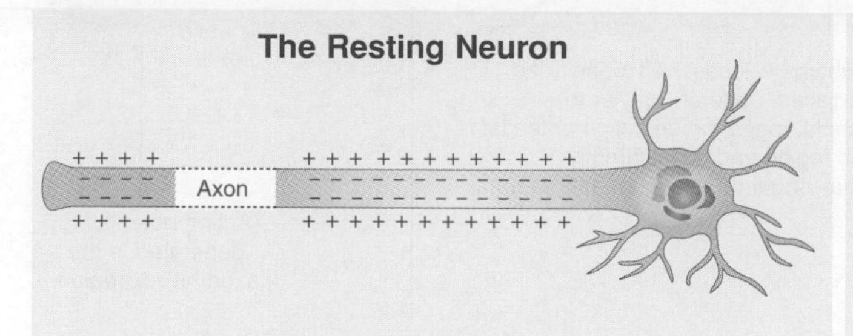

When a neuron is not transmitting an impulse, the inside of the cell is negatively charged relative to the outside and the cell is said to be electrically polarized. The potential difference (voltage) across the membrane is called the **resting potential**. For most nerve cells this is about -70 mV. Nerve transmission is possible because this membrane potential exists.

The Nerve Impulse

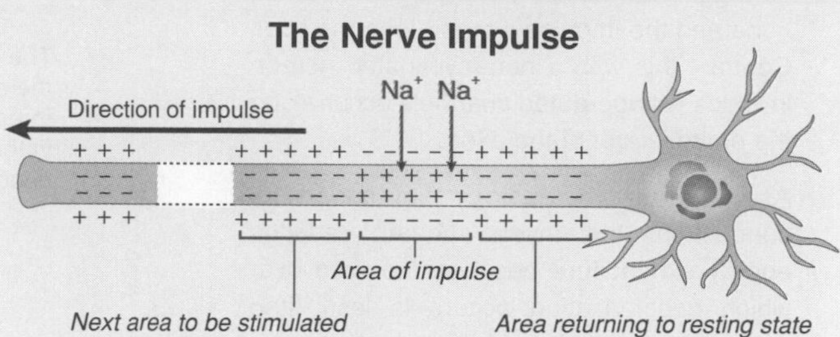

When a neuron is stimulated, the distribution of charges on each side of the membrane briefly reverses. This process of **depolarization** causes a burst of electrical activity to pass along the axon of the neuron as an **action potential**. As the charge reversal reaches one region, local currents depolarize the next region and the impulse spreads along the axon.

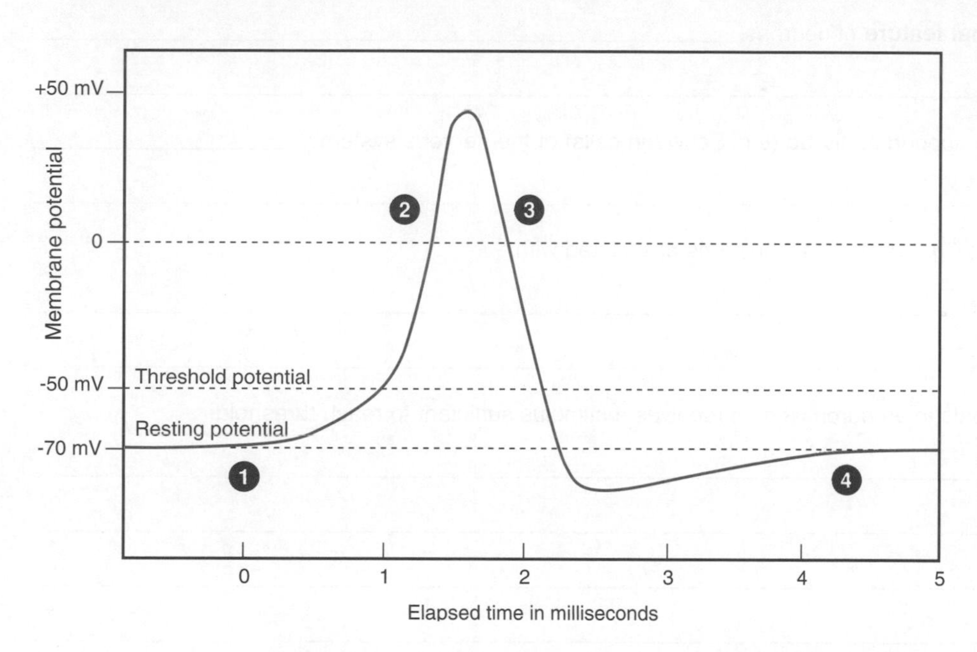

The depolarization in an axon can be shown as a change in membrane potential (in millivolts). A stimulus must be strong enough to reach the **threshold potential** before an action potential is generated. This is the voltage at which the depolarization of the membrane becomes unstoppable.

The action potential is **all or nothing** in its generation and because of this, impulses (once generated) always reach threshold and move along the axon without attenuation. The resting potential is restored by the movement of potassium ions (K^+) out of the cell. During this **refractory period**, the nerve cannot respond, so nerve impulses are discrete.

Voltage-Gated Ion Channels and the Course of an Action Potential

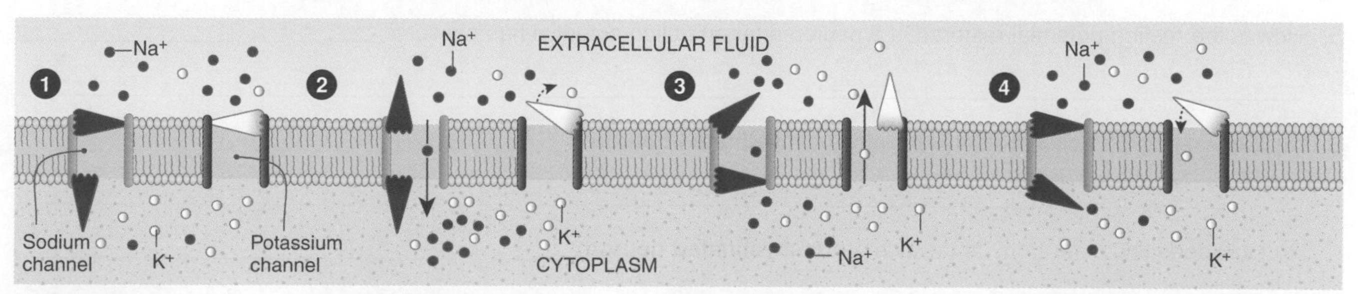

Resting state:
Voltage activated Na^+ and K^+ channels are closed.

Depolarization:
Voltage activated Na^+ channels open and there is a rapid influx of Na^+ ions. The interior of the neuron becomes positive relative to the outside.

Repolarization:
Voltage activated Na^+ channels close and the K^+ channels open; K^+ moves out of the cell, restoring the negative charge to the cell interior.

Returning to resting state:
Voltage activated Na^+ and K^+ channels close to return the neuron to the resting state.

ISBN: 978-1-92717357-2

Periodicals: *Refractory period*

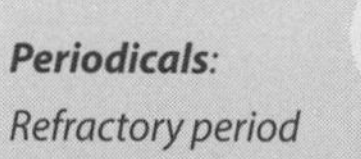

Related activities: *Chemical Synapses*
Weblinks: *Nerve Action Potential, Neurobiology*

Axon myelination is a feature of vertebrate nervous systems and it enables them to achieve very rapid speeds of nerve conduction. Myelinated neurons conduct impulses by **saltatory conduction**, a term that describes how the impulse jumps along the fiber. In a myelinated neuron, **action potentials are generated only at the nodes**, which is where the voltage gated channels occur. The axon is insulated so the action potential at one node is sufficient to trigger an action potential in the next node and the impulse jumps along the fiber. Contrast this with a non-myelinated neuron in which voltage-gated channels occur along the entire length of the axon.

As well as increasing the speed of conduction, the myelin sheath reduces energy expenditure because the area over which depolarization occurs is less (and therefore also the number of sodium and potassium ions that need to be pumped to restore the resting potential).

Saltatory Conduction in Myelinated Axons

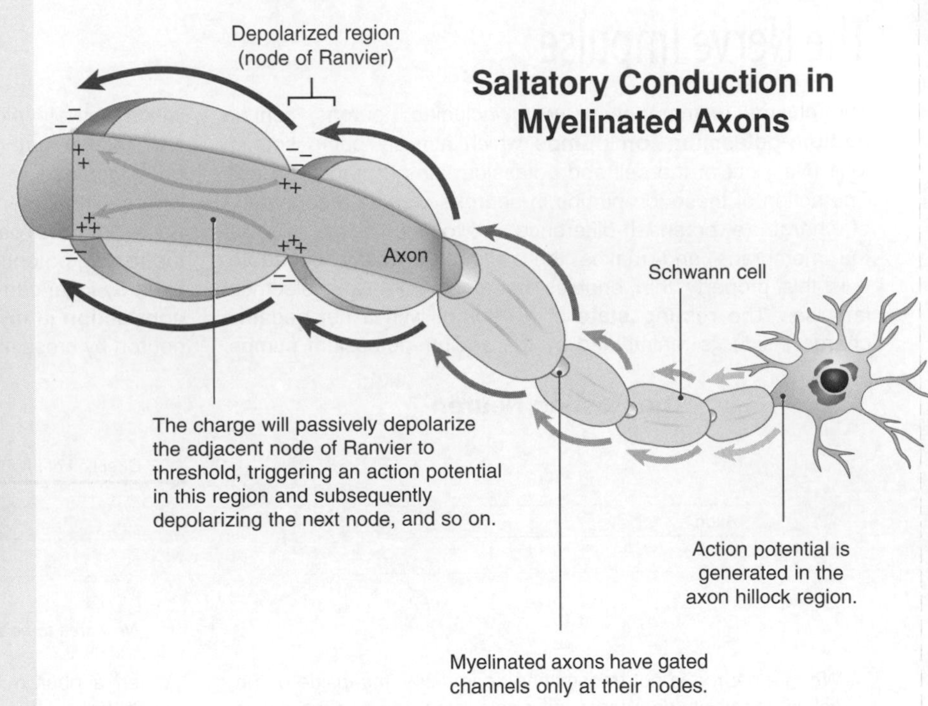

1. In your own words, define what an **action potential** is: ____________________

2. (a) Identify the defining **functional feature** of neurons: ____________________

   (b) How does this differ from the supporting tissue (e.g. Schwann cells) of the nervous system? ____________________

3. Describe the movement of voltage-gated channels and ions associated with:

   (a) Depolarization of the neuron: ____________________

   (b) Repolarization of the neuron: ____________________

4. Summarize the sequence of events in a neuron when it receives a stimulus sufficient to reach threshold:

5. How is the resting potential restored in a neuron after an action potential has passed? ____________________

6. (a) Explain how an action potential travels in a **myelinated neuron**: ____________________

   (b) How does this differ from its travel in a **non-myelinated neuron**? ____________________

7. Explain how the **refractory period** influences the direction in which an impulse will travel: ____________________

ISBN: 978-1-92717357-2

Neurotransmitters

Neurotransmitters are chemicals that allow the transmission of signals between neurons. They are found in the axon endings of neurons and are released into the space between one neuron and the next (the **synaptic cleft**) after a depolarization or hyperpolarization of the nerve ending. Neurotransmitters can be classified into amino acids, peptides, or monoamines. The many neurotransmitters produce various responses depending on their location in the body. They can be excitatory (likely to cause an action potential in the receiving neuron) or inhibitory (causing hyperpolarization) depending on the receptor they activate.

Neurotransmitters Carry Signals Between Neurons

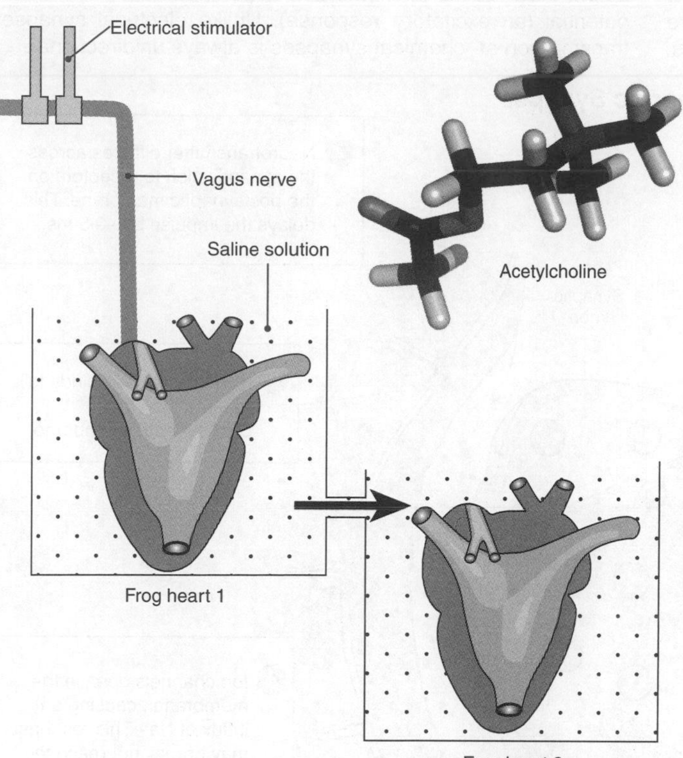

Chemical signaling between neurons was first demonstrated in 1921 by Otto Loewi. In his experiment, the still beating hearts of two frogs were placed in connected flasks filled with saline solution. The vagus nerve (parasympathetic) of the first heart was still attached and was stimulated by electricity to reduce its rate of beating. After a delay, the rate of beating in the second heart also slowed. Increasing the beating rate in the first heart caused an increase in the beating rate in the second heart, showing electrical stimulus of the first heart caused it to release a chemical into the saline solution that then affected the heartbeat of the second heart. The chemical was found to be **acetylcholine**.

Neurotransmitters

Name	Postsynaptic effect	
Acetylcholine	Excitatory/ inhibitory	Responsible for the stimulation of muscles. Found in sensory neurons and the autonomic nervous system.
Norepinephrine	Excitatory	Brings the nervous system into high alert. Increases heart rate and blood pressure.
Dopamine	Excitatory/ inhibitory	Associated with reward mechanisms in the brain. Produces the "feel good" feeling.
Gamma amino butyric acid (GABA)	Inhibitory	Inhibits excitatory neurotransmitters that can cause anxiety.
Glutamate	Excitatory	Found in the central nervous system and concentrated in the brain.
Serotonin	Inhibitory	Serotonin is strongly involved in regulation of mood and perception.
Endorphin	Excitatory	Involved in pain reduction and pleasure.

1. Describe the purpose of a neurotransmitter: ______

2 (a) Explain why stimulating the first frog heart with electricity caused it to change its beating rate:

______

(b) Explain why the second heart in the experiment reduced its beating rate after a delay:

______

3. Why can some neurotransmitters be both excitatory and inhibitory? ______

ISBN: 978-1-92717357-2

Related activities: *Chemical Synapses*

Chemical Synapses

Action potentials are transmitted between neurons across **synapses**: junctions between the end of one axon and the dendrite or cell body of a receiving neuron. Electrical synapses, where cells are electrically coupled through gap junctions between cells, occur in heart muscle and in the cerebral cortex, but they are relatively uncommon elsewhere. Most synapses in the nervous system are **chemical synapses**. In these, the axon terminal is a swollen knob, and a small gap, the **synaptic cleft**, separates it from the receiving neuron. The synaptic knobs are filled with tiny packets of chemicals called **neurotransmitters**.

Nerve transmission involves the diffusion of the neurotransmitter across the cleft, where it interacts with the receiving membrane and causes an electrical response. The response of a receiving (post-synaptic) cell to the arrival of a neurotransmitter depends on the nature of the cell itself, on its location in the nervous system, and on the neurotransmitter involved. Synapses that release acetylcholine (ACh) are termed **cholinergic**. In the example below, ACh results in membrane depolarization and an action potential (an excitatory response). Unlike electrical synapses, transmission at chemical synapses is always unidirectional.

A Cholinergic Synapse

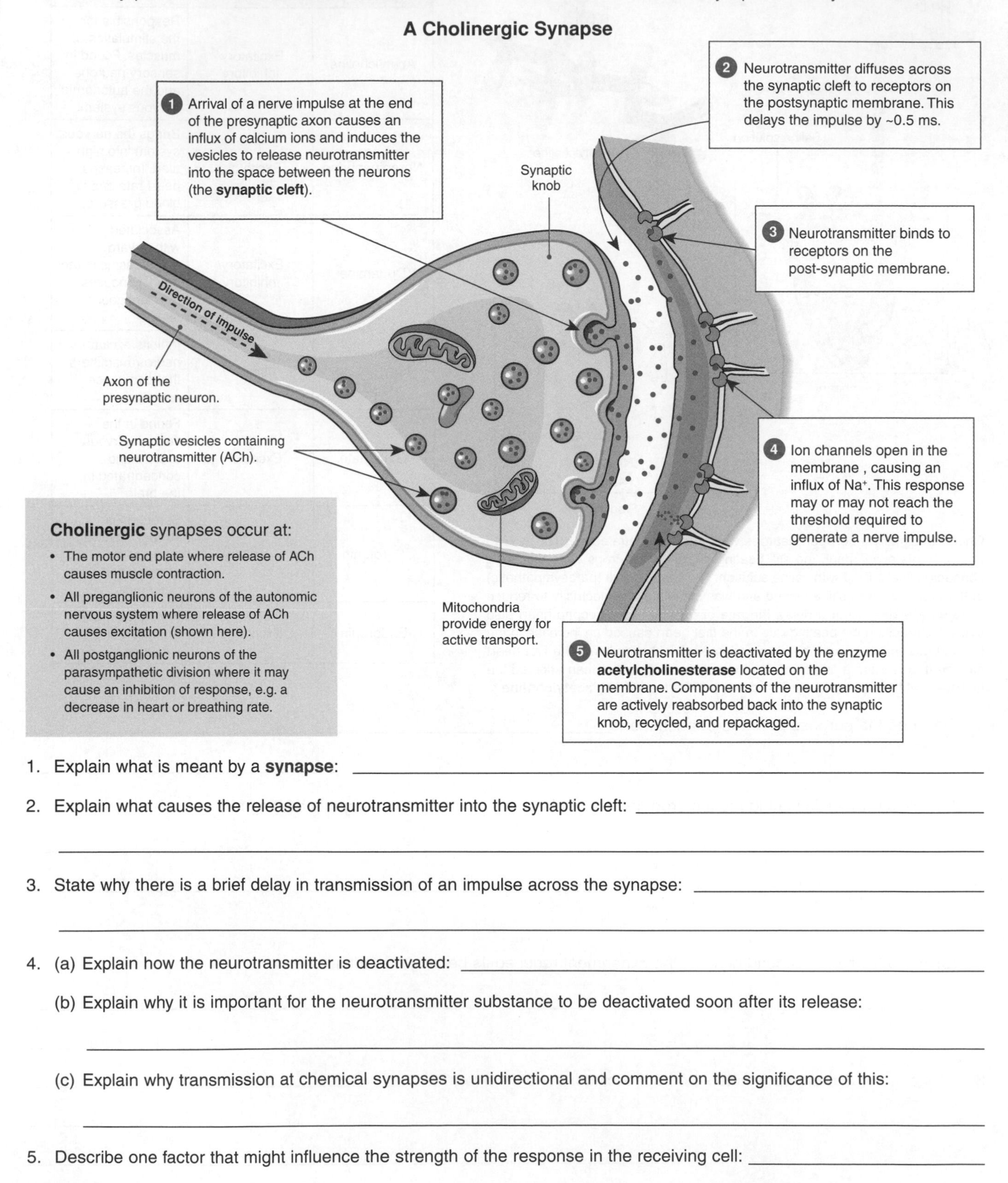

Cholinergic synapses occur at:

- The motor end plate where release of ACh causes muscle contraction.
- All preganglionic neurons of the autonomic nervous system where release of ACh causes excitation (shown here).
- All postganglionic neurons of the parasympathetic division where it may cause an inhibition of response, e.g. a decrease in heart or breathing rate.

1. Explain what is meant by a **synapse**: ______

2. Explain what causes the release of neurotransmitter into the synaptic cleft: ______

3. State why there is a brief delay in transmission of an impulse across the synapse: ______

4. (a) Explain how the neurotransmitter is deactivated: ______

   (b) Explain why it is important for the neurotransmitter substance to be deactivated soon after its release:

   ______

   (c) Explain why transmission at chemical synapses is unidirectional and comment on the significance of this:

   ______

5. Describe one factor that might influence the strength of the response in the receiving cell: ______

ISBN: 978-1-92717357-2

Integration at Synapses

Synapses play a pivotal role in the ability of the nervous system to respond appropriately to stimulation and to adapt to change. The nature of synaptic transmission allows the **integration** (interpretation and coordination) of inputs from many sources. These inputs need not be just excitatory (causing depolarization). Inhibition results when the neurotransmitter released causes negative chloride ions (rather than sodium ions) to enter the postsynaptic neuron. The postsynaptic neuron then becomes more negative inside (hyperpolarized) and an action potential is less likely to be generated. At synapses, it is the sum of **all** inputs (excitatory and inhibitory) that leads to the final response in a postsynaptic cell. Integration at synapses makes possible the various responses we have to stimuli. It is also the most probable mechanism by which learning and memory are achieved.

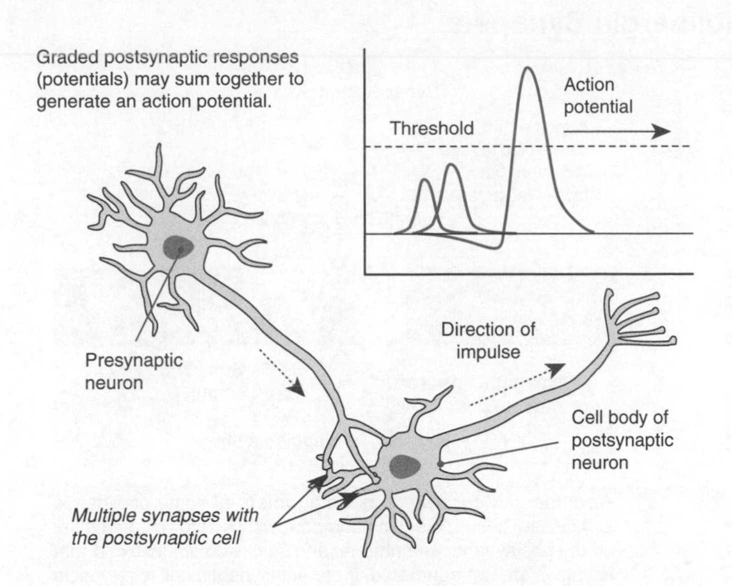

Synapses and Summation

Nerve transmission across chemical synapses has several advantages, despite the delay caused by neurotransmitter diffusion. Chemical synapses transmit impulses in one direction to a precise location and, because they rely on a limited supply of neurotransmitter, they are subject to fatigue (inability to respond to repeated stimulation). This protects the system against overstimulation.

Synapses also act as centers for the **integration** of inputs from many sources. The response of a postsynaptic cell is often graded; it is not strong enough on its own to generate an action potential. However, because the strength of the response is related to the amount of neurotransmitter released, subthreshold responses can sum to produce a response in the post-synaptic cell. This additive effect is termed **summation**. Summation can be **temporal** or **spatial** (below). A neuromuscular junction (photo below) is a specialized form of synapse between a motor neuron and a skeletal muscle fiber. Functionally, it is similar to any excitatory cholinergic synapse.

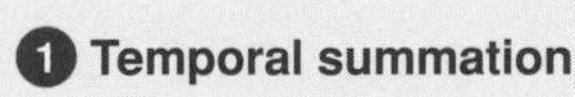

1 Temporal summation

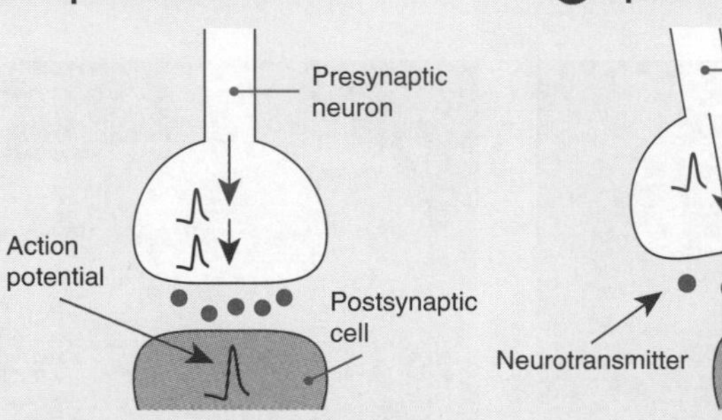

Several impulses may arrive at the synapse in quick succession from a single axon. The individual responses are so close together in time that they sum to reach threshold and produce an action potential in the postsynaptic neuron.

2 Spatial summation

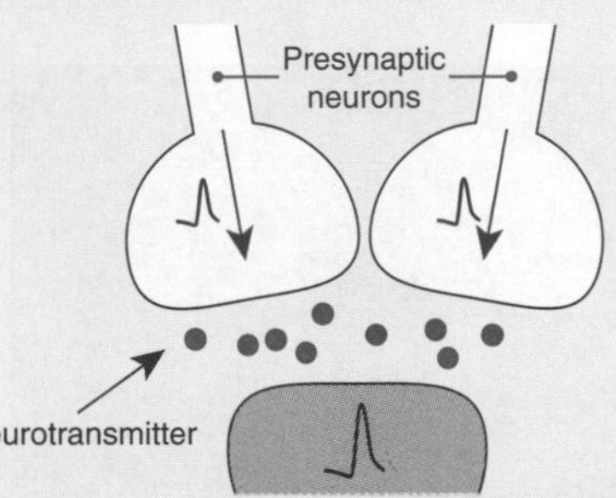

Individual impulses from spatially separated axon terminals may arrive **simultaneously** at different regions of the same postsynaptic neuron. The responses from the different places sum to reach threshold and produce an action potential.

3 Neuromuscular junction

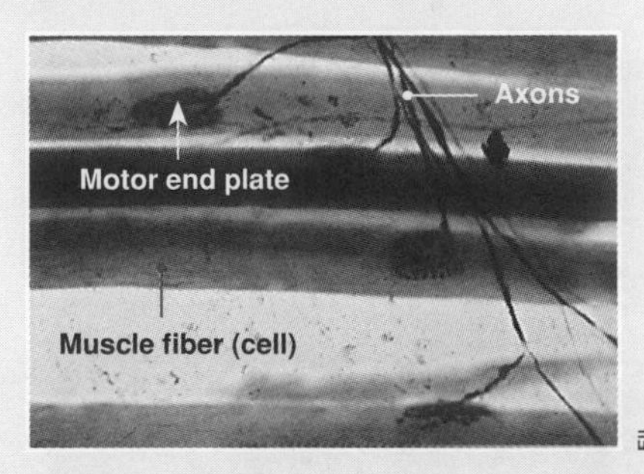

EII

The arrival of an impulse at the neuromuscular junction causes the release of acetylcholine from the synaptic knobs. This causes the muscle cell membrane (sarcolemma) to depolarize, and an action potential is generated in the muscle cell.

1. Explain the purpose of nervous system integration: ______

2. (a) Explain what is meant by **summation**: ______

   (b) In simple terms, distinguish between temporal and spatial summation: ______

3. Describe two ways in which a neuromuscular junction is similar to any excitatory cholinergic synapse:

   (a) ______

   (b) ______

ISBN: 978-1-92717357-2

Related activities: *Chemical Synapses, Skeletal Muscle Structure and Function*

Drugs at Synapses

Synapses in the peripheral nervous system are classified by the type of neurotransmitter they release. **Cholinergic** synapses release **acetylcholine** (**Ach**), while adrenergic synapses release **epinephrine** (adrenaline) or norepinephrine (noradrenaline). Postsynaptic receptors are also classified by the type of neurotransmitters they bind. Cholinergic receptors all bind acetylcholine, but they can also bind **drugs** that mimic Ach. Drugs act on the nervous system by mimicking (**agonists**) or blocking (**antagonists**) the activity of neurotransmitters. Because of the small amounts of chemicals involved in synaptic transmission, drugs that affect the activity of neurotransmitters, or their binding sites, can have powerful effects in small doses.

Drugs at Cholinergic Synapses

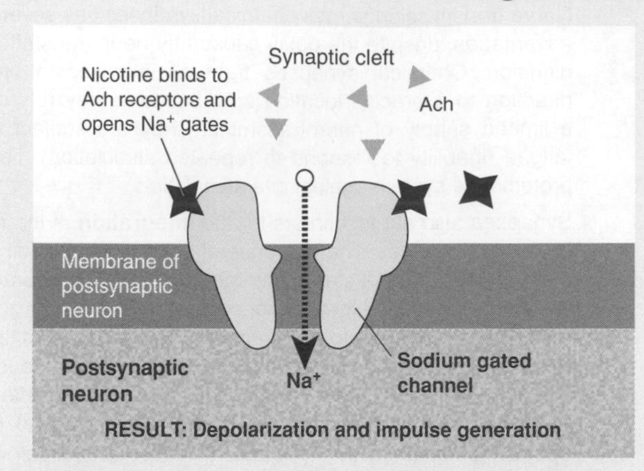

Nicotine acts as a **direct agonist** at nicotinic synapses. Nicotine binds to and activates acetylcholine (Ach) receptors on the postsynaptic membrane. This opens sodium gates, leading to a sodium influx and membrane depolarization. Some agonists work indirectly by preventing Ach breakdown. Such drugs are used to treat elderly patients with Alzheimer's.

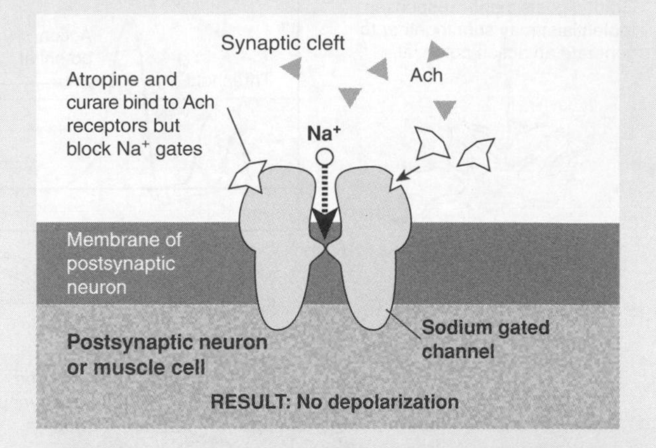

Atropine and **curare** act as **antagonists** at some cholinergic synapses. These molecules compete with Ach for binding sites on the postsynaptic membrane, and block sodium influx so that impulses are not generated. If the postsynaptic cell is a muscle cell, muscle contraction is prevented. In the case of curare, this causes death by flaccid paralysis.

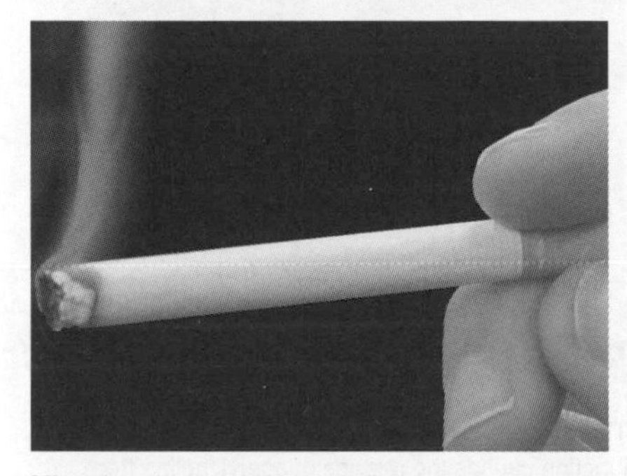

Nicotine is the highly addictive substance in cigarettes. It acts on the nicotinic acetylcholine receptors, increasing the levels of several neurotransmitters, including dopamine. Dopamine produces feelings of euphoria and relaxation. These feelings reinforce nicotine consumption, and create nicotine addiction.

Muscarine, a compound found in several types of mushrooms, binds to muscarinic acetylcholine receptors. Muscarine is used to treat a number of medical conditions (e.g. glaucoma), but consumption of the mushrooms can deliver a fatal overdose of muscarine.

Mamba snake venom contains a number of neurotoxins including **dendrotoxins**. These small peptide molecules act as acetylcholine receptor antagonists (blocking muscarinic receptors). They have many effects including disrupting muscle contraction.

1. Providing an example of each, outline two ways in which drugs can act at a cholinergic synapse:

   (a) ______________________________

   (b) ______________________________

2. Explain why atropine and curare are described as direct antagonists: ______________________________

3. Suggest why curare (carefully administered) is used during abdominal surgery: ______________________________

ISBN: 978-1-92717357-2

Chemical Imbalances in the Brain

The brain uses chemicals (**neurotransmitters**) to transmit messages between nerve cells. Neurotransmitters are released from presynaptic neurons and diffuse across the synaptic cleft to postsynaptic neurons to cause a specific effect. Many brain disorders result from disturbances to natural levels of specific neurotransmitters, and can lead to the failure of specific neural pathways. Sometimes the pathways can be restored using drugs that either replace or boost levels of specific neurotransmitters.

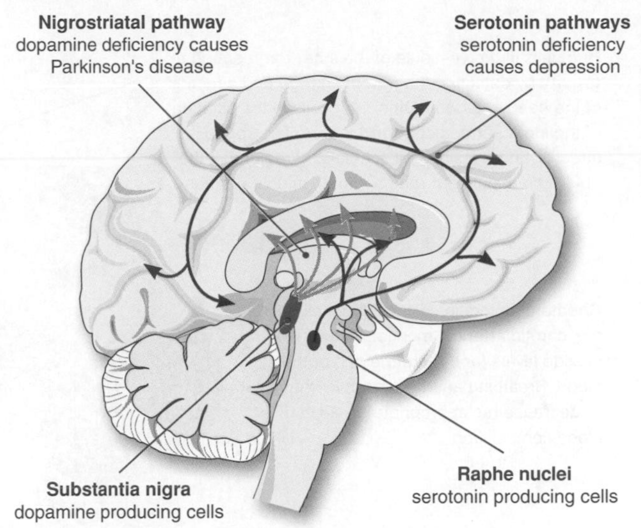

Parkinson's Disease

Patients with **Parkinson's disease** show decreased stimulation in the motor cortex of the brain. This results from reduced dopamine production in the substantia nigra region (right) where dopamine is produced. This is usually the result of the death of nerve cells. Symptoms, slow physical movement and spasmodic tremors, often don't begin to appear until a person has lost 70% of their dopamine-producing cells.

Treating Parkinson's Disease

Parkinson's disease is caused by reduced dopamine production and low dopamine levels in the brain pathways involved with movement. Treatments for Parkinson's have focused on increasing the body's dopamine levels. Dopamine is unable to cross the blood-brain barrier, so cannot be administered as a treatment. However, **L-dopa** is a dopamine precursor that can cross the blood-brain barrier and enter the brain. Once in the brain, it is converted to dopamine. L-dopa has been shown to reduce some of the symptoms of Parkinson's disease.

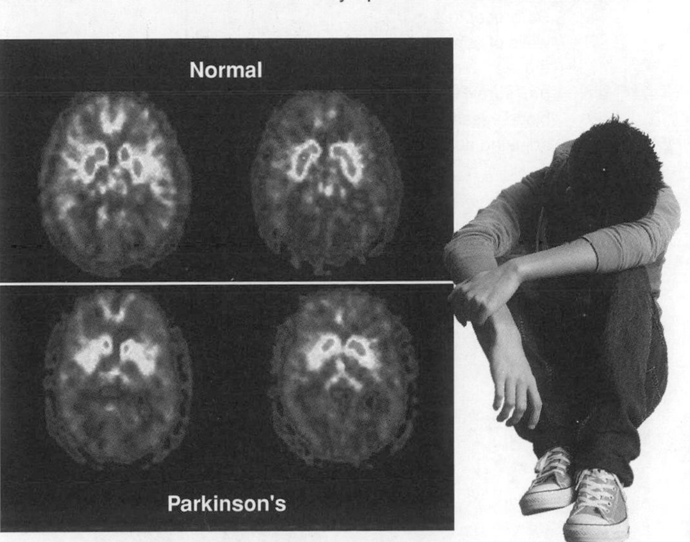

Positron emission tomography (PET) measures the activity of dopamine neurons in the substantia nigra area of the brain. Parkinson's patients (lower panel) show reduced activity in the dopamine neurons compared with normal patients.

Depression

A person with **depression** (left) experiences prolonged periods of extremely low mood, including low self esteem, regret, guilt, and feelings of hopelessness. Depression may be caused by a mixture of environmental factors (e.g. stress) and biological factors (e.g. low **serotonin** production by the raphe nuclei in the brain, above).

Treating Depression

Recognition of the link between **serotonin** and **depression** has resulted in the development of **antidepressant drugs** that alter serotonin levels. Monoamine oxidase inhibitors (MAOI) are commonly used antidepressants that increase serotonin levels by preventing its breakdown in the brain. Newer drugs, called Selective Serotonin Re-uptake Inhibitors (SSRIs), stop serotonin re-uptake by presynaptic cells. This increases the levels of extracellular serotonin, making more available to bind to the postsynaptic cells, and stabilizing serotonin levels in the brain. SSRIs have fewer side effects than other antidepressants because they specifically target serotonin and no other neurotransmitters.

1. Describe the function of a neurotransmitter: ______________________________

____

____

____

2. Describe the pharmacological cause of the following diseases and identify the major symptom of each:

(a) Parkinson's disease: ______________________________

____

____

____

(b) Depression: ______________________________

____

____

ISBN: 978-1-92717357-2

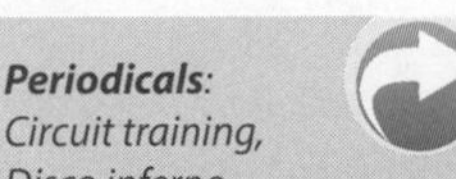

Detecting Changing States

A **stimulus** is any physical or chemical change in the environment capable of provoking a response in an organism. All organisms respond to stimuli in order to survive. This response is adaptive; it acts to maintain the organism's state of homeostasis. Stimuli may be either external (outside the organism) or internal (within its body). Some of the stimuli to which humans respond are described below, together with the sense organs that detect and respond to these stimuli. Note that sensory receptors respond only to specific stimuli, so the sense organs we possess determine how we perceive the world.

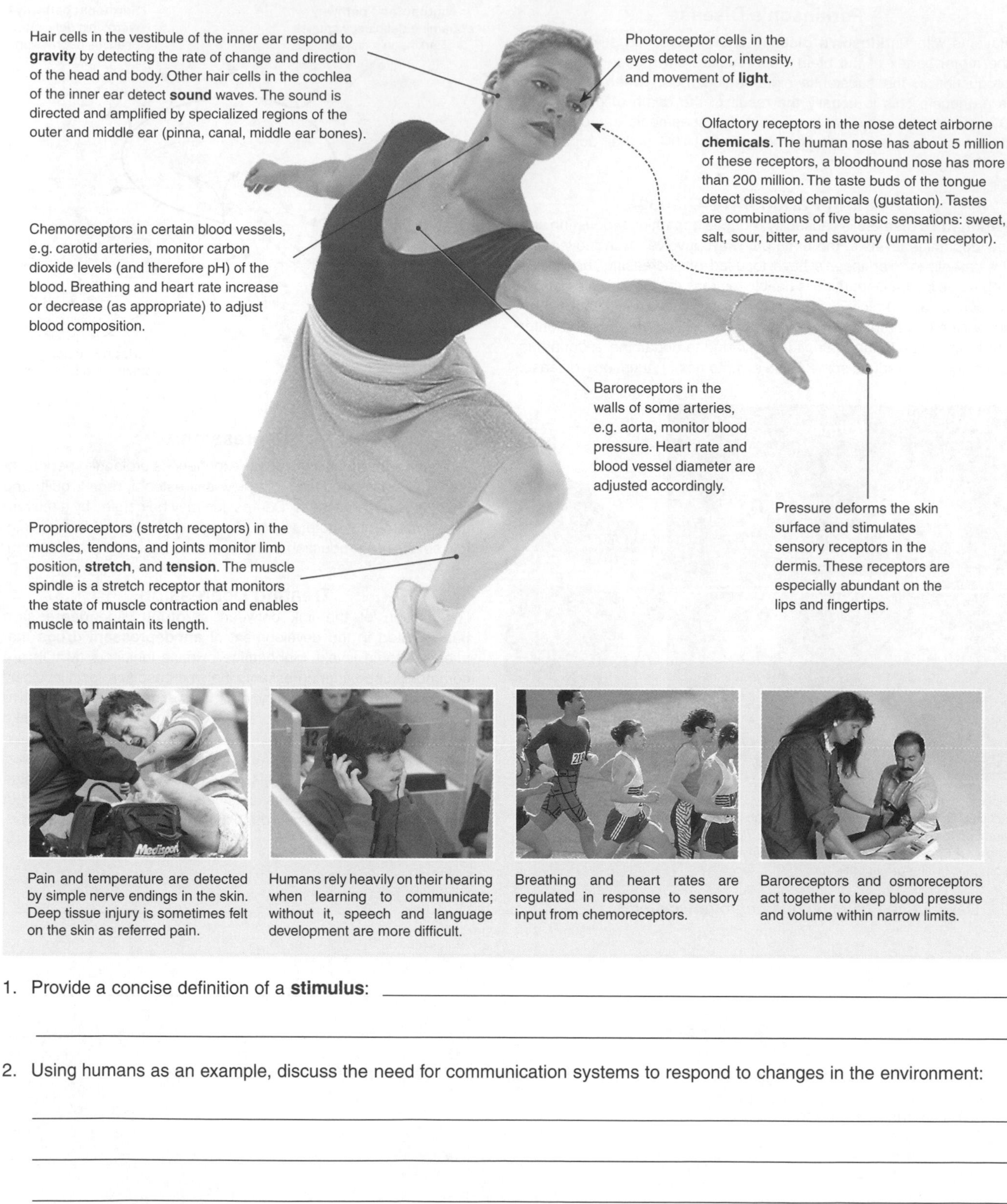

Pain and temperature are detected by simple nerve endings in the skin. Deep tissue injury is sometimes felt on the skin as referred pain.

Humans rely heavily on their hearing when learning to communicate; without it, speech and language development are more difficult.

Breathing and heart rates are regulated in response to sensory input from chemoreceptors.

Baroreceptors and osmoreceptors act together to keep blood pressure and volume within narrow limits.

1. Provide a concise definition of a **stimulus**: ______________________________

____

2. Using humans as an example, discuss the need for communication systems to respond to changes in the environment:

____

____

____

3. (a) Name one internal stimulus and its sensory receptor: ______________________________

(b) Describe the role of this sensory receptor in contributing to **homeostasis**: ______________________________

____

____

Related activities: *The Basis of Sensory Perception*

***Periodicals**:*
Sense and sense ability

ISBN: 978-1-92717357-2

The Basis of Sensory Perception

Sensory receptors are specialized to detect stimuli and respond by producing an electrical (or chemical) discharge. In this way they act as **biological transducers**, converting the energy from a stimulus into an electrochemical signal. They can do this because the stimulus opens (or closes) ion channels and leads to localized changes in membrane potential called **receptor potentials**. Receptor potentials are graded and not self-propagating, but sense cells can amplify them, generating action potentials directly or inducing the release of a neurotransmitter. Whether or not the sensory cell itself fires action potentials, ultimately the stimulus is transduced into action potentials whose frequency is dependent on stimulus strength. The simplest sensory receptors consist of a single sensory neuron (e.g. nerve endings). More complex sense cells form synapses with their sensory neurons (e.g. taste buds). Sensory receptors are classified according to the stimuli to which they respond (e.g. photoreceptors respond to light).

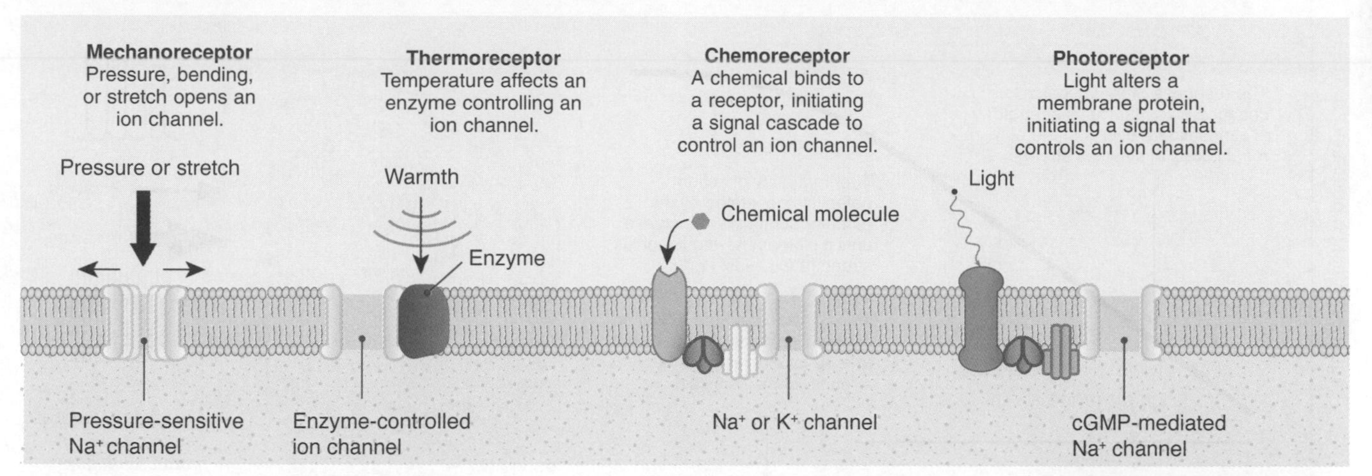

Signal Transduction

Sensory cells convert one type of stimulus energy (e.g. pressure) into an electrical signal by altering the flow of ions across the plasma membrane and generating receptor potentials. In many cases (as in the Pacinian corpuscle), this leads directly to action potentials which are generated in the voltage-gated region of the sensory cell.

In some receptor cells, the receptor potential leads to neurotransmitter release, which then leads directly or indirectly to action potentials in a post-synaptic cell.

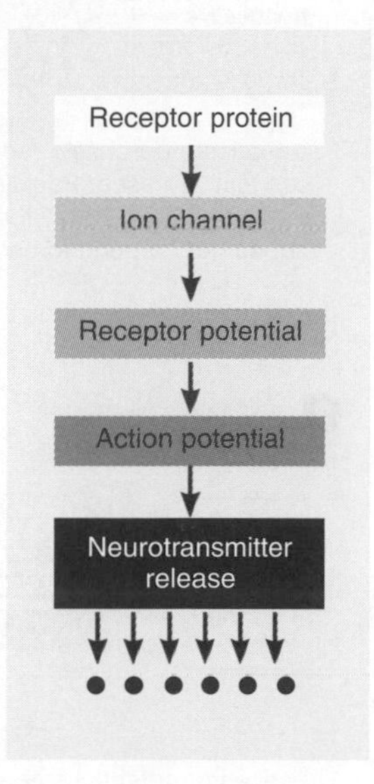

The Pacinian Corpuscle

Pacinian corpuscles are pressure receptors in deep subcutaneous tissues of the body. They are relatively large but structurally simple, consisting of a sensory nerve ending (dendrite) surrounded by a capsule of connective tissue layers. Pressure deforms the capsule, stretching the nerve ending and leading to a localized depolarization called a **receptor potential**. Receptor potentials are graded and do not spread far, although they may sum together and increase in amplitude.

The sense cell converts the receptor potentials to action potentials in the spike generating zone at the start of the axon (where there are voltage-gated channels). The action potential is then propagated along the axon.

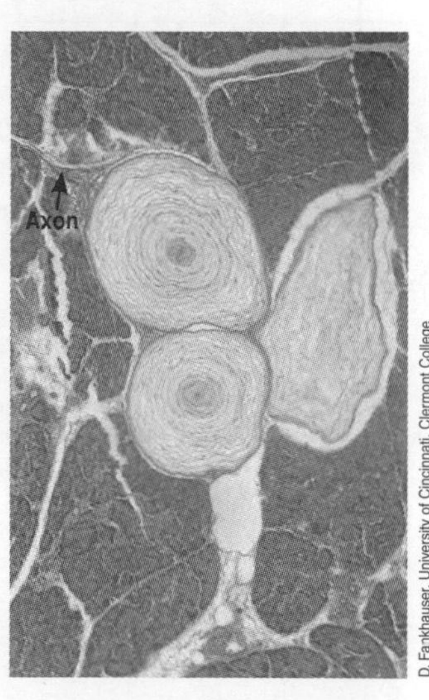

D. Fankhauser, University of Cincinnati, Clermont College

1. Explain why sensory receptors are termed 'biological transducers': ______

2. Identify one feature that all sensory receptors have in common: ______

3. Explain how a stimulus received by a sensory receptor is converted into an electrical response: ______

4. (a) Describe the properties of receptor potentials: ______

(b) Explain how summation of receptor potentials leads to an increased frequency of action potentials in the axon: ______

ISBN: 978-1-92717357-2

Encoding Information

A receptor must do more than simply record a stimulus. It is important that it also provides information about the stimulus strength. Action potentials obey the 'all or none law' and are always the same size, so stimulus strength cannot be encoded by varying the amplitude of the action potentials. Instead, the frequency of impulses conveys information about the stimulus intensity; the higher the frequency of impulses, the stronger the stimulus. This encoding method is termed **frequency modulation**, and it the way that receptors inform the brain about stimulus strength. In the Pacinian corpuscle, described below and earlier in this chapter, frequency modulation is possible because a stronger pressure produces larger receptor potentials, which depolarize the first node of Ranvier to threshold more rapidly and results in a more rapid volley of action potentials. Sensory receptors also show **sensory adaptation** and will cease responding to a stimulus of the same intensity.

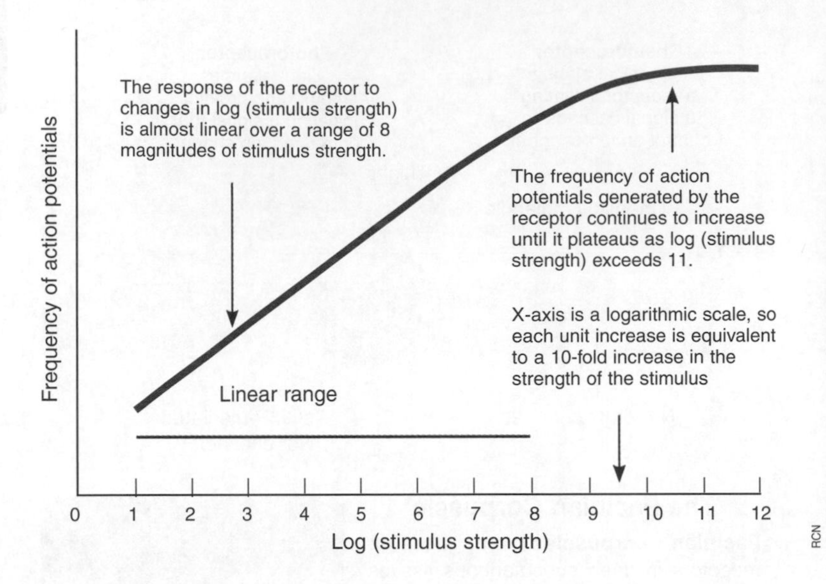

Receptors can use variation in action potential frequency to encode stimulus strengths that vary by nearly 11 orders of magnitude.

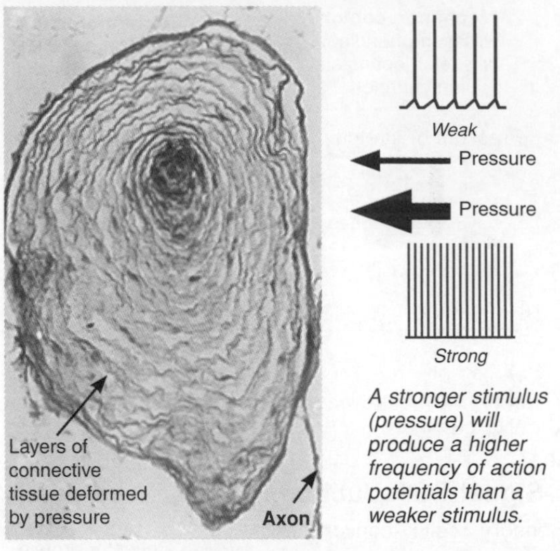

A Pacinian corpuscle (above), illustrating the many layers of connective tissue. Pacinian corpuscles are rapidly adapting receptors; they fire at the beginning and end of a stimulus, but do not respond to unchanging pressure.

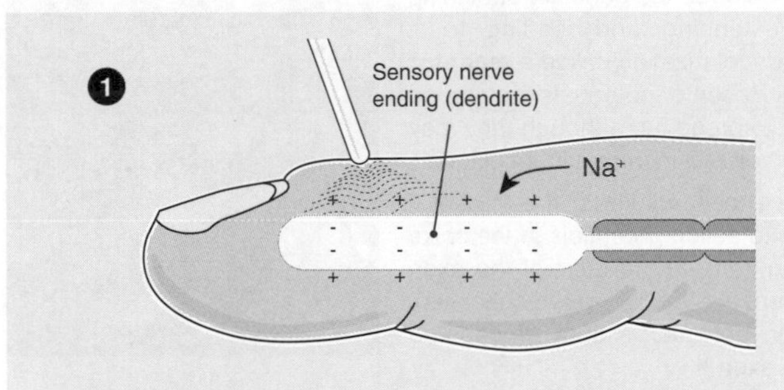

Deforming the corpuscle leads to an increase in the permeability of the nerve to sodium. Na^+ diffuses into the nerve ending creating a localised depolarization. This depolarization is called a **receptor potential**.

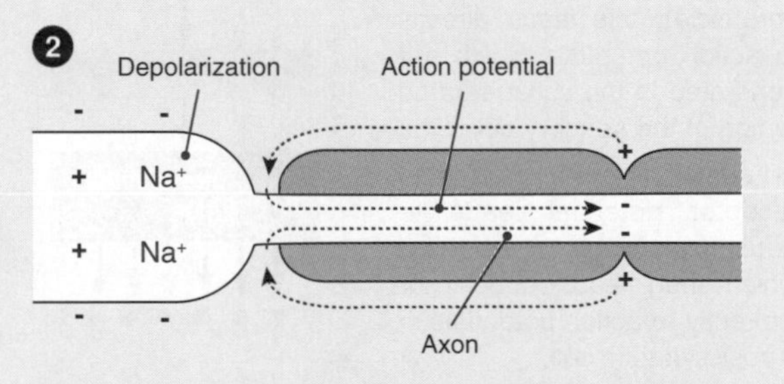

The receptor potential spreads to the first node of Ranvier, which is the first region with voltage-gated Na^+ channels. It depolarizes the node to threshold and generates an action potential, which propagates along the axon.

1. (a) Explain how the strength of a stimulus is encoded by the nervous system: ______

   (b) Explain the significance of encoding information in this way: ______

2. Using the example of the Pacinian corpuscle, explain how stimulus strength is linked to frequency of action potentials: ______

3. Why is sensory adaptation important? ______

Related activities: *The Basis of Sensory Perception*
Weblinks: *Neuron Information Coding and Transfer*

Periodicals: *What is a Pacinian corpuscle?*

ISBN: 978-1-92717357-2

The Structure of the Eye

The eye is a complex and highly sophisticated sense organ specialized to detect light. The adult eyeball is about 25 mm in diameter. Only the anterior one-sixth of its total surface area is exposed; the rest lies recessed and protected by the **orbit** into which it fits. The eyeball is protected and given shape by a fibrous tunic. The posterior part of this structure is the **sclera** (the white of the eye), while the anterior transparent portion is the **cornea**, which covers the colored iris.

The Structure and Function of the Human Eye

The human eye is essentially a three layered structure comprising an outer fibrous layer (the sclera and cornea), a middle vascular layer (the choroid, ciliary body, and iris), and inner **retina** (neurons and **photoreceptor cells**). The shape of the eye is maintained by the fluid filled cavities (aqueous and vitreous humors), which also assist in light refraction. Eye color is provided by the pigmented iris. The iris also regulates the entry of light into the eye through the contraction of circular and radial muscles.

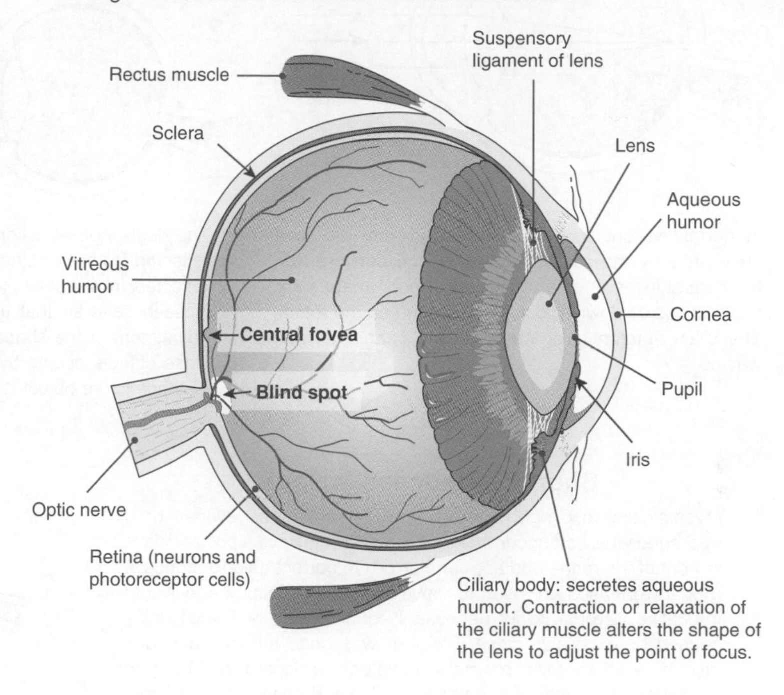

Forming a Visual Image

Before light can reach the photoreceptor cells of the retina, it must pass through the cornea, aqueous humor, pupil, lens, and vitreous humor. For vision to occur, light reaching the photoreceptor cells must form an image on the retina. This requires **refraction** of the incoming light, **accommodation** of the lens, and **constriction** of the pupil.

The anterior of the eye is concerned mainly with **refracting** (bending) the incoming light rays so that they focus on the retina. Most refraction occurs at the cornea. The lens adjusts the degree of refraction to produce a sharp image. **Accommodation** adjusts the eye for near or far objects. Constriction of the pupil narrows the diameter of the hole through which light enters the eye, preventing light rays entering from the periphery.

The point at which the nerve fibers leave the eye as the optic nerve, is the **blind spot** (the point at which there are no photoreceptor cells). Nerve impulses travel along the optic nerves to the visual processing areas in the cerebral cortex. Images on the retina are inverted and reversed by the lens but the brain interprets the information it receives to correct for this image reversal.

1. Identify the function of each of the structures of the eye listed below:

   (a) Cornea: ______________________________

   ____

   (b) Ciliary body: ______________________________

   ____

   (c) Iris: ______________________________

   ____

2. (a) The first stage of vision involves forming an image on the retina. In simple terms, explain what this involves:

   ____

   ____

   (b) Explain how accommodation is achieved: ______________________________

   ____

   ____

   ____

   ____

   ____

ISBN: 978-1-92717357-2

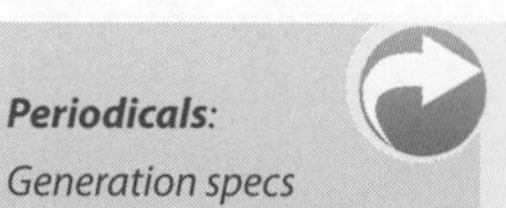

The lens of the eye has two convex surfaces (biconvex). When light enters the eye, the lens bends the incoming rays towards each other so that they intersect at the focal point on the central fovea of the retina. By altering the curvature of the lens, the focusing power of the eye can be adjusted. This adjustment of the eye for near or far vision is called **accommodation** and it is possible because of the elasticity of the lens. For some people, the shape of the eyeball or the lens prevents convergence of the light rays on the central fovea, and images are focused in front of, or behind, the retina. Such visual defects (below) can be corrected with specific lenses. As we age, the lens loses some of its elasticity and, therefore, its ability to accommodate. This inability to focus on nearby objects due to loss of lens elasticity is a natural part of **aging** and is called **far sight**.

Normal vision

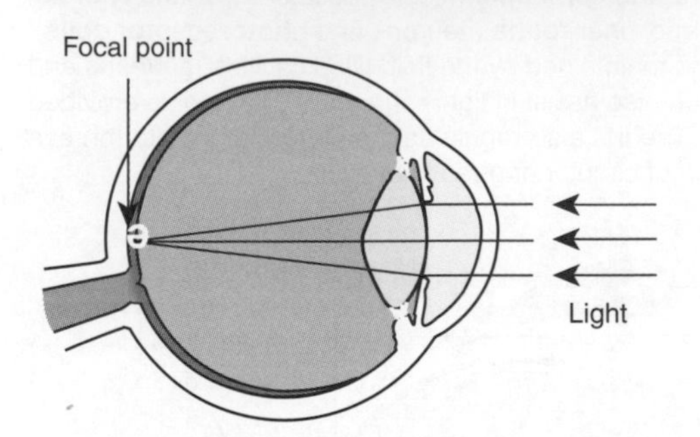

In normal vision, light rays from an object are bent sufficiently by the cornea and lens, and converge on the central fovea. A clear image is formed. Images are focused upside down and mirror reversed on the retina. The brain automatically interprets the image as right way up.

Accommodation for near and distant vision

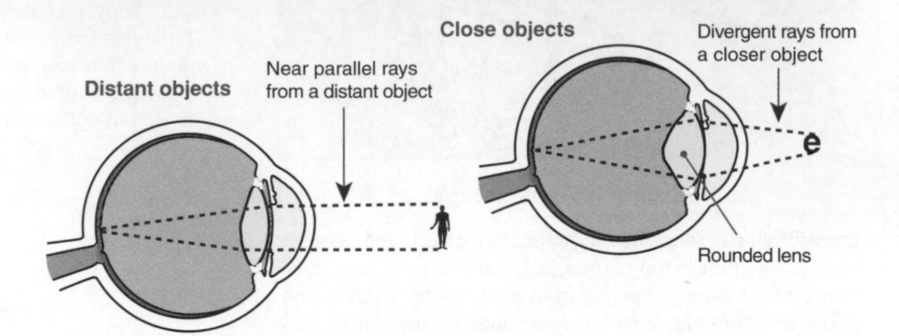

The degree of refraction occurring at each surface of the eye is precise. The light rays reflected from an object 6 m or more away are nearly parallel to one another. Those reflected from near objects are divergent. The light rays must be refracted differently in each case so that they fall exactly on the central fovea. This is achieved through adjustment of the shape of the lens (accommodation). Accommodation from distant to close objects occurs by rounding the lens to shorten its focal length, since the image distance to the object is essentially fixed.

Short sightedness (myopia)

Myopia (top row, right) results from an elongated eyeball or a thickened lens. Left uncorrected, distant objects have a point of focus in front of the retina and appear blurred. To correct myopia, concave (negative) lenses are used to move the point of focus backward to the retina. Myopia is not necessarily genetic, nor is it necessarily caused by excessive close work, as was once thought, although myopia does seem to be more prevalent amongst those living in very confined spaces (e.g. people working and living in submarines).

Long sightedness (hypermetropia)

Long sightedness (bottom row, right) results from a shortened eyeball or from a lens that is too thin. Left uncorrected, light is focused at a point that would be behind the retina and near objects appear blurred. Mild or moderate hypermetropia, which occurs naturally in young children, may be overcome by **accommodation**. In more severe cases, corrective lenses are used to bring the point of focus forward to produce a clear image. This is achieved using a convex (positive) lens.

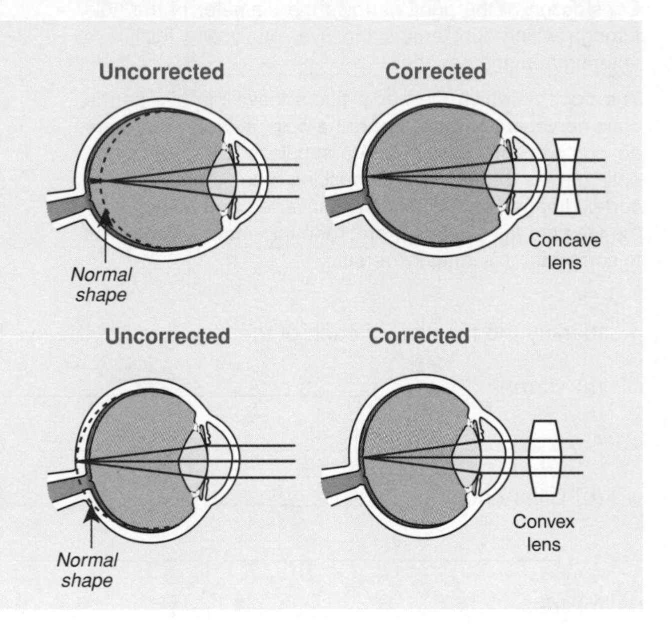

3. (a) Describe the function of the pupil:

   (b) Suggest why control of pupil diameter would be under reflex control:

4. With respect to formation of the image, describe what is happening in:

   (a) Short sighted people:

   (b) Long sighted people:

5. In general terms, describe how the use of lenses corrects the following problems associated with vision:

   (a) Myopia:

   (b) Hypermetropia:

ISBN: 978-1-92717357-2

The Physiology of Vision

Sense organs are collections of sensory receptors with their accessory structures. They are transducers of physical or chemical stimuli. The eye is a sensory organ that converts light into nerve impulses resulting in the formation of a visual image. In the mammalian eye this is achieved by focusing the light through a lens to form an image on the back of the eye (the **retina**). When light reaches the retina, it is absorbed by the photosensitive pigments associated with the membranes of the **photoreceptor cells** (the rods and cones). The pigment molecules are altered by the absorption of light in such a way as to lead to the generation of nerve impulses. The electrical signals are transmitted from the eye via the **optic nerve**, along the visual pathway to the visual cortex of the brain, where the information is interpreted. The retina is not uniform. The area called the central fovea is where there is a high density of cones and virtually no rods. It is the region of highest acuity.

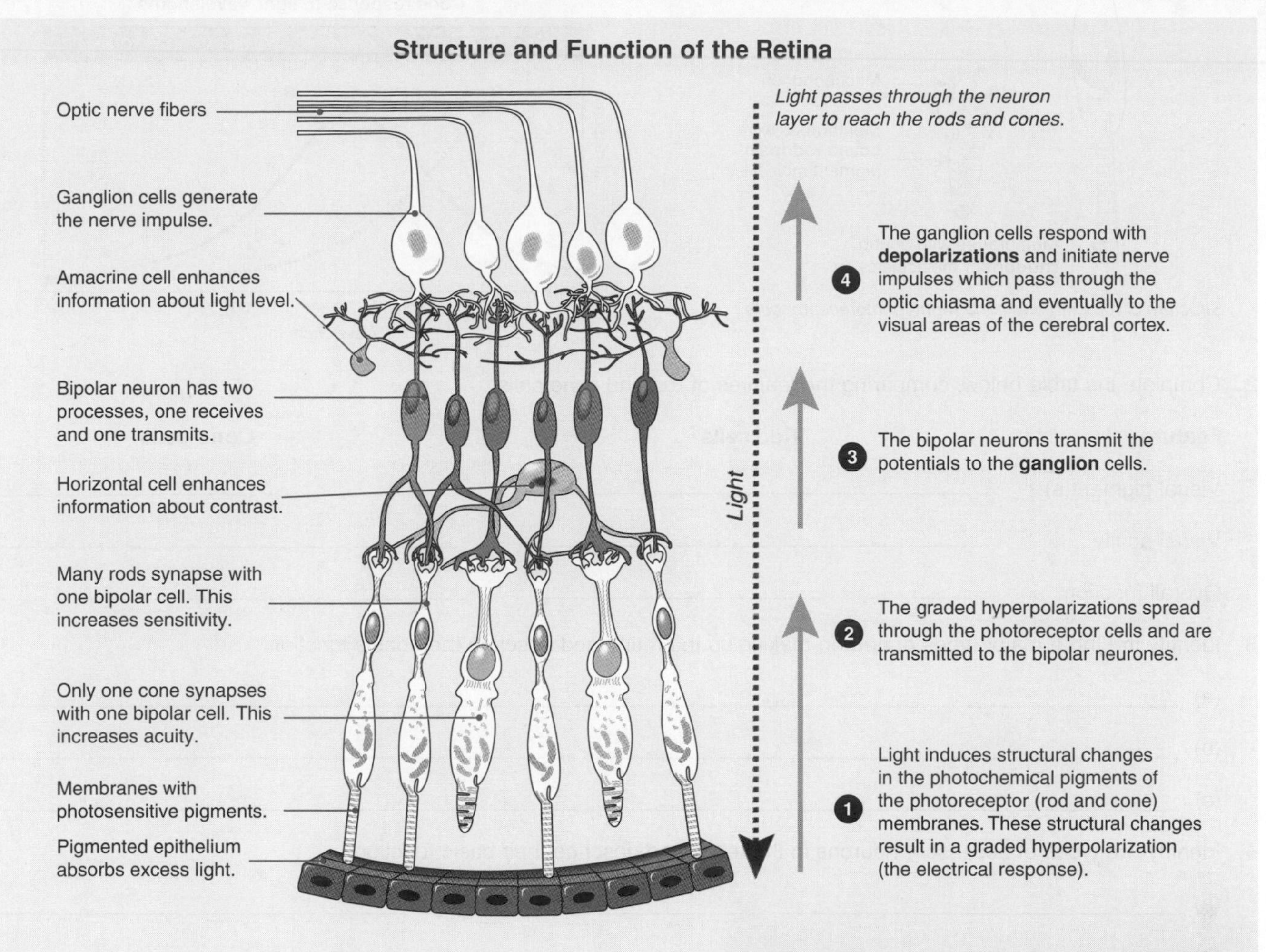

1. Describe the role of each of the following in human vision:

(a) Retina: ______________________________

____

____

____

(b) Optic nerve: ______________________________

____

____

____

(c) Central fovea: ______________________________

____

____

____

Mikael Häggström, Wiki PD

Photograph through the eye of a normal retina. The blind spot, where the where ganglion cell axons exit the eye to form the optic nerve, is seen as the bright area to the left of the image. The central fovea, where cone density is highest, is in the darker region at the centre of the image. Note the rich blood supply.

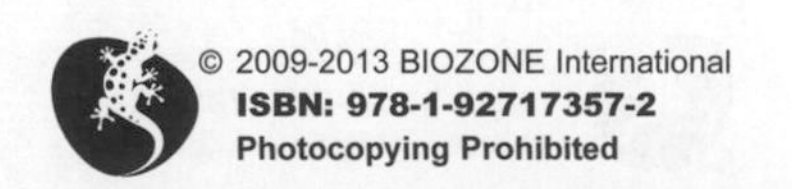

ISBN: 978-1-92717357-2

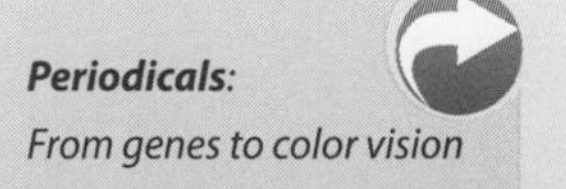
Periodicals:
From genes to color vision

Related activities: *The Structure of the Eye*
Weblinks: *Eye Structure and Function*

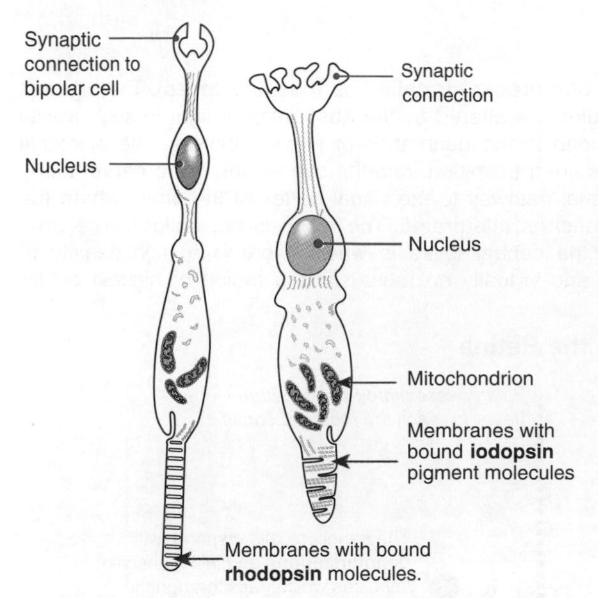

Structure of rod (left) and cone (right) photoreceptor cells

The Basis of Trichromatic Vision.

There are three classes of cones, each with a maximal response in either short (blue), intermediate (green) or long (yellow-green) wavelength light (below). The yellow-green cone is also sensitive to the red part of the spectrum and is often called the red cone. The differential responses of the cones to light of different wavelengths provides the basis of trichromatic color vision.

Cone response to light wavelengths

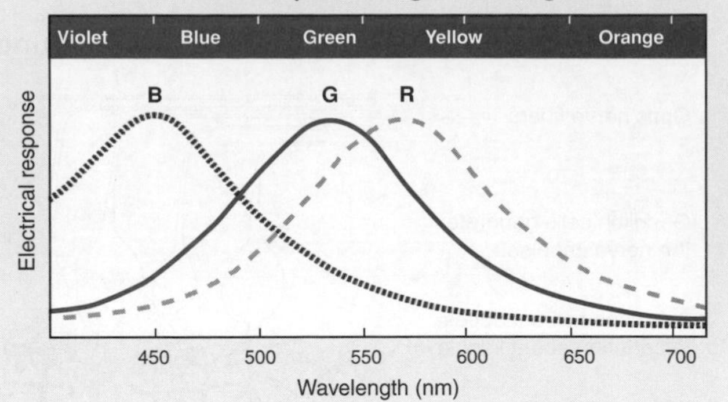

2. Complete the table below, comparing the features of rod and cone cells:

Feature	Rod cells	Cone cells
Visual pigment(s):		
Visual acuity:		
Overall function:		

3. Identify the three major types of neuron making up the retina and describe their basic function:

(a)

(b)

(c)

4. Identify two types of accessory neurons in the retina and describe their basic function:

(a)

(b)

5. Account for the differences in acuity and sensitivity between rod and cone cells:

6. (a) What is meant by the term **photochemical pigment** (photopigment)?

(b) Identify two photopigments and their location:

7. In your own words, explain how light is able to produce a nerve impulse in the ganglion cells:

ISBN: 978-1-92717357-2

Skin Senses

The skin is an important sensory organ, with receptors for pain, pressure, touch, and temperature. While some are specialized receptor structures, many are simple unmyelinated nerves. Tactile (touch) and pressure receptors are **mechanoreceptors** and are stimulated by mechanical distortion. In the **Pacinian corpuscle**, the layers of tissue comprising the sensory structure are pushed together with pressure, stimulating the axon. Views of human skin and its structures are illustrated below. Human skin is fairly uniform in its basic structure, but the density and distribution of glands, hairs, and receptors varies according to the region of the body. **Meissner's corpuscles**, for example, are concentrated in areas sensitive to light touch and, in hairy skin, tactile receptors are clustered into specialized epithelial structures called touch domes or hair disks.

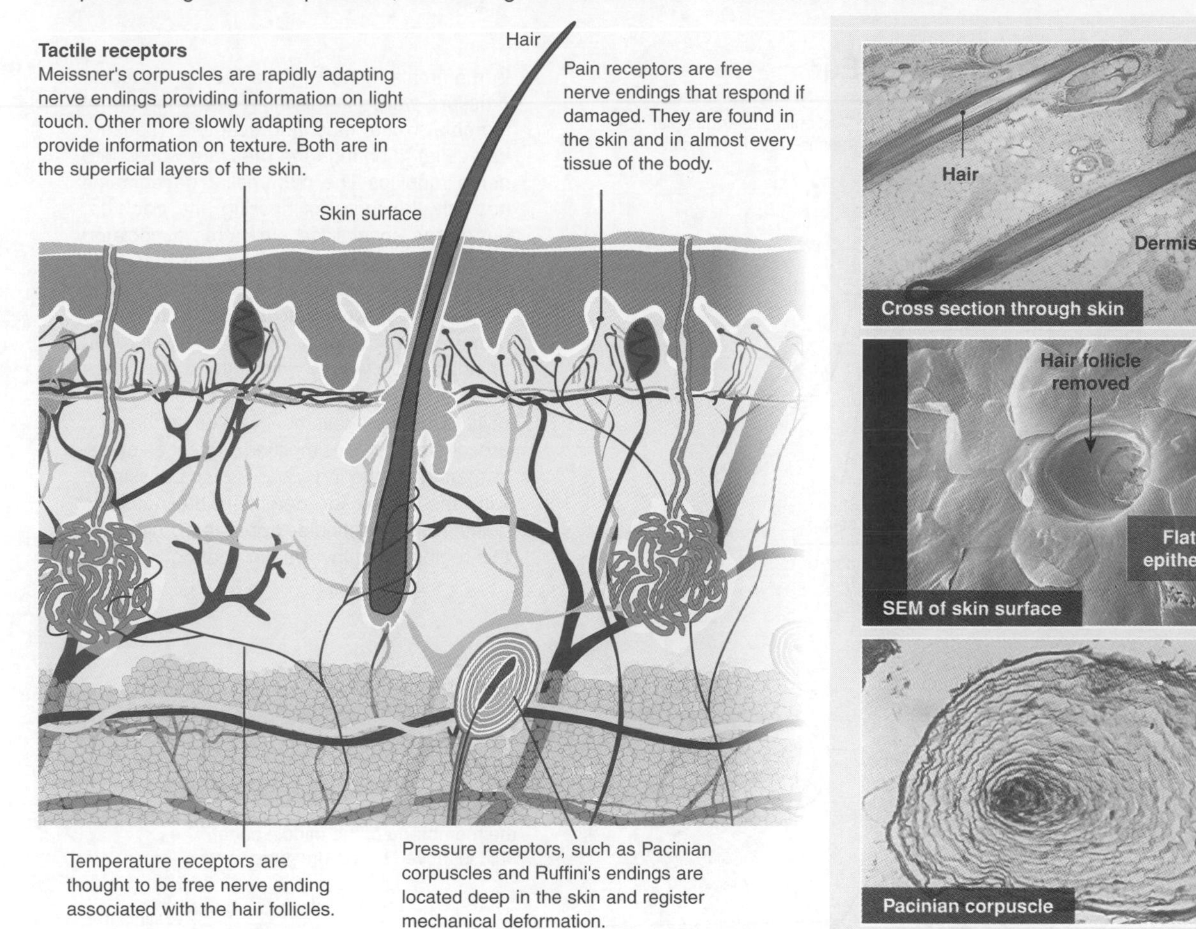

Testing the Distribution of Touch Receptors

The receptors in the skin are more concentrated in some parts of the body than in others. The distribution of receptors can be tested by using the **two point touch test**. This involves finding the smallest distance at which someone can correctly distinguish two point stimuli.

Method

Repeatedly touch your lab partner's skin lightly with either one or two points of fine scissors or tweezers. Your partner's eyes should be closed. At each touch, they should report the sensation as "one" or "two", depending on whether they perceive one or two touches.

Begin with the scissor points far apart and gradually reduce the separation until only about 8 in 10 reports are correct.

This separation distance (in mm) is called the **two point threshold**. When the test subject can feel only one receptor (when there are two) it means that only one receptor is being stimulated. A large two point threshold indicates a low receptor density, a low one indicates a high receptor density.

Repeat this exercise for: the forearm, the back of the hand, the palm of the hand, the fingertip, and the lips, and then complete the table provide below:

Area of skin	Two point threshold (in mm)
Forearm	
Back of hand	
Palm of hand	
Fingertip	
Lips	

1. Name the region with the greatest number of touch receptors:

2. Name the region with the least number of touch receptors:

3. Explain why there is a difference between these two regions:

ISBN: 978-1-92717357-2

Hearing

In humans the receptors for detecting sound waves are organized into hearing organs called **ears**. Sound is produced by the vibration of particles in a medium and it travels in waves that can pass through solids, liquids, or gases. The distance between wave 'crests' determines the frequency (pitch) of the sound. The absolute size (amplitude) of the waves determines the intensity or loudness of the sound. Sound reception in humans is the role of **mechanoreceptors**: tiny **hair cells** in the cochlea of the inner ear. The hair cells are very sensitive and are easily damaged by prolonged exposure to high intensity sounds. Gradual hearing loss with age is often caused by the cumulative loss of sensory hair cell function, especially at the higher frequencies. The ear also houses the vestibular apparatus, which is sensitive to the body's equilibrium.

The Human Ear

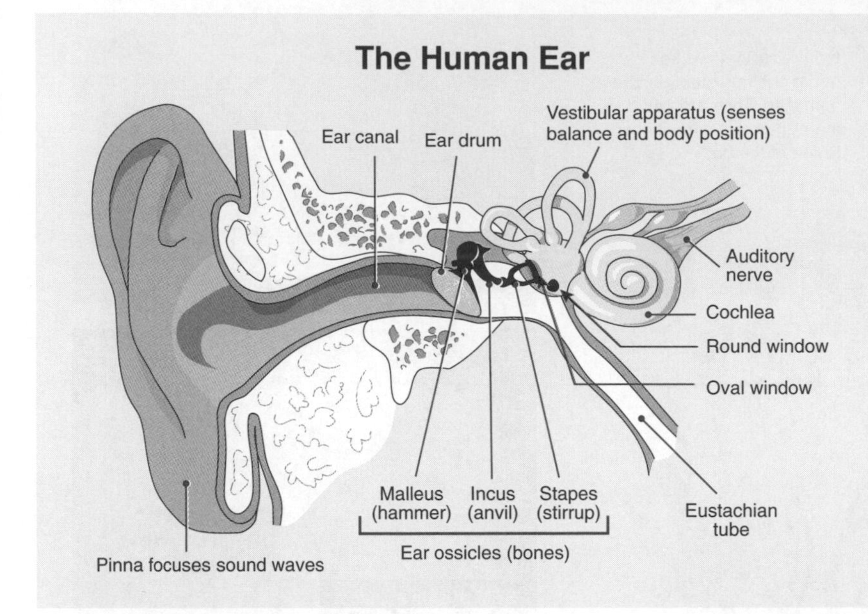

In mammals, sound waves are converted to pressure waves in the inner ear. The ears of mammals use mechanoreceptors (sensory hair cells) to change the pressure waves into nerve impulses. The mammalian ear contains not only the organ of hearing, the **cochlea**, but all the specialized structures associated with gathering, directing, and amplifying the sound. The cochlea is a tapered, coiled tube, divided lengthwise into **three fluid filled canals**. The cochlea is shown below, unrolled to indicate the way in which sound waves are transmitted through the canals to the sensory cells. The mechanisms involved in hearing are outlined in a simplified series of steps. In mammals, the inner ear is also associated with the organ for detecting balance and position (the vestibular apparatus), although this region is not involved in hearing.

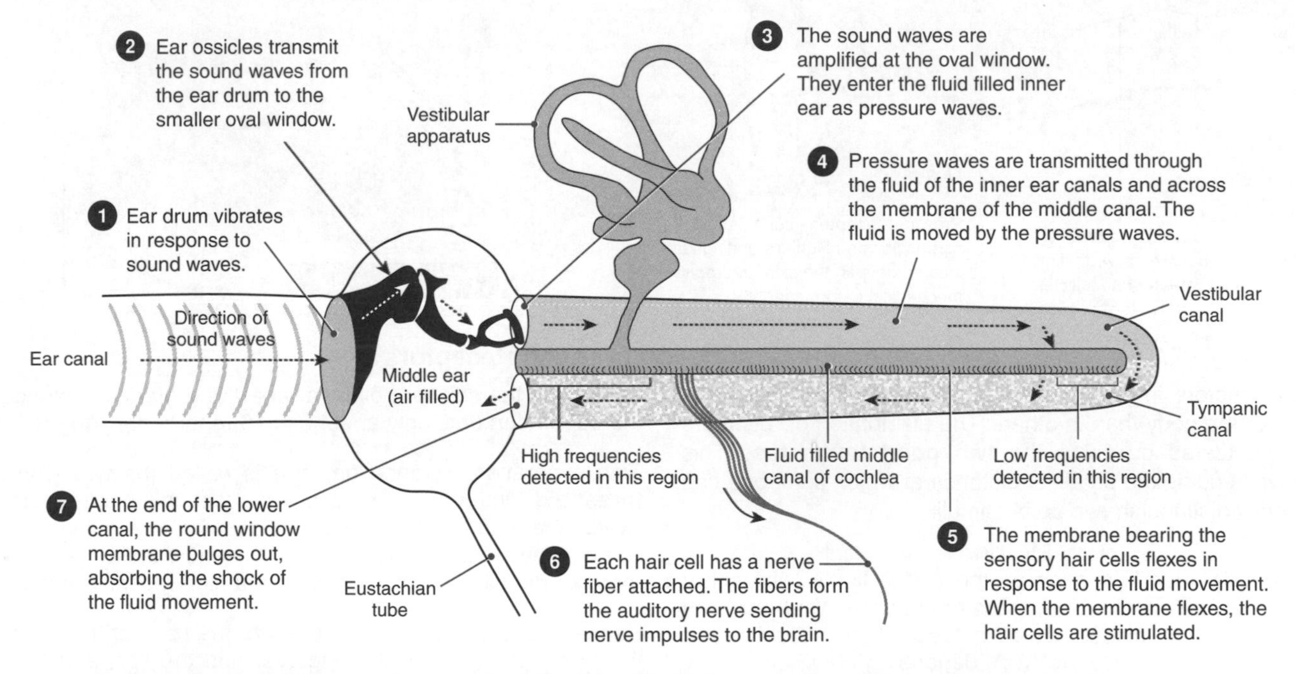

1. In a short sentence, outline the role of each of the following in the reception and response to sound:

   (a) The ear drum: ______________________________

   (b) The ear ossicles: ______________________________

   (c) The oval window: ______________________________

   (d) The sensory hair cells: ______________________________

   (e) The auditory nerve: ______________________________

2. Explain the significance of the inner ear being fluid filled: ______________________________

____

____

Related activities: *The Basis of Sensory Perception*
Weblinks: *Hearing, Sound Waves and the Cochlea*

ISBN: 978-1-92717357-2

Taste and Smell

Chemosensory receptors are responsible for our sense of smell (**olfaction**) and taste (**gustation**). The receptors for smell and taste both respond to chemicals, either carried in the air (smell) or dissolved in a fluid (taste). In humans and other mammals, these are located in the nose and tongue respectively.

Each receptor type is basically similar: they are collections of receptor cells equipped with chemosensory microvilli or cilia. When chemicals stimulate their membranes, the cells respond by producing nerve impulses that are transmitted to the appropriate region of the cerebral cortex for interpretation.

Taste (Gustation)

The organs of taste are the **taste buds** of the tongue. Most of the taste buds on the tongue are located on raised protrusions of the tongue surface called **papillae**. Each bud is flask-like in shape, with a pore opening to the surface of the tongue enabling molecules and ions dissolved in saliva to reach the receptor cells inside. Each taste bud is an assembly of 50-150 taste cells. These connect with nerves that send messages to the gustatory region of the brain. There are five basic taste sensations. **Salty** and **sour** operate through ion channels, while **sweet**, **bitter**, and **umami** (savory) operate through membrane signaling proteins. These taste senations are found on all areas of the tongue although some regions are more sensitive than others.

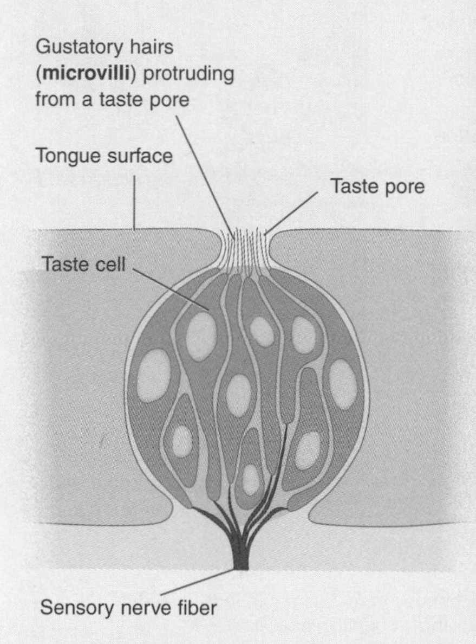

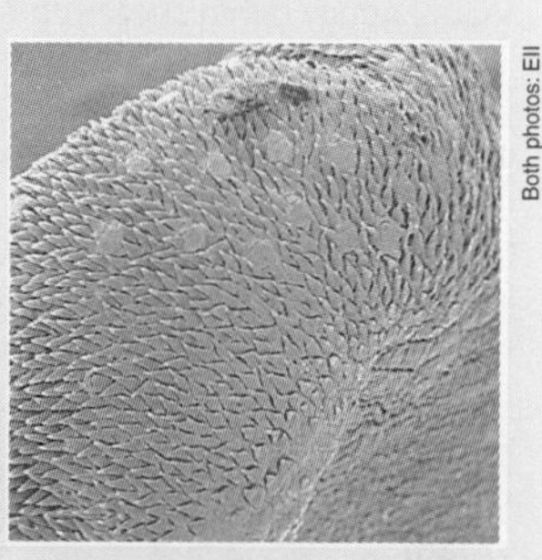

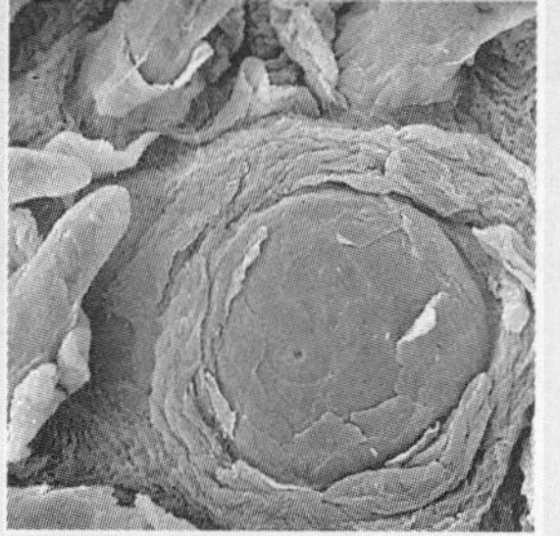

Both photos: Ell

Note that taste also relies heavily on smell because odors from food also stimulate olfactory receptors.

Above: SEMs of the surface of the tongue (top) and close up of one of the papillae (below).

Smell (Olfaction)

In humans, the receptors for smell are located at the top of the nasal cavity. The receptors are specialized hair cells that detect airborne molecules and respond by sending nerve impulses to the olfactory centre of the brain. Unlike taste receptors, olfactory receptors can detect many different odors. However, they quickly adapt to the same smell and will cease to respond to it. This phenomenon is called **sensory adaptation**.

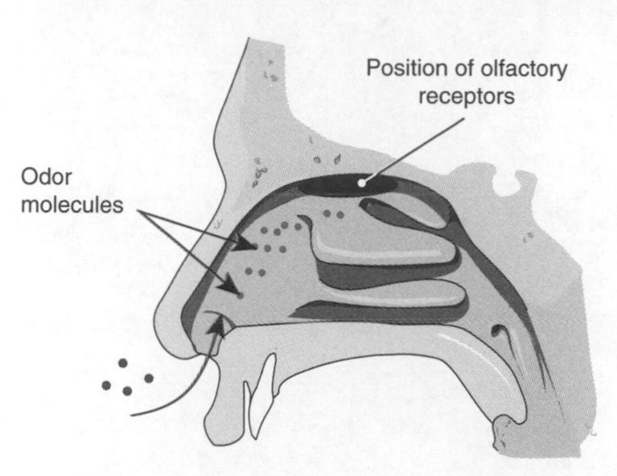

Detail of olfactory membrane

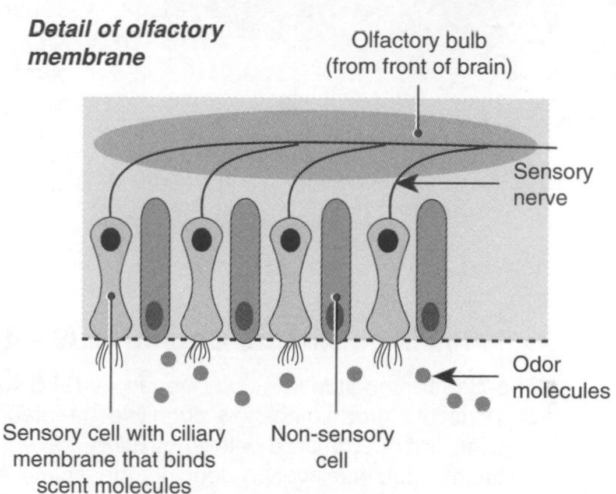

1. Describe the basic mechanism by which chemical sense operates: ______

2. Take a deep breath of a non-toxic, pungent substance such as perfume. Take a sniff of the substance at 10 second intervals for about 1 minute. Make a record of how strongly you perceived the smell to be at each time interval. Use a scale of 1 to 6: **1.** *Very strong;* **2.** *Quite strong;* **3.** *Noticeable;* **4.** *Weak;* **5.** *Very faint;* **6.** *Could not detect.*

Time	Strength	Time	Strength	Time	Strength
10 s		30 s		50 s	
20 s		40 s		1 min	

3. (a) Explain what happened to your sense of smell over the time period: ______

   (b) State the term that describes this phenomenon: ______

   (c) Describe the adaptive advantage of this phenomenon: ______

ISBN: 978-1-92717357-2

Related activities: *The Basis of Sensory Perception*
Weblinks: *Sense of Taste, Sense of Smell, Olfactory Receptor Stimulation*

Aging and the Nervous System

The aging process affects all body systems, including the nervous system. Neuron loss begins around age 30, and accumulates over time, which is why the changes are often more obvious in the elderly. Common changes include impaired (diminished) hearing and vision, short term memory loss, slower reaction times, and loss of fine motor skills. Performing mental and physical exercise, slows down the loss of neurons in the areas of the brain associated with memory, and helps the remaining neurons to function properly. Lack of mental and physical stimulation, a poor diet, and the consumption of two or more alcoholic drinks a day can increase the rate of neuron loss in the brain.

The Effects of Aging on the Nervous System

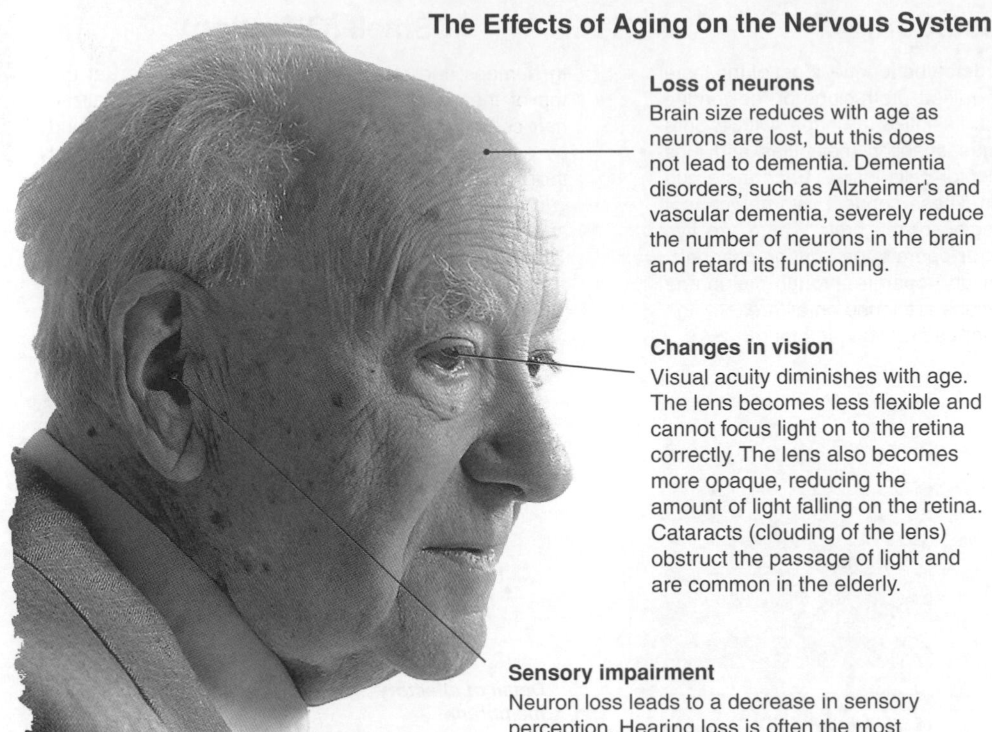

Loss of neurons
Brain size reduces with age as neurons are lost, but this does not lead to dementia. Dementia disorders, such as Alzheimer's and vascular dementia, severely reduce the number of neurons in the brain and retard its functioning.

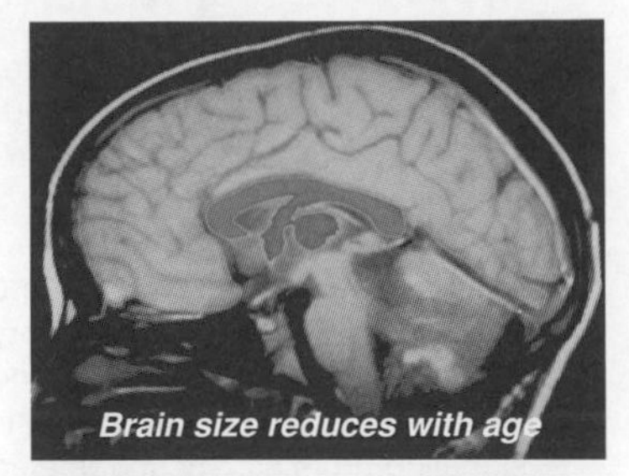

Brain size reduces with age

Changes in vision
Visual acuity diminishes with age. The lens becomes less flexible and cannot focus light on to the retina correctly. The lens also becomes more opaque, reducing the amount of light falling on the retina. Cataracts (clouding of the lens) obstruct the passage of light and are common in the elderly.

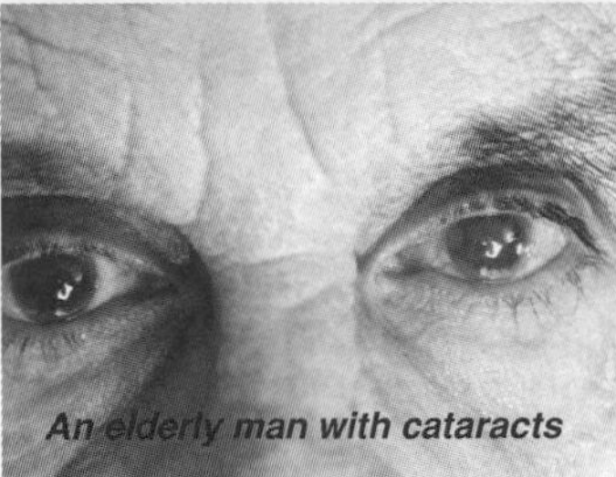

An elderly man with cataracts

Sensory impairment
Neuron loss leads to a decrease in sensory perception. Hearing loss is often the most obvious sensory impairment in elderly people and usually begins with inability to hear high pitched sounds. Hearing aids are often worn to correct the problem.

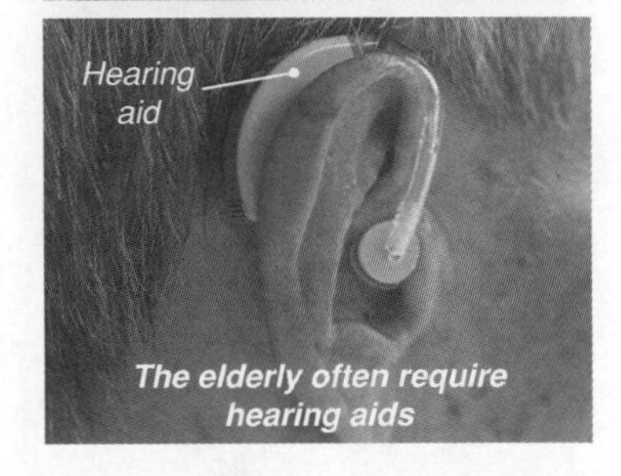

The elderly often require hearing aids

How Age Affects Cognitive Ability

The Seattle longitudinal study began in 1956 with the purpose of determining how cognitive (mental) ability and intelligence change with age. Every seven years, additional subjects were added to the study, and all participants undertook a series of cognitive tests and psychological questioning. Approximately 6,000 people have been tested.

The graph (right) summarizes some of the results to date. Some cognitive abilities (perceptual speed and numeric ability), begin to decrease from early maturity, while others, such as verbal memory, do not begin to deteriorate until much later in life (60 years old). The study also showed that training (use of specific mental techniques) could slow the decline in cognitive ability.

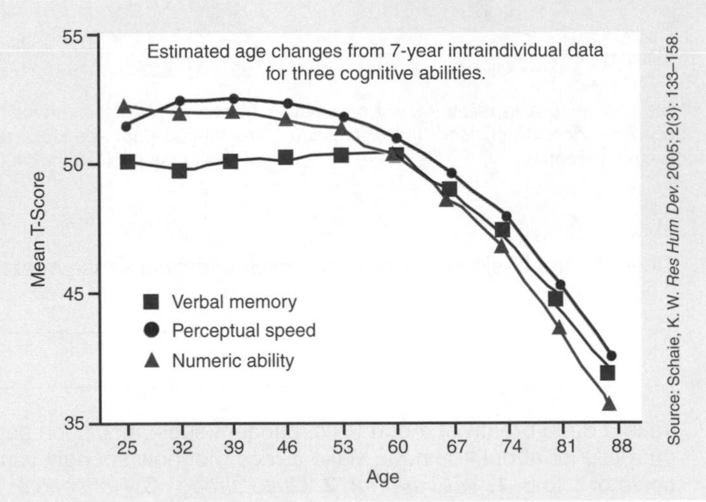

1. (a) Why do many cognitive abilities diminish with age? ____________________

(b) What steps can be taken to reduce the rate of cognitive decline? ____________________

ISBN: 978-1-92717357-2

Alzheimer's and the Brain

Alzheimer's disease is a disabling neurological disorder affecting about 5% of the population over 65. Although its causes are largely unknown, people with a family history of Alzheimer's have a greater risk, implying that a genetic factor is involved. Some of the cases of Alzheimer's with a familial (inherited) pattern involve a mutation of the gene for amyloid precursor protein (APP), found on chromosome 21 and nearly all people with Down syndrome (trisomy 21) who live into their 40s develop the disease. The gene for the protein apoE, which has an important role in lipid transport, degeneration and regulation in nervous tissue, is also a risk factor that may be involved in modifying the age of onset. Sufferers of Alzheimer's have trouble remembering recent events and they become confused and forgetful. In the later stages of the disease, people with Alzheimer's become very disorientated, lose past memories, and may become paranoid and moody. Dementia and loss of reason occur at the end stages of the disease. The effects of the disease are irreversible and it has no cure.

The Malfunctioning Brain: The Effects of Alzheimer's Disease

Alzheimer's is associated with accelerated loss of neurons, particularly in regions of the brain that are important for memory and intellectual processing, such as the cerebral cortex and hippocampus. The disease has been linked to abnormal accumulations of protein-rich **amyloid** plaques and tangles, which invade the brain tissue and interfere with synaptic transmission.

Cerebral cortex: Conscious thought, reasoning, and language. Alzheimer's sufferers show considerable loss of function from this region.

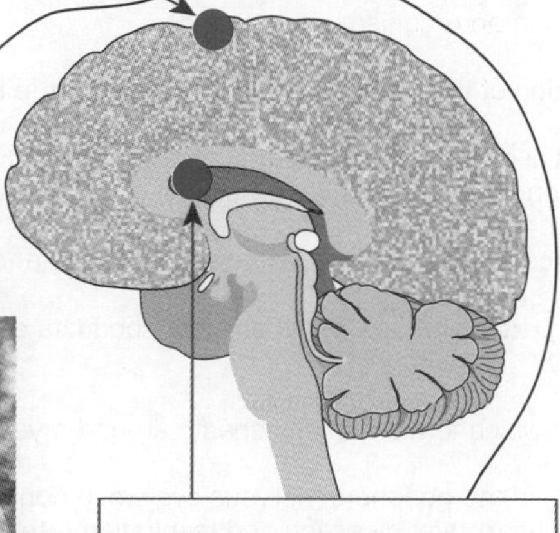

Hippocampus: A swelling in the floor of the lateral ventricle. It contains complex foldings of the cortical tissue and is involved in the establishment of memory patterns. In Alzheimer's sufferers, it is one of the first regions to show loss of neurons and accumulation of amyloid.

It is not uncommon for Alzheimer's sufferers to wander and become lost and disorientated.

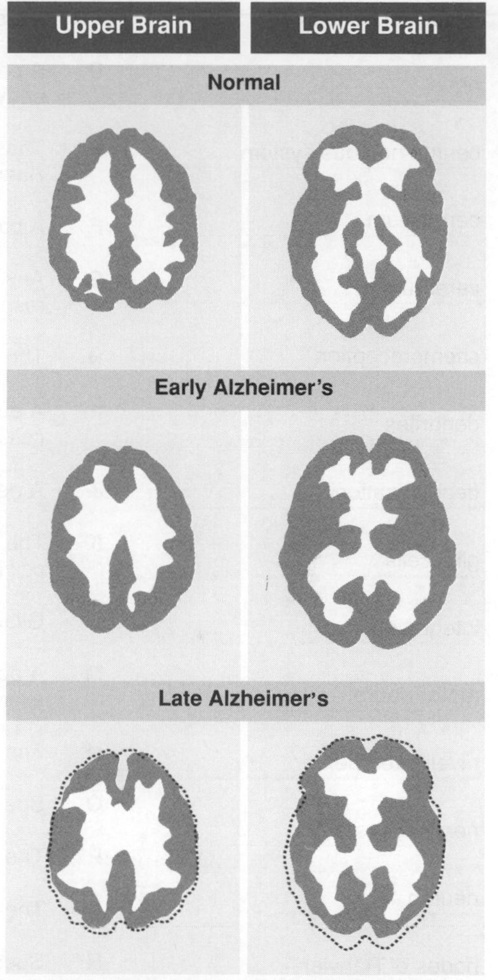

The brain scans above show diminishing brain function in certain areas of the brain in Alzheimer's sufferers. Note, particularly in the two lower scans, how much the brain has shrunk (original size indicated by the dotted line). Light areas indicate brain activity.

1. Describe the biological basis behind the degenerative changes associated with Alzheimer's disease:

2. Describe the evidence for the Alzheimer's disease having a genetic component in some cases:

3. Some loss of neuronal function occurs normally as a result of aging. Identify the features distinguishing Alzheimer's disease from normal age related loss of neuronal function:

ISBN: 978-1-92717357-2

KEY TERMS: Mix and Match

INSTRUCTIONS: Test your vocabulary by matching each term to its definition, as identified by its preceding letter code.

action potential ....

autonomic nervous system ....

axon ....

biological transducer ....

brain ....

central nervous system ....

cerebellum ....

cerebrum ....

chemoreception ....

dendrites ....

depolarization ....

glial cells ....

integration ....

motor neuron ....

myelinated nerve ....

nervous system ....

neuron ....

nodes of Ranvier ....

peripheral nervous system ....

saltatory conduction ....

Schwann cells ....

sense organ ....

sensory receptor ....

somatic nervous system ....

spinal cord ....

stimulus ....

synapse ....

A The highly organised system of neurons that generates and conveys signals in the form of electrical impulses.

B A sensory receptor that is specialized to detect stimuli and respond by producing an electrical discharge.

C A cylinder of nervous tissue extending from the base of the brain down the back. It transmits messages to and from the brain, and controls spinal reflexes.

D A part of the peripheral nervous system associated with the voluntary control of body movements via skeletal muscles.

E The propagation of action potentials along myelinated axons from one node of Ranvier to the next.

F A part of the hindbrain that coordinates body movements, posture, and balance.

G Any physical or chemical change in the environment capable of provoking a response in an organism.

H The portion of the nervous system comprising the brain and spinal cord.

I A potential difference produced across the plasma membrane of nerve or muscle cells when they are stimulated.

J A nerve cell, made up of a cell body, dendrites, and an axon.

K The long extension of a nerve cell that conducts electrical impulses away from the cell body.

L Glial cell which forms the fatty sheath around myelinated nerve fibers.

M A portion of the peripheral nervous system. It controls visceral functions, for example heart rate, digestion and respiration rate.

N The junction between two neurons or between a neuron and an effector.

O Specialized cells that support and protect neurons.

P The gaps occurring at intervals along the myelin sheath.

Q The reduction or neutralization in electrical potential difference across a membrane.

R Specialized nerve cells that detect stimuli and respond by producing an electrical discharge.

S An organ containing sensory neurons that is able to respond to external stimuli by sending nerve impulses to the brain.

T The control centre for the body, consisting of billions of interconnected neurons.

U The physiological reception of chemical stimuli.

V Neuron with axons sheathed within a lipid and protein rich substance called myelin.

W The part of the nervous system that comprises all the nerves and sensory receptors outside of the central nervous system.

X A nerve cell that carries impulses away from the central nervous system to an effector muscle. May be non-myelinated or myelinated.

Y The largest region of the brain. It controls and integrates motor, sensory, and higher mental functions (e.g reason and emotion).

Z The reception, interpretation, and coordination of inputs from many sources by a neuron.

AA Thin processes from the neuron cell body that receive stimuli.

ISBN: 978-1-92717357-2

The Endocrine System

Key terms

blood glucose
cell signaling
cyclic AMP
diabetes mellitus
endocrine gland
exocrine gland
glucagon
hormone
hypothalamus
insulin
menopause
negative feedback
neurosecretory cell
pancreas
pituitary gland
positive feedback
second messenger
signal molecule
signal transduction pathway
target cell

Key concepts

- Endocrine glands produce hormones that are carried in the blood to affect target cells and tissues.
- Endocrine regulation is slower and longer lasting that nervous regulation.
- Endocrine secretions are regulated primarily by negative feedback mechanisms.
- Disruption to normal endocrine regulation can result in disease.

Learning Objectives

☐ 1. Use the **KEY TERMS** to compile a glossary for this topic.

The Basis of Endocrine Control pages 35-36, 116-118

☐ 2. Outline the general roles of the two regulatory systems (hormonal and nervous) with which humans achieve homeostasis. Appreciate that they are interdependent.

☐ 3. Define the terms **endocrine gland**, **hormone**, and **target cell** (tissue or organ). Describe the general organization of the endocrine system and outline the general role of hormones in the maintenance of homeostasis.

☐ 4. Explain how hormones bring about their effects (also see #5). Explain how endocrine activity is regulated primarily through **negative feedback**. Recognize the few situations where **positive feedback** plays a role in endocrine regulation.

☐ 5. Describe types of **cell signaling** and identify the involvement of **signal molecules** in a **signal transduction pathway**. Contrast the action of steroid hormones (direct activation of genes) and peptide hormones (action by **second messenger**).

The Endocrine System pages 119-129

☐ 6. Identify the major endocrine glands and tissues of the body. Name the hormones produced and their functions, and explain how their release is regulated.

☐ 7. Describe the structural and functional relationship between the **hypothalamus** and the **pituitary gland**. Discuss the central role of the hypothalamus in regulating the secretions of other endocrine glands, as illustrated by the role of the hypothalamus in the stress response.

☐ 8. Distinguish between the anterior and posterior pituitary and identify the position and role of the portal vein and **neurosecretory cells**. Describe the function and regulation of the pituitary hormones, as follows:
(a) Anterior pituitary (adenohypophysis): GH, TSH, ACTH, MSH, LH, FSH, prolactin
(b) Posterior pituitary (neurohypophysis): ADH and oxytocin

Case study: Blood glucose regulation

☐ 9. Describe the factors that lead to variation in **blood glucose** levels and understand the normal range over which blood glucose levels fluctuate. Describe the general structure of the pancreas and outline its role as both an **exocrine** and an endocrine gland. Explain how blood glucose level is regulated, including reference to:
(a) Negative feedback mechanisms.
(b) The hormones **insulin** and **glucagon**.
(c) The role of the liver in glucose-glycogen conversions.
(d) Causes and effects of type 1 and type 2 **diabetes mellitus**.
(e) The effect of drugs, e.g. nicotine and alcohol, on blood glucose.

The Endocrine System page 130

☐ 10. Describe degenerative changes occurring in endocrine function, with particular reference to **menopause**. Describe the physiological changes associated with menopause and relate these to increased risk of some diseases, e.g. osteoporosis and heart disease.

Periodicals:
Listings for this chapter are on page 279

Weblinks:
www.thebiozone.com/weblink/AnaPhy-3572.html

BIOZONE APP:
Student Review Series
The Endocrine System

Cell Signaling

Cells use **signals** (chemical messengers) to gather information about, and respond to, changes in their cellular environment and for communication between cells. The signaling and response process is called the **signal transduction pathway**. It often involves a number of enzymes and molecules in a **signal cascade** which causes a large response in the target cell. Cell signaling pathways are categorized primarily on the distance over which the signal molecule travels to reach its target cell. They generally fall into three categories. The **endocrine** pathway involves the transport of hormones over large distances through the circulatory system. During **paracrine** signaling, the signal travels an intermediate distance to act upon neighboring cells. **Autocrine** signaling involves a cell producing and reacting to its own signal. These three pathways are illustrated below.

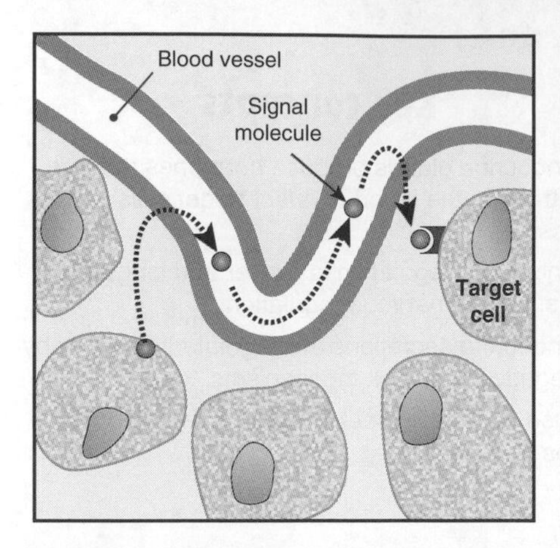

Endocrine signaling: Hormone signals are released by ductless endocrine glands and carried through the body by the circulatory system to target cells. Examples include sex hormones, growth factors, and neurohormones such as dopamine.

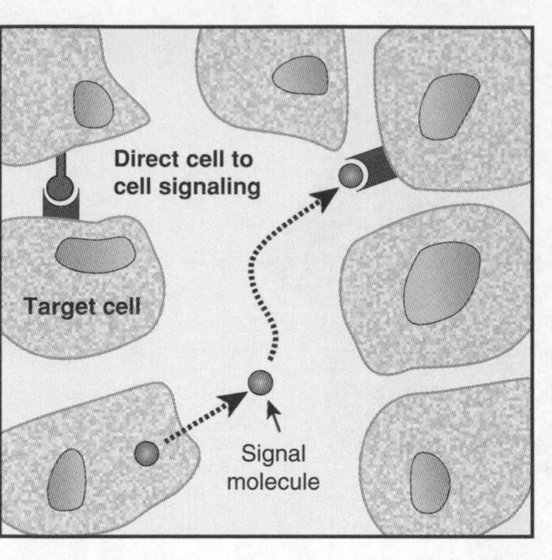

Paracrine signaling: Signals released from a cell act on target cells in the immediate vicinity. The chemical messenger can be transferred through the extracellular fluid (e.g. at synapses) or directly between cells, which is important during embryonic development.

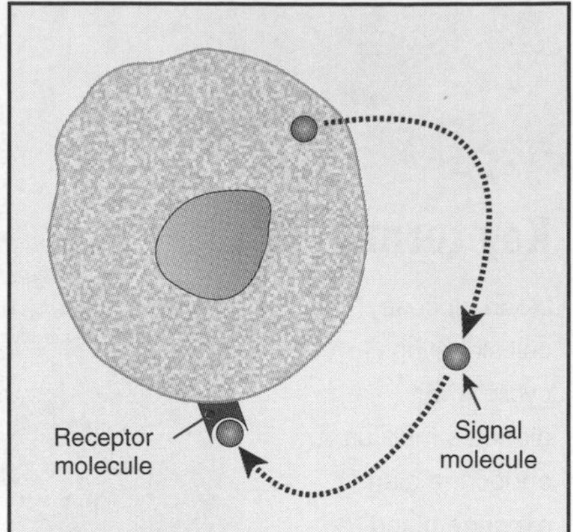

Autocrine signaling: Cells produce and react to their own signals. For example, when a foreign protein enters the body, some T-cells (lymphocytes) produce a growth factor to stimulate their own production. The increased number of T-cells helps to fight the infection.

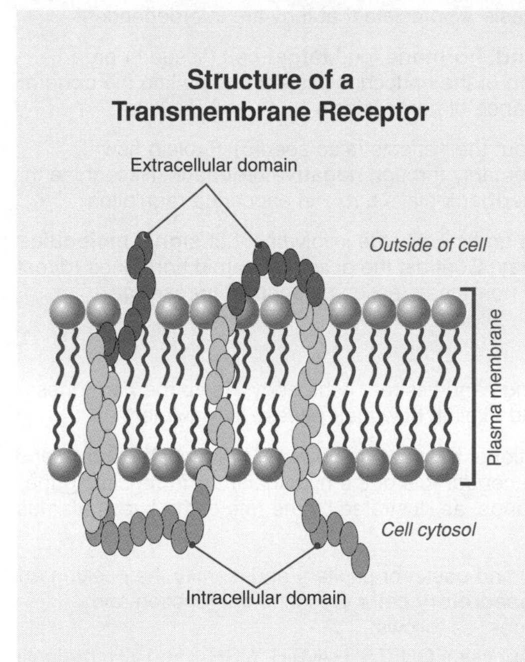

Insulin like growth factor 1 (protein)

Epinephrine (peptide)

Progesterone (steroid)

Examples of cell signaling molecules

The binding sites of cell receptors are very specific: they only bind certain **ligands** (signal molecules). This stops them from reacting to every signal bombarding the cell. Receptors fall into two main categories:

▶ **Cytoplasmic receptors**: These are located within the cell cytoplasm, and bind ligands which are able to cross the plasma membrane unaided.

▶ **Transmembrane receptors**: These span the cell membrane and bind ligands which cannot cross the plasma membrane on their own. They have an extracellular domain outside the cell, and an intracellular domain within the cell cytosol (left).

G-protein linked receptors, ion channels, and protein kinases are examples of transmembrane receptors.

1. Briefly describe the three types of cell signaling:

   (a) ______________________________

   (b) ______________________________

   (c) ______________________________

2. Identify the components that all three cell signaling types have in common: ______________________________

____

____

____

ISBN: 978-1-92717357-2

Related activities: *Hormonal Regulation, Signal Transduction*

Weblinks: *Hormones, Receptors, and Target Cells*

Signal Transduction

Once a hormone signal is released, it is carried in the blood to target cells that respond specifically to that hormone. For example, **epinephrine** binds to a family of membrane receptors called **G protein-coupled receptors**. Once the epinephrine has bound, the G protein becomes activated and regulates several activities, many involved with the **flight or fight response**. There are two main groups of receptors for epinephrine: α and β receptors. In both instances the epinephrine acts as a **first messenger**. It does not cross the plasma membrane, but binds and activates a transmembrane receptor to produce a **second messenger**. In the case of α receptors, the activated G protein reacts directly with an enzyme to cause a cellular response. When β receptors are activated, the G protein interacts with an intermediate enzyme, **adenylate cyclase** which produces cyclic AMP **(cAMP).** This in turn interacts with and regulates other proteins and enzymes to initiate a cellular response. Once the target cells respond, the response is recognized by the hormone-producing cell through a feedback signal and the hormone is degraded.

Cell Signaling with Epinephrine

Binding of the first messenger, epinephrine, to α receptors (below) activates the G protein component of the G protein-coupled receptor. The G protein travels into the cell cytoplasm where it binds specific enzymes and alters cellular function.

Beta (β) receptors (below) use cyclic AMP as a second messenger to link the hormone to the cellular response. Cellular concentration of cAMP increases markedly once a hormone binds and the cascade of enzyme-driven reactions is initiated.

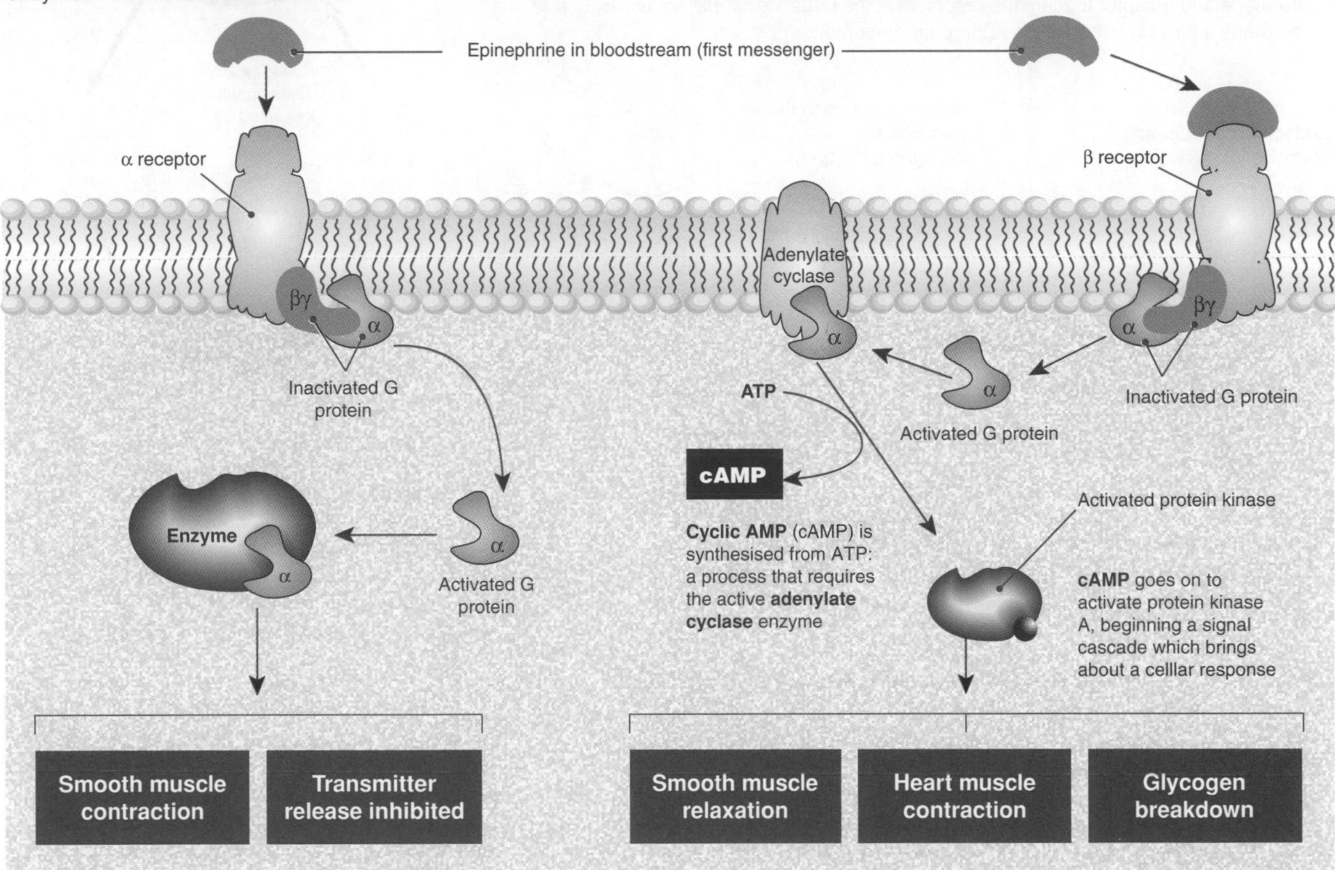

1. Explain how differences in the membrane receptor are related to how epinephrine brings about its intracellular effects:

2. Explain how a very small amount of hormone is able to exert a disproportionately large effect on a target cell:

3. Explain how the binding of a hormone to a target cell can be likened to an enzyme-substrate reaction:

ISBN: 978-1-92717357-2

Periodicals: *The ups and downs of hormones*

Related activities: *Hormonal Regulation, Cell Signaling*
Weblinks: *Signal Transduction, Peptide Hormone Action*

Hormonal Regulation

The endocrine system regulates the body's processes by releasing chemical messengers called **hormones** into the bloodstream. Hormones are potent chemical regulators: they are produced in minute quantities yet can have a large effect on metabolism. The endocrine system is made up of endocrine cells glands and the hormones they produce. Unlike exocrine glands (e.g. sweat and salivary glands), endocrine glands are **ductless glands**, secreting hormones directly into the bloodstream rather than through a duct or tube. Some organs (e.g. the pancreas) have both endocrine and exocrine regions, but these are structurally and functionally distinct. The stimulus for hormone release may be **hormonal** (e.g. action of hypothalamic releasing hormones), a **humoral** trigger (such as high blood glucose), or **neural** (e.g. sympathetic nervous system stimulation).

The Mechanism of Hormone Action

Endocrine cells produce hormones and secrete them into the bloodstream where they are distributed throughout the body. Although hormones are broadcast throughout the body, they affect only specific target cells. These target cells have receptors on the plasma membrane which recognize and bind the hormone (see inset, below right). The binding of hormone and receptor triggers the response in the target cell. Cells are unresponsive to a hormone if they do not have the appropriate receptors.

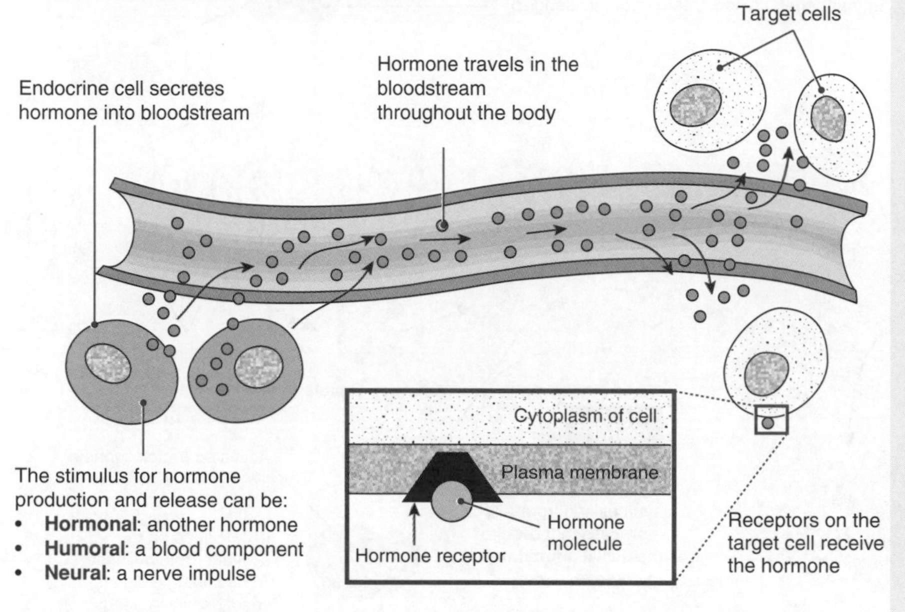

Antagonistic Hormones

Insulin secretion

Blood glucose rises: insulin is released

Raises blood glucose level

Lowers blood glucose level

Blood glucose falls: glucagon is released

Glucagon secretion

The effects of one hormone are often counteracted by an opposing hormone. Feedback mechanisms adjust the balance of the two hormones to maintain a physiological function. Example: insulin acts to decrease blood glucose and glucagon acts to raise it.

1. (a) What are **antagonistic hormones**? Describe an example of how two such hormones operate:

____

____

Example: ______________________________

____

(b) Explain the role of feedback mechanisms in adjusting hormone levels (explain using an example if this is helpful):

____

____

2. Explain how a hormone can bring about a response in target cells even though all cells may receive the hormone:

____

____

3. Explain why hormonal control differs from nervous system control with respect to the following:

(a) The speed of hormonal responses is slower: ______________________________

____

(b) Hormonal responses are generally longer lasting: ______________________________

____

Related activities: *Nervous Regulatory Systems, Control of Blood Glucose*
Weblinks: *Control of Endocrine Activity*

ISBN: 978-1-92717357-2

The Endocrine System

The endocrine glands are distributed throughout the body, frequently associated with the organs of other body systems. Under appropriate stimulation they secrete **hormones**, which are carried in the blood to exert a specific metabolic effect on target cells. They are then broken down and excreted. Hormones may be amino acids, peptides, proteins (often modified), fatty acids, or steroids. Some basic features of the endocrine system are shown below. The hypothalamus, although part of the brain and not strictly an endocrine gland, contains neurosecretory cells, and links the nervous and endocrine systems.

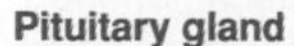

Hypothalamus (neural)
Coordinates nervous and endocrine systems. Secretes releasing hormones, which regulate the hormones of the anterior pituitary. Produces oxytocin and ADH, which are released from the posterior pituitary.

Pineal
This small gland in the brain secretes melatonin, which regulates the sleep-wake cycle. Melatonin secretion follows a circadian rhythm and coordinates reproductive hormones too.

Thyroid gland
Secretes thyroxine, an iodine containing hormone which stimulates metabolism and growth via protein synthesis.

Pancreatic islets
Specialised α and β endocrine cells in the pancreas produce glucagon and insulin. Together, these control blood sugar levels.

Ovaries (in females)
Produce estrogen and progesterone. These hormones control and maintain female characteristics, stimulate the menstrual cycle, maintain pregnancy, and prepare the mammary glands for lactation.

Pituitary gland
The pituitary is located below the hypothalamus. It secretes at least nine hormones that regulate the activities of other endocrine glands.

Parathyroid glands
On the surface of the thyroid, they secrete PTH (parathyroid hormone), which regulates blood calcium levels and promotes the release of calcium from bone. High levels of calcium in the blood inhibit PTH secretion.

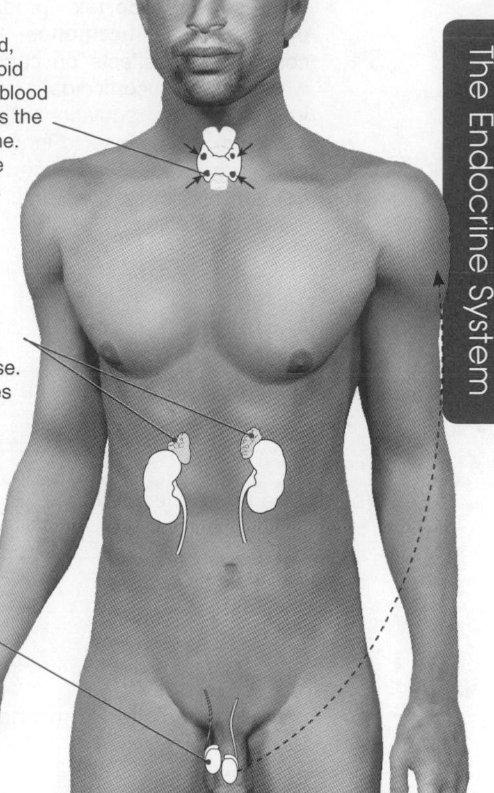

Adrenal glands
The adrenal medulla produces epinephrine and norepinephrine responsible for the fight or flight response. The adrenal cortex produces various steroid hormones, including cortisol (response to stress) and aldosterone (sodium regulation).

Testes (in males)
Produce testosterone, which controls and maintains male features and promotes sperm production.

1. Explain how a hormone is different from a neurotransmitter: ______________________________

____

2. Using ruled lines, connect each of the following endocrine glands with its correct role in the body

(a) Pituitary gland	The hormone from this gland regulates the levels of calcium in the blood
(b) Ovaries	Master gland secreting at least nine hormones, including growth hormone and TSH
(c) Pineal gland	Produces hormones involved in the regulation of metabolic rate
(d) Parathyroid glands	Secretes melatonin to regulate sleep patterns and cycles of reproductive hormones
(e) Thyroid	Produce estrogen and progesterone in response to hormones from the pituitary

3. Review the three types of stimuli for hormone release and describe a specific example of each:

(a) Hormonal stimulus: ______________________________

(b) Humoral stimulus: ______________________________

(c) Neural stimulus: ______________________________

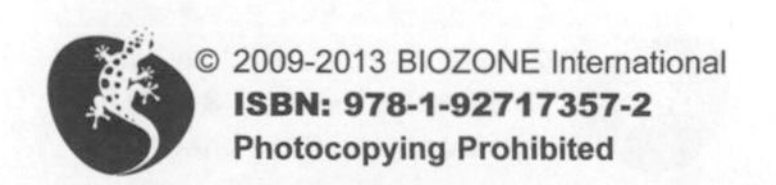

ISBN: 978-1-92717357-2

Related activities: *Neurohormones, The Stress Response*
Weblinks: *Drag and Drop Hormone Match, Hormones and their Effects*

The Adrenal Glands

The **adrenals** sit above the superior tip of the kidneys. Each adrenal gland has two structurally and functionally distinct regions; an outer cortex and an inner medulla.

- The **adrenal medulla** releases two catecholamine hormones: **epinephrine** and **norepinephrine**. Their release is controlled by the sympathetic nervous system and they are responsible for the 'fight or flight' response to stress, which includes increased breathing and heart rates, and paling of skin.
- The **adrenal cortex** produces a number of **corticosteroid hormones**. Glucocorticoids (e.g. cortisol) have effects on carbohydrate metabolism, while mineralocorticoids, (e.g. aldosterone) are involved in ion regulation. Glucocorticoids are also secreted in response to long term stress. Hormone release from the adrenal cortex is controlled by the hormone ACTH from the anterior pituitary gland.

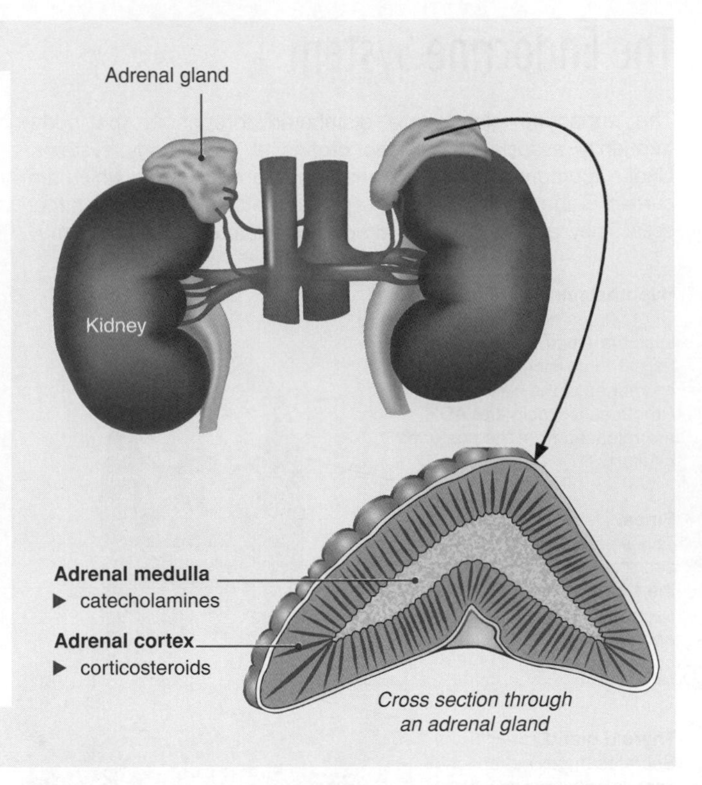

Cross section through an adrenal gland

4. (a) Identify a **target tissue** and a hormone that controls its activity: ______

   (b) Outline the homeostatic function of this hormone and explain how it controls the activity of the target tissue:

   (c) Explain the regulation of this hormone, including the role of nerves, hormones, and negative feedback as appropriate:

5. (a) Describe the structure of the adrenal glands: ______

   (b) Relate the structural differentiation of the adrenal glands to the functional role of each distinct region:

   (c) Explain the relationship between the pituitary gland and the adrenals: ______

   (d) Explain the relationship between the sympathetic nervous system and the adrenals: ______

ISBN: 978-1-92717357-2

Neurohormones

The **hypothalamus** is located below the thalamus, just above the brain stem and the pituitary gland, with which it has a close structural and functional relationship. Information comes to the hypothalamus through sensory pathways from sensory receptors. On the basis of this information, the hypothalamus controls and integrates many basic physiological activities (e.g. temperature regulation, food and fluid intake, and sleep), including the reflex activity of the **autonomic nervous system**.

One of the most important functions of the hypothalamus is to link the nervous system to the endocrine system (via the pituitary). The hypothalamus contains **neurosecretory cells**. These are specialized secretory neurons, which are both nerve cells and endocrine cells. They produce hormones (usually peptides) in the cell body, which are packaged into droplets and transported along the axon. At the axon terminal, the **neurohormone** is released into the blood in response to nerve impulses.

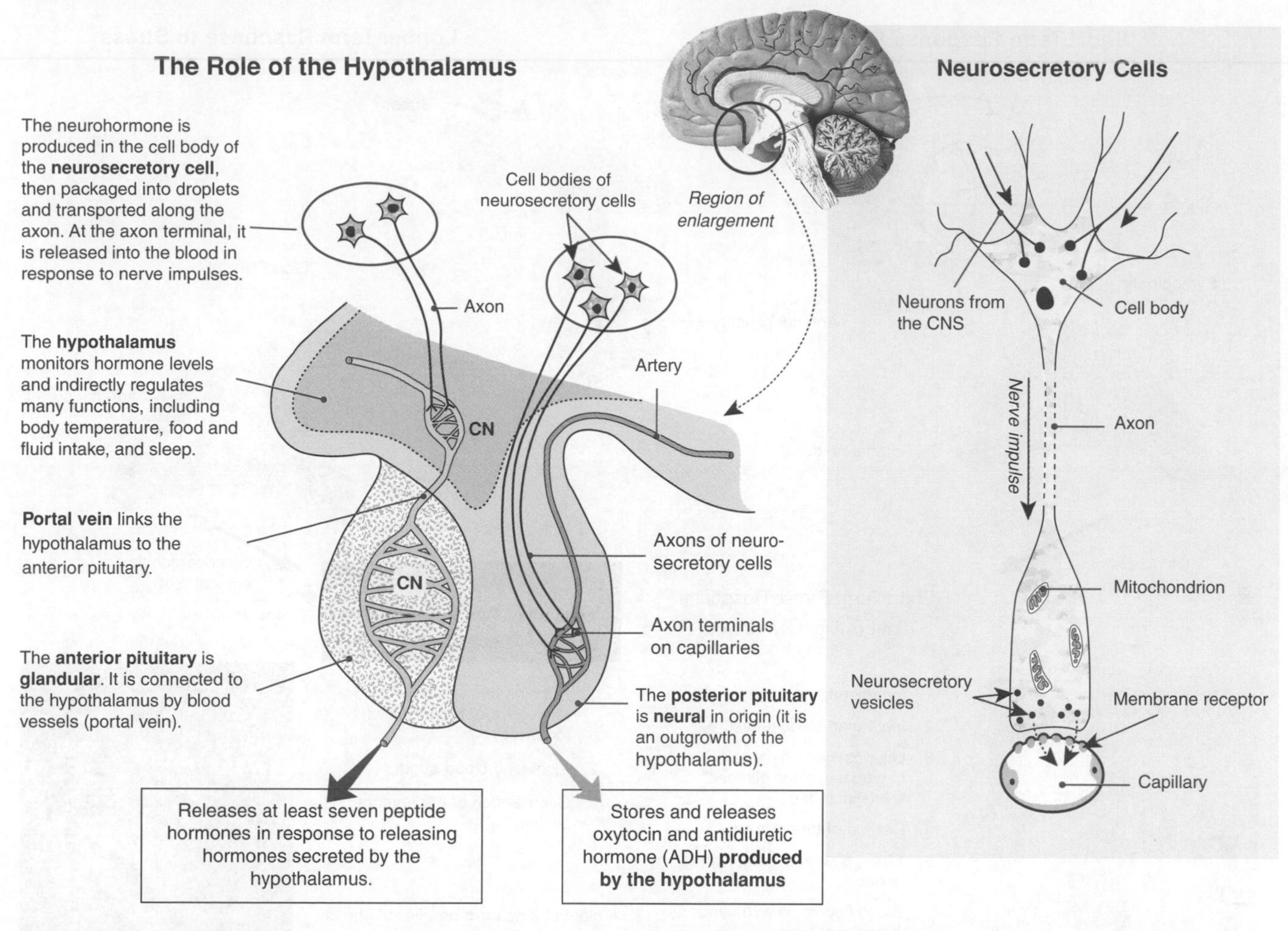

1. (a) Explain how the anterior and posterior pituitary differ with respect to their relationship to the hypothalamus:

   (b) Explain how these differences relate to the nature of the hormonal secretions for each region:

2. Describe the role of the neurohormones released by the hypothalamus:

3. Explain why the adrenal and thyroid glands atrophy if the pituitary gland ceases to function:

4. Although the anterior pituitary is often called the master gland, the hypothalamus could also claim that title. Explain:

ISBN: 978-1-92717357-2

Related activities: Hormones of the Pituitary, The Stress Response

The Stress Response

The interactions of the **hypothalamus, pituitary** and **adrenal glands** together constitute the hypothalamic-pituitary-adrenal axis, which is responsible for controlling the body's reactions to stress and regulating many of the body's processes, including digestion, immune function, mood, sexuality, and energy storage and expenditure. The **stress response** is triggered through sympathetic stimulation of the central medulla region of the adrenal glands, in what is popularly know as the **fight or flight syndrome**. This stimulation causes the release of catecholamines (epinephrine and norepinephrine). These hormones help to prepare the body to cope with short-term stressful situations. Continued stress results in release of glucocorticoids (especially **cortisol**) from the outer cortex of the adrenals. These hormones help the body to resist longer term stress. Their secretion is a normal part of what is called the **general adaptation syndrome**, but continued unrelieved stress can be damaging, or even fatal.

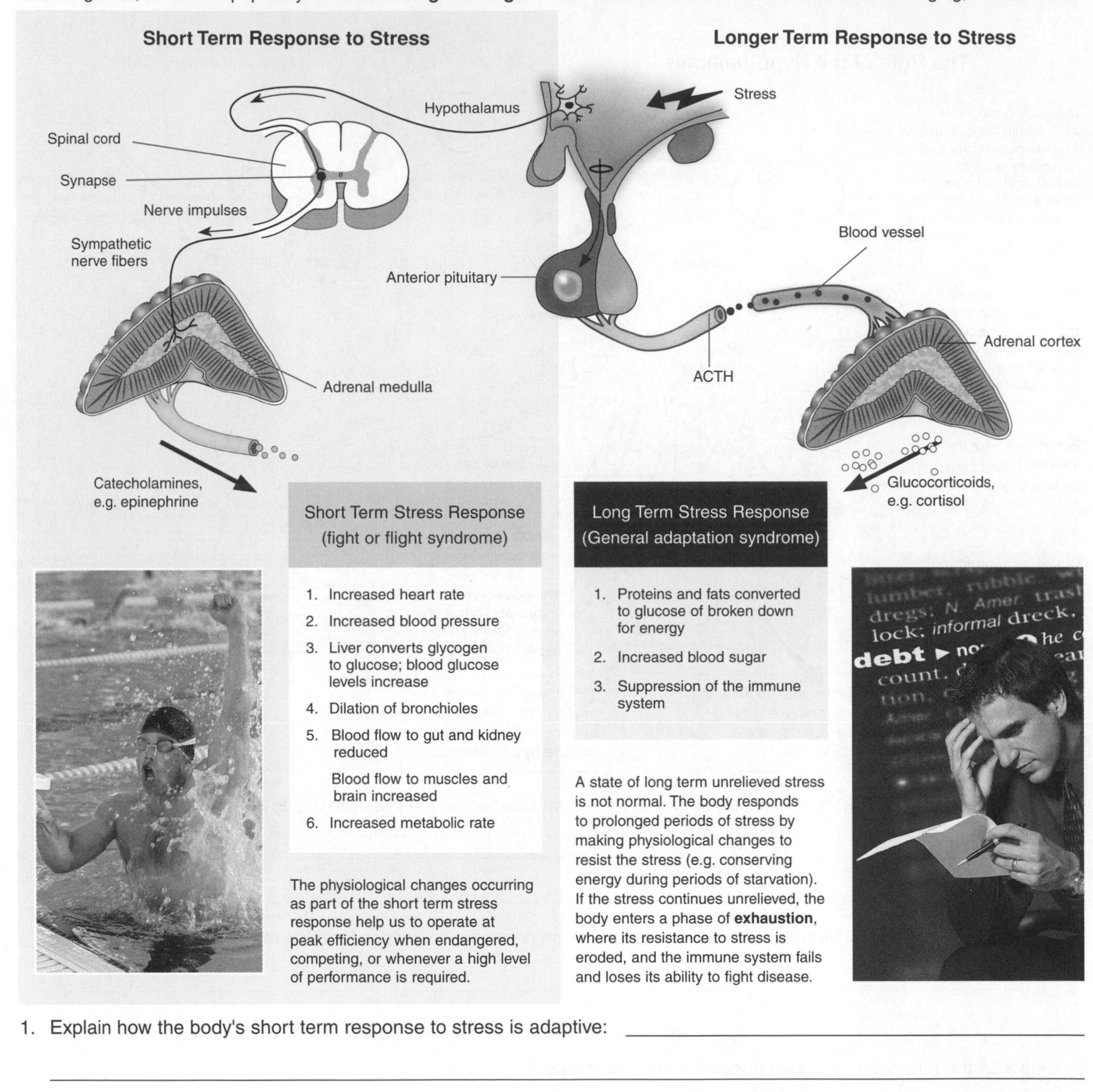

1. Explain how the body's short term response to stress is adaptive: ______

2. (a) Describe features of the long term stress response that help to maintain activity through the period of the stress:

______

(b) Describe how these responses could be damaging if the stress is unrelieved for prolonged periods (e.g. months):

______

Related activities: Neurohormones, Hormones of the Pituitary
Weblinks: Action of ACTH

ISBN: 978-1-92717357-2

Hormones of the Pituitary

The **pituitary gland** (or hypophysis) is a tiny endocrine gland, about the size of a pea, hanging from the inferior surface of the hypothalamus. It has two regions or lobes, each with different structure and origin. The **posterior pituitary** is neural (nervous) in origin and is essentially an extension of the hypothalamus. Its neurosecretory cells have their cell bodies in the hypothalamus, and release oxytocin and ADH directly into the bloodstream in response to nerve impulses. The **anterior pituitary** is connected to the hypothalamus by blood vessels and receives releasing and inhibiting hormones (factors) from the hypothlamus via a capillary network. These releasing hormones regulate the secretion of the anterior pituitary's hormones.

Hormones of the Anterior Pituitary

The anterior pituitary releases at least seven **peptide hormones** (below) into the blood from simple secretory cells. The release of these hormones is regulated by releasing and inhibiting hormones from the hypothalamus.

Hormones of the Posterior Pituitary

The posterior pituitary develops as an extension of the hypothalamus. The release of its two hormones, oxytocin and antidiuretic hormone, occurs directly as a result of nervous input to the hypothalamus.

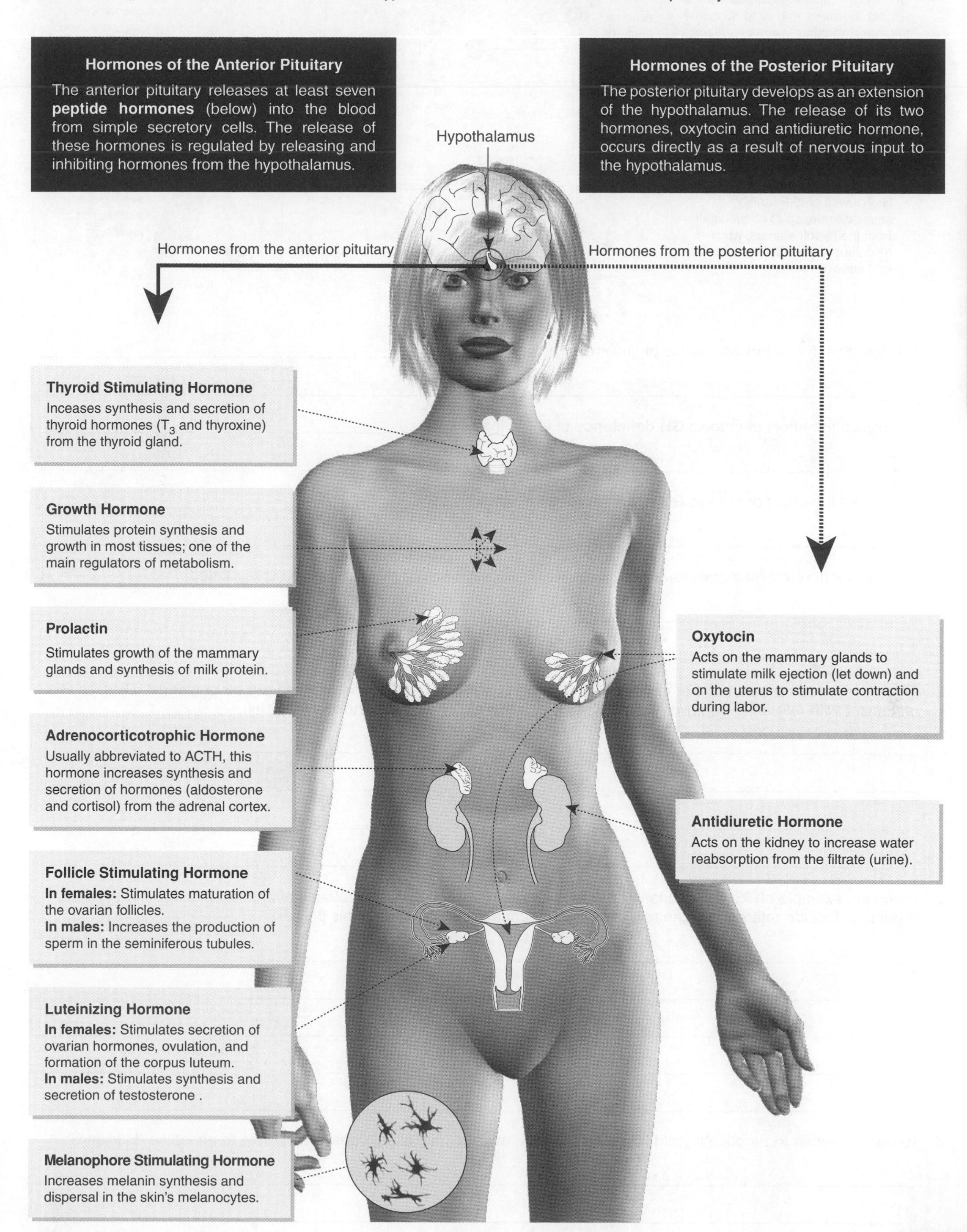

Thyroid Stimulating Hormone
Inceases synthesis and secretion of thyroid hormones (T_3 and thyroxine) from the thyroid gland.

Growth Hormone
Stimulates protein synthesis and growth in most tissues; one of the main regulators of metabolism.

Prolactin
Stimulates growth of the mammary glands and synthesis of milk protein.

Adrenocorticotrophic Hormone
Usually abbreviated to ACTH, this hormone increases synthesis and secretion of hormones (aldosterone and cortisol) from the adrenal cortex.

Follicle Stimulating Hormone
In females: Stimulates maturation of the ovarian follicles.
In males: Increases the production of sperm in the seminiferous tubules.

Luteinizing Hormone
In females: Stimulates secretion of ovarian hormones, ovulation, and formation of the corpus luteum.
In males: Stimulates synthesis and secretion of testosterone .

Melanophore Stimulating Hormone
Increases melanin synthesis and dispersal in the skin's melanocytes.

Oxytocin
Acts on the mammary glands to stimulate milk ejection (let down) and on the uterus to stimulate contraction during labor.

Antidiuretic Hormone
Acts on the kidney to increase water reabsorption from the filtrate (urine).

ISBN: 978-1-92717357-2

Related activities: *The Endocrine System, Neurohormones*
Weblinks: *Growth Hormone: An Inside Look*

Effects of Growth Hormone

Growth hormone (GH) is released in response to GHRH (growth hormone releasing hormone) from the hypothalamus. GH acts directly and indirectly to affect metabolic activities associated with growth.

GH directly stimulates metabolism of fat, but its major role is to stimulate the liver and other tissues to secrete IGF-I (Insulin-like Growth Factor) and through this stimulate bone and muscle growth. GH secretion is regulated is via negative feedback:

- High levels of IGF-1 suppress GHRH secretion from the hypothalamus.
- High levels of IGF-1 also stimulate release of somatostatin from the hypothalamus, which also suppresses GH secretion (not shown).

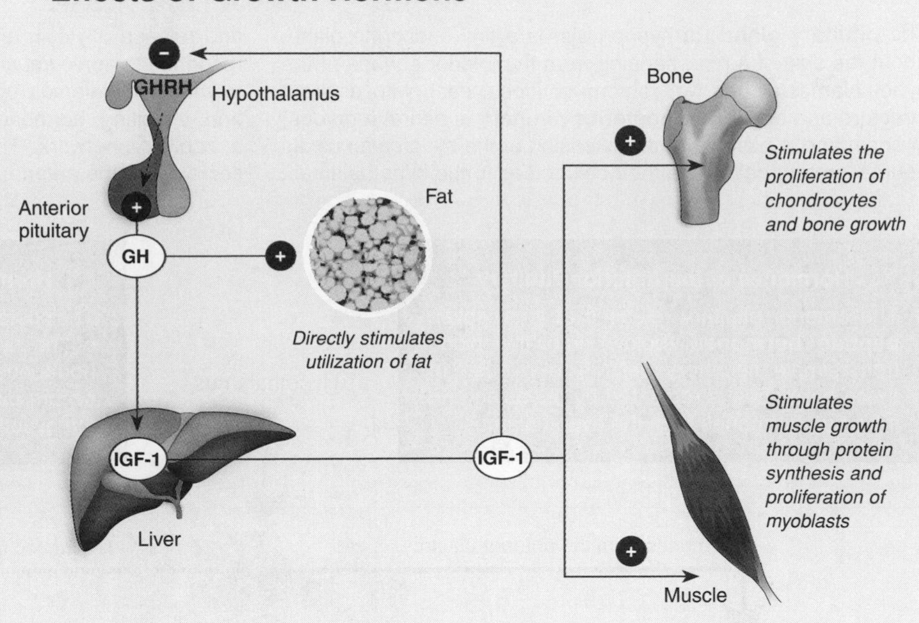

1. (a) Describe the metabolic effects of growth hormone: ______

   (b) Predict the effect of chronic **GH deficiency** of GH in infancy: ______

   (c) Predict the effect of chronic **GH hypersecretion** in infancy: ______

   (d) Describe the two main mechanisms through which the secretion of growth hormone is regulated: ______

2. "The pituitary releases a number of hormones that regulate the secretion of hormones from other glands". Discuss this statement with reference to growth hormone (GH) and **thyroid stimulating hormone** (TSH): ______

3. Using the example of TSH and its target tissue (the thyroid), explain how the release of anterior pituitary hormones is regulated. Include reference to the role of negative feedback mechanisms in this process: ______

4. Iodine is needed to produce thyroid hormones. Explain why the thyroid enlarges in response to an iodine deficiency: ______

ISBN: 978-1-92717357-2

Control of Blood Glucose

The endocrine portion of the **pancreas** (the α and β cells of the **islets of Langerhans**) produces two hormones, **insulin** and **glucagon**, which maintain blood glucose at a steady state through **negative feedback**. Insulin promotes a decrease in blood glucose by promoting cellular uptake of glucose and synthesis of glycogen. **Glucagon** promotes an increase in blood glucose through the breakdown of glycogen and the synthesis of glucose from amino acids. When normal blood glucose levels are restored, negative feedback stops hormone secretion. Regulating blood glucose to within narrow limits allows energy to be available to cells as needed. Extra energy is stored as glycogen or fat, and is mobilized to meet energy needs as required. The liver is pivotal in these carbohydrate conversions. One of the consequences of a disruption to this system is the disease **diabetes mellitus**. In type 1 diabetes, the insulin-producing β cells are destroyed as a result of autoimmune activity and insulin is not produced. In type 2 diabetes, the pancreatic cells produce insulin, but the body's cells become increasingly resistant to it.

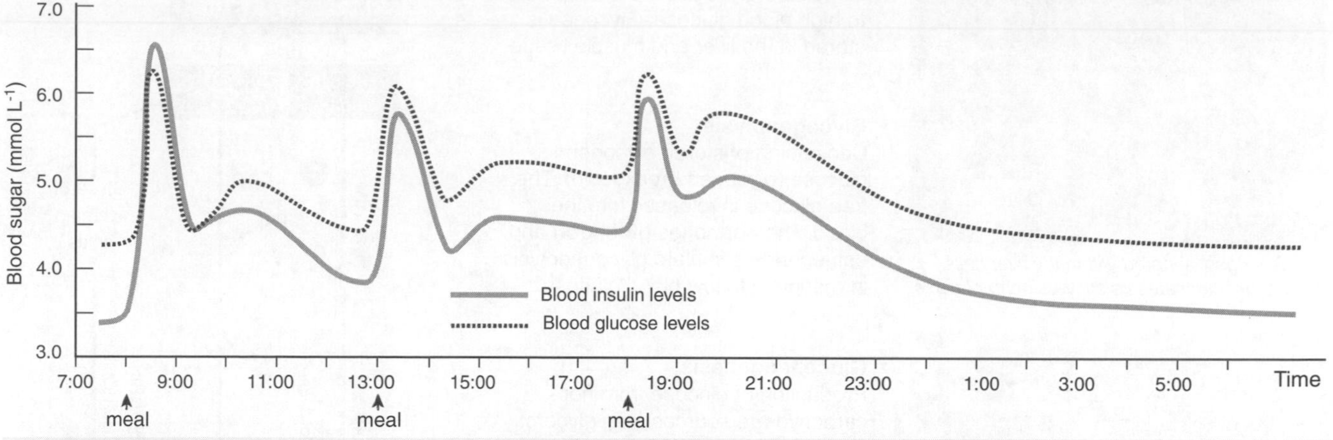

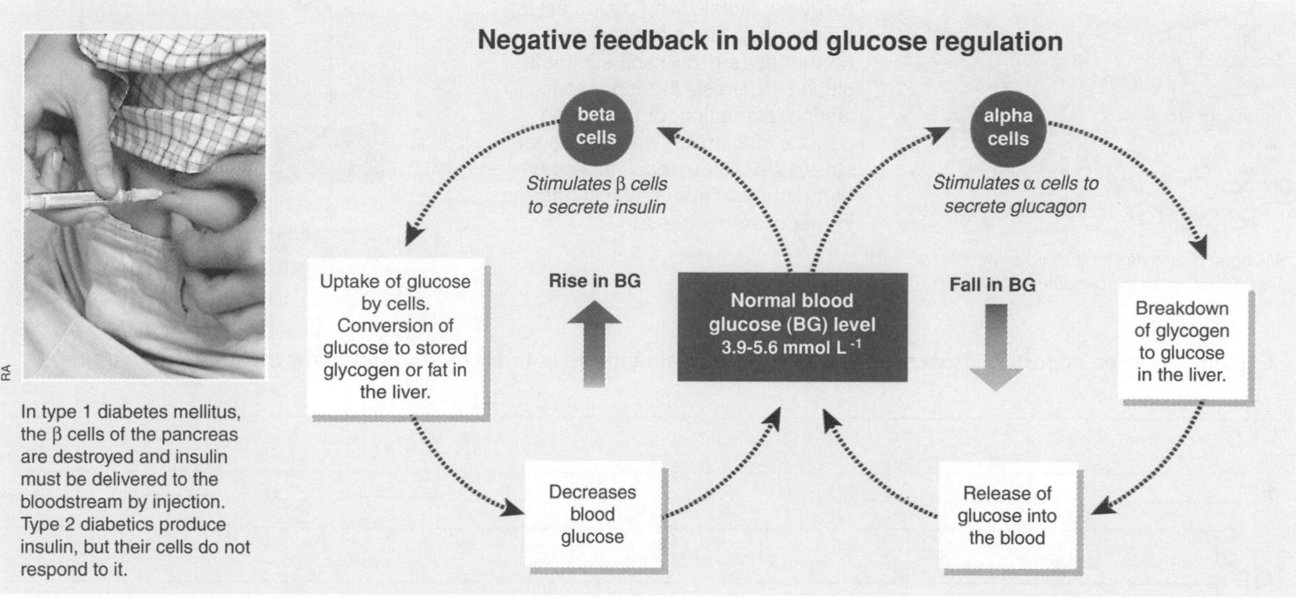

In type 1 diabetes mellitus, the β cells of the pancreas are destroyed and insulin must be delivered to the bloodstream by injection. Type 2 diabetics produce insulin, but their cells do not respond to it.

1. (a) Identify the stimulus for the release of insulin: ______________________

   (b) Identify the stimulus for the release of glucagon: ______________________

   (c) Explain how glucagon brings about an increase in blood glucose level: ______________________

   ____

   (d) Explain how insulin brings about a decrease in blood glucose level: ______________________

   ____

2. Explain the pattern of fluctuations in blood glucose and blood insulin levels in the graph above:

   ____

   ____

   ____

   ____

3. Identify the mechanism regulating insulin and glucagon secretion (humoral, hormonal, neural): ______________________

ISBN: 978-1-92717357-2

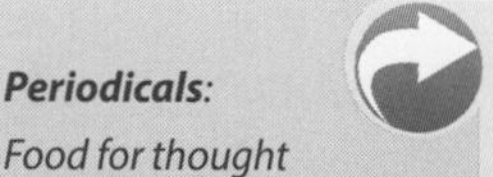
Periodicals: *Food for thought*

Related activities: *The Role of the Liver in Carbohydrate Metabolism* ***Weblinks:*** *Insulin and Glucose Regulation*

Carbohydrate Metabolism in the Liver

The liver has a central role in carbohydrate metabolism, specifically the production of glucose from non-carbohydrate sources, and the interconversion of glucose and glycogen (a glucose polysaccharide). These processes ensure that carbohydrate is stored or made available to cells as required and are regulated by hormones, principally insulin and glucagon, but also epinephrine and glucocorticoids.

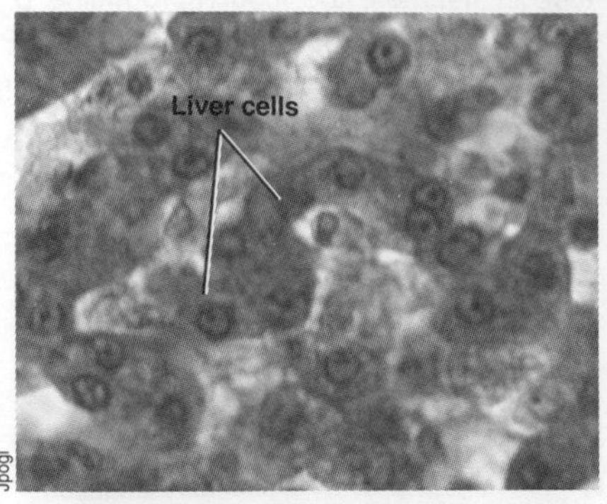

Glycogen is stored within the liver cells. Glucagon stimulates its conversion to glucose.

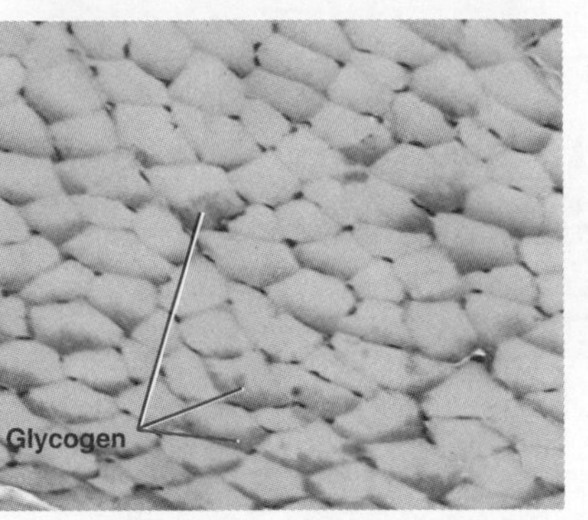

Glycogen is also stored in muscle, where it is squeezed out to the periphery of the cells.

- **Glycogenesis**
  Excess glucose in the blood is converted to **glycogen**. **Insulin** stimulates glycogenesis in response to high blood glucose. Glycogen is stored in the liver and muscle tissue.

- **Glycogenolysis**
  Conversion of stored glycogen to glucose (glycogen breakdown). The free glucose is released into the blood. The hormones **glucagon** and epinephrine stimulate glycogenolysis in response to low blood glucose.

- **Gluconeogenesis**
  Production of glucose from non-carbohydrate sources (e.g. glycerol, pyruvate, lactate, and amino acids). Epinephrine and glucocorticoid hormones (e.g. cortisol) stimulate gluconeogenesis in response to fasting, starvation, or prolonged periods of exercise when glycogen stores are exhausted. It is also part of the general adaptation syndrome in response to stress.

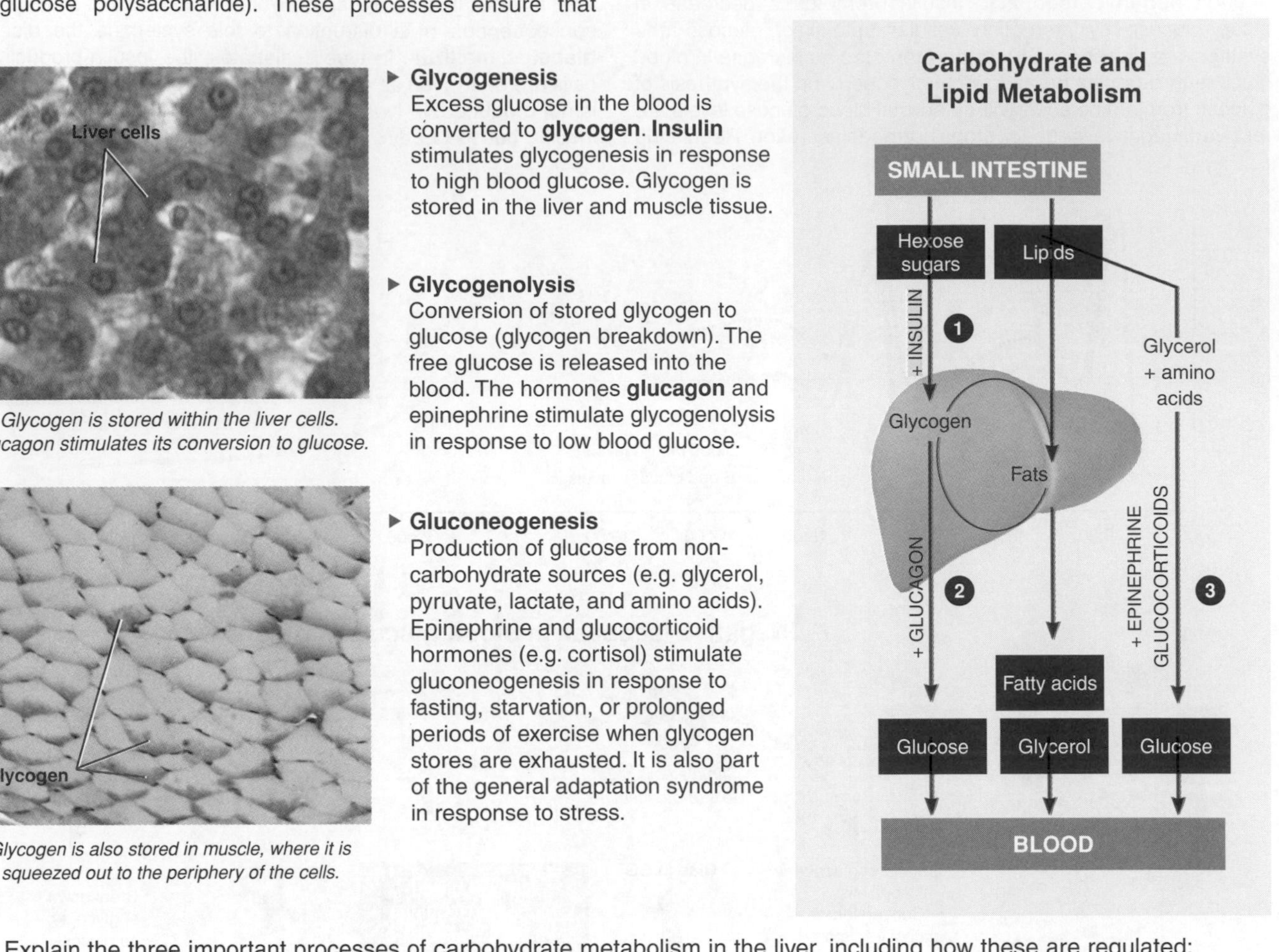

1. Explain the three important processes of carbohydrate metabolism in the liver, including how these are regulated:

   (a) ______

   (b) ______

   (c) ______

2. Identify the processes occurring at each numbered stage on the diagram above, right:

   (a) Process occurring at point 1: ______

   (b) Process occurring at point 2: ______

   (c) Process occurring at point 3: ______

3. Explain why it is important that the body can readily convert and produce different forms of carbohydrates:

   ______

ISBN: 978-1-92717357-2

Type 1 Diabetes Mellitus

Diabetes is a general term for a range of disorders sharing two common symptoms: production of large amounts of urine and excessive thirst. Other symptoms depend on the type of diabetes. **Diabetes mellitus** is the most common form of diabetes and is characterized by **hyperglycemia** (high blood sugar). **Type 1 diabetes** is characterized by **absolute insulin deficiency**. It usually begins in childhood as a result of autoimmune destruction of the insulin-producing cells of the pancreas. For this reason, it was once called juvenile-onset diabetes. It is a severe, incurable condition, which is treated primarily with insulin injections.

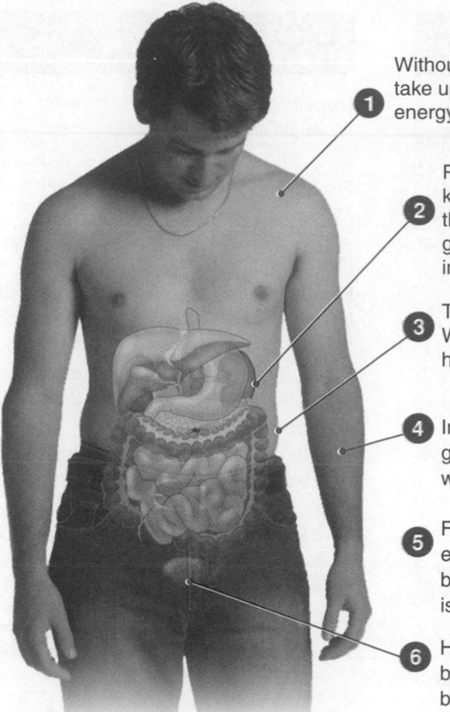

1. Without insulin, cells cannot take up glucose and so lack an energy source for metabolism.
2. Production of urine from the kidneys increases to clear the body of excess blood glucose. Glucose is present in the urine.
3. There is constant thirst. Weight is lost despite hunger and overeating.
4. Inability to utilize glucose leads to muscle weakness and fatigue.
5. Fats are metabolized for energy leading to a fall in blood pH (ketosis). This is potentially fatal.
6. High sugar levels in blood and urine promote bacterial and fungal infections of the bladder and urino-genital tract.

Cause of Type 1 Diabetes Mellitus

Incidence: About 10-15% of all diabetics.

Age at onset: Early; often in childhood.

Symptoms: Hyperglycemia (high blood sugar), excretion of glucose in the urine (glucosuria), increased urine production, excessive thirst and hunger, weight loss, and ketosis.

Cause: Absolute deficiency of insulin due to lack of insulin production (pancreatic beta cells are destroyed in an autoimmune reaction). There is a genetic component but usually a childhood viral infection triggers the development of the disease. Mumps, coxsackie, and rubella are implicated.

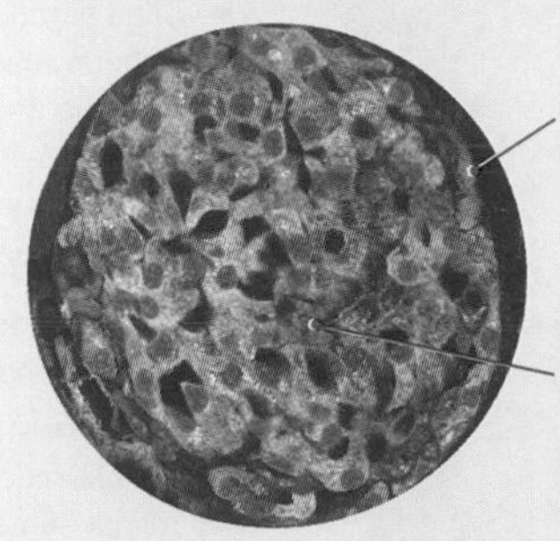

α **cells** produce glucagon, which promotes glucose release from the liver.

β **cells** (most of the cells in this field of view) produce insulin, the hormone promoting cellular uptake of glucose. β cells are destroyed in type 1 diabetes mellitus.

Cell types in the endocrine region of a normal pancreas

Image: Solimena Lab, Uni of Tech, Germany

The Endocrine System

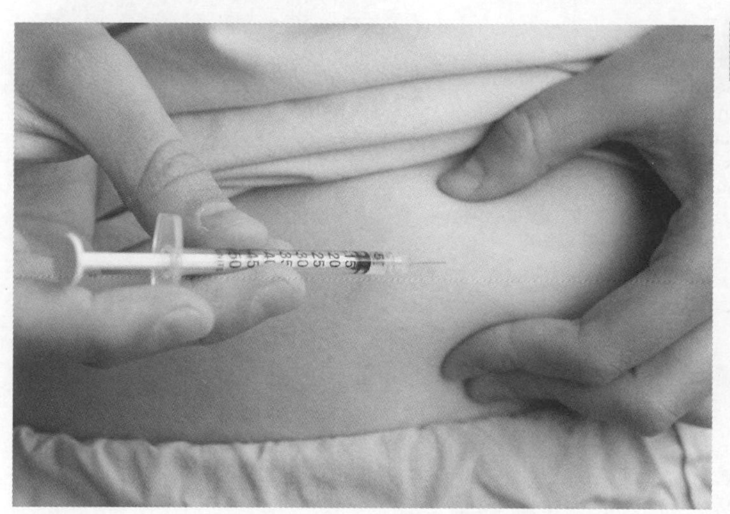

Treatments of Type 1 Diabetes Mellitus

Present treatments: Regular insulin injections are combined with dietary management to keep blood sugar levels stable. Blood glucose levels are monitored regularly with testing kits to guard against sudden, potentially fatal, falls in blood glucose (hypoglycemia).

Insulin was once extracted from dead animals, but animal-derived insulin produces many side effects. Genetically engineered microbes now provide abundant, low cost human insulin, without the side effects associated with animal insulin.

Newer treatments: Cell therapy involves transplanting islet cells into the patient where they produce insulin and regulate blood sugar levels. The islet cells may be derived from stem cells or from the pancreatic tissue of pigs. New technology developed by New Zealand company Living Cell Technologies Ltd, encapsulates the pig islet cells within microspheres so they are protected from destruction by the patient's immune system.

Cell therapy promises to be an effective way to provide sustained relief for diabetes.

1. Describe the **symptoms** of type 1 diabetes mellitus and relate these to the physiological cause of the disease:

2. Explain how regular insulin injections assist the type 1 diabetic to maintain their blood glucose homeostasis:

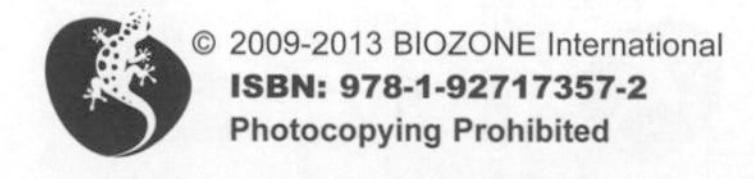

ISBN: 978-1-92717357-2

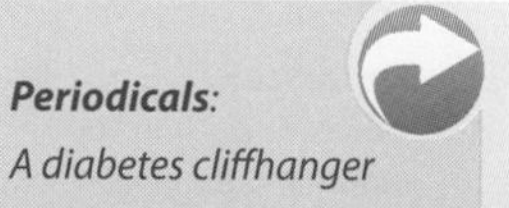

Periodicals:
A diabetes cliffhanger

Related activities: *Control of Blood Glucose*
Weblinks: *Type 1 Diabetes*

Type 2 Diabetes Mellitus

Type 2 diabetes mellitus is characterized by a resistance to insulin's effects and relative (rather than absolute) insulin deficiency. The pancreas produces insulin, but the body's cells cease to respond to it and glucose levels in the blood remain high.

Type 2 diabetes is a chronic, progressive disease, and becomes worse with time if not managed. Its long-term effects include heart disease, strokes, blindness, and kidney failure. However, ketoacidosis (a feature of type 1 diabetes), is uncommon.

Symptoms of Type 2 Diabetes Mellitus

a Symptoms may be mild at first. The body's cells do not respond appropriately to the insulin present and blood glucose levels become elevated. Normal blood glucose level is 3.3-6.1 mmol L^{-1} (60-110 mg dL^{-1}). In diabetics, fasting blood glucose level is 7 mmol L^{-1} (126 mg dL^{-1}) or higher.

b Symptoms occur with varying degrees of severity:

- Cells are starved of fuel. This can lead to increased appetite and overeating and may contribute to an existing obesity problem.
- Urine production increases to rid the body of the excess glucose. Glucose is present in the urine and patients are frequently very thirsty.
- The body's inability to use glucose properly leads to muscle weakness and fatigue, irritability, frequent infections, and poor wound healing.

c Uncontrolled elevated blood glucose eventually results in damage to the blood vessels and leads to:

- coronary artery disease
- peripheral vascular disease
- retinal damage, blurred vision and blindness
- kidney damage and renal failure
- persistent ulcers and gangrene

Risk Factors

Obesity: BMI greater than 27. Distribution of weight is also important.

Age: Risk increases with age, although the incidence of type 2 diabetes is increasingly reported in obese children.

Sedentary lifestyle: Inactivity increases risk through its effects on bodyweight.

Family history: There is a strong genetic link for type 2 diabetes. Those with a family history of the disease are at greater risk.

Ethnicity: Certain ethnic groups are at higher risk of developing of type 2 diabetes.

High blood pressure: Up to 60% of people with undiagnosed diabetes have high blood pressure.

High blood lipids: More than 40% of people with diabetes have abnormally high levels of cholesterol and similar lipids in the blood.

Treating Type 2 Diabetes

Diabetes is not curable but can be managed to minimize the health effects:

- Regularly check blood glucose level
- Manage diet to reduce fluctuations in blood glucose level
- Take regular exercise
- Reduce weight
- Reduce blood pressure
- Reduce or stop smoking
- Take prescribed anti-diabetic drugs
- In time, insulin therapy may be required

Cellular uptake of glucose is impaired and glucose enters the bloodstream instead. Type 2 diabetes is sometimes called **insulin resistance**.

Fat cell

Insulin

The **beta cells** of the pancreatic islets (above) produce insulin, the hormone responsible for the cellular uptake of glucose. In type 2 diabetes, the body's cells do not utilize the insulin properly.

1. Distinguish between type 1 and type 2 diabetes, relating the differences to the different methods of treatment:

2. Explain what dietary advice you would give to a person diagnosed with type 2 diabetes:

3. Explain why the increase in type 2 diabetes is considered epidemic in the developed world:

ISBN: 978-1-92717357-2

***Related activities**: Control of Blood Glucose*
***Weblinks**: Cellular Mechanisms of Diabetes*
***Periodicals**: Glucose - getting the balance right*

Alcohol, Nicotine, and Blood Glucose

The amount of glucose in the blood (the blood glucose level) is tightly regulated by negative feedback. Two commonly used substances, **alcohol** and **nicotine,** both affect blood glucose levels and its regulation. Nicotine is the highly addictive component of tobacco, and a potent carcinogen (cancer causing agent). It is also responsible for depression of appetite in smokers, partly through its effect on blood glucose. The alcohol in alcoholic beverages acts as a preservative and adds to its flavor. Alcohol is toxic and its metabolism reduces the body's capacity to regulate blood glucose levels.

Alcohol

The liver metabolizes alcohol, so while there is alcohol in the blood, it has less capacity to regulate blood glucose. Alcohol lowers blood glucose levels by stimulating insulin production. Low to moderate alcohol intakes don't affect the body's response to insulin. However, long-term alcohol abuse damages the liver and interferes with many aspects of metabolism, including metabolism of glucose.

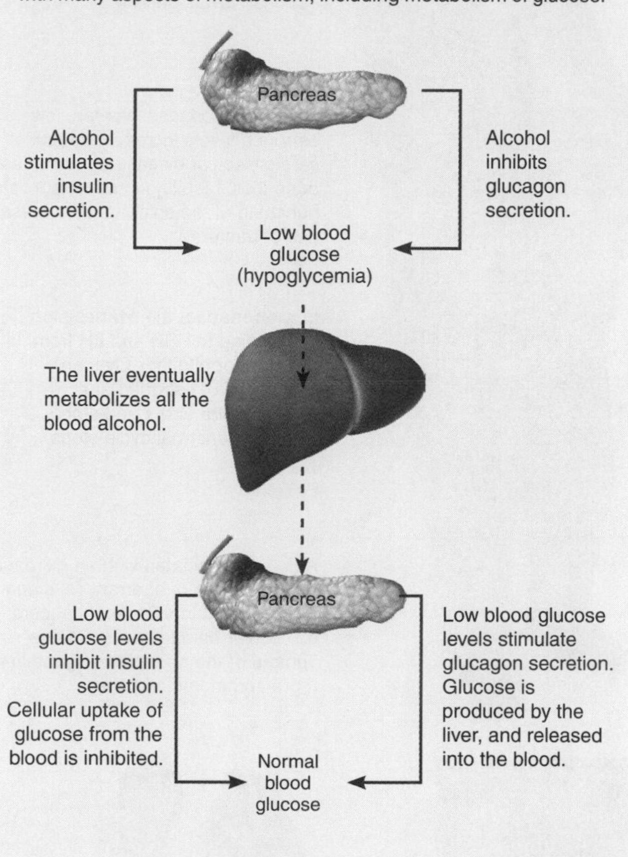

Nicotine

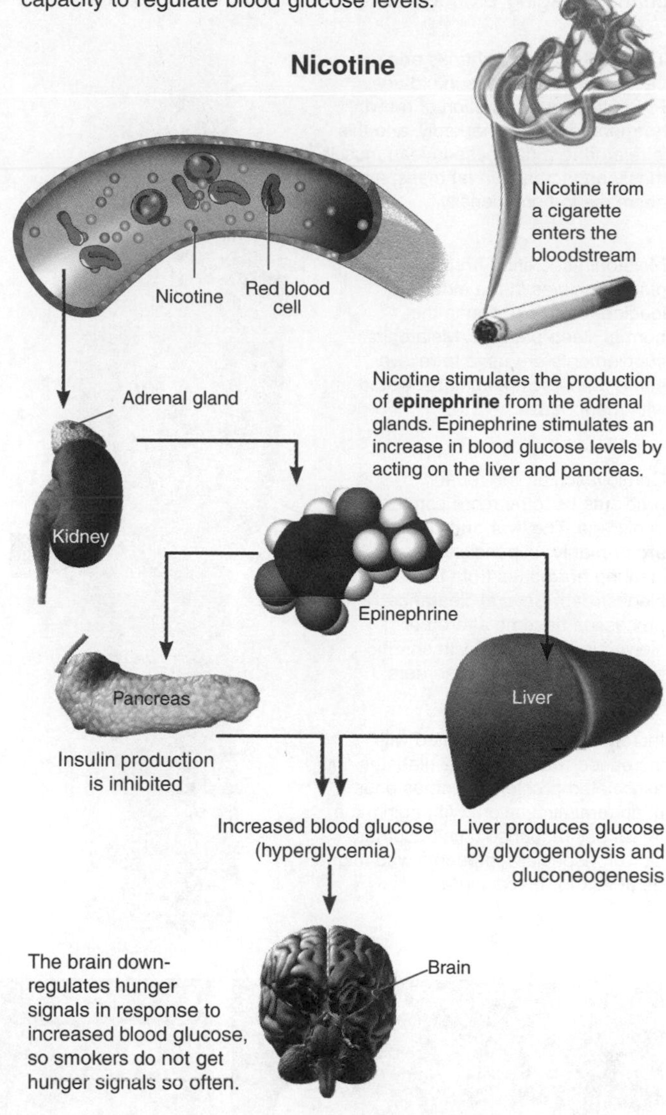

1. (a) How does drinking alcohol result in low blood glucose? ______________________

   (b) The liver prioritizes the metabolism of alcohol before restoring blood glucose. Suggest why this is the case: ______________________

2. (a) Explain how nicotine reduces appetite: ______________________

   (b) Explain why people who stop smoking often put on weight: ______________________

ISBN: 978-1-92717357-2

Aging and the Endocrine System

Despite age-related changes, the endocrine system functions well in most older people. However, some endocrine changes do occur because of normal cellular damage accumulating as a result of the aging process and genetically programmed cellular changes. Aging produces changes in hormone production and secretion, hormone metabolism (how quickly excess hormones are broken down and leave the body, for example, through urination), levels of circulating hormones, target tissue response to hormones, and biological rhythms such as sleep and the menstrual cycle.

The pituitary gland shrinks and becomes more fibrous in old age. Production and secretion of growth hormone declines markedly, and this is related to a decrease in lean muscle mass, an increase in fat mass, and a decrease in bone density.

Melatonin secretion from the pineal declines in the elderly, leading to a disruption in the normal sleep patterns. Melatonin supplements are used to relieve some of the symptoms associated with menopause.

Cardiovascular and renal problems become more common in old age. The liver and kidneys are primarily responsible for clearing hormones from the bloodstream Several clearance processes become altered or slowed in individuals with chronic heart, liver, or kidney disorders.

Increasing age is correlated with increased rates of type 2 diabetes and associated problems such as poor peripheral circulation. With aging, the target cell response time becomes slower, especially in people who might be at risk for this disorder.

Age-related thinning and loss of hair is related to hormonal changes as circulating levels of a testosterone derivative (DHT) increase and damage the hair follicles. In women, hair loss and thinning is the result of DHT sensitivity increasing as estrogen levels decline.

In post-menopausal women, low estrogen levels increase the risk of cardiovascular disease and accelerate bone loss, resulting in **osteoporosis**, hunching of the spine, and increased risk of fractures.

In **menopause**, the ovaries stop responding to FSH and LH from the anterior pituitary. Ovarian production of estrogen and progesterone slows and stops, and the menstrual cycle stops.

Aging is associated with an increase in the number of aberrant cells and an increased incidence of cancers as cellular damage accumulates. Tumors of the endocrine organs are more common in old age.

Menopause is a stage of life and not a disease. The symptoms of approaching menopause, which include hot flushes of the skin, mood swings, and irregular menstrual periods, can be alleviated with attention to a healthy diet and lifestyle. Menopause is determined retrospectively, after 12 months without menstruation.

Physical exercise can help slow or reverse some of the physical changes that occur with aging as a result of hormonal changes. Regular exercise helps to regulate body weight and insulin levels, reduces insulin resistance, improves cardiovascular efficiency and muscle strength, and slows the loss of bone mass.

Production and secretion of growth hormone (GH) is maximal during adolescence and declines thereafter. When GH is administered to adults with GH deficiency, body fat declines and muscle mass and bone density increase. However, GH does not reverse all indications of aging and can even increase mortality in the elderly.

1. Describe two consequences of declining output of growth hormone in old age:

   (a) ______________________________

   (b) ______________________________

2. Explain why the levels of LH and FSH increase in post-menopausal women: ______________________________

3. Identify one process that accelerates in elderly women as a result of loss of estrogen: ______________________________

ISBN: 978-1-92717357-2

KEY TERMS: Mix and Match

INSTRUCTIONS: Test your vocabulary by matching each term to its definition, as identified by its preceding letter code.

Term		Definition
blood glucose	**A**	The hormone that lowers blood glucose, primarily through the cellular uptake of glucose, but also by promoting storage of glucose as glycogen.
cell signaling	**B**	Chemical messenger that induces a specific physiological response.
cyclic AMP	**C**	Communication between cells involving a chemical messenger or signal molecule (ligand) and a receptor molecule (on the target cells).
diabetes mellitus	**D**	A pathway that converts a mechanical or chemical stimulus to a cell into a specific cellular response.
endocrine gland	**E**	The region of the brain which coordinates the nervous and endocrine systems via the pituitary gland.
exocrine gland	**F**	An abdominal organ with both endocrine and exocrine functions.
glucagon	**G**	An important second messenger derived from ATP.
hormone	**H**	The permanent end of menstruation.
hypothalamus	**I**	The amount of glucose present in the blood.
insulin	**J**	Any chemical that transfers information between cells and causes a response in a target cell.
menopause	**K**	Ductless gland secreting hormones directly into the blood.
negative feedback	**L**	Cell that responds to a specific messenger.
neurosecretory cell	**M**	A molecule that relays signals from receptors on the cell surface to a target molecule inside the cell.
pancreas	**N**	An endocrine gland located below the hypothalamus. It produces hormones that control other glands and many body functions.
pituitary gland	**O**	The hormone that brings about physiological processes to elevate blood glucose levels if they become too low.
positive feedback	**P**	Specialized secretory neurons, which are both nerve cells and endocrine cells (they secrete neurohormones).
second messenger	**Q**	Mechanism in which the result of a process (the response) influences the system so that changes are reduced. Leads to self-regulation and stability.
signal molecule	**R**	A gland that delivers secretions via a duct.
signal transduction pathway	**S**	A mechanism by which the body can speed up or escalate a physiological response.
target cell	**T**	A disease caused by the body's inability to produce or react to insulin.

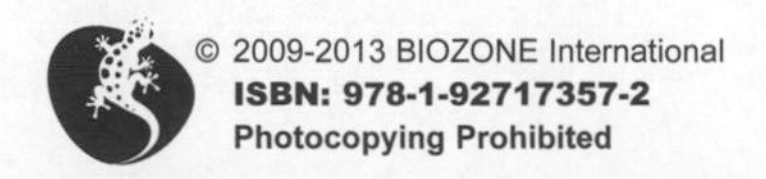

ISBN: 978-1-92717357-2

Interacting Systems

Cardiovascular and Lymphatic Systems

Endocrine system

- Lymph and blood distribute hormones.
- A number of hormones, including epinephrine and ADH, influence blood pressure.
- The thymus (a lymphoid organ with endocrine function) produces hormones for the development of the lymphatic organs and T cell maturation.
- Blood pressure and fluid volume are regulated via the renin-angiotensin-aldosterone system.

Cardiovascular system

- Lymphatic vessels pick up leaked tissue fluid and proteins and returns them to general circulation.
- Spleen destroys RBCs, removes cellular debris from the blood, and stores iron
- Blood is the source of lymph and circulates antibodies and immune system cells.

Digestive system

- Products of digestion are transported in the blood and lymph.
- Lymphoid nodules in the intestinal wall protect against invasion by pathogens.
- Gastric acidity destroys pathogens.

Skeletal system

- Bones are the site of blood cell production (hematopoiesis) for both cardiovascular and lymphatic systems.
- Bones protect cardiovascular and lymphatic organs and provide a store of calcium, which is needed for regulating blood volume.

Integumentary system

- Blood vessels in the skin provide a reservoir of blood and are a site of heat loss.
- Skin provides a physical and chemical barrier to pathogens.

Nervous system

- The nervous system innervates lymphatic organs and vessels. The brain helps to regulate the immune response.
- The autonomic ANS regulates cardiac output; the sympathetic division regulates blood pressure and distribution.

Lymphatic system and immunity

- The lymphatic system returns tissue fluid and proteins to the cardiovascular system.
- Lymph forms from blood. Blood and lymph transport antibodies and immune cells.

Respiratory system

- Gas exchange loads O_2 and unloads CO_2 to and from the blood. Respiration aids return of venous blood and lymph to the heart.
- IgA secreted by plasma cells protects the respiratory mucosa against pathogens.
- The pharynx houses lymphoid tissues.

Urinary system

- Urinary system eliminates wastes transported in the blood and maintains water, electrolyte, and acid-base balance.
- The urinary system helps to regulate blood pressure and volume by altering urine output and via release of renin.
- Urine flushes some pathogens from the body.

Reproductive system

- Slightly acidic vaginal secretions in women prevent the growth of bacteria and fungi.
- Estrogen maintains cardiovascular health in women (and in men as a result of conversion from testosterone).

Muscular system

- Activity of skeletal muscles aids return of blood and lymph to the heart.
- Muscles protect superficial lymph nodes.
- Aerobic activity improves cardiovascular health and efficiency.

General Functions and Effects on all Systems

The cardiovascular system delivers oxygen and nutrients to organs and tissues and removes carbon dioxide and other waste products of metabolism. Lymphatic vessels pick up leaked tissue fluid and proteins and return them to general circulation. The immune system protects the body's tissues and organs against pathogens.

Disease

Symptoms of disease
- Pain (moderate to severe)
- Loss of function (e.g. cardiac arrest)
- Swelling and immunodeficiency
- Tissue loss and gangrene

Diseases of the heart and blood vessels
- Congenital heart disorders
- Atherosclerosis and heart disease
- Peripheral vascular disease & stroke
- Cancers (e.g. leukemia)

Lymphatic system disorders
- Cancers (e.g. lymphomas)
- Infectious diseases (e.g. HIV)
- Inherited disorders (e.g. SCID)
- Autoimmune disorders

Medicine & Technology

Diagnosis of disorders
- Electrocardiography
- X-rays, MRI and CT scans
- Ultrasound
- Blood and DNA tests

Treatment of CVD
- Surgery
- Diet and lifestyle management
- Drug therapies

Treatment of immune disorders
- Surgery
- Radiotherapy (for cancers)
- Drug and physical therapies
- Immunotherapy (desensitization)
- Gene and cell therapy (SCID)

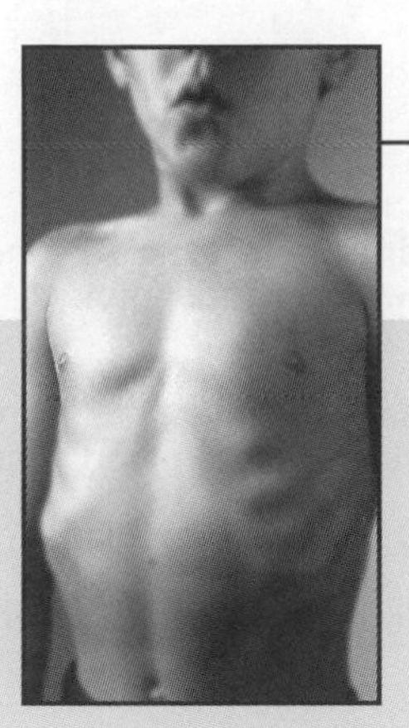

- Asthma
- Allergy
- Atherosclerosis
- Myocardial infarction
- HIV

- ECGs
- Coronary bypass
- Valve replacement
- Asthma treatment
- Organ transplants

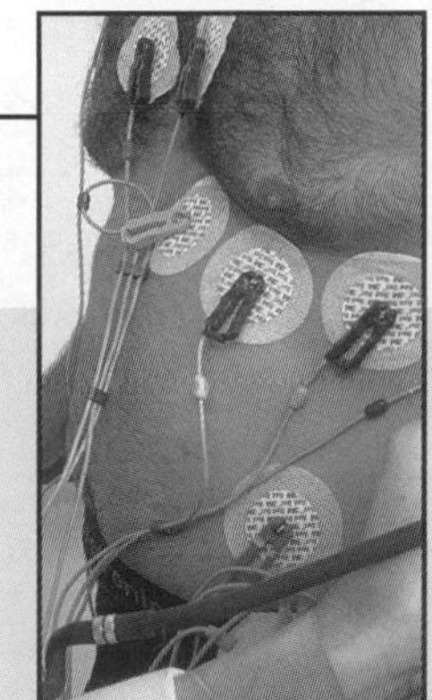

Internal Transport

The Cardiovascular and Lymphatic Systems

The cardiovascular and lymphatic systems are linked in a communication network extending throughout the body.

Exercise and medical therapies can alleviate the symptoms of some age and disease related changes to these systems.

- CVD risk factors
- Cardiac pacemakers
- Stroke
- Cancer
- Rheumatoid arthritis

Effects of exercise on:
- stroke volume
- cardiac output
- pulse rate
- blood pressure
- venous return

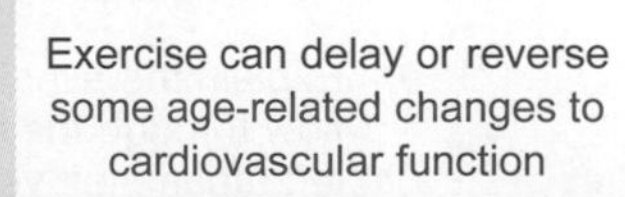

Exercise can delay or reverse some age-related changes to cardiovascular function

Cpl. Earnest J. Barnes

Effects of aging on heart and immune function
- Higher rates of autoimmune disease
- Increased susceptibility to disease
- Decline in cardiovascular performance
- Higher risk of stroke and CVD
- Increased risk of hypertension

Effects of exercise on heart and immune function
- Improved immune function
- Improved cardiovascular performance
- Increase in cardiac output
- Reduced risk of CVD and stroke
- Lowered blood pressure

The Effects of Aging

Exercise

The Cardiovascular System

Key terms

aorta
artery (*pl.* arteries)
atherosclerosis
atrioventricular (AV) node
atrioventricular valves
atrium (*pl.* atria)
blood
blood vessel
capillary (pl. capillaries)
capillary network
cardiac cycle
cardiac output
cardiovascular disease
cardiovascular system
diastole
erythrocytes
granulocytes
heart
hematopoiesis
leukocytes
lymph
myogenic
pacemaker (=sinoatrial node)
plasma
plasma
platelets
pulmonary arter
pulmonary circulation
pulmonary vein
Purkyne fibers
semilunar valves
systemic circulation
systole
tissue fluid
vein
vena cava
ventricles

Periodicals:
Listings for this chapter are on page 279

Weblinks:
www.thebiozone.com/weblink/AnaPhy-3572.html

BIOZONE APP:
Student Review Series
The Cardiovascular System

Key concepts

- The cardiovascular system transports blood to all parts of the body within closed vessels.
- Blood carries oxygen, nutrients, and waste products and has a defensive role in the body.
- The heart is a muscular pump with its own intrinsic rate of beating. This rate is influenced by the autonomic nervous system.
- Cardiovascular diseases affect the heart and blood vessels. Many are treatable.

Learning Objectives

☐ 1. Use the **KEY TERMS** to compile a glossary for this topic.

The Cardiovascular System

pages 135-143

☐ 2. Describe the primary components of the **cardiovascular system**, identifying the location of the **heart** and major vessels. Distinguish between the **pulmonary circulation** and the **systemic circulation**.

☐ 3. Use annotated diagrams to describe the structure of **arteries**, **capillaries**, and **veins** and explain the relationship between structure and function of these **blood vessels**.

☐ 4. Draw a diagram to show the relative positions of blood vessels in a **capillary network** and their relationship to the lymphatic vessels. Explain the functional relationship between **blood**, **lymph**, **plasma**, and **tissue fluid**, including reference to how the tissue fluid is formed.

☐ 5. Describe the process of **hematopoiesis**, the formation of blood cells from stem cells in the red bone marrow.

☐ 6. Recognize the transport functions of the lymphatic system, identifying how lymph is returned to the blood circulatory system.

☐ 7. Describe the composition of blood, including the functional role of the **plasma** and the cellular components (**erythrocytes**, **leukocytes**, **granulocytes**, **platelets**).

Heart Structure and Function

pages 29, 145-153

☐ 8. Recall the features of cardiac muscle. Use a diagram to describe the internal and external gross structure of the heart, including **atria**, **ventricles**, **atrioventricular valves**, **semilunar valves**, major vessels, and coronary circulation. Relate the differences in the thickness of the heart chambers to their functions.

☐ 9. Explain the events of the **cardiac cycle**, relating these to the maintenance of blood flow through the heart. Analyze data showing pressure and volume changes in the left atrium, left ventricle, and the aorta during the cardiac cycle.

☐ 10. Describe the intrinsic control of heart beat, including the role of the pacemaker (**sinoatrial node**) in initiating the heartbeat and the role of the heart's internal conduction system.

☐ 11. Describe the extrinsic regulation of the intrinsic heart rate through the autonomic nervous system. Identify the role of the medulla, baroreceptors, and chemoreceptors in the response of the heart to changing demands.

Cardiovascular Health

pages 144, 154-160

☐ 12. Describe and explain the short and long term responses of the cardiovascular system to aerobic exercise. Include reference to resting pulse, heart rate, stroke volume, and **cardiac output**.

☐ 13. Recognize **cardiovascular disease** (CVD) as a term encompassing a range of diseases. Describe **atherosclerosis** and its relationship to myocardial infarction. Describe risk factors in the development of CVD.

The Human Transport System

The blood vessels of the circulatory system form a vast network of tubes that carry blood away from the heart, transport it to the tissues of the body, and then return it to the heart. The arteries, arterioles, capillaries, venules, and veins are organized into specific routes to circulate the blood throughout the body. The figure below shows a number of the basic **circulatory routes** through which the blood travels. Mammals have a **double circulatory system**: a **pulmonary system** (or circulation), which carries blood between the heart and lungs, and a **systemic system** (circulation), which carries blood between the heart and the rest of the body. The systemic circulation has many subdivisions. Two important subdivisions are the coronary (cardiac) circulation, which supplies the heart muscle, and the **hepatic portal circulation**, which runs from the gut to the liver.

Schematic Overview of the Human Circulatory System

Deoxygenated blood (coloured grey below) travels to the right side of the heart via the vena cavae. The heart pumps the deoxygenated blood to the lungs where it releases carbon dioxide and receives oxygen. The oxygenated blood (coloured white below) travels via the pulmonary vein back to the heart from where it is pumped to all parts of the body. The **venous system** (figure, left) returns blood from the capillaries to the heart. The **arterial system** (figure right) carries blood from the heart to the capillaries. **Portal systems** carry blood between two capillary beds.

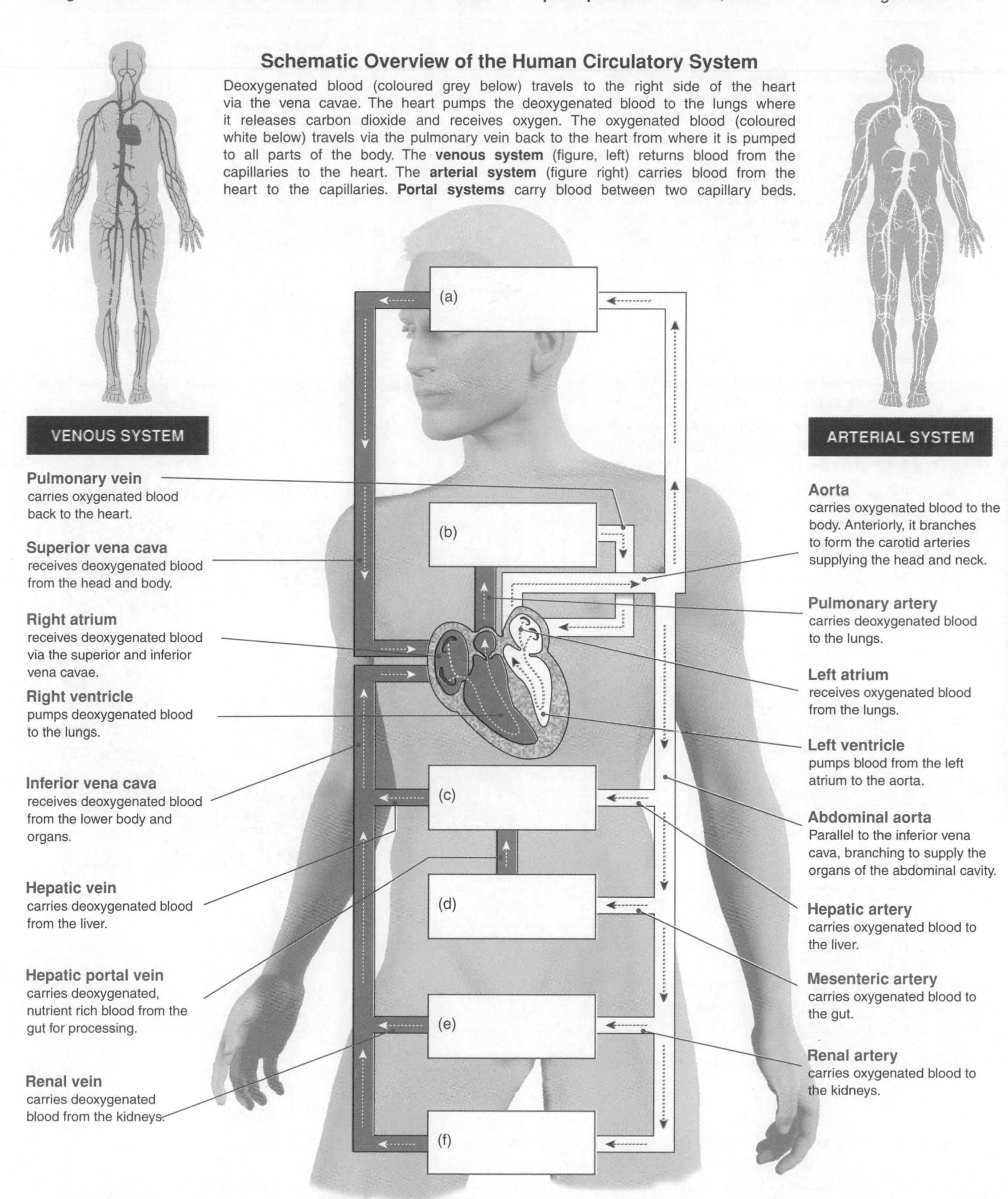

1. Complete the diagram above by labeling the boxes with the organs or structures they represent.

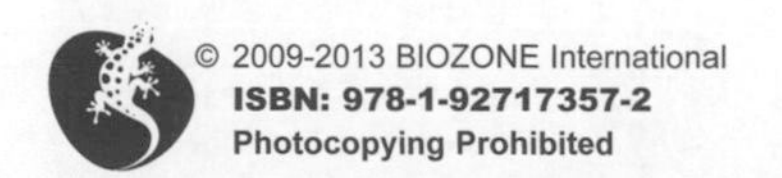

ISBN: 978-1-92717357-2

Arteries

Arteries are the blood vessels that carry blood away from the heart to the capillaries within the tissues. The large arteries that leave the heart divide into medium-sized (distributing) arteries. Within the tissues and organs, these distribution arteries branch to form very small vessels called **arterioles**, which deliver blood to capillaries. Arterioles lack the thick layers of arteries and consist only of an endothelial layer wrapped by a few smooth muscle fibers at intervals along their length. Resistance to blood flow is altered by contraction (**vasoconstriction**) or relaxation (**vasodilation**) of the blood vessel walls, especially in the arterioles. Vasoconstriction increases resistance and leads to an increase in blood pressure whereas vasodilation has the opposite effect. This mechanism is important in regulating the blood flow into tissues.

Arteries

Arteries have an elastic, stretchy structure that gives them the ability to withstand the high pressure of blood being pumped from the heart. At the same time, they help to maintain pressure by having some contractile ability themselves (a feature of the central muscle layer). Arteries nearer the heart have more elastic tissue, giving greater resistance to the higher blood pressures of the blood leaving the left ventricle. Arteries further from the heart have more muscle to help them maintain blood pressure. Between heartbeats, the arteries undergo elastic recoil and contract. This tends to smooth out the flow of blood through the vessel.

Arteries comprise three main regions (right):

1. A thin inner layer of epithelial cells called the **endothelium** lines the artery.
2. A central layer (the **tunica media**) of elastic tissue and smooth muscle that can both stretch and contract.
3. An outer connective tissue layer (the **tunica externa**) has a lot of elastic tissue.

Artery Structure

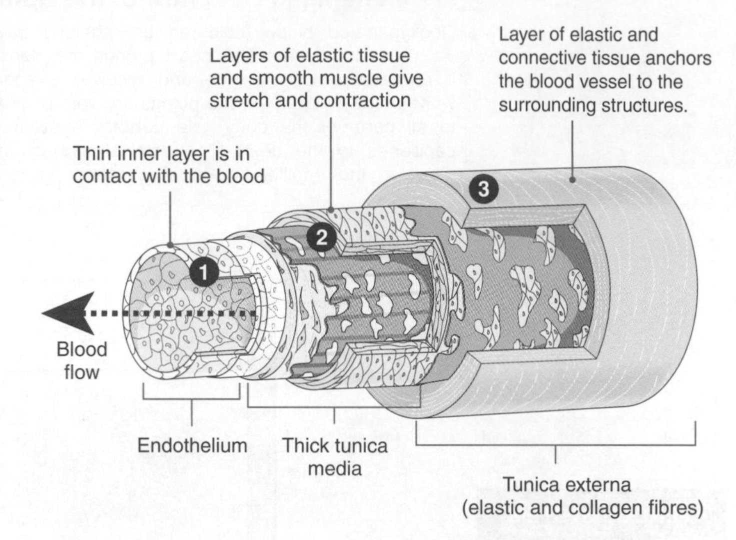

Cross section through a large artery

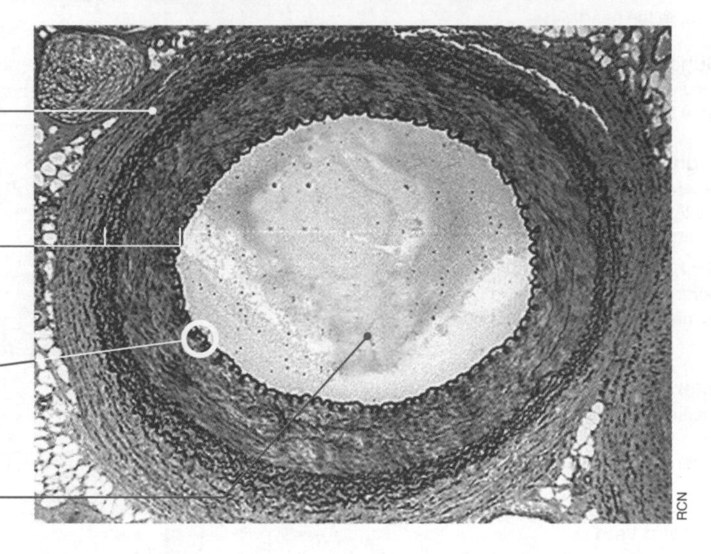

(a)

(b)

(c)

(d)

1. Using the diagram to help you, label the photograph of the cross section through an artery (above).
2. (a) Explain why the walls of arteries need to be thick with a lot of elastic tissue: ______

   (b) Explain why arterioles lack this elastic tissue layer: ______
3. Explain the purpose of the smooth muscle in the artery walls: ______
4. (a) Describe the effect of vasodilation on the diameter of an arteriole: ______

   (b) Describe the effect of vasodilation on blood pressure: ______

Related activities: *Veins, Capillaries*
Weblinks: *Arteries*

Periodicals:
Cunning plumbing

ISBN: 978-1-92717357-2

Veins

Veins are the blood vessels that return blood to the heart from the tissues. The smallest veins (**venules**) return blood from the capillary beds to the larger veins. Veins and their branches contain about 59% of the blood in the body. The structural differences between veins and arteries are mainly associated with differences in the relative thickness of the vessel layers and the diameter of the lumen. These, in turn, are related to the vessel's functional role.

Veins

When several capillaries unite, they form small veins called **venules**. The venules collect the blood from capillaries and drain it into **veins**. Veins are made up of essentially the same three layers as arteries but they have less elastic and muscle tissue and a larger **lumen**. The venules closest to the capillaries consist of an **endothelium** and a tunica externa of connective tissue. As the venules approach the veins, they also contain the tunica media characteristic of veins (right). Although veins are less elastic than arteries, they can still expand enough to adapt to changes in the pressure and volume of the blood passing through them. Blood flowing in the veins has lost a lot of pressure because it has passed through the narrow capillary vessels. The low pressure in veins means that many veins, especially those in the limbs, need to have valves to prevent back-flow of the blood as it returns to the heart.

Vein Structure

Inner thin layer of simple squamous epithelium lines the vein (**endothelium** or **tunica intima**).

Central thin layer of elastic and muscle tissue (**tunica media**). The smaller venules lack this inner layer.

Layer of elastic connective tissue (**tunica externa**)

One-way valves are located along the length of veins to prevent the blood from flowing backwards.

Blood flow

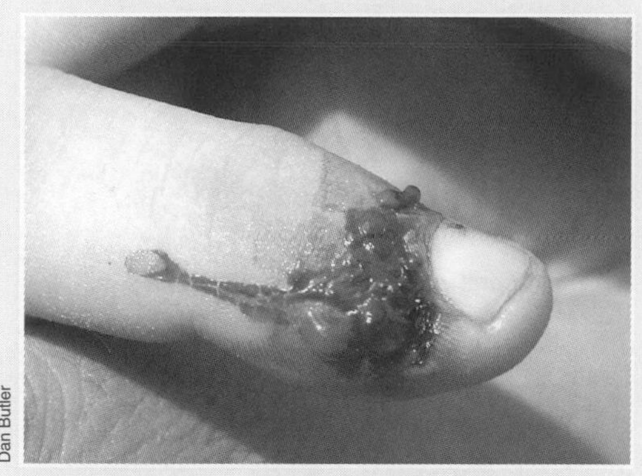

If a vein is cut, as is shown in this severe finger wound, the blood oozes out slowly in an even flow, and usually clots quickly as it leaves. In contrast, arterial blood spurts rapidly and requires pressure to staunch the flow.

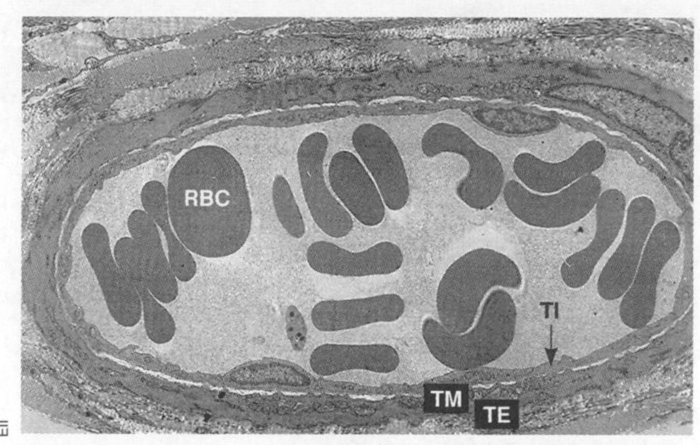

EII

Above: TEM of a vein showing red blood cells (RBC) in the lumen, and the tunica intima (TI), tunica media (TM), and tunica externa (TE).

1. Contrast the structure of veins and arteries for each of the following properties:

   (a) Thickness of muscle and elastic tissue: ______________________________

   (b) Size of the lumen (inside of the vessel): ______________________________

2. With respect to their functional roles, give a reason for the difference you have described above: ______________________________

3. Explain the role of the valves in assisting the veins to return blood back to the heart: ______________________________

4. Blood oozes from a venous wound, rather than spurting as it does from an arterial wound. Account for this difference: ______________________________

ISBN: 978-1-92717357-2

Related activities: *Arteries, Capillaries*
Weblinks: *Veins*

Capillaries

Capillaries are very small vessels that connect arterial and venous circulation and allow efficient exchange of nutrients and wastes between the blood and tissues. Capillaries form networks called capillary beds and are abundant where metabolic rates are high. Fluid that leaks out of the capillaries has an essential role in bathing the tissues.

Exchanges in Capillaries

Blood passes from the arterioles into the capillaries. Capillaries are small blood vessels with a diameter of just 4-10 µm. The only tissue present is an **endothelium** of squamous epithelial cells. Capillaries are so numerous that no cell is more than 25 µm from any capillary. It is in the capillaries that the exchange of materials between the body cells and the blood takes place.

Blood pressure causes fluid to leak from capillaries through small gaps where the endothelial cells join. This fluid bathes the tissues, supplying nutrients and oxygen, and removing wastes (right).The density of capillaries in a tissue is an indication of that tissue's metabolic activity. For example, cardiac muscle relies heavily on oxidative metabolism. It has a high demand for blood flow and is well supplied with capillaries. Smooth muscle is far less active than cardiac muscle, relies more on anaerobic metabolism, and does not require such an extensive blood supply.

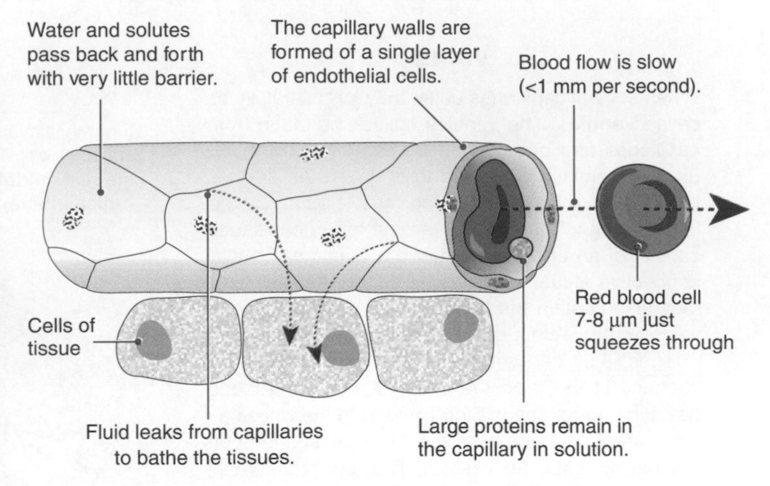

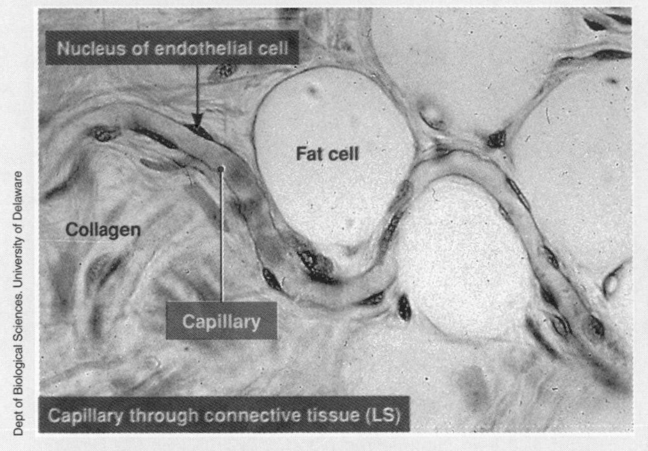

Capillaries are found near almost every cell in the body. In many places, the capillaries form extensive branching networks. In most tissues, blood normally flows through only a small portion of a capillary network when the metabolic demands of the tissue are low. When the tissue becomes active, the entire capillary network fills with blood.

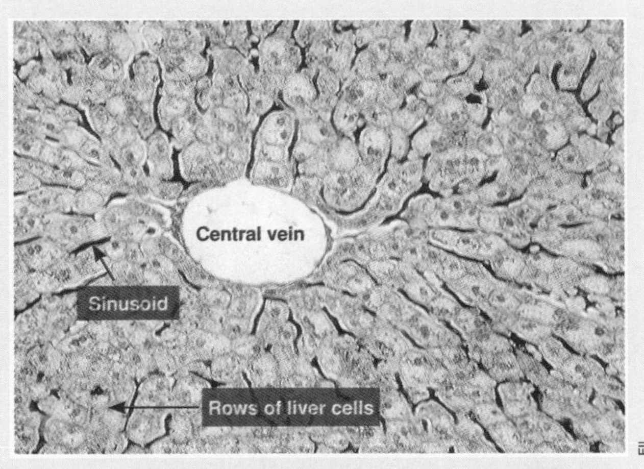

Microscopic blood vessels in some dense organs, such as the liver (above), are called **sinusoids**. They are wider than capillaries and follow a more convoluted path through the tissue. Instead of the usual endothelial lining, they are lined with phagocytic cells. Like capillaries, sinusoids transport blood from arterioles to venules.

1. Describe the structure of a capillary, contrasting it with the structure of a vein and an artery:

2. Sinusoids provide a functional replacement for capillaries in some organs:

   (a) How do sinusoids differ structurally from capillaries?

   (b) In what way are capillaries and sinusoids similar?

ISBN: 978-1-92717357-2

Related activities: Arteries, Veins, Capillary Networks
Weblinks: Microcirculation

Capillary Networks

Capillaries form branching networks where exchanges between the blood and tissues take place. The flow of blood through a capillary bed is called **microcirculation**. In most parts of the body, there are two types of vessels in a capillary bed: the **true capillaries**, where exchanges take place, and a vessel called a **vascular shunt**, which connects the arteriole and venule at either end of the bed. The shunt diverts blood past the true capillaries when the metabolic demands of the tissue are low (e.g. vasoconstriction in the skin when conserving body heat). When tissue activity increases, the entire network fills with blood.

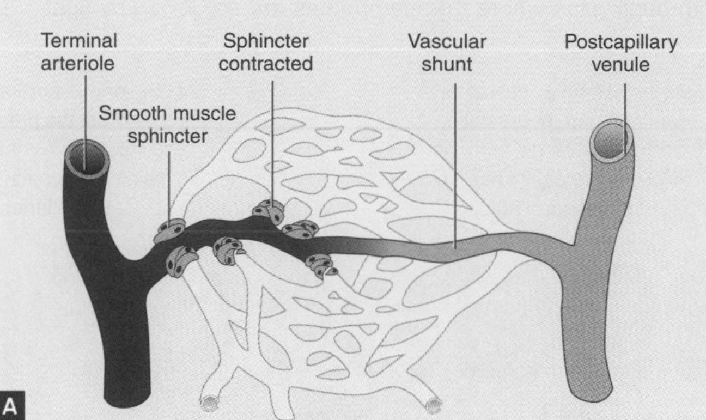

When the sphincters contract (close), blood is diverted via the vascular shunt to the post-capillary venule, bypassing the exchange capillaries.

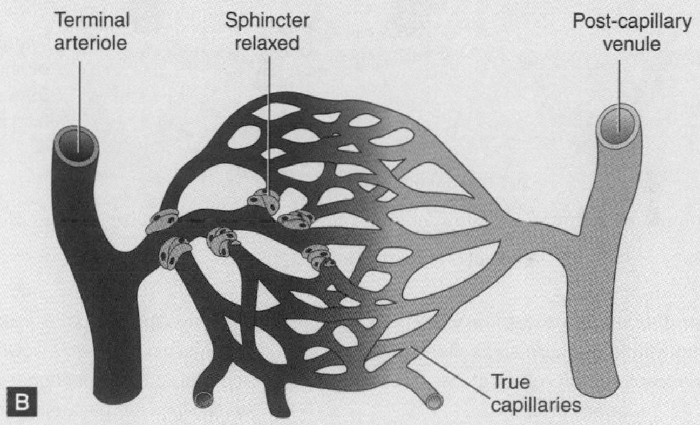

When the sphincters are relaxed (open), blood flows through the entire capillary bed allowing exchanges with the cells of the surrounding tissue.

Connecting Capillary Beds

The role of portal venous systems

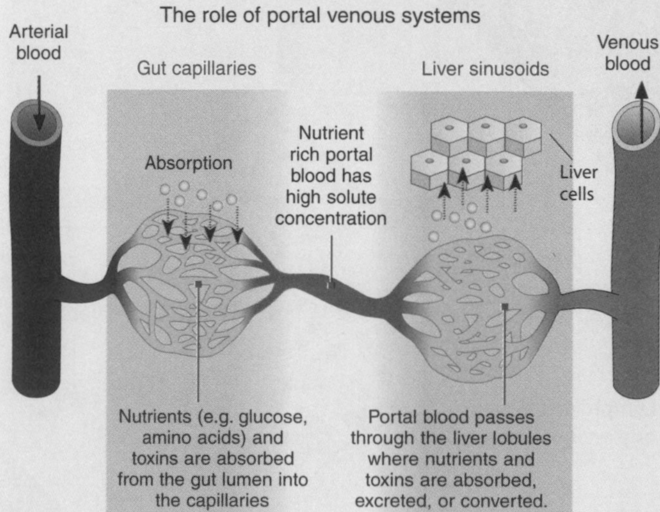

A portal venous system occurs when a capillary bed drains into another capillary bed through veins, without first going through the heart. Portal systems are relatively uncommon; most capillary beds drain into veins which then drain into the heart, not into another capillary bed. The diagram above depicts the hepatic portal system, which includes both capillary beds and the blood vessels connecting them.

1. Describe the structure of a capillary network:

2. Explain the role of the smooth muscle sphincters and the vascular shunt in a capillary network:

3. (a) Describe a situation where the capillary bed would be in the condition labelled **A**:

   (b) Describe a situation where the capillary bed would be in the condition labelled **B**:

4. How does a portal venous system differ from other capillary systems?

ISBN: 978-1-92717357-2

Related activities: *Capillaries*
Weblinks: *Microcirculation*

The Formation of Tissue Fluid

The network of capillaries supplying the body's tissues ensures that no substance has to diffuse far to enter or leave a cell. Substances exchanged first diffuse through the interstitial fluid (or tissue fluid), which surrounds and bathes the cells. As with all cells, substances can move into and out of the endothelial cells of the capillary walls in several ways; by diffusion, by cytosis, and through gaps where the membranes are not joined by tight junctions. Some fenestrated capillaries are also more permeable than others. These specialised capillaries are important where absorption or filtration occurs (e.g. in the intestine or the kidney). Capillaries are leaky, so fluid flows across their plasma membranes. Whether fluid moves into or out of a capillary depends on the balance between the blood (hydrostatic) pressure and the concentration of solutes at each end of a capillary bed.

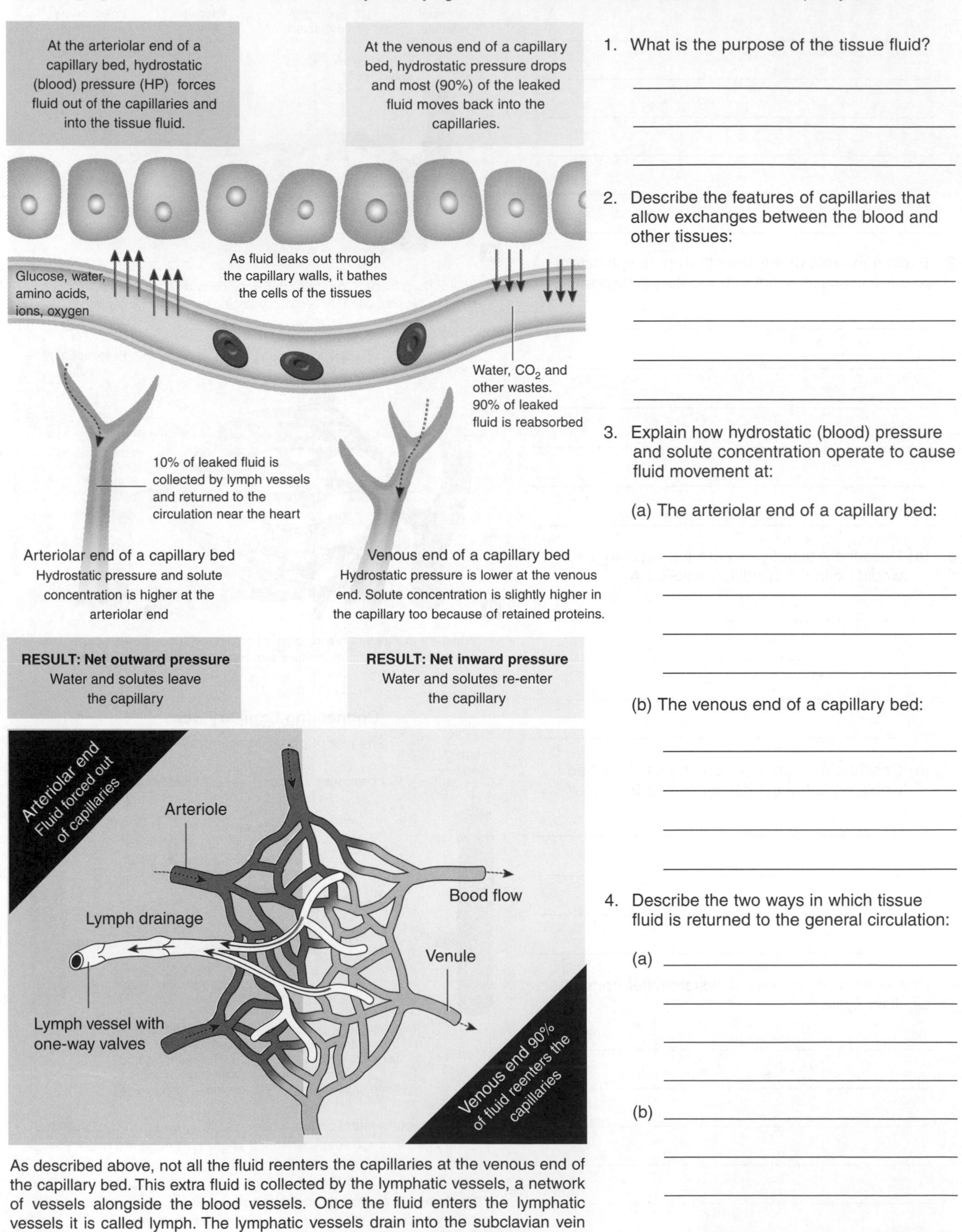

As described above, not all the fluid reenters the capillaries at the venous end of the capillary bed. This extra fluid is collected by the lymphatic vessels, a network of vessels alongside the blood vessels. Once the fluid enters the lymphatic vessels it is called lymph. The lymphatic vessels drain into the subclavian vein near the heart. Lymph is similar to tissue fluid but has more lymphocytes.

1. What is the purpose of the tissue fluid?

2. Describe the features of capillaries that allow exchanges between the blood and other tissues:

3. Explain how hydrostatic (blood) pressure and solute concentration operate to cause fluid movement at:

   (a) The arteriolar end of a capillary bed:

   (b) The venous end of a capillary bed:

4. Describe the two ways in which tissue fluid is returned to the general circulation:

   (a)

   (b)

ISBN: 978-1-92717357-2

Blood

Blood is a complex liquid tissue comprising cellular components suspended in plasma. It makes up about 8% of body weight. If a blood sample is taken, the cells can be separated from the plasma by centrifugation. The cells (formed elements) settle as a dense red pellet below the transparent, straw-colored **plasma**. Blood performs many functions. It transports nutrients, respiratory gases, hormones, and wastes and has a role in thermoregulation through the distribution of heat. Blood also defends against infection and its ability to clot protects against blood loss. The examination of blood is also useful in diagnosing disease. The cellular components of blood are normally present in particular specified ratios. A change in the morphology, type, or proportion of different blood cells can therefore be used to indicate a specific disorder or infection (see the next page).

Non-Cellular Blood Components

The non-cellular blood components form the plasma. Plasma is a watery matrix of ions and proteins and makes up 50-60% of the total blood volume.

Water
The main constituent of blood and lymph.
Role: Transports dissolved substances. Provides body cells with water. Distributes heat and has a central role in thermoregulation. Regulation of water content helps to regulate blood pressure and volume.

Mineral ions
Sodium, bicarbonate, magnesium, potassium, calcium, chloride.
Role: Osmotic balance, pH buffering, and regulation of membrane permeability. They also have a variety of other functions, e.g. Ca^{2+} is involved in blood clotting.

Plasma proteins
7-9% of the plasma volume.
Serum albumin
Role: Osmotic balance and pH buffering, Ca^{2+} transport.
Fibrinogen and prothrombin
Role: Take part in blood clotting.
Immunoglobulins
Role: Antibodies involved in the immune response.
α-globulins
Role: Bind/transport hormones, lipids, fat soluble vitamins.
β-globulins
Role: Bind/transport iron, cholesterol, fat soluble vitamins.
Enzymes
Role: Take part in and regulate metabolic activities.

Substances transported by non-cellular components
Products of digestion
Examples: sugars, fatty acids, glycerol, and amino acids.
Excretory products
Example: urea
Hormones and vitamins
Examples: insulin, sex hormones, vitamins A and B_{12}.
Importance: These substances occur at varying levels in the blood. They are transported to and from the cells dissolved in the plasma or bound to plasma proteins.

Cellular Blood Components

The cellular components of the blood (also called the formed elements) float in the plasma and make up 40-50% of the total blood volume.

Erythrocytes (red blood cells or RBCs)
5-6 million per mm^3 blood; 38-48% of total blood volume.
Role: RBCs transport oxygen (O_2) and a small amount of carbon dioxide (CO_2). The oxygen is carried bound to hemoglobin (Hb) in the cells. Each Hb molecule can bind four molecules of oxygen.

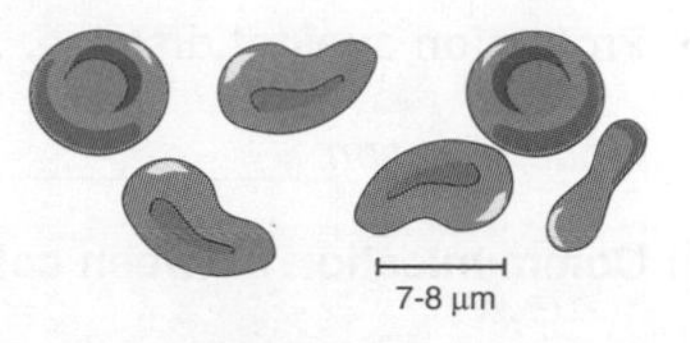

2 μm

Platelets
Small, membrane bound cell fragments derived from bone marrow cells; about 1/4 the size of RBCs.
0.25 million per mm^3 blood.
Role: To start the blood clotting process.

Leukocytes (white blood cells)
5-10 000 per mm^3 blood
2-3% of total blood volume.
Role: Involved in internal defense. There are several types of white blood cells (see below)..

Lymphocytes
T and B cells.
24% of the white cell count.
Role: Antibody production and cell mediated immunity.

Neutrophils
Phagocytes.
70% of the white cell count.
Role: Engulf foreign material.

Eosinophils
Rare leukocytes; normally 1.5% of the white cell count.
Role: Mediate allergic responses such as hayfever and asthma.

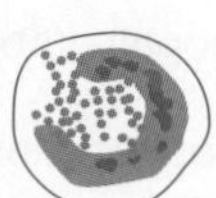

Basophils
Rare leukocytes; normally 0.5% of the white cell count.
Role: Produce heparin (an anti-clotting protein), and histamine. Involved in inflammation.

The Cardiovascular System

ISBN: 978-1-92717357-2

Related activities: *Gas Transport in Humans, The Body's Defenses, Blood Clotting and Defense*

The Examination of Blood

Different types of microscopy give different information about blood. A SEM (right) shows the detailed external morphology of the blood cells. A fixed smear of a blood sample viewed with a light microscope (far right) can be used to identify the different blood cell types present, and their ratio to each other. Determining the types and proportions of different white blood cells in blood is called a **differential white blood cell count.** Elevated counts of particular cell types indicate allergy or infection.

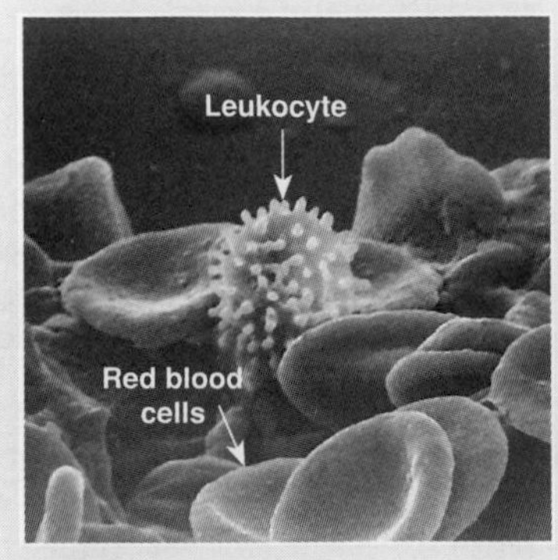

SEM of red blood cells and a leukocytes.

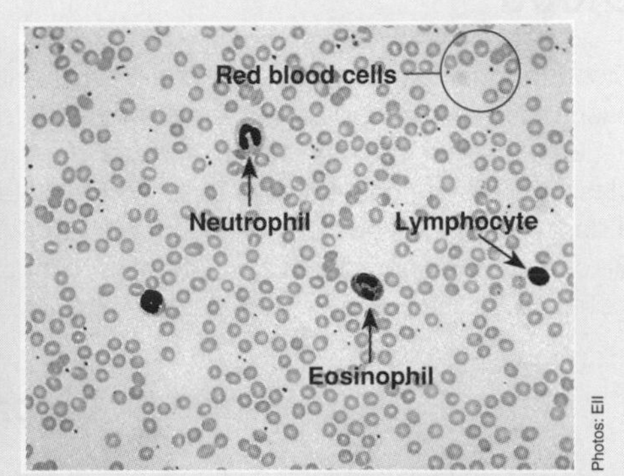

Light microscope view of a fixed blood smear.

1. For each of the following blood functions, identify the component (or components) of the blood responsible and state how the function is carried out (the mode of action). The first one is done for you:

   (a) **Temperature regulation**. *Blood component*: Water component of the plasma

   Mode of action: Water absorbs heat and dissipates it from sites of production (e.g. organs)

   (b) **Protection against disease.** *Blood component*: ____________________

   Mode of action: ____________________

   (c) **Communication between cells, tissues, and organs.** *Blood component*: ____________________

   Mode of action: ____________________

   (d) **Oxygen transport.** *Blood component*: ____________________

   Mode of action: ____________________

   (e) **CO_2 transport.** *Blood components*: ____________________

   Mode of action: ____________________

   (f) **Buffer against pH changes**. *Blood components*: ____________________

   Mode of action: ____________________

   (g) **Nutrient supply**. *Blood component*: ____________________

   Mode of action: ____________________

   (h) **Tissue repair**. *Blood components*: ____________________

   Mode of action: ____________________

   (i) **Transport of hormones, lipids, and fat soluble vitamins**. *Blood component*: ____________________

   Mode of action: ____________________

2. Identify a feature that distinguishes red and white blood cells: ____________________

3. Explain two physiological advantages of red blood cell structure (lacking nucleus and mitochondria):

   (a) ____________________

   (b) ____________________

4. Suggest what each of the following results from a differential white blood cell count would suggest:

   (a) Elevated levels of eosinophils (above the normal range): ____________________

   (b) Elevated levels of neutrophils (above the normal range): ____________________

   (c) Elevated levels of basophils (above the normal range): ____________________

   (d) Elevated levels of lymphocytes (above the normal range): ____________________

ISBN: 978-1-92717357-2

Hematopoiesis

Hematopoiesis (also called hemopoiesis) refers to the formation of blood cells. All cellular blood components are derived from **hematopoietic stem cells** (HSCs). In a healthy adult person, approximately 10^{11}-10^{12} new blood cells are produced every day in order to maintain homeostasis of the peripheral circulation. Before birth, hematopoiesis occurs in aggregates of blood cells in the yolk sac, then the liver, and eventually the bone marrow. In normal adults, HSCs reside in the red bone marrow and have the ability to give rise to all of the different mature blood cell types. Like all stem cells, HSCs can divide many times while remaining unspecialized and, when given the right signals, they can differentiate into other cell types. When they proliferate, some of the daughter cells remain as HSCs and some give rise to progenitor cells. The progenitor cells then each commit to any of the alternative differentiation pathways that lead to the production of one or more types of blood cells. Blood cells are divided into lineages. **Erythroid cells** are the oxygen carrying red blood cells. **Lymphoid cells** are the white blood cells of the adaptive immune system. They are derived from common lymphoid progenitors. **Myeloid cells** are derived from common myeloid progenitors, and are involved in many diverse roles within the body's defense system.

Stem Cells and Blood Cell Production

New blood cells are produced in the red bone marrrow, which becomes the main site of blood production after birth, taking over from the fetal liver. All types of blood cells develop from a single cell type: called a **multipotent stem cell** or hemocytoblast. These cells are capable of mitosis and of differentiation into 'committed' precursors of each of the main types of blood cell.

Each of the different cell lines is controlled by a specific growth factor. When a stem cell divides, one of its daughters remains a stem cell, while the other becomes a precursor cell, either a lymphoid cell or myeloid cell. These cells continue to mature into the various type of blood cells, developing their specialized features and characteristic roles as they do so.

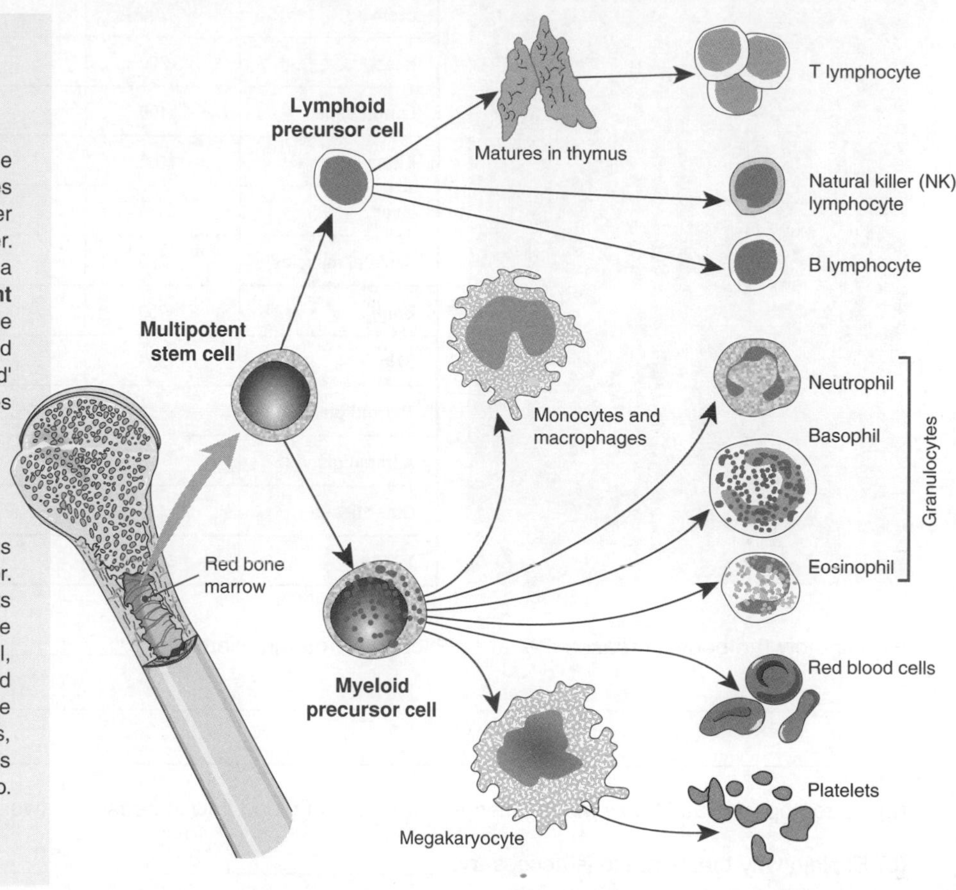

1. Where are new blood cells produced in the body:

   (a) Before birth? ______

   (b) After birth? ______

2. Identify the blood cell types arising from each of the progenitor cell types:

   (a) Myeloid progenitor cells: ______

   (b) Lymphoid progenitor cells: ______

3. (a) Using an example, explain the purpose of stem cells in an adult: ______

   (b) Identify where else in the body, apart from the red bone marrow, you might find stem cells: ______

   (c) Explain why blood cells are constantly being produced, when some other cells (e.g. neurons) are not:

ISBN: 978-1-92717357-2

Related activities: Blood, The Body's Defenses, Stem Cell Technology

Exercise and Blood Flow

Exercise promotes health by improving the rate of blood flow back to the heart (venous return). This is achieved by strengthening all types of muscle and by increasing the efficiency of the heart. During exercise blood flow to different parts of the body changes in order to cope with the extra demands of the muscles, the heart and the lungs.

1. The following table gives data for the **rate** of blood flow to various parts of the body at rest and during strenuous exercise. **Calculate** the **percentage** of the total blood flow that each organ or tissue receives under each regime of activity.

Organ or tissue	At rest		Strenuous exercise	
	$cm^3\ min^{-1}$	% of total	$cm^3\ min^{-1}$	% of total
Brain	700	14	750	4.2
Heart	200		750	
Lung tissue	100		200	
Kidneys	1100		600	
Liver	1350		600	
Skeletal muscles	750		12 500	
Bone	250		250	
Skin	300		1900	
Thyroid gland	50		50	
Adrenal glands	25		25	
Other tissue	175		175	
TOTAL	5000	**100**	17 800	**100**

2. Explain how the body increases the rate of blood flow during exercise: ______

3. (a) State approximately how many times the total rate of blood flow increases between rest and exercise: ______

   (b) Explain why the increase is necessary: ______

4. (a) Identify which organs or tissues show no change in the rate of blood flow with exercise: ______

   (b) Explain why this is the case: ______

5. (a) Identify the organs or tissues that show the most change in the rate of blood flow with exercise: ______

   (b) Explain why this is the case: ______

ISBN: 978-1-92717357-2

Related activities: *The Effects of Aerobic Training*

The Human Heart

The heart is at the center of the human cardiovascular system. It is a hollow, muscular organ, weighing on average 342 grams. Each day it beats over 100,000 times to pump 3780 litres of blood through 100,000 kilometers of blood vessels. It comprises a system of four muscular chambers (two **atria** and two **ventricles**) that alternately fill and empty of blood, acting as a double pump. The left side pumps blood to the body tissues and the right side pumps blood to the lungs. The heart lies between the lungs, to the left of the body's midline, and it is surrounded by a double layered **pericardium** of tough fibrous connective tissue. The pericardium prevents over-distension of the heart and anchors the heart within the **mediastinum**.

Human heart structure

(sectioned, anterior view)

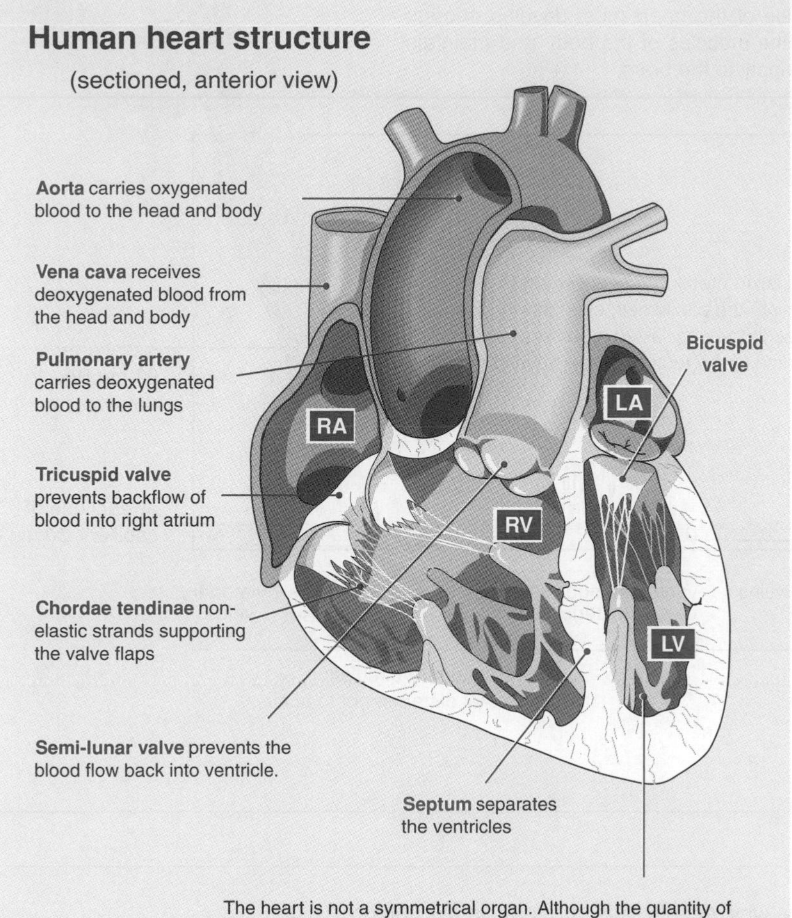

The heart is not a symmetrical organ. Although the quantity of blood pumped by each side is the same, the walls of the left ventricle are thicker and more muscular than those of the right ventricle. The difference affects the shape of the ventricular cavities, so the right ventricle is twisted over the left.

Key to abbreviations

- **RA** Right atrium: receives deoxygenated blood via the anterior and posterior vena cava
- **RV** Right ventricle: pumps deoxygenated blood to the lungs via the pulmonary artery
- **LA** Left atrium: receives blood returning to the heart from the lungs via the pulmonary veins
- **LV** Left ventricle: pumps oxygenated blood to the head and body via the aorta

Top view of a heart in section, showing valves

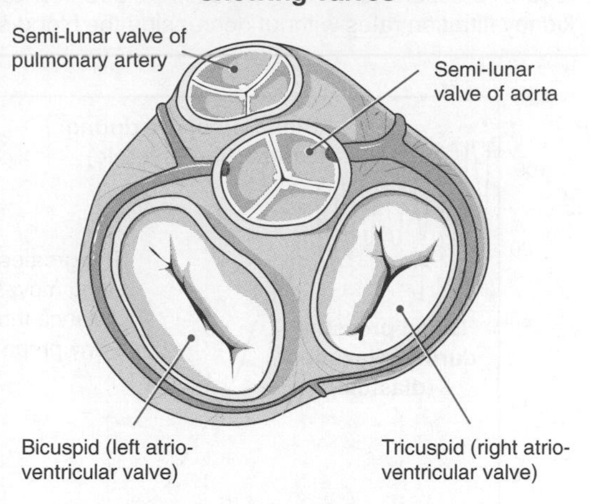

Posterior view of heart

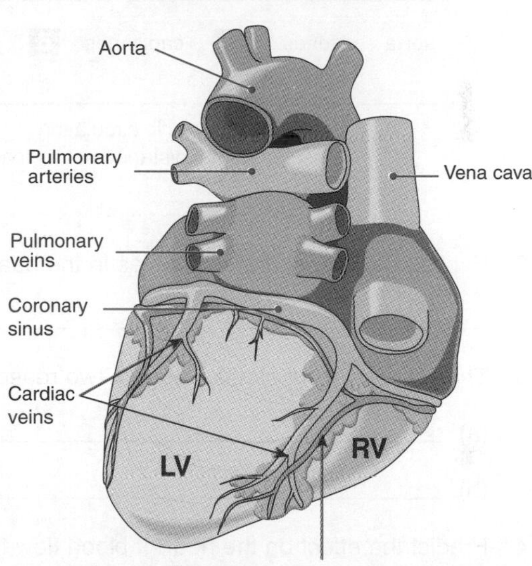

Coronary arteries: The high oxygen demands of the heart muscle are met by a dense capillary network. Coronary arteries arise from the aorta and spread over the surface of the heart supplying the cardiac muscle with oxygenated blood. Deoxygenated blood is collected by cardiac veins and returned to the right atrium via a large coronary sinus.

1. In the schematic diagram of the heart, below, label the four chambers and the main vessels entering and leaving them. The arrows indicate the direction of blood flow. Use large colored circles to mark the position of each of the four valves.

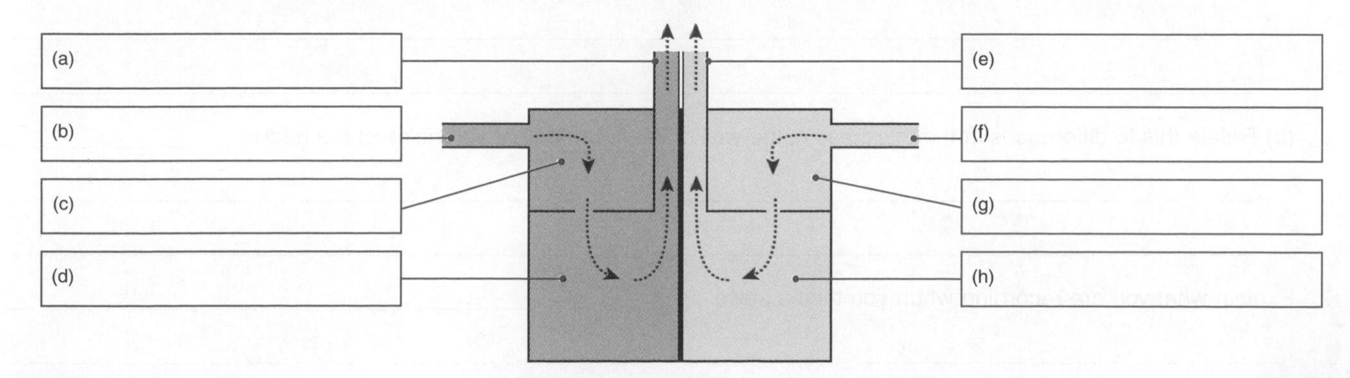

ISBN: 978-1-92717357-2

Periodicals: *The heart, Keeping pace*

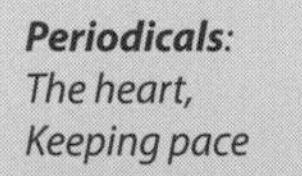

Related activities: *Dissecting a Mammalian Heart*
Weblinks: *Anatomy of the Heart, How the Heart Works*

Pressure Changes and the Asymmetry of the Heart

The heart is not a symmetrical organ. The left ventricle and its associated arteries are thicker and more muscular than the corresponding structures on the right side. This asymmetry is related to the necessary pressure differences between the pulmonary (lung) and systemic (body) circulations (not to the distance over which the blood is pumped *per se*). The graph below shows changes in blood pressure in each of the major blood vessel types in the systemic and pulmonary circuits (the horizontal distance not to scale). The pulmonary circuit must operate at a much lower pressure than the systemic circuit to prevent fluid from accumulating in the alveoli of the lungs. The left side of the heart must develop enough "spare" pressure to enable increased blood flow to the muscles of the body and maintain kidney filtration rates without decreasing the blood supply to the brain.

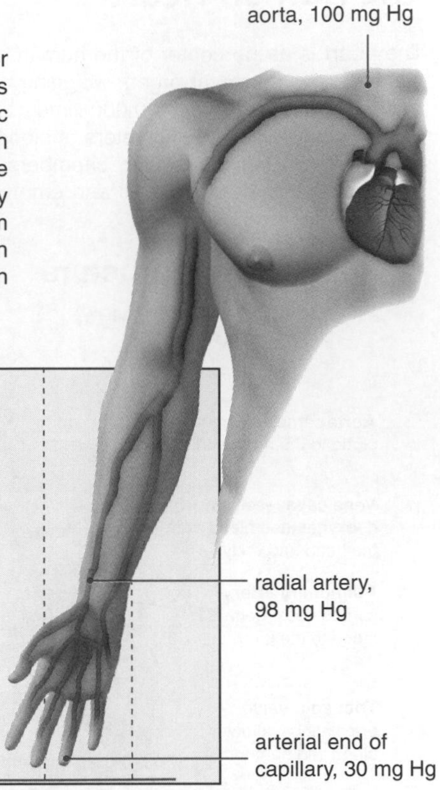

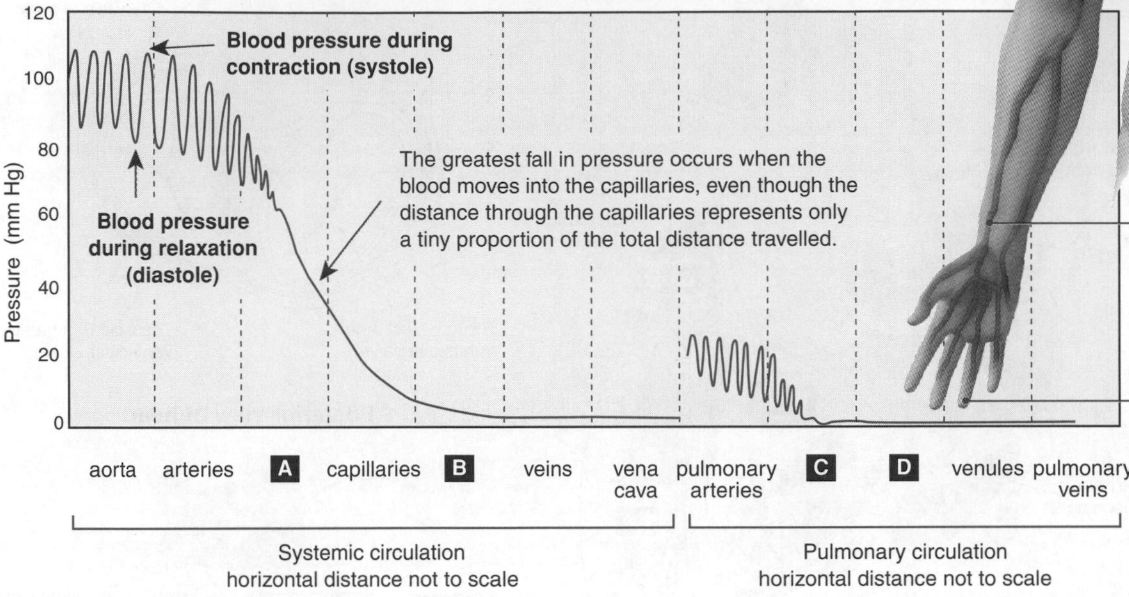

2. Explain the purpose of the valves in the heart: ______________________________

3. The heart is full of blood. Suggest two reasons why, despite this, it needs its own blood supply:

   (a) ______________________________

   (b) ______________________________

4. Predict the effect on the heart if blood flow through a coronary artery is restricted or blocked: ______________________________

5. Identify the vessels corresponding to the letters **A-D** on the graph above:

   A: ____________ B: ____________ C: ____________ D: ____________

6. (a) Explain why the pulmonary circuit must operate at a lower pressure than the systemic system: ______________________________

   (b) Relate this to differences in the thickness of the wall of the left and right ventricles of the heart: ______________________________

7. Explain what you are recording when you take a pulse: ______________________________

ISBN: 978-1-92717357-2

The Cardiac Cycle

The heart pumps with alternate contractions (**systole**) and relaxations (**diastole**). The **cardiac cycle** refers to the sequence of events of a heartbeat and involves three main stages: atrial systole, ventricular systole, and complete cardiac diastole. Pressure changes within the heart's chambers generated by the cycle of contraction and relaxation are responsible for blood movement and cause the heart valves to open and close, preventing the back-flow of blood. The noise of the blood when the valves open and close produces the heartbeat sound (**lubb-dupp**). The heart beat occurs in response to electrical impulses, which can be recorded as a trace, called an **electrocardiogram** or **ECG**. The ECG pattern is the result of the different impulses produced at each phase of the cardiac cycle, and each part is identified with a letter code. An ECG provides a useful method of monitoring changes in heart rate and activity and detection of heart disorders. The electrical trace is accompanied by volume and pressure changes (below).

The Cardiac Cycle

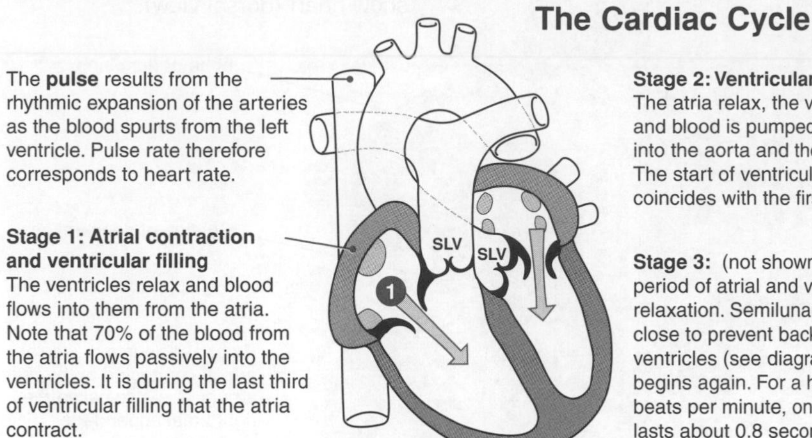

The **pulse** results from the rhythmic expansion of the arteries as the blood spurts from the left ventricle. Pulse rate therefore corresponds to heart rate.

Stage 1: Atrial contraction and ventricular filling
The ventricles relax and blood flows into them from the atria. Note that 70% of the blood from the atria flows passively into the ventricles. It is during the last third of ventricular filling that the atria contract.

Heart during ventricular filling

Stage 2: Ventricular contraction
The atria relax, the ventricles contract, and blood is pumped from the ventricles into the aorta and the pulmonary artery. The start of ventricular contraction coincides with the first heart sound.

Stage 3: (not shown) There is a short period of atrial and ventricular relaxation. Semilunar valves (**SLV**) close to prevent backflow into the ventricles (see diagram, left). The cycle begins again. For a heart beating at 75 beats per minute, one cardiac cycle lasts about 0.8 seconds.

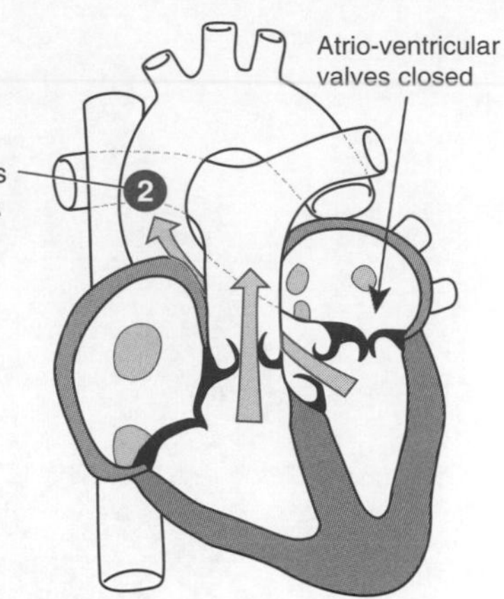

Heart during ventricular contraction

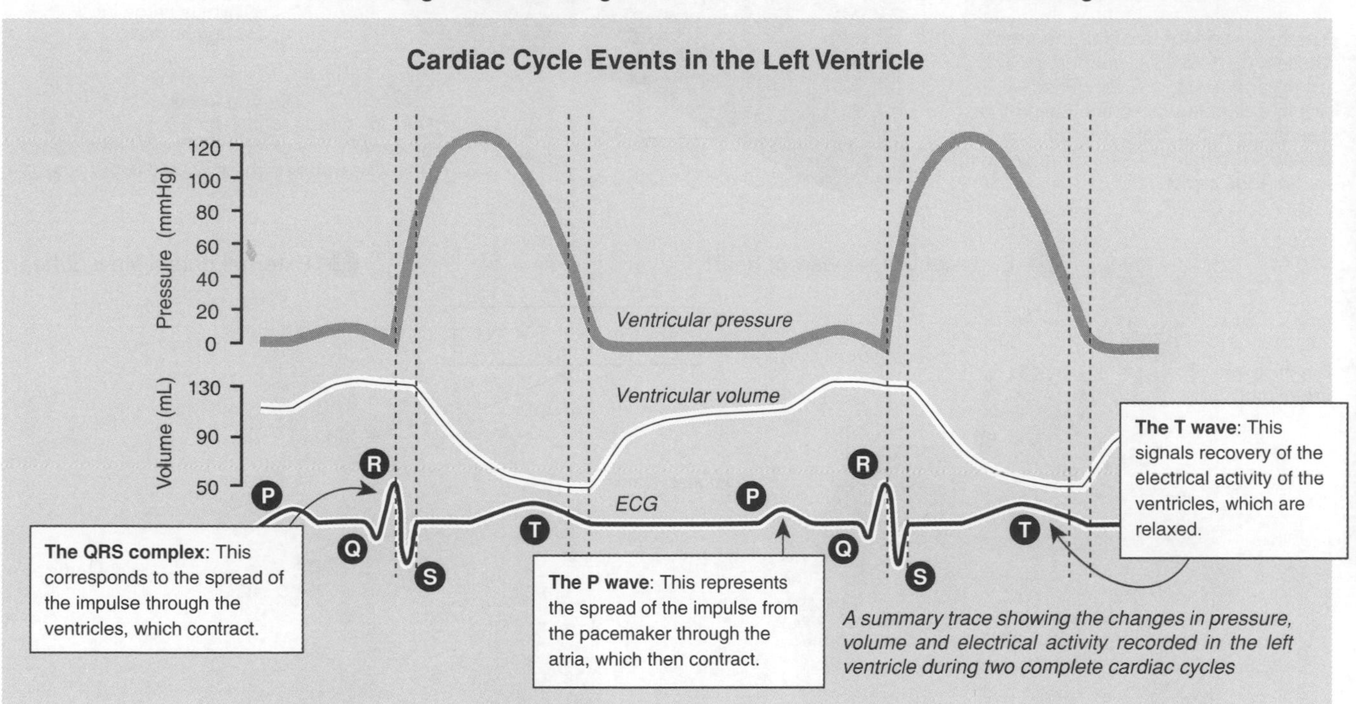

A summary trace showing the changes in pressure, volume and electrical activity recorded in the left ventricle during two complete cardiac cycles

1. Identify each of the following phases of an ECG by its international code:

   (a) Excitation of the ventricles and ventricular systole: __________

   (b) Electrical recovery of the ventricles and ventricular diastole: __________

   (c) Excitation of the atria and atrial systole: __________

2. Suggest the physiological reason for the period of electrical recovery experienced each cycle (the T wave):

   ____

3. Using the letters indicated, mark the points on trace above corresponding to each of the following:

   (a) E: Ejection of blood from the ventricle

   (b) AVC: Closing of the atrioventricular valve

   (c) FV: Filling of the ventricle

   (d) AVO: Opening of the atrioventricular valve

ISBN: 978-1-92717357-2

Dissecting a Mammalian Heart

The dissection of a sheep's heart is a common practical activity and allows hands-on exploration of the physical appearance and structure of a mammalian heart. A diagram of a heart is an idealized representation of an organ that may look quite different in reality. You must learn to transfer what you know from a diagram to the interpretation of the real organ. If you are lucky, you will have access to a fresh specimen, with most of the large vessels intact. The aim of this activity is to provide photographs of some aspects of a heart dissection to help you in identifying specific structures during your own dissection. It may also serve, in part, as a dissection replacement for those who, for personal reasons, do not wish to perform the dissection themselves.

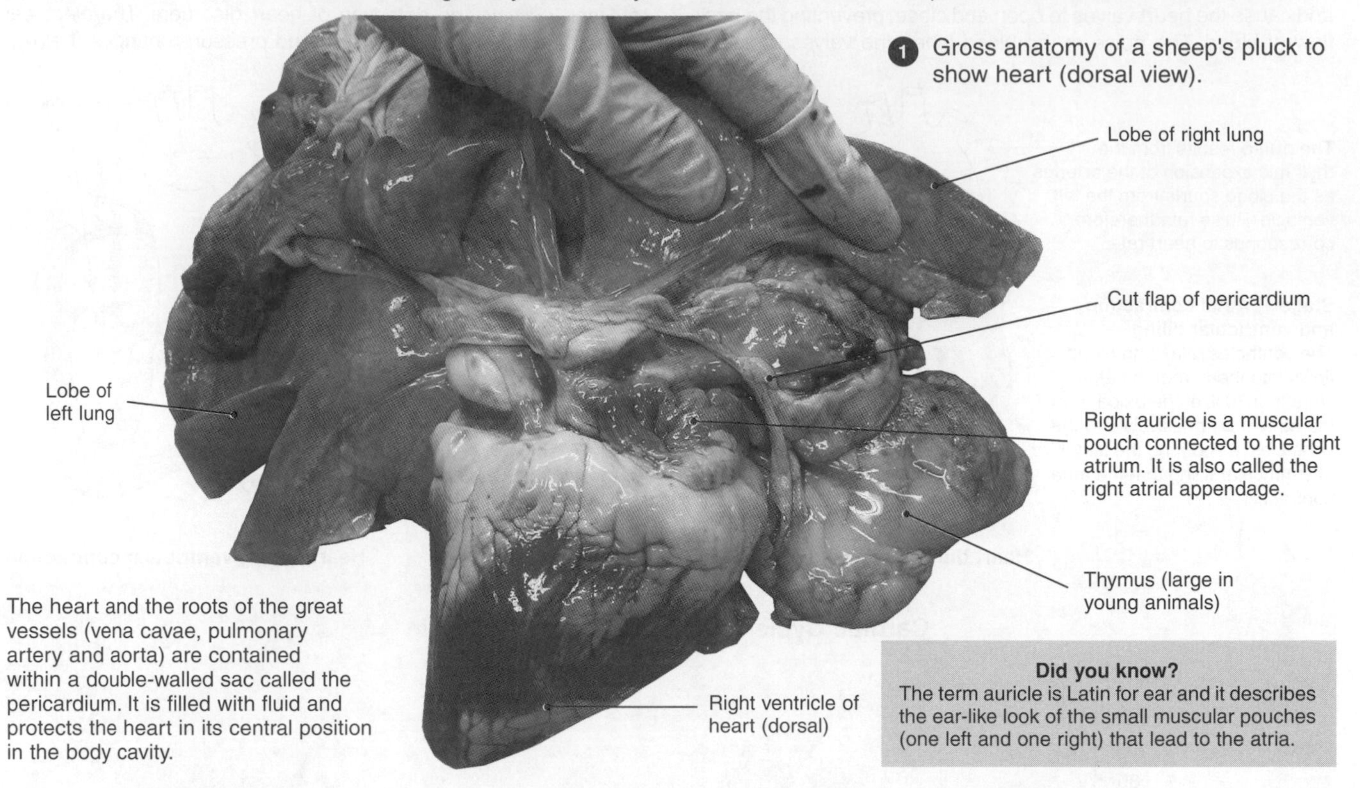

The heart and the roots of the great vessels (vena cavae, pulmonary artery and aorta) are contained within a double-walled sac called the pericardium. It is filled with fluid and protects the heart in its central position in the body cavity.

Did you know?
The term auricle is Latin for ear and it describes the ear-like look of the small muscular pouches (one left and one right) that lead to the atria.

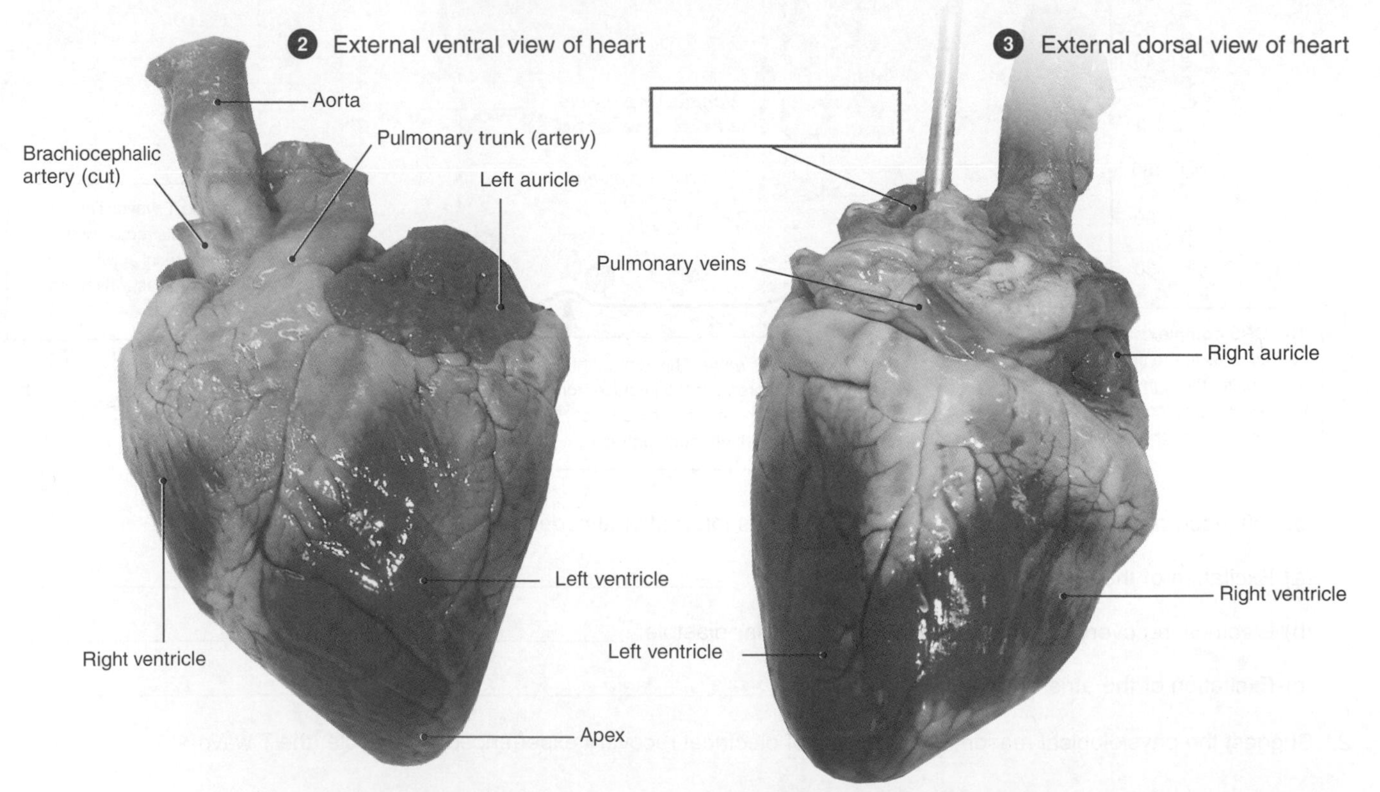

Note the main surface features of an isolated heart. The narrow pointed end forms the **apex** of the heart, while the wider end, where the blood vessels enter is the **base**. The ventral surface of the heart (above) is identified by a groove, the **interventricular sulcus**, which marks the division between the left and right ventricles.

1. Use colored lines to indicate the interventricular sulcus and the base of the heart. Label the coronary arteries.

On the dorsal surface of the heart, above, locate the large thin-walled **vena cavae** and **pulmonary veins**. You may be able to distinguish between the anterior and posterior vessels. On the right side of the dorsal surface (as you look at the heart) at the base of the heart is the **right atrium**, with the **right ventricle** below it.

2. On this photograph, label the vessel indicated by the probe.

ISBN: 978-1-92717357-2

Related activities: The Human Heart
Weblinks: Anatomy of the Heart

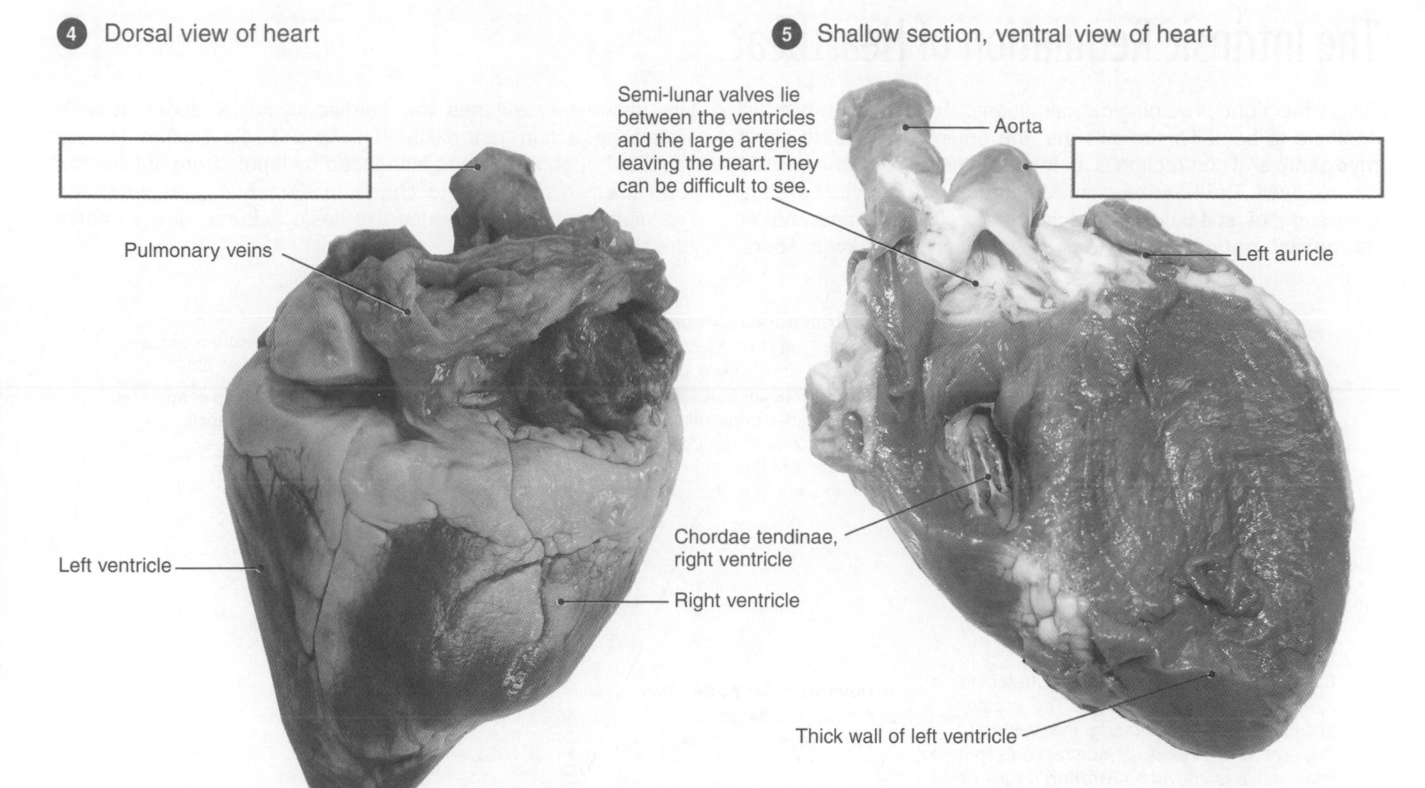

3. On this **dorsal view**, label the vessel indicated. Palpate the heart and feel the difference in the thickness of the left and right ventricle walls.

4. This photograph shows a shallow section to expose the right ventricle. Label the vessel in the box indicated.

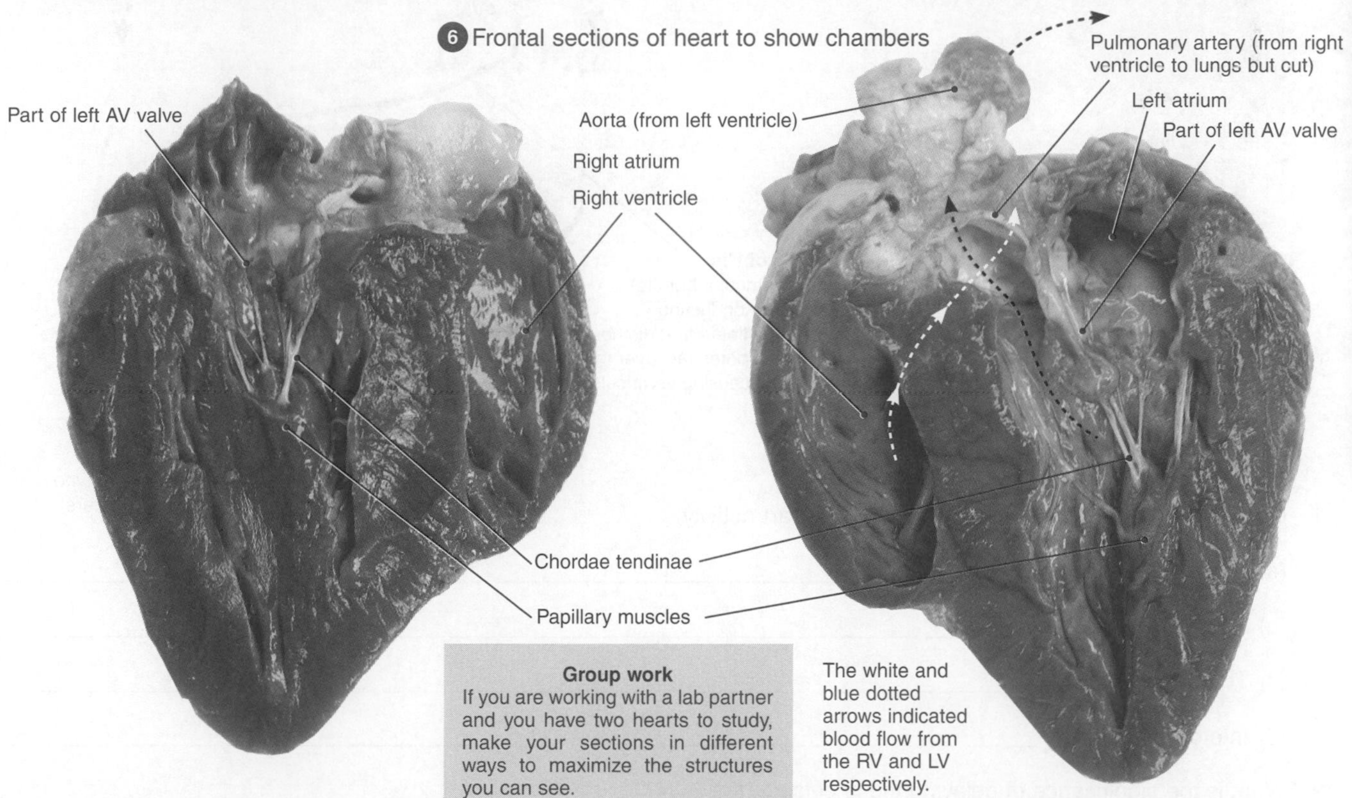

Group work
If you are working with a lab partner and you have two hearts to study, make your sections in different ways to maximize the structures you can see.

The white and blue dotted arrows indicated blood flow from the RV and LV respectively.

If the heart is sectioned and the two halves opened, the valves of the heart can be seen. Each side of the heart has a one-way valve between the atrium and the ventricle known as the **atrioventricular valve**. They close during ventricular contraction to prevent back flow of the blood into the lower pressure atria.

5. Judging by their position and structure, what do you suppose is the function of the chordae tendinae?

The atrioventricular (AV) valves of the two sides of the heart are similar in structure except that the right AV valve has three cusps (tricuspid) while the left atrioventricular valve has two cusps (bicuspid or mitral valve). Connective tissue (**chordae tendineae**) run from the cusps to **papillary muscles** on the ventricular wall.

6. What feature shown here most clearly distinguishes the left and right ventricles?.

ISBN: 978-1-92717357-2

The Intrinsic Regulation of Heartbeat

Given the right physiological conditions, an isolated heart will continue to beat. This shows that the origin of the heartbeat is **myogenic** and contraction is an **intrinsic property** of the cardiac muscle itself. The heartbeat is regulated by a conduction system consisting of specialized muscle cells called the **pacemaker** (**sinoatrial node**) and a tract of conducting Purkyne fibers. The pacemaker initiates the cardiac cycle by spontaneously generating action potentials. It sets a basic rhythm for the heart, although this rate is influenced by inputs from outside the heart itself in response to changing demands (see opposite). The diagram below illustrates the basic features of the heart's intrinsic control system.

Generation of the Heartbeat

The basic rhythmic heartbeat is **myogenic**. The nodal cells (SAN and atrioventricular node) spontaneously generate rhythmic action potentials without neural stimulation. The normal resting rate of self-excitation of the SAN is about 50 beats per minute.

The amount of blood ejected from the left ventricle per minute is called the **cardiac output**. It is determined by the **stroke volume** (the volume of blood ejected with each contraction) and the **heart rate** (number of heart beats per minute).

Cardiac muscle responds to stretching by contracting more strongly. The greater the blood volume entering the ventricle, the greater the force of contraction. This relationship is known as **Starling's Law of the heart** and it is important in regulating stroke volume in response to demand.

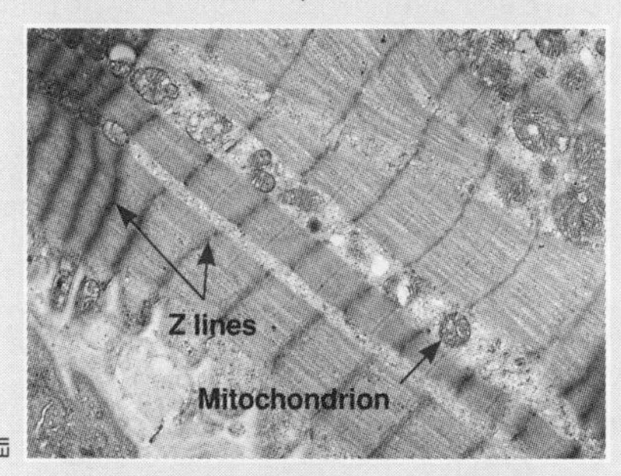

TEM of cardiac muscle showing striations in a fiber (muscle cell). Each fiber has one or two nuclei and many large mitochondria. Note the Z lines that delineate the contractile units (or sarcomeres) of the rod-like units (myofibrils) of the fiber. The fibers are joined by specialized electrical junctions called Intercalated discs, which allow impulses to spread rapidly through the heart muscle.

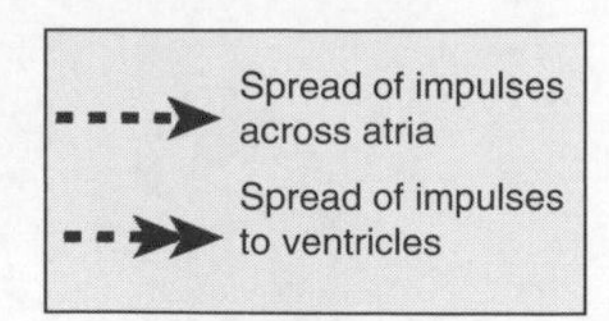

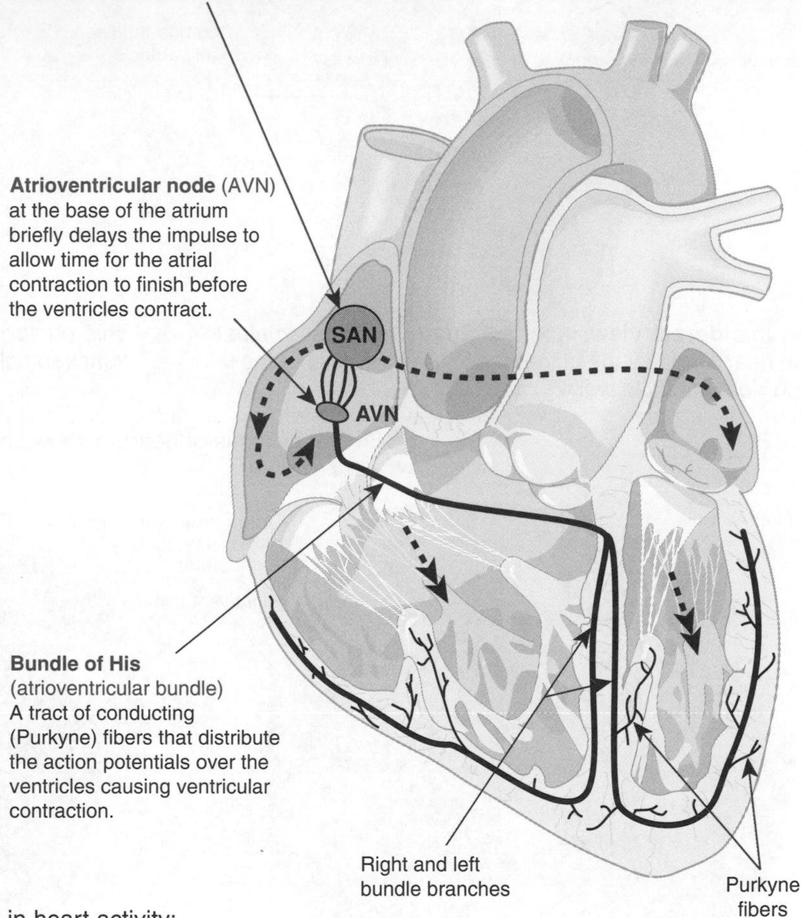

1. Describe the role of each of the following in heart activity:

   (a) The sinoatrial node: ______________________

   (b) The atrioventricular node: ______________________

   (c) The bundle of His: ______________________

   (d) Intercalated discs: ______________________

2. What is the significance of delaying the impulse at the AVN? ______________________

   ____

   ____

3. (a) What is the physiological response of cardiac muscle to stretching? ______________________

   ____

   (b) What is the physiological advantage of this response? ______________________

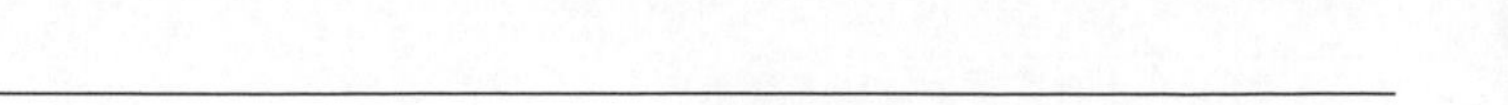

ISBN: 978-1-92717357-2

Extrinsic Control of Heart Rate

The pacemaker sets the basic rhythm of the heart, but this rate is influenced by the **cardiovascular control center**, primarily in response to sensory information from pressure receptors in the walls of the blood vessels entering and leaving the heart. Control of heart rate is complex and has some similarities to control of breathing rate. This is not surprising given the importance of both systems in supplying oxygen to the tissues. However, the main trigger for changing the basic heart rate is change in blood pressure, whereas the main trigger for changing the basic rhythm of breathing is change in blood CO_2.

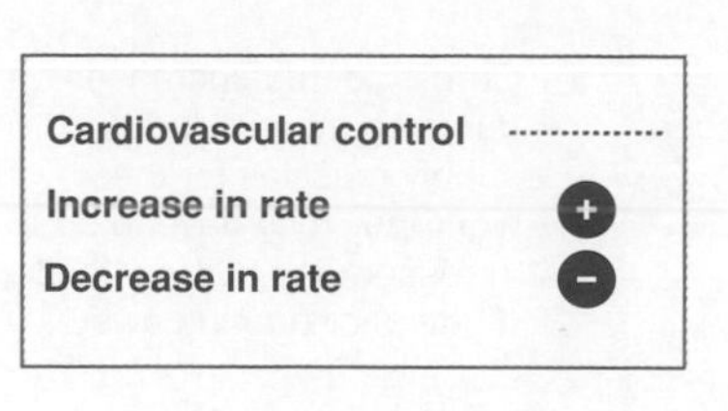

Higher brain centers influence the cardiovascular center, e.g. excitement or anticipation of an event.

Baroreceptors in aorta, carotid arteries, and vena cava give feedback to cardiovascular center on **blood pressure**. Blood pressure is directly related to the pumping action of the heart.

+ or −

Cardiovascular center responds directly to norepinephrine and to low pH (high CO_2). It sends output to the sinoatrial node (SAN) to increase heart rate. Changing the rate and force of heart contraction is the main mechanism for controlling cardiac output in order to meet changing demands.

Sympathetic output to heart via **cardiac nerve** increases heart rate. Sympathetic output predominates during exercise or stress.

Parasympathetic output to heart via **vagus nerve** decreases heart rate. Parasympathetic (vagal) output predominates during rest.

Extrinsic input to SAN

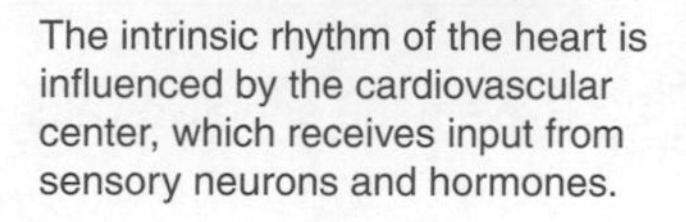

The intrinsic rhythm of the heart is influenced by the cardiovascular center, which receives input from sensory neurons and hormones.

Influences on Heart Rate

Increase	Decrease
Increased physical activity	Decreased physical activity
Decrease in blood pressure	Increase in blood pressure
Secretion of epinephrine or norepinephrine	Re-uptake and metabolism of epinephrine or norepinephrine
Increase in H^+ or CO_2 concentrations in blood	Decrease in H^+ or CO_2 concentrations in blood

Reflex Responses to Changes in Blood Pressure

Reflex	Receptor	Stimulus	Response
Bainbridge reflex	Pressure receptors in vena cava and atrium	Stretch caused by increased venous return	Increase heart rate
Carotid reflex	Pressure receptors in the carotid arteries	Stretch caused by increased arterial flow	Decrease heart rate
Aortic reflex	Pressure receptors in the aorta	Stretch caused by increased arterial flow	Decrease heart rate

Opposing actions keep blood pressure within narrow limits

1. Explain how each of the following extrinsic factors influences the basic intrinsic rhythm of the heart:

   (a) Increased venous return: ______

   (b) Release of epinephrine in anticipation of an event: ______

   (c) Increase in blood CO_2: ______

2. How do these extrinsic factors bring about their effects? ______

3. What type of activity might cause increased venous return? ______

4. (a) Identify the nerve that brings about **increased** heart rate: ______

   (b) Identify the nerve that brings about **decreased** heart rate: ______

5. Account for the different responses to stretch in the vena cava and the aorta: ______

The Cardiovascular System

ISBN: 978-1-92717357-2

Related activities: Stress and Heart Rate
Weblinks: Baroreceptor Reflex Control of Blood Pressure

Stress and Heart Rate

The heart's intrinsic rate of beating is influenced by higher brain function and by the cardiovascular control center in the medulla in response to sensory information. Thus, when the body needs more oxygen, heart rate (as well as lung ventilation rate) will increase. Changes in the rate and force of heart contraction alter the **cardiac output**, i.e. the amount of blood pumped by the heart. When the body needs to prepare for physical exertion, cardiac output increases to meet the greater demand for gas exchange. Both nervous and endocrine controls are involved in regulating these changes.

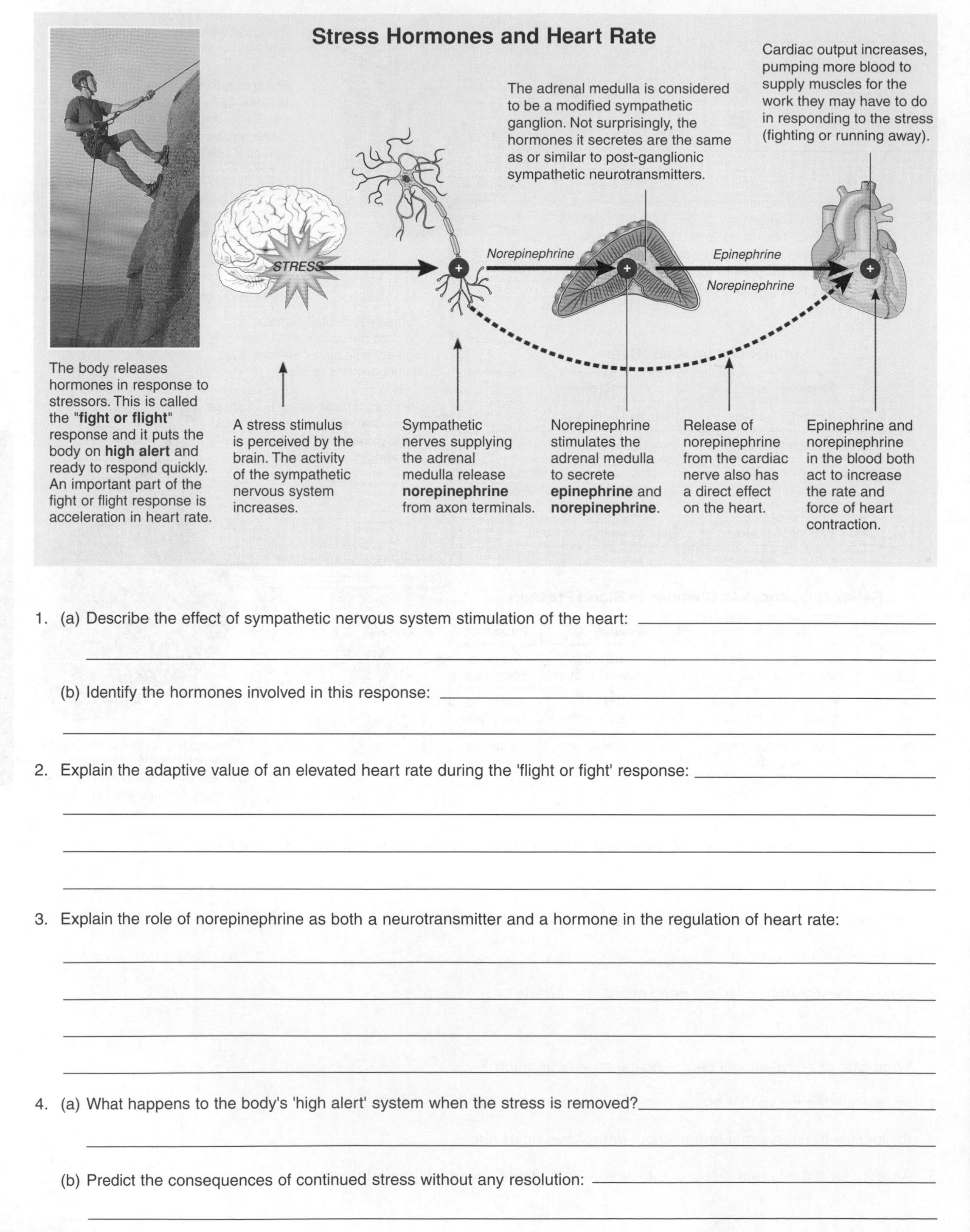

1. (a) Describe the effect of sympathetic nervous system stimulation of the heart:

   (b) Identify the hormones involved in this response:

2. Explain the adaptive value of an elevated heart rate during the 'flight or fight' response:

3. Explain the role of norepinephrine as both a neurotransmitter and a hormone in the regulation of heart rate:

4. (a) What happens to the body's 'high alert' system when the stress is removed?

   (b) Predict the consequences of continued stress without any resolution:

ISBN: 978-1-92717357-2

Related activities: Extrinsic Control of Heart Rate, The Stress Response

Review of the Human Heart

A circulatory system is required to transport materials because diffusion is too inefficient and slow to supply all the cells of the body adequately. The circulatory system in humans transports nutrients, respiratory gases, wastes, and hormones, aids in regulating body temperature and maintaining fluid balance, and has a role in internal defense. The circulatory system comprises a network of vessels, a circulatory fluid (blood), and a heart. This activity summarizes key features of the structure and function of the human heart. The necessary information can be found in earlier activities in this topic.

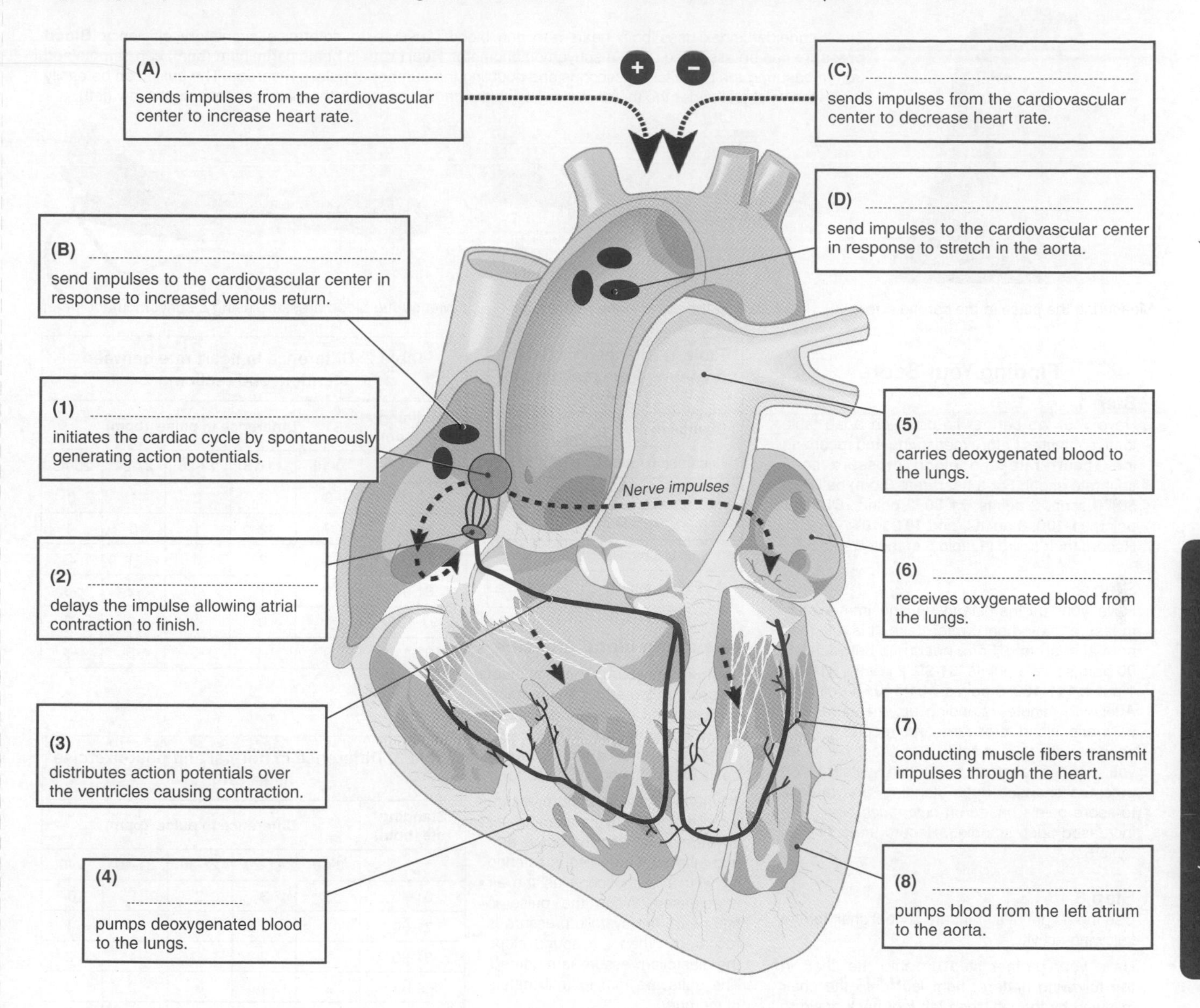

The Cardiovascular System

1. On the diagram above, label the identified components of heart structure and intrinsic control (**1-8**), and the components involved in extrinsic control of heart rate (**A-D**).

2. An **ECG** is the result of different impulses produced at each phase of the **cardiac cycle** (the sequence of events in a heartbeat). For each electrical event indicated in the ECG below, describe the corresponding event in the cardiac cycle:

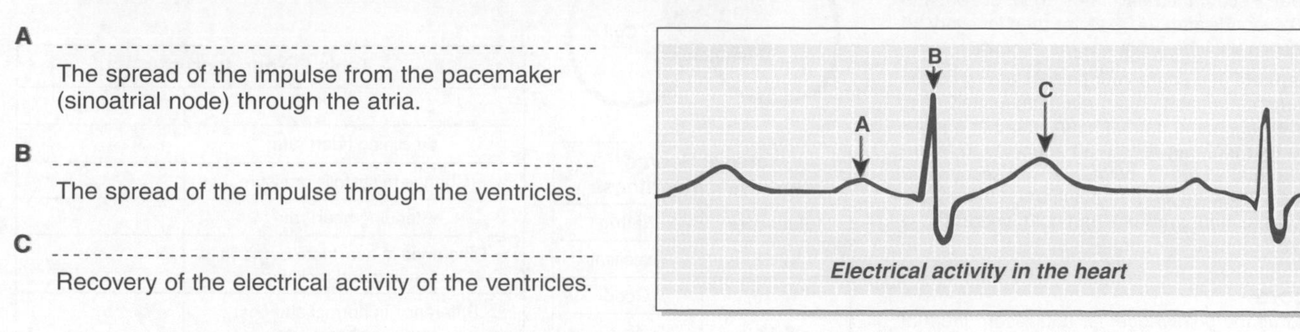

Electrical activity in the heart

3. (a) On the trace above, mark the region where the ventricular pressure is highest.

   (b) What is happening to the ventricular volume at this time?

ISBN: 978-1-92717357-2

Related activities: *The Human Heart, The Intrinsic Regulation of Heartbeat, Extrinsic Control of Heart Rate, The Cardiac Cycle*

Investigating Cardiovascular Physiology

Cardiovascular fitness can be evaluated by measuring responses to changes in body position. When lying down, heart rate decreases and blood vessels dilate. As you stand up there is a sudden decrease in blood pressure as gravity pulls blood towards your feet. The cardiovascular system must compensate for this by increasing heart rate and constricting the blood vessels. Its ability to do this can be measured using the **Schneider index,** a test of circulatory efficiency, which gives a score out of 18 (table 4 below). Perform the following activity in the classroom in pairs to measure your own cardiovascular fitness.

The Schneider index uses both heart rate and blood pressure to determine circulatory efficiency. **Blood pressure** can be assessed using a sphygmomanometer. **Heart rate** in beats per minute (bpm) can be obtained from measuring the pulse for 30 seconds and doubling the number of pulses recorded. The pulse can be easily felt on the wrist just under the thumb (below) or on the carotid artery on the neck just beneath the jaw (left).

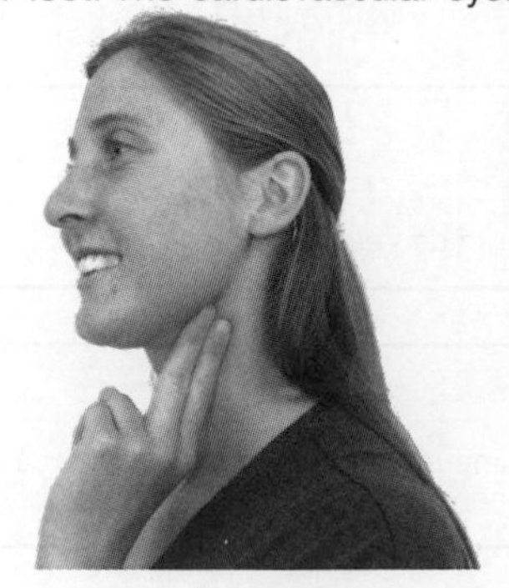

Measuring the pulse in the carotid artery

Measuring the pulse in the radial artery.

Measuring blood pressure using a sphygmomanometer.

Finding Your Score

Step 1

Have your lab partner lie down on a lab table for five minutes before measuring and recording their **heart rate** and **blood pressure** on a separate sheet. For a heart rate (bpm) between **50-70** score 3 points, **71-80**: 2 points, **81-90**: 1 point, **91-100**: 0 points, and **101-110+**: -1 point. Record their score in table 5 of their workbook.

Step 2

Have your partner stand up and immediately measure their heart rate. Take this as their normal heart rate. For a heart rate between **61-80** bpm score 3 points, **81-90**: 2 points, **91-110**: 1 point, **111-130**, 0 points, **131-140**: -1 point.

After two minutes standing up, measure your partner's blood pressure and record it on a separate sheet. Use **table 1** to score points for your partner based on how much systolic blood pressure changed upon standing, and **table 2** to score points based on how much heart rate increased upon standing. Record the scores in table 5.

Step 3

Use the height of a standard school chair for the following activity.

Have your partner step up onto the chair in the following pattern: right foot onto the chair followed by the left, then left foot back down to the floor followed by the right. Repeat this five times allowing no more than three seconds for each repetition.

Immediately after the exercise, record your partner's heart rate for 15 seconds and multiply by four. Record the heart rate on a separate sheet. Record the heart rates at 30, 60, 90, and 120 seconds after the exercise then for every 30 seconds until the pulse returns to normal.

Use **table 3** to score your partner based on the difference between heart rate immediately after the exercise and normal heart rate. For a heart rate that took between **0-30 seconds** to return to normal score 3 points, **31-60** seconds: 2 points: **61-90** seconds: 1 point, **91-120** seconds: 0 points, and **greater than 120 seconds**: -1 point.

Score

Add all the points together and record them at the bottom of table 5 in your partner's workbook. Match this score to the scores in table 4 and rate your partner's cardiovascular fitness.

Table 1: Change in systolic pressure from reclining to standing

Change in pressure	Points
Increase 8+ mmHg	3
Increase 2-7 mm Hg	2
0 (+/- 1 mmHg)	1
Fall 2-5 mmHg	0
Fall 6+ mmHg	-1

Measuring blood pressure

Systolic pressure: The maximum pressure in the artery as the heart contracts and forces blood through. **Diastolic pressure**: The lowest pressure in the artery between beats (when the heart is resting).

To measure the systolic pressure the cuff of the sphygmomanometer is wrapped around the upper arm and inflated. The pulse is listened for with a stethoscope as the air is released. When the pulse is first heard the systolic pressure is recorded. When the sound stops the diastolic pressure is recorded. The units are mmHg (millimeters of mercury).

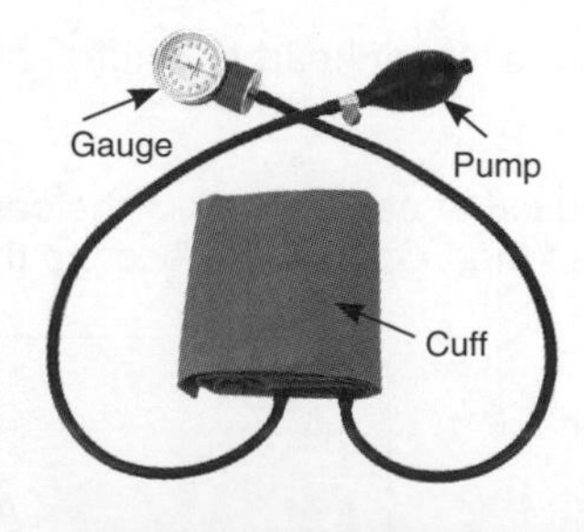

Table 2: Difference in heart rate between standing and reclining

Reclining rate (bpm)	Difference in pulse (bpm)				
	0-10	11-18	19-26	27-34	25-43
50-60	3	3	2	1	0
61-70	3	2	1	0	-1
71-80	3	2	0	-1	-2
81-90	2	1	-1	-2	-3
91-100	1	0	-2	-3	-3
101-110	0	-1	-3	-3	-3

Table 3: Difference in normal and post-exercise heart rate

Standing rate (bpm)	Difference in pulse (bpm)				
	0-10	11-20	21-30	31-40	>40
61-70	3	3	2	1	0
71-80	3	2	1	0	-1
81-90	3	2	1	-1	-2
91-100	2	1	0	-2	-3
101-110	1	0	-1	-3	-3
111-120	1	-1	-2	-3	-3
121-130	0	-2	-3	-3	-3
131-140	0	-3	-3	-3	-3

Table 4: Scores cardiovascular fitness

Total score	Rating
17-18	Excellent
14-16	Good
8-13	Fair
0-7	Poor
Negative value	!!!!!

Table 5: Your scores

	Points
Reclining heart rate	
Change in systolic pressure	
Standing heart rate	
Difference in heart rate between standing and reclining	
Difference in normal and post-exercise heart rate	
Time for pulse to return to normal	
Score:	

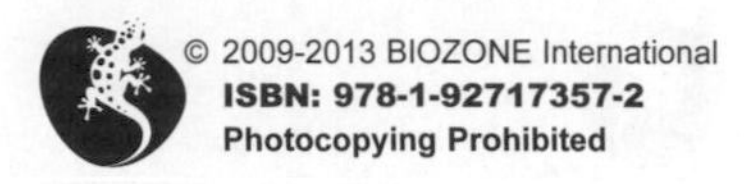

ISBN: 978-1-92717357-2

Related activities: *Extrinsic Control of Heart Rate*

The Effects of Aerobic Training

The body has an immediate response to exercise, but over time responds to the stress of repeated exercise (or **training**) by adapting and improving both its capacity for exercise and the efficiency with which it performs. Regular, intense exercise causes predictable physiological changes in muscular, cardiovascular, and respiratory performance. The heart adapts so that it can pump more blood per stroke. This increase in **cardiac output** is brought about not only by an increase in heart rate, but also by a training-induced increase in the stroke volume. The circulatory system also adapts to repeated exercise, with changes in blood flow facilitating an increased flow of blood to the muscles and skin and an increased rate of gas exchange. The pulmonary system adjusts accordingly, with greater efficiencies in ventilation rate and breathing rhythm.

The Physiological Effects of Aerobic Training

The Pulmonary System

- Improvement in lung ventilation rate. The rate and depth of breathing increases during exercise but for any given level of exercise, the ventilation response is reduced with training.
- Improvement in ventilation rhythm, so that breathing is in tune with the exercise rhythm. This promotes efficiency.

Overall result
Improved exchange of gases.

The Muscular System

- Improvement in aerobic generation of ATP
- Larger mitochondria
- More mitochondria
- Increase in muscle myoglobin
- Greater Krebs cycle enzyme activity
- Improved ability to use fats as fuels
- Increased capillary density

Overall result
Improved function of the oxidative system and better endurance.

The Cardiovascular System

- Exercise lowers blood plasma volume by as much as 20% and the cellular portion of the blood becomes concentrated. With training, blood volume at rest increases to compensate.
- **Heart rate** increases during exercise but aerobic training leads to a lower steady state heart rate overall for any given level of work.
- Increase in **stroke volume** (the amount of blood pumped with each beat). This is related to an increased heart capacity, an increase in the heart's force of contraction, and an increase in venous return.
- Increased cardiac output as a result of the increase in stroke volume.
- During exercise, systolic blood pressure increases as a result of increased cardiac output. In response to training, the resting systolic blood pressure is lowered.
- Blood flow changes during exercise so that more blood is diverted to working muscles and less is delivered to the gut.

Overall result
Meets the increased demands of exercise most efficiently.

1. (a) State what you understand by the term training: ______________________

   (b) In general terms, explain how training forces a change in physiology: ______________________

2. With respect to increasing functional efficiency, describe the role of each of the following effects of aerobic training:

   (a) Increase in stroke volume and cardiac output: ______________________

   (b) Increased ventilation efficiency: ______________________

   (c) Increase in capillary density in the muscle tissue: ______________________

ISBN: 978-1-92717357-2

Related activities: Energy and Exercise, Exercise and Blood Flow

Endurance refers to the ability to carry out sustained activity. Muscular strength and short term muscular endurance allows sprinters to run fast for a short time or body builders and weight lifters to lift an immense weight and hold it. Cardiovascular and respiratory endurance refer to the body as a whole: the ability to endure a high level of activity over a prolonged period. This type of endurance is seen in marathon runners, and long distance swimmers and cyclists.

Sprint-focused sports demand quite different training to that required for endurance sports, and the physiologies and builds of the athletes are quite different. A body builder ready for a competition would be ill-equipped to complete a 90 km cycle race!

2008's Mr Olympia winner, Dexter Jackson: high muscular development and endurance.

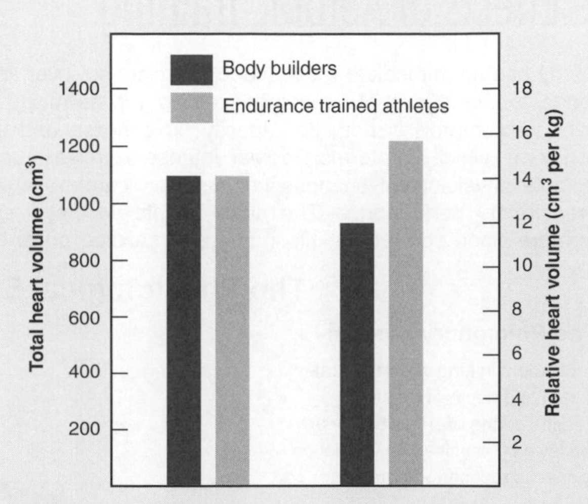

Difference in heart size of highly trained body builders and endurance athletes. Total heart volume is compared to heart volume as related to body weight. Average weights as follows: Body builders = 90.1 kg. Endurance athletes = 68.7 kg.

Weightlifters have high muscular strength and short term muscular endurance; they can lift extremely heavy weights and hold them for a short time. Typical sports requiring these attributes are sprinting, weight lifting, body building, boxing, and wrestling.

Distance runners have very good cardiovascular, respiratory, and muscular endurance; they sustain high intensity exercise for a long time. Typical sports needing overall endurance are distance running, cycling, and swimming (triathletes combine all three).

3. Explain why heart size increases with endurance activity: ______

4. In the graph above right, explain why the relative heart volume of endurance athletes is greater than that of body builders, even though their total heart volumes are the same:

5. Heart stroke volume increases with endurance training. Explain how this increases the efficiency of the heart as a pump:

6. The **resting pulse** is much lower in trained athletes compared with non-active people. Explain the health benefits of a lower resting pulse:

ISBN: 978-1-92717357-2

Cardiovascular Disease

Cardiovascular disease (CVD) describes all diseases affecting the heart and blood vessels. It includes coronary heart disease, atherosclerosis (hardening of the arteries), hypertension (high blood pressure), peripheral vascular disease, stroke (reduced blood supply to the brain), and congenital heart disorders. CVD causes 20% of all deaths worldwide and is the principal cause of deaths in developed countries. Most CVD develops as a result of lifestyle factors, but a small proportion of the population (< 1%), are born with a CVD. This is called a **congenital disorder**. There are many types of congenital heart defects. Some obstruct blood flow in the heart or vessels near it, whereas others cause blood to flow through the heart in an abnormal pattern.

Types of Cardiovascular Disease

Atherosclerosis (hardening of the arteries) is caused by deposits of fats and cholesterol in the inner walls of the arteries. Blood flow becomes restricted and increases the risk of blood clots (thrombosis). Complications arising as a result of atherosclerosis include heart attack (infarction), gangrene, and stroke. A stroke is the rapid loss of brain function due to a disturbance in the blood supply to the brain, and may result in death if the damage is severe. Speech, or vision and movement on one side of the body is often affected.

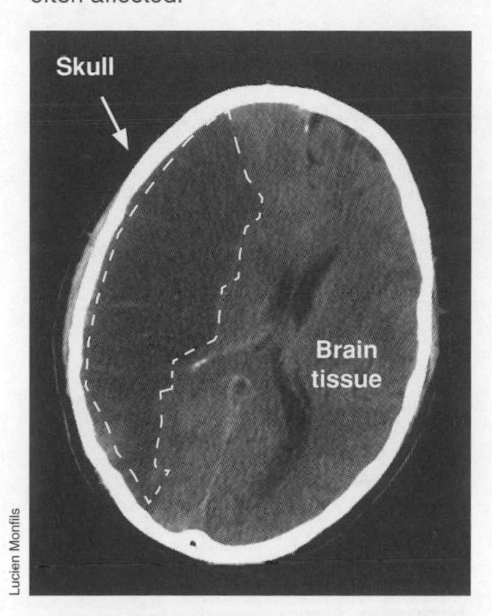

Lucien Monfils

The CT scan (above) shows a brain affected by a severe cerebral infarction or ischaemic stroke. The loss of blood supply results in tissue death (outlined area). Blood clots resulting from atherosclerosis are a common cause of ischaemic stroke.

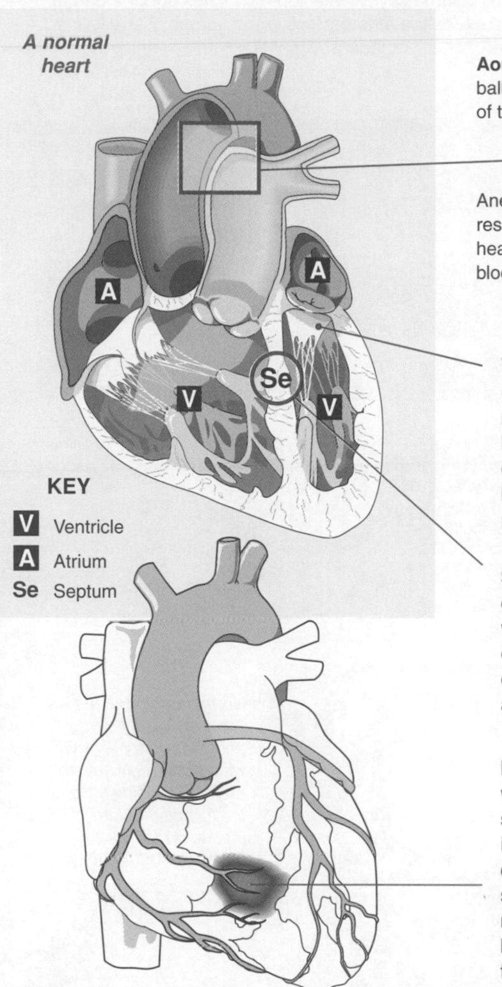

Restricted supply of blood to heart muscle resulting in myocardial infarction

Aortic aneurysm: A ballooning and weakening of the wall of the aorta.

Aneurysms usually result from generalized heart disease and high blood pressure.

Valve defects: Unusual heart sounds (murmurs) can result when a valve (often the mitral valve) does not close properly, allowing blood to bubble back into the atria. Valve defects may be congenital (present at birth) but they can also occur as a result of rheumatic fever.

Septal defects: These hole-in-the-heart congenital defects occur where the dividing wall (**septum**) between the left and right sides of the heart is not closed. These defects may occur between the atria or the ventricles, and are sometimes combined with valve problems.

Myocardial infarction *(heart attack):* Occurs when an area of the heart is deprived of blood supply resulting in tissue damage or death. It is the major cause of death in developed countries. Symptoms of infarction include a sudden onset of chest pain, breathlessness, nausea, and cold clammy skin. Damage to the heart may be so severe that it leads to heart failure and even death (myocardial infarction is fatal within 20 days in 40 to 50% of all cases).

1. Define the term cardiovascular disease (CVD): ______________________

____

____

2. Suggest why CVD is the principal cause of death in developed countries? ______________________

____

____

____

3. Explain the difference between a congenital cardiovascular defect and a defect that develops later in life:

____

____

____

ISBN: 978-1-92717357-2

Atherosclerosis

Atherosclerosis is a disease of the arteries caused by **atheromas** (fatty deposits) on the inner arterial walls. An atheroma is made up of cells (mostly macrophages) or cell debris, with associated fatty acids, cholesterol, calcium, and varying amounts of fibrous connective tissue. The accumulation of fat and plaques causes the lining of the arteries to degenerate. Atheromas weaken the arterial walls and eventually restrict blood flow through the arteries, increasing the risk of **aneurysm** (swelling of the artery wall) and **thrombosis** (blood clots). Complications arising as a result of atherosclerosis include heart attacks, strokes, and gangrene. A typical progression for the formation of an atheroma is illustrated below.

Initial lesion	Fatty streak	Intermediate lesion	Atheroma	Fibroatheroma	Complicated plaque
Atherosclerosis is triggered by damage to an artery wall caused by blood borne chemicals or persistent **hypertension**.	Low density lipoproteins (LDLs) accumulate beneath the endothelial cells. Macrophages follow and absorb them, forming foam cells.	Foam cells accumulate forming greasy yellow lesions called atherosclerotic plaques.	A core of extracellular lipids under a cap of fibrous tissue forms.	Lipid core and fibrous layers. Accumulated smooth muscle cells die. Fibres deteriorate and are replaced with scar tissue.	Calcification of plaque. Arterial wall may ulcerate. Hypertension may worsen. Plaque may break away and blood cells may then collect at the damaged site, forming a clot.

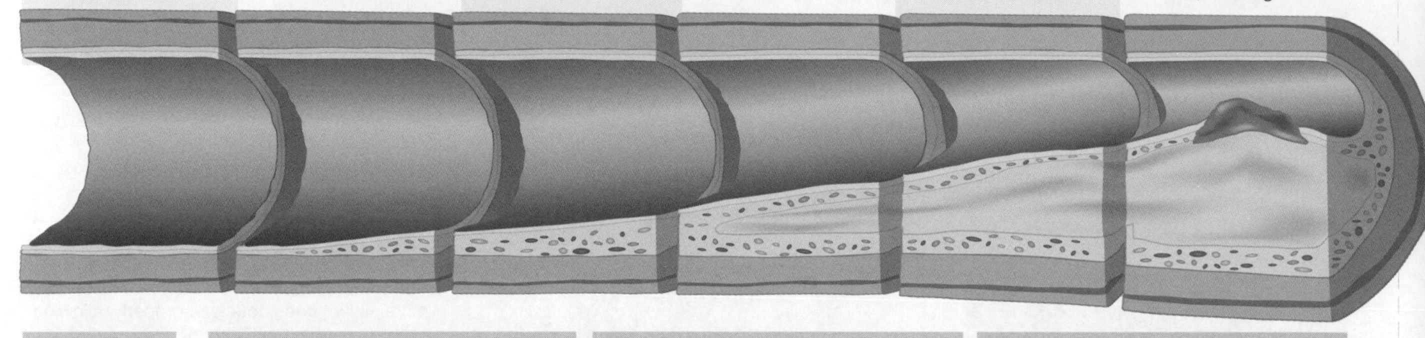

Earliest onset	From first decade	From third decade	From fourth decade	
Growth mechanism	Growth mainly by lipid accumulation		Smooth muscle/ collagen increase	Thrombosis, hematoma
Clinical correlation	Clinically silent	Clinically silent or overt		

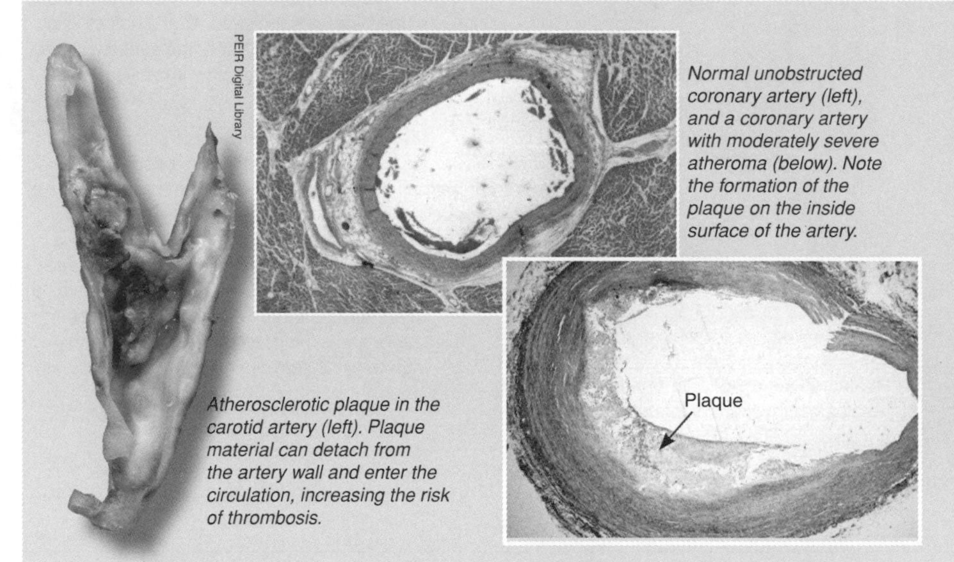

Atherosclerotic plaque in the carotid artery (left). Plaque material can detach from the artery wall and enter the circulation, increasing the risk of thrombosis.

Normal unobstructed coronary artery (left), and a coronary artery with moderately severe atheroma (below). Note the formation of the plaque on the inside surface of the artery.

Recent studies indicate that most heart attacks are caused by the body's **inflammatory response** to a plaque. The inflammatory process causes young, soft, cholesterol-rich plaques to rupture and break into pieces. If these block blood vessels they can cause lethal heart attacks, even in previously healthy people.

Aorta opened lengthwise (above), with extensive atherosclerotic lesions (arrowed).

1. Explain why most people are unlikely to realize they are developing atherosclerosis until serious complications arise:

2. Explain how an atherosclerotic plaque changes over time:

3. Describe some of the consequences of developing atherosclerosis:

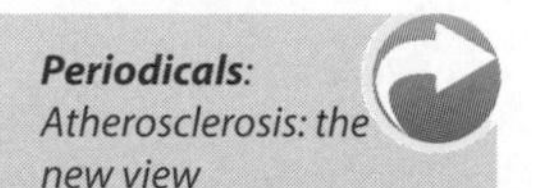

ISBN: 978-1-92717357-2

CVD Risk Factors

A **risk factor** is a variable that increases the risk of a certain disease developing. Several risk factors increase the likelihood of a person developing CVD. Some risk factors are **controllable** in that they can be modified by lifestyle changes. Controllable risk factors include diet, cigarette smoking, obesity, high blood cholesterol, high blood pressure, diabetes, and physical inactivity.

Uncontrollable risk factors (advancing age, gender, and heredity) cannot be modified, but overall risk can minimized by reducing the number of controllable risk factors. The more risk factors a person has, the greater the likelihood they will develop CVD (below). Increased levels of education and awareness about CVD and its risk factors have helped to reduce levels of the disease.

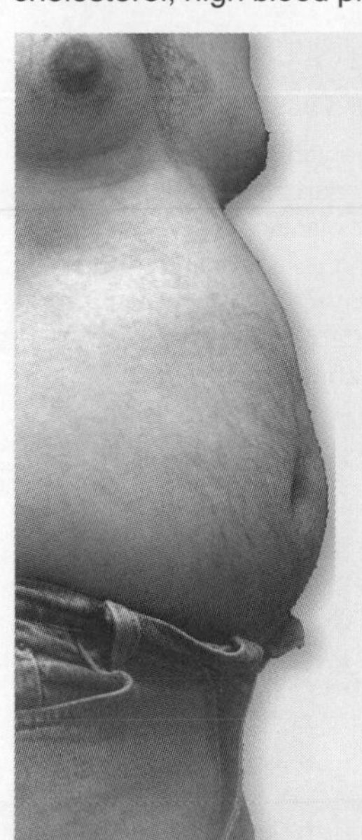

Cardiovascular Disease: Who is at Risk?

Controlled risk factors for cardiovascular disease

- High blood pressure
- Cigarette smoking
- High blood cholesterol
- High LDL:HDL ratio
- Obesity
- Type 2 diabetes mellitus
- High achiever personality
- Environmental stress
- Sedentary lifestyle

A person's risk of CVD increases markedly with an increase in the number of risk factors. This is particularly so for smoking, because smoking acts synergistically with other risk factors, particularly hypertension (high blood pressure) and high blood lipids. This means that any given risk factor has a proportionately greater effect in a smoker than in a non-smoker.

Estimated coronary heart disease rate according to various combinations of risk factors over 10 years (source: International Diabetes Foundation, 2001)

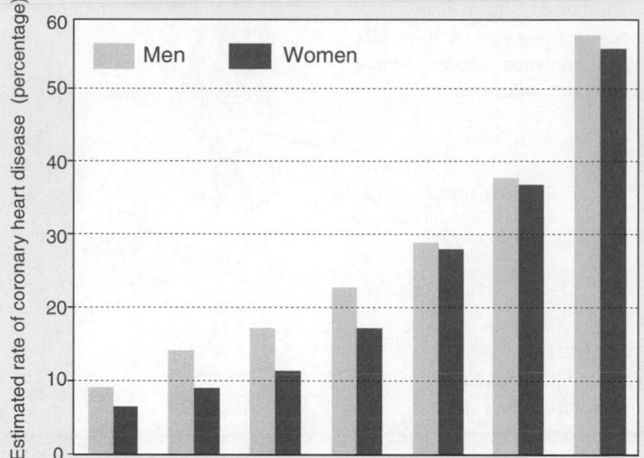

Risk factors							
Systolic blood pressure (mm Hg)	120	160	160	160	160	160	160
Cholesterol (mg 100cm^{-3})	220	220	259	259	259	259	259
HDL cholesterol (mg 100cm^{-3})	50	50	50	35	35	35	35
Diabetes	–	–	–	–	+	+	+
Cigarette smoking	–	–	–	–	–	+	+
Enlargement of left ventricle	–	–	–	–	–	–	+

1. (a) Distinguish between **controllable** and **uncontrollable** risk factors in the development of cardiovascular disease:

(b) Suggest why some controllable risk factors often occur together:

(c) Evaluate the evidence supporting the observation that patients with several risk factors are at a higher risk of CVD:

ISBN: 978-1-92717357-2

Correcting Heart Problems

Medical technology now provides the means to correct many heart problems, even if only temporarily. Some symptoms of CVD, arising as a result of blockages to the coronary arteries, are now commonly treated using techniques such as coronary bypass surgery and angioplasty. Other cardiac disorders, such as disorders of heartbeat, are frequently treated using cardiac pacemakers. Valve defects, which are often congenital, can be successfully corrected with surgical valve replacement. The latest technology involves non-surgical replacement of aortic valves. The procedure, known as percutaneous (through the skin) heart valve replacement, will greatly reduce the trauma associated with correcting these particular heart disorders.

Coronary Bypass Surgery

This is a now commonly used surgery to bypass blocked coronary arteries with blood vessels from elsewhere in the body (e.g. leg vein or mammary artery). Sometimes, double or triple bypasses are performed.

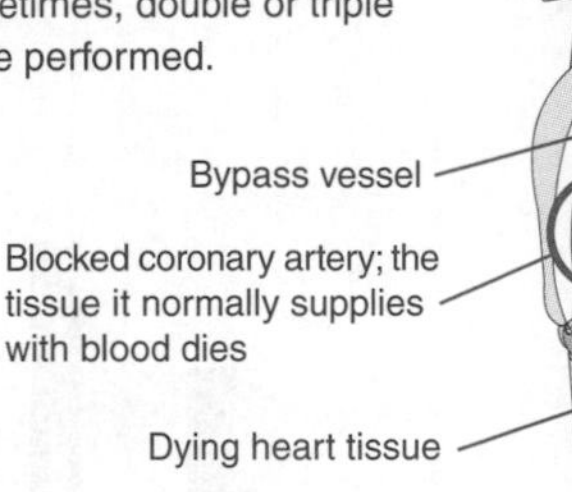

Angioplasty

Angioplasty (right) is an alternative procedure used for some patients with coronary artery disease. A balloon tipped catheter is placed via the aorta into the coronary artery. The balloon is inflated to reduce the blockage of the artery and later removed. Heparin (an anticlotting agent) is given to prevent the formation of blood clots. The death rate from complications is about 1%.

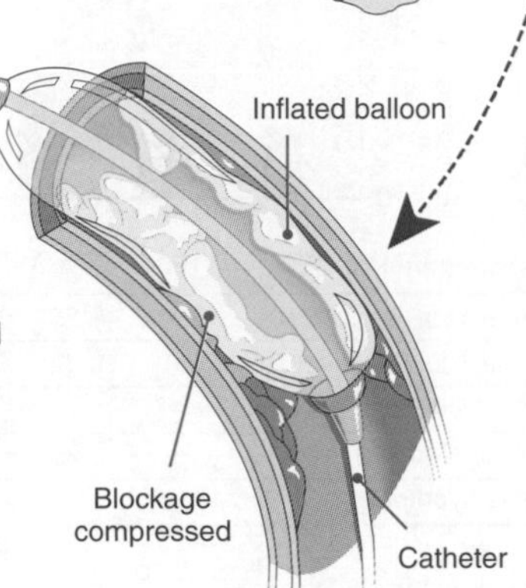

Heart Valve Replacement

Heart valves can be replaced with either **biological** (tissue) **valves** or **synthetic valves.** Tissue valves are sourced from animal (e.g. pig) or human donors. They last only 7-10 years, but there are relatively few blood clotting and tissue rejection problems associated with them. For these reasons, they are often used in older patients. Synthetic ball or disc valves are constructed from non-biological materials. They last a long time but tend to create blood clots (raising the risk of stroke). They are used on younger patients, who must take long-term anti-clotting drugs.

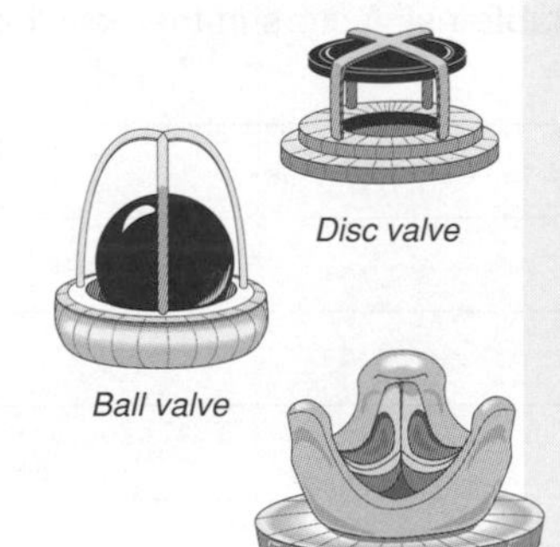

Cardiac Pacemakers

A cardiac pacemaker is sometimes required to maintain an effective heart rate in cases where the heart beats irregularly or too slowly. Pacemakers provide regular electrical stimulation of the heart muscle so that it contracts and relaxes with a normal rhythm. They stand by until the heart rate falls below a pre-set rate. **Temporary pacemakers** are often used after cardiac surgery or heart attacks, while **permanent pacemakers** are required for patients with ongoing problems. Pacemakers allow a normal (even strenuous) lifestyle.

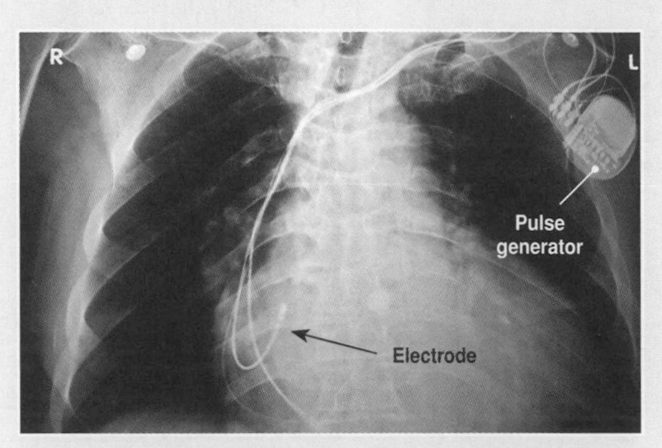

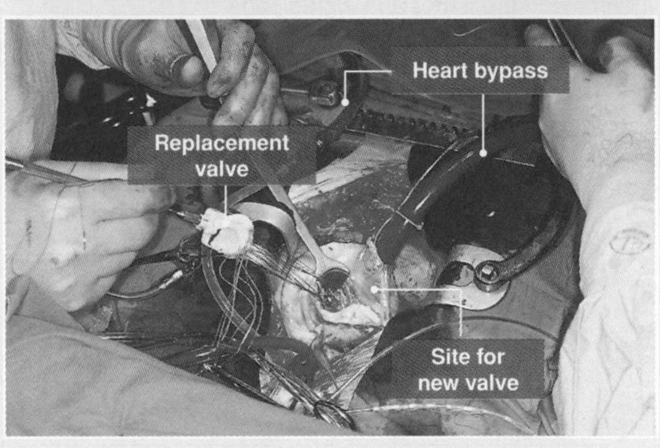

Above: Valve replacement operation in progress. The valve can be seen threaded up and ready for placement. Two large tubes bypass the heart so that circulation to the lungs and rest of the body is maintained.

1. Describe the problems associated with the use of each of the following types of replacement heart valve:

   (a) Tissue valves: ____________________

   (b) Synthetic valves: ____________________

2. Suggest why tissue valves are usually a preferred option for use in elderly patients: ____________________

3. Explain why patients who have undergone coronary bypass surgery or angioplasty require careful supervision of their diet and lifestyle following the operation, even though their problem has been alleviated:

ISBN: 978-1-92717357-2

Related activities: *Cardiovascular Disease*

KEY TERMS: Mix and Match

INSTRUCTIONS: Test your vocabulary by matching each term to its definition, as identified by its preceding letter code.

aorta ....

artery (*pl. arteries*) ....

atrioventricular (AV) node ....

atrioventricular valves ....

atrium (*pl. atria*) ....

blood ....

capillary (*pl. capillaries*) ....

cardiac cycle ....

cardiac output ....

cardiovascular disease ....

cardiovascular system ....

diastole ....

erythrocytes ....

heart ....

hematopoiesis ....

leukocytes ....

lymph ....

myogenic ....

pacemaker (=sinoatrial node) ....

plasma ....

pulmonary artery ....

pulmonary circulation ....

pulmonary vein ....

systemic circulation ....

systole ....

vein ....

vena cava ....

ventricle ....

A Muscular organ responsible for pumping blood around the body.

B Chamber of the heart that pumps blood into the arteries.

C Part of the double circulatory system in air breathing vertebrates that transports blood from the heart to the lungs and back to the heart.

D The smallest blood vessel, with an endothelium only one cell thick.

E Clear fluid contained within the lymphatic system. Similar in composition to the interstitial fluid.

F A blood vessel which carries oxygenated blood from the lungs back to the heart.

G The large vein which receives deoxygenated blood from the head and body and transports it to the heart.

H Artery which carries oxygenated blood away from the heart to the head and body.

I A general term used to describe diseases of the heart and blood vessels.

J White blood cell. Includes lymphocytes, macrophages and other phagocytic cells.

K The sequence of events of a heartbeat, and involves three main stages: atrial systole, ventricular systole and complete cardiac diastole.

L Chamber of the heart that receives blood from the body or lungs.

M The contraction of the heart muscle during the cardiac cycle.

N An artery which carries deoxygenated blood to the lungs from the heart.

O The period of time in which the heart relaxes and fills with blood.

P The specialized cardiac cells that initiate the cardiac cycle, and sets the basic heart rate.

Q A large blood vessel with a thick, muscled wall which carries blood away from the heart.

R The body system comprising the heart and blood vessels.

S Contractions initiated by the heart muscle cells spontaneously and independently of nervous stimulation.

T Blood cells that carries oxygen around the body. Also called red blood cells.

U Part of a double circulatory system that transports blood from the heart to the body and back to the heart.

V Valves located between the atria and the ventricles. They prevent back-flow so blood flows only in one direction through the heart.

W Specialized tissue between the right atrium and the right ventricle of the heart. It delays the delivery of impulse between the atria and ventricles.

X The non-cellular portion of the blood.

Y The volume of blood pumped by the heart per minute.

Z Large blood vessels that return blood to the heart.

AA Circulatory fluid comprising numerous cell types, which transports respiratory gases, nutrients, and wastes.

BB The formation of blood cells.

ISBN: 978-1-92717357-2

The Lymphatic System and Immunity

Key concepts

- The body can distinguish self from non-self.
- The body can defend itself against pathogens.
- Non-specific defenses target any foreign material.
- The immune response targets specific antigens and has a memory for antigens previously encountered.
- The properties of immune cells can be used to target specific antigens.

Key terms

ABO blood groups
active immunity
allergies
antibody
antigen
artificially acquired immunity
B-cells (B lymphocytes)
cell-mediated immunity
clonal selection
fever
gene therapy
HIV
humoral immunity
hypersensitivity
immune system
immunological memory
inflammatory response
lymphatic system
major histocompatibility complex (MHC)
monoclonal antibodies
naturally acquired immunity
non-specific defenses
non-specific resistance
passive immunity
phagocytes
primary response
Rh blood groups
secondary response
specific resistance
T-cells (T lymphocytes)
vaccination
vaccine

Periodicals:
Listings for this chapter are on page 280

Weblinks:
www.thebiozone.com/weblink/AnaPhy-3572.html

Learning Objectives

☐ 1. Use the **KEY TERMS** to compile a glossary for this topic.

Recognizing Self and Non-self

pages 163-165

☐ 2. Explain how a body distinguishes its own cells and tissues from those that are foreign. Include reference to the role of the **major histocompatibility complex** (MHC).

☐ 3. Explain the basis of the **Rh** and **ABO blood group** systems in humans. Explain the consequences of blood type incompatibility in blood transfusions.

Defense Mechanisms

pages 166-178

☐ 4. Describe **blood clotting**, including the role of clotting factors, thrombin, and fibrin. Appreciate the role of blood clotting as a defense against the spread of infection.

☐ 5. Describe **non-specific defenses**, distinguishing the first and second lines of defense. Describe the nature and role of each of the following in protecting against pathogens:
(a) The skin and mucus-secreting and ciliated membranes.
(b) Body secretions and natural anti-bacterial and anti-viral proteins such as interferon.
(c) The **inflammatory response**, **fever**, and **phagocytosis** by phagocytes.

☐ 6. Identify the role of **specific resistance** in the body's resistance to infection. Explain how the **immune system** recognizes and responds to foreign material. Explain the significance of the immune system having both specificity and memory.

☐ 7. Distinguish between **naturally acquired** and **artificially acquired immunity**, and between **active** and **passive immunity**.

☐ 8. Describe the structure and role of the **lymphatic system**.

☐ 9. Distinguish between **cell-mediated immunity** (involving **T-cells**) and **humoral** (antibody-mediated) **immunity** (involving **B-cells**) and their roles.

☐ 10. Describe **clonal selection** and the basis of **immunological memory**. Explain how the immune system is able to respond to a large, unpredictable range of **antigens**. Understand the concept of that self-tolerance and how it develops.

☐ 11. Explain the role of antigens in provoking a specific immune response. Describe the structure of an **antibody** and relate the structure to function.

☐ 12. Describe B-cell activation and differentiation. Contrast the roles of **plasma cells** and **memory cells**. Explain the role of **immunological memory** in long term immunity.

Disease, Treatment, and Prevention

pages 164, 179-192

☐ 13. Outline the principle of **vaccination**. Identify the role of the **primary response** and the **secondary response** to infection and their significance in the success of **vaccines**.

☐ 14. Describe examples of **autoimmune disease** and their causes. Outline the role of the immune system in **allergies** and other **hypersensitivity** reactions.

☐ 15. Discuss the effects of **HIV** (Human Immunodeficiency Virus) on immune system function. Include reference to both the long and short term effects of HIV infection.

☐ 16. Describe the production, mode of action, and applications of **monoclonal antibodies**.

☐ 17. Describe the use of **gene therapy** to restore immune system function in patients with the heritable immunodeficiency disease, SCID (Severe Combine Immune Deficiency).

Targets for Defense

In order for the body to present an effective defense against pathogens, it must first be able to recognize its own tissues (self) and ignore the body's normal microflora (e.g. the bacteria of the skin and gastrointestinal tract). In addition, the body needs to be able to deal with abnormal cells which, if not eliminated, may become cancerous. Failure of self/non-self recognition can lead to autoimmune disorders, in which the immune system mistakenly attacks its own tissues. The body's ability to recognize its own molecules has implications for procedures such as tissue grafts, organ transplants, and blood transfusions. Incompatible tissues (identified as foreign) are attacked by the body's immune system (**rejected**). Even a healthy pregnancy involves suppression of specific features of the self recognition system, allowing the mother to tolerate a nine month gestation with the fetus.

The Body's Natural Microbiota

After birth, normal and characteristic microbial populations begin to establish themselves on and in the body. A typical human body contains 1 X 10^{13} body cells, yet harbors 1 X 10^{14} bacterial cells. These microorganisms establish more or less permanent residence but, under normal conditions, do not cause disease. In fact, this normal microflora can benefit the host by preventing the overgrowth of harmful pathogens. They are not found throughout the entire body, but are located in certain regions.

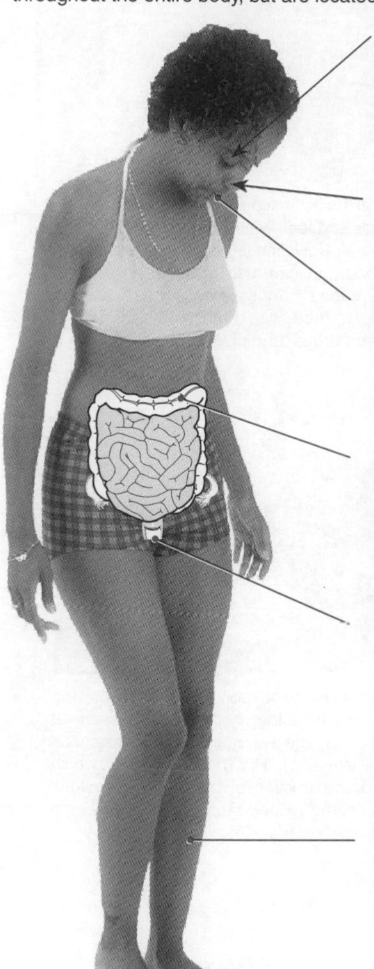

Eyes
The conjuctiva, a continuation of the skin or mucous membrane, contains a similar microbiota to the skin.

Nose and throat
Harbors a variety of microorganisms, e.g. *Staphylococcus* spp.

Mouth
Supports a large and diverse microbiota. It is an ideal microbial environment; high in moisture, warmth, and nutrient availability.

Large intestine
Contains the body's largest resident population of microbes because of its available moisture and nutrients.

Urinary and genital systems
The lower urethra in both sexes has a resident population; the vagina has a particular acid-tolerant population of microbes because of the low pH nature of its secretions.

Skin
Skin secretions prevent most of the microbes on the skin from becoming residents.

The Major Histocompatibility Complex

The human immune system achieves self-recognition through the **major histocompatibility complex** (MHC). This is a cluster of tightly linked genes on chromosome 6 in humans. These genes code for protein molecules (MHC antigens) that are attached to the surface of body cells. They are used by the immune system to recognize its own or foreign material. **Class I MHC** antigens are located on the surface of virtually all human cells, but **Class II MHC** antigens are restricted to macrophages and the antibody-producing B-lymphocytes.

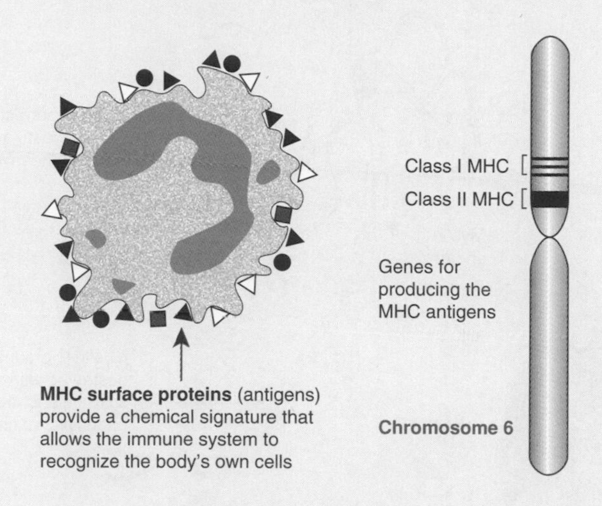

Tissue Transplants

The MHC is responsible for the rejection of tissue grafts and organ transplants. Foreign MHC molecules are antigenic, causing the immune system to respond in the following way:

- T cells directly lyse the foreign cells
- Macrophages are activated by T cells and engulf foreign cells
- Antibodies are released that attack the foreign cell
- The complement system injures blood vessels supplying the graft or transplanted organ

To minimize this rejection, attempts are made to match the MHC of the donor to that of the recipient as closely as possible.

1. Explain why it is healthy to have a natural population of microbes on and inside the body: ______

2. (a) Explain the nature and purpose of the **major histocompatibility complex** (MHC): ______

(b) Explain the importance of such a self-recognition system: ______

3. Identify two situations when the body's recognition of 'self' is undesirable: ______

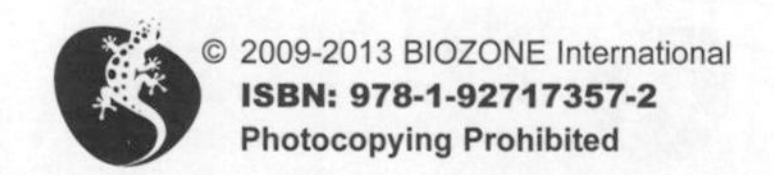

ISBN: 978-1-92717357-2

Periodicals: *What is the human microbiome?*
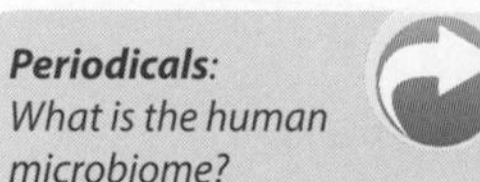

Related activities: *The Body's Defenses*

Autoimmune Diseases

Any of numerous disorders, including **rheumatoid arthritis**, insulin dependent (type 1) **diabetes mellitus**, and **multiple sclerosis**, are caused by an individual's immune system reaction to their own cells or tissues. The immune system normally distinguishes self from non-self. Some lymphocytes are capable of reacting against self, but these are generally suppressed. **Autoimmune diseases** occur when there is some interruption of the normal control process, allowing lymphocytes to escape from suppression, or when there is an alteration in some body tissue so that it is no longer recognized as self. The exact mechanisms behind autoimmune malfunctions are not fully understood but pathogens or drugs may play a role in triggering an autoimmune response in someone who already has a genetic predisposition. The reactions are similar to those that occur in allergies, except that in autoimmune disorders, the hypersensitivity response is to the body itself, rather than to an outside substance.

Multiple Sclerosis

Multiple sclerosis (MS) is a progressive inflammatory disease of the central nervous system in which scattered patches of **myelin** (white matter) in the brain and spinal cord are destroyed. Myelin is the fatty connective tissue sheath surrounding conducting axons. When it is destroyed, the axons cannot transmit impulses effectively and this leads to the symptoms of MS: numbness, tingling, muscle weakness and **paralysis**.

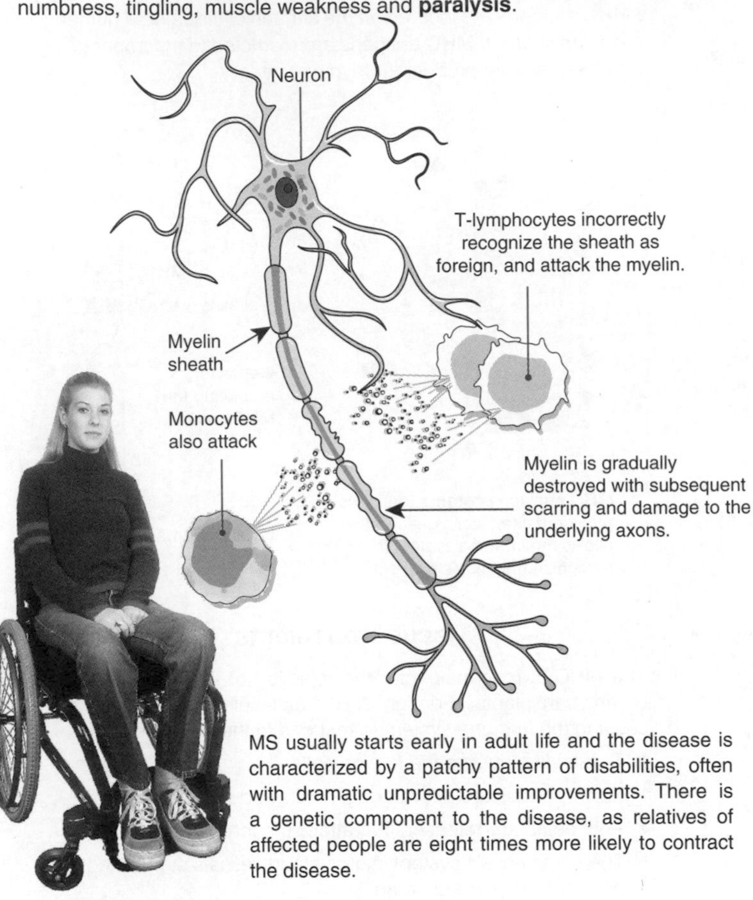

MS usually starts early in adult life and the disease is characterized by a patchy pattern of disabilities, often with dramatic unpredictable improvements. There is a genetic component to the disease, as relatives of affected people are eight times more likely to contract the disease.

Other Immune System Disorders

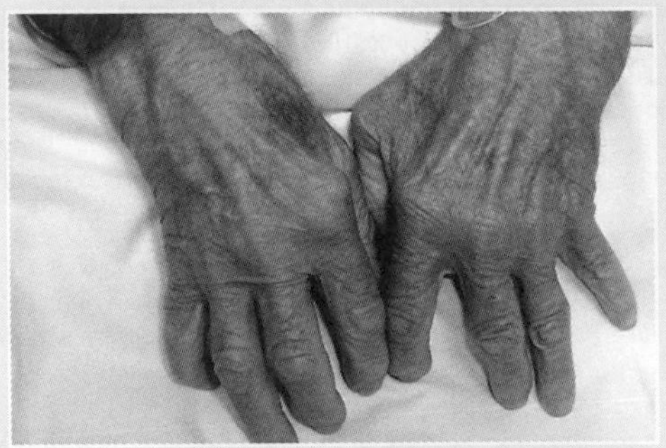

Rheumatoid arthritis is a type of joint inflammation, usually in the hands and feet, which results in destruction of cartilage and painful, swollen joints. The disease often begins in adulthood, but can also occur in children or the elderly. Rheumatoid arthritis affects more women than men and is treated with anti-inflammatory and immunosuppressant drugs, and physiotherapy.

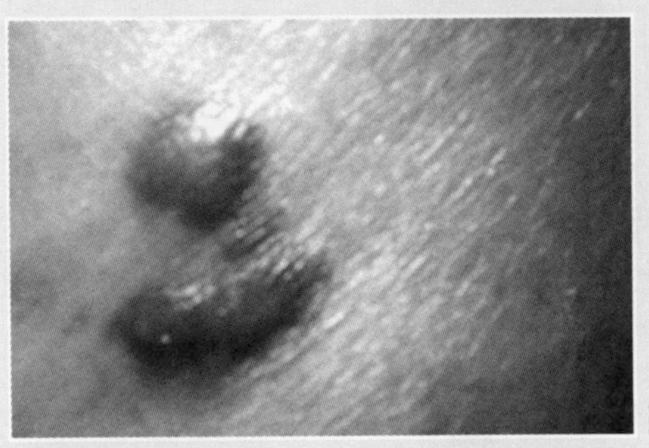

Lacking a sufficient immune response is called **immune deficiency**, and may be either **congenital** (present at birth) or **acquired** as a result of drugs, cancer, or infectious agents (e.g. HIV infection). HIV causes AIDS, which results in a steady destruction of the immune system. Sufferers then succumb to opportunistic infections and rare cancers such as Kaposi's sarcoma (above).

1. Explain the basis of the following autoimmune diseases:

   (a) Multiple sclerosis: ______

   (b) Rheumatoid arthritis: ______

2. Suggest why autoimmune diseases are difficult to treat effectively: ______

3. Explain why sufferers of immune deficiencies, such as AIDS, develop a range of debilitating infections: ______

ISBN: 978-1-92717357-2

Related activities: Neuron Structure and Function, The Immune System

Blood Group Antigens

Blood is classified into groups according to the different marker molecules (**antigens**) on the surface of red blood cells (RBCs). The type of antigens present determines an individual's blood type. **ABO blood group** antigens (below) and Rh antigens are the most important in the blood typing system because they are strongly immunogenic (cause a strong immune response). Blood must be checked for compatibility before a patient can receive donated blood. Transfusion of incompatible blood may cause a fatal transfusion reaction in which RBCs from the donated blood clump together (agglutinate), block capillaries, and rupture (hemolysis). To prevent this occurring, blood is carefully matched before transfusion. Although human RBCs have more than 500 known antigens, fewer than 30 are regularly tested for and they are not common enough to be used for cross-matching. Blood typing can also be used forensically to eliminate suspects in a crime or to predict the possibilities of paternity.

	Blood type A	Blood type B	Blood type AB	Blood type O
Antigens* present on the *red blood cells	antigen ***A***	antigen ***B***	antigens ***A*** and ***B***	Neither antigen ***A*** nor ***B***
Anti-bodies* present in the *plasma	Contains **anti-B** antibodies; but no antibodies that would attack its own antigen ***A***	Contains **anti-A** antibodies; but no antibodies that would attack its own antigen ***B***	Contains neither **anti-A** nor **anti-B** antibodies	Contains both **anti-A** and **anti-B** antibodies

Blood Type	Freq. in US		Antigen	Antibody	Can donate blood to:	Can receive blood from:
	Rh+	Rh−				
A	34%	6%	A	anti-B	A, AB	A, O
B	9%	2%				
AB	3%	1%				
O	38%	7%				

1. Complete the table above to show the antibodies and antigens in each blood group, and donor/recipient blood types:

2. Explain why blood from an incompatible donor causes a transfusion reaction in the recipient: ______________________

3. Why is blood type O− sometimes called the universal donor? ______________________

4. Why is blood type AB+ sometimes called the universal recipient? ______________________

5. Why was the discovery of the ABO system such a significant medical breakthrough? ______________________

The Lymphatic System and Immunity

ISBN: 978-1-92717357-2

Related activities: *Antibodies, Blood*
Weblinks: *The Blood Typing Game*

Blood Clotting and Defense

Apart from its transport role, **blood** has a role in the body's defense against infection and **hemostasis** (the prevention of bleeding and maintenance of blood volume). The tearing or puncturing of a blood vessel initiates **clotting**. Clotting is normally a rapid process that seals off the tear, preventing blood loss and the invasion of bacteria into the site. Clot formation is triggered by the release of clotting factors from the damaged cells at the site of the tear or puncture. A hardened clot forms a scab, which acts to prevent further blood loss and acts as a mechanical barrier to the entry of pathogens.

Blood Clotting

When tissue is wounded, the blood quickly coagulates to prevent further blood loss and maintain the integrity of the circulatory system. For external wounds, clotting also prevents the entry of pathogens. Blood clotting involves a cascade of reactions involving at least twelve clotting factors in the blood. The end result is the formation of an insoluble network of fibers, which traps red blood cells and seals the wound.

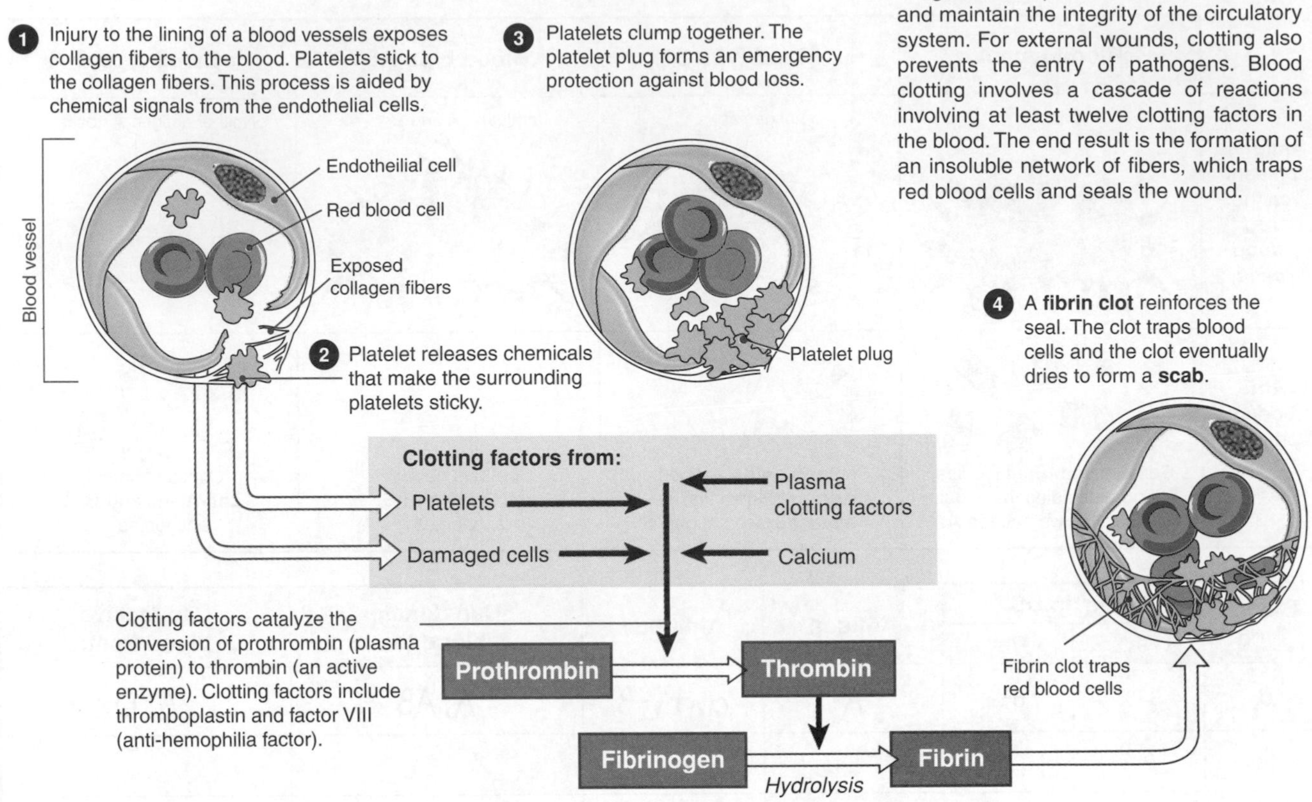

1. Explain two roles of the blood clotting system in internal defense and hemostasis:

   (a) ______

   (b) ______

2. Explain the role of each of the following in the sequence of events leading to a blood clot:

   (a) Injury: ______

   (b) Release of chemicals from platelets: ______

   (c) Clumping of platelets at the wound site: ______

   (d) Formation of a fibrin clot: ______

3. (a) Explain the role of clotting factors in the blood in formation of the clot: ______

   (b) Explain why these clotting factors are not normally present in the plasma: ______

4. (a) Name one inherited disease caused by the absence of a clotting factor: ______

   (b) Name the clotting factor involved: ______

ISBN: 978-1-92717357-2

The Body's Defenses

The human body has a tiered system of defenses to prevent or limit infection by pathogens. The first line of defense has a role in keeping microorganisms from entering the body. If this fails, a second line of defense targets any foreign bodies (including microbes) that manage to get inside. If these defenses fail, the body's immune system provides a third line of (specific) defense. The ability to ward off disease through the various defense mechanisms is called **resistance**. **Non-specific resistance** is provided by the first and second lines of defense. **Specific resistance** (the immune response) is specific to particular pathogens. Part of the immune response involves the production of **antibodies**, which are large proteins that identify and neutralize foreign material such as microorganisms. Antibodies recognize and respond to specific parts of the microbes called **antigens**. Antigens are often proteins or carbohydrates such as fragments of cell wall. The name comes from **anti**body **gen**erator.

1st Line of Defense

The skin provides a physical barrier to the entry of pathogens. Healthy skin is rarely penetrated by microorganisms. Its low pH is unfavorable to the growth of many bacteria and its chemical secretions (e.g. sebum, antimicrobial peptides) inhibit growth of bacteria and fungi. Tears, mucus, and saliva also help to wash bacteria away.

2nd Line of Defense

A range of defense mechanisms operate inside the body to inhibit or destroy pathogens. These responses react to the presence of any pathogen, regardless of which species it is. White blood cells are involved in most of these responses.

It includes the **complement system** whereby plasma proteins work together to bind pathogens and induce an inflammatory responses to help fight infection.

3rd Line of Defense

Once the pathogen has been identified by the immune system, **lymphocytes** launch a range of specific responses to the pathogen, including the production of **antibodies**. Each type of antibody is produced by a B-cell clone and is specific against a particular antigen.

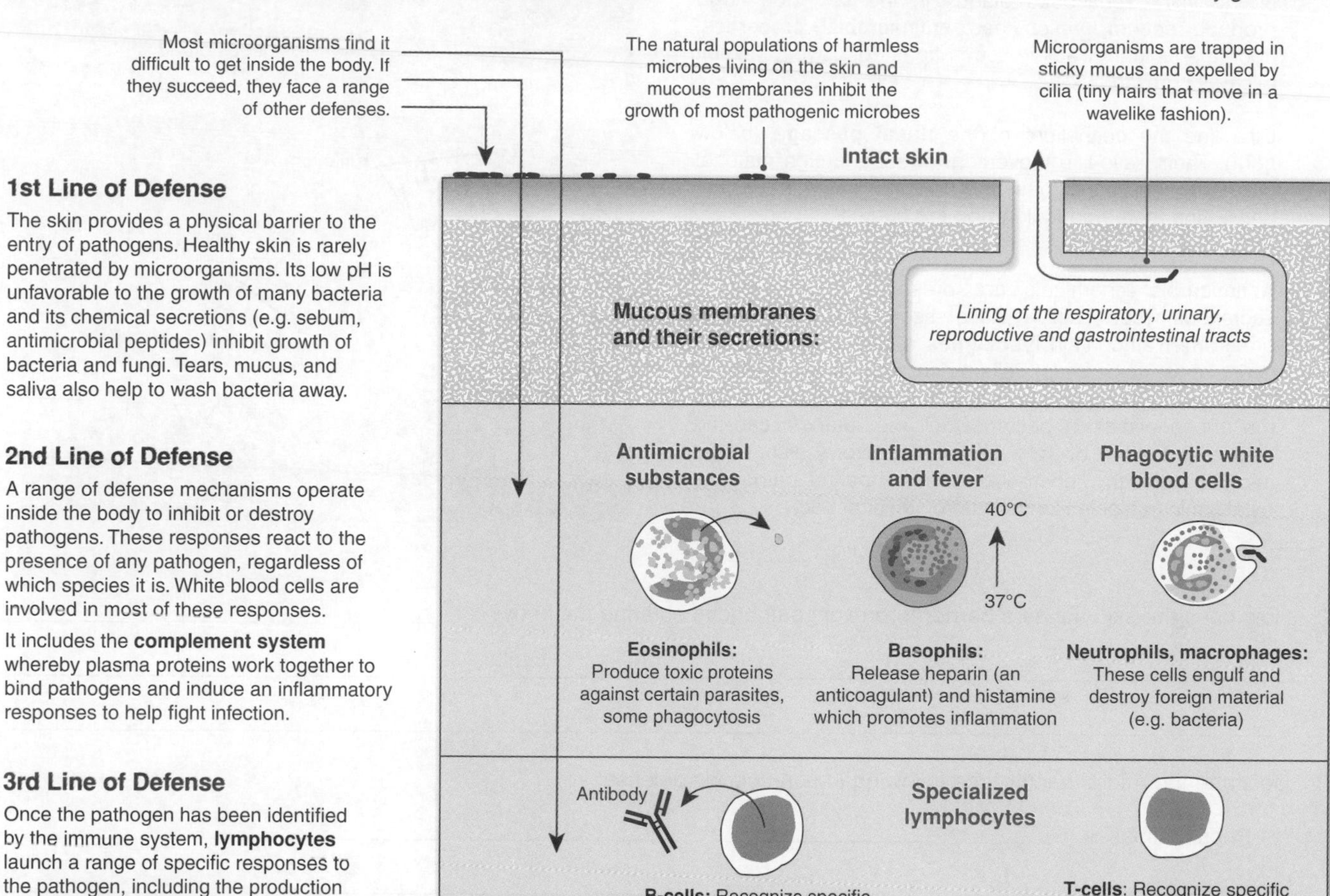

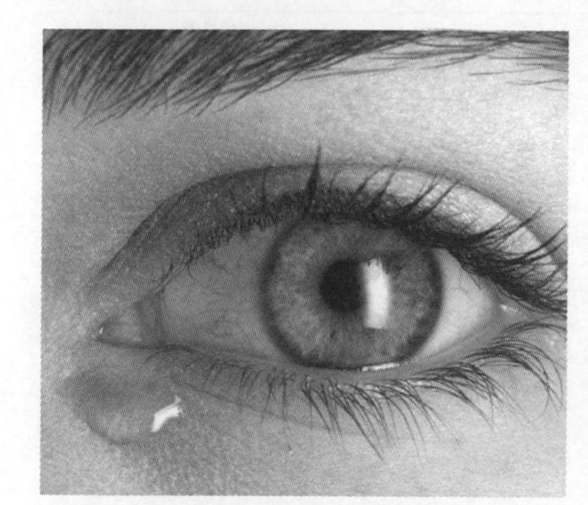

Tears contain antimicrobial substances as well as washing contaminants from the eyes.

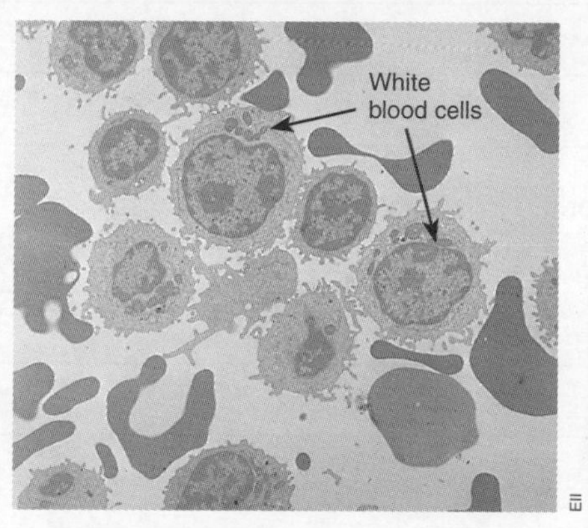

A range of white blood cells (the larger cells in the photograph) form the second line of defense.

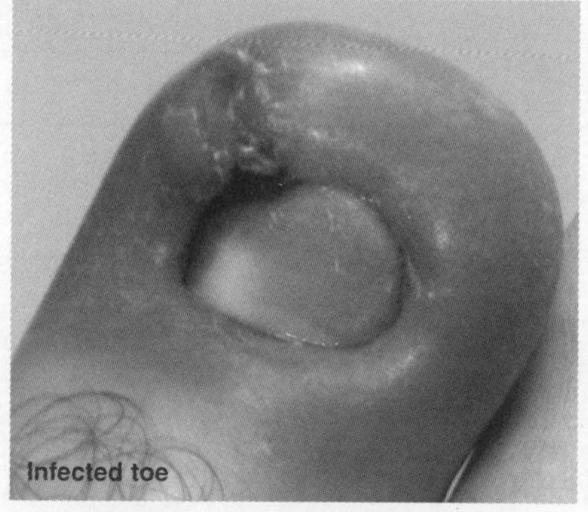

Inflammation is a localized response to infection characterized by swelling, pain, and redness.

1. Distinguish between specific and non-specific resistance:

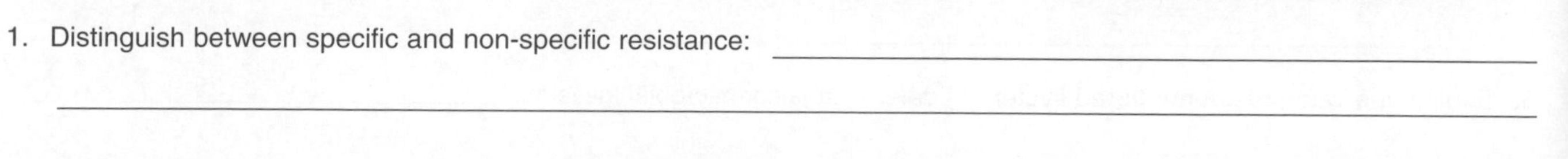

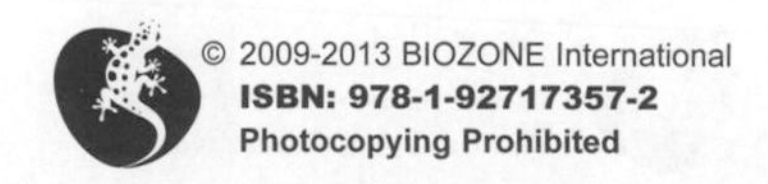

ISBN: 978-1-92717357-2

Periodicals: *Skin, scabs and scars, Fight for your life*

Related activities: *Targets for Defense, The Immune System*

Weblinks: *Immunoanimations*

The Importance of the First Line of Defense

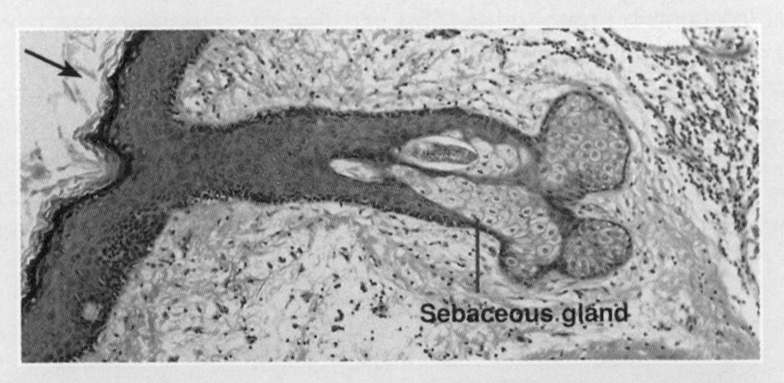

The skin is the largest organ of the body. It forms an important physical barrier against the entry of pathogens into the body. A natural population of harmless microbes live on the skin, but most other microbes find the skin inhospitable. The continual shedding of old skin cells (arrow, right) physically removes bacteria from the surface of the skin. Sebaceous glands in the skin (top right) produce sebum, which has antimicrobial properties, and the slightly acidic secretions of sweat inhibit microbial growth.

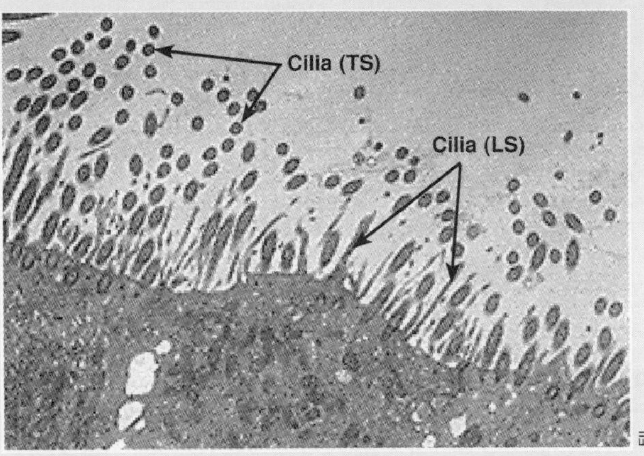

Cilia line the epithelium of the **nasal passage** (below right). Their wave-like movement sweeps foreign material out and keeps the passage free of microorganisms, preventing them from colonizing the body.

Antimicrobial chemicals are present in many bodily secretions. Tears, saliva, nasal secretions, and human breast milk all contain **lysozymes** and **phospholipases**. Lysozymes kill bacterial cells by catalyzing the hydrolysis of cell wall linkages, whereas phospholipases hydrolyze the phospholipids in bacterial cell membranes, causing bacterial death. Low pH gastric secretions also inhibit microbial growth, and reduce the number of pathogens establishing colonies in the gastrointestinal tract.

2. How does the skin act as a barrier to prevent pathogens entering the body? ______

3. Describe the role of each of the following in non-specific defense:

(a) Phospholipases: ______

(b) Cilia: ______

(c) Sebum: ______

4. Describe the functional role of each of the following defense mechanisms:

(a) Phagocytosis by white blood cells: ______

(b) Antimicrobial substances: ______

(c) Antibody production: ______

5. Explain the value of a three tiered system of defense against microbial invasion: ______

ISBN: 978-1-92717357-2

The Action of Phagocytes

Human cells that ingest microbes and digest them by the process of **phagocytosis** are called **phagocytes**. All are types of white blood cells. During many kinds of infections, especially bacterial infections, the total number of white blood cells increases by two to four times the normal number. The ratio of various white blood cell types changes during the course of an infection.

How a Phagocyte Destroys Microbes

1 Detection
Phagocyte detects microbes by the chemicals they give off (chemotaxis) and sticks the microbes to its surface.

2 Ingestion
The microbe is engulfed by the phagocyte wrapping pseudopodia around it to form a vesicle.

3 Phagosome forms
A phagosome (phagocytic vesicle) is formed, which encloses the microbes in a membrane.

4 Fusion with lysosome
Phagosome fuses with a lysosome (which contains powerful enzymes that can digest the microbe).

5 Digestion
The microbes are broken down by enzymes into their chemical constituents.

6 Discharge
Indigestible material is discharged from the phagocyte cell.

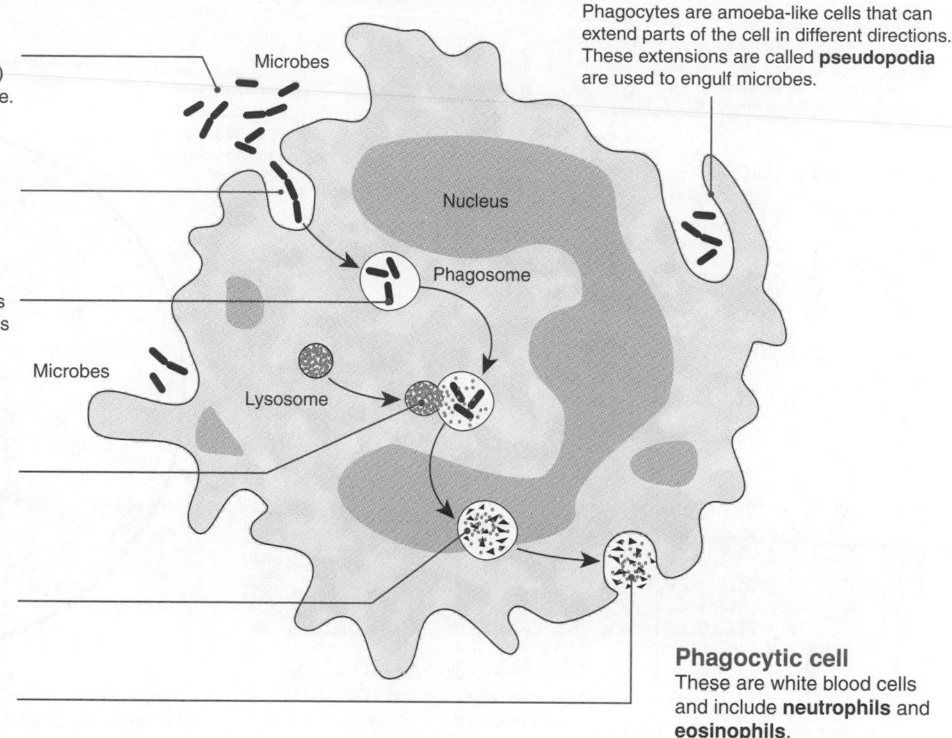

Phagocytic cell
These are white blood cells and include **neutrophils** and **eosinophils**.

The Interaction of Microbes and Phagocytes

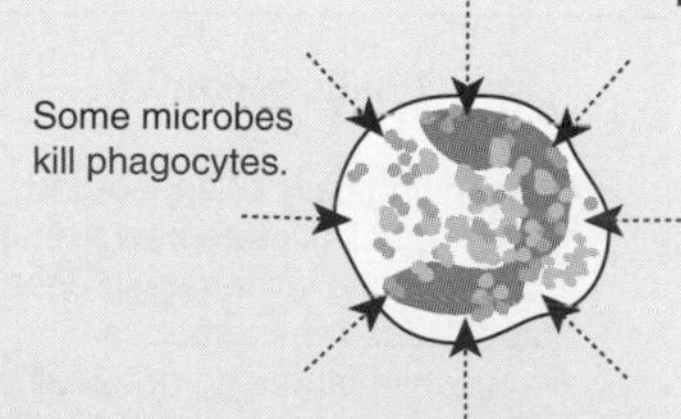

Some microbes kill phagocytes

Some microbes produce toxins that can actually kill phagocytes, e.g. toxin-producing staphylococci and the dental plaque-forming bacteria *Actinobacillus*.

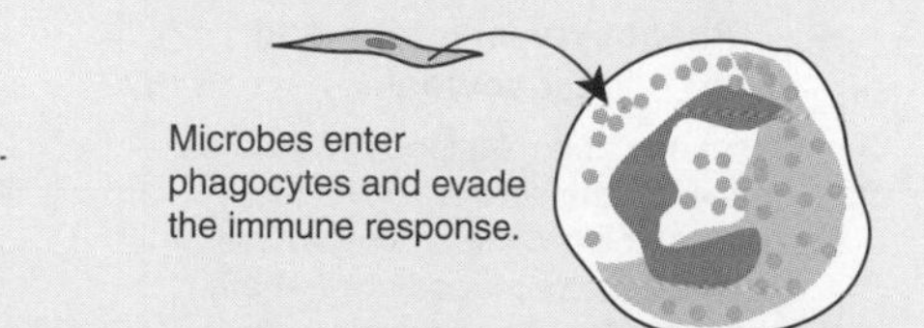

Microbes evade immune system

Some microbes can evade the immune system by entering phagocytes. The microbes prevent fusion of the lysosome with the phagosome and multiply inside the phagocyte, almost filling it. Examples include *Chlamydia*, *Mycobacterium tuberculosis*, *Shigella*, and malarial parasites.

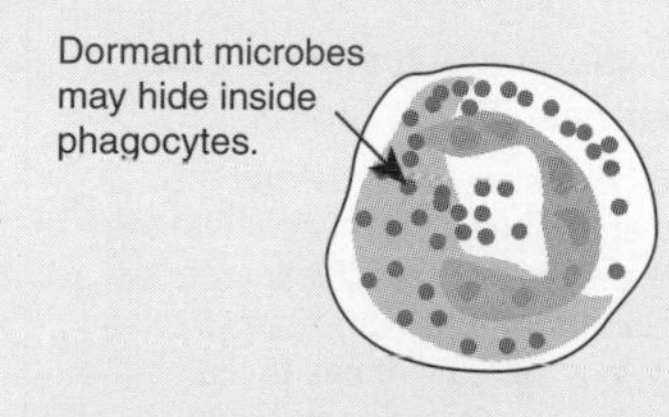

Dormant microbes hide inside

Some microbes can remain dormant inside the phagocyte for months or years at a time. Examples include the microbes that cause brucellosis and tularemia.

1. Identify the white blood cells capable of phagocytosis: ______________________________

2. Describe how a blood sample from a patient may be used to determine whether they have a microbial infection (without looking for the microbes themselves):

3. Explain how some microbes are able to overcome phagocytic cells and use them to their advantage:

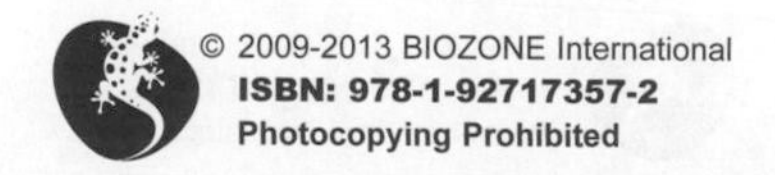

ISBN: 978-1-92717357-2

Inflammation

Damage to the body's tissues can be caused by physical agents (e.g. sharp objects, heat, radiant energy, or electricity), microbial infection, or chemical agents (e.g. gases, acids and bases). The damage triggers a defensive response called **inflammation**. It is usually characterized by four symptoms: pain, redness, heat and swelling. The inflammatory response is beneficial and has the following functions: (1) to destroy the cause of the infection and remove it and its products from the body; (2) if this fails, to limit the effects on the body by confining the infection to a small area; (3) replacing or repairing tissue damaged by the infection. The process of inflammation can be divided into three distinct stages. These are described below.

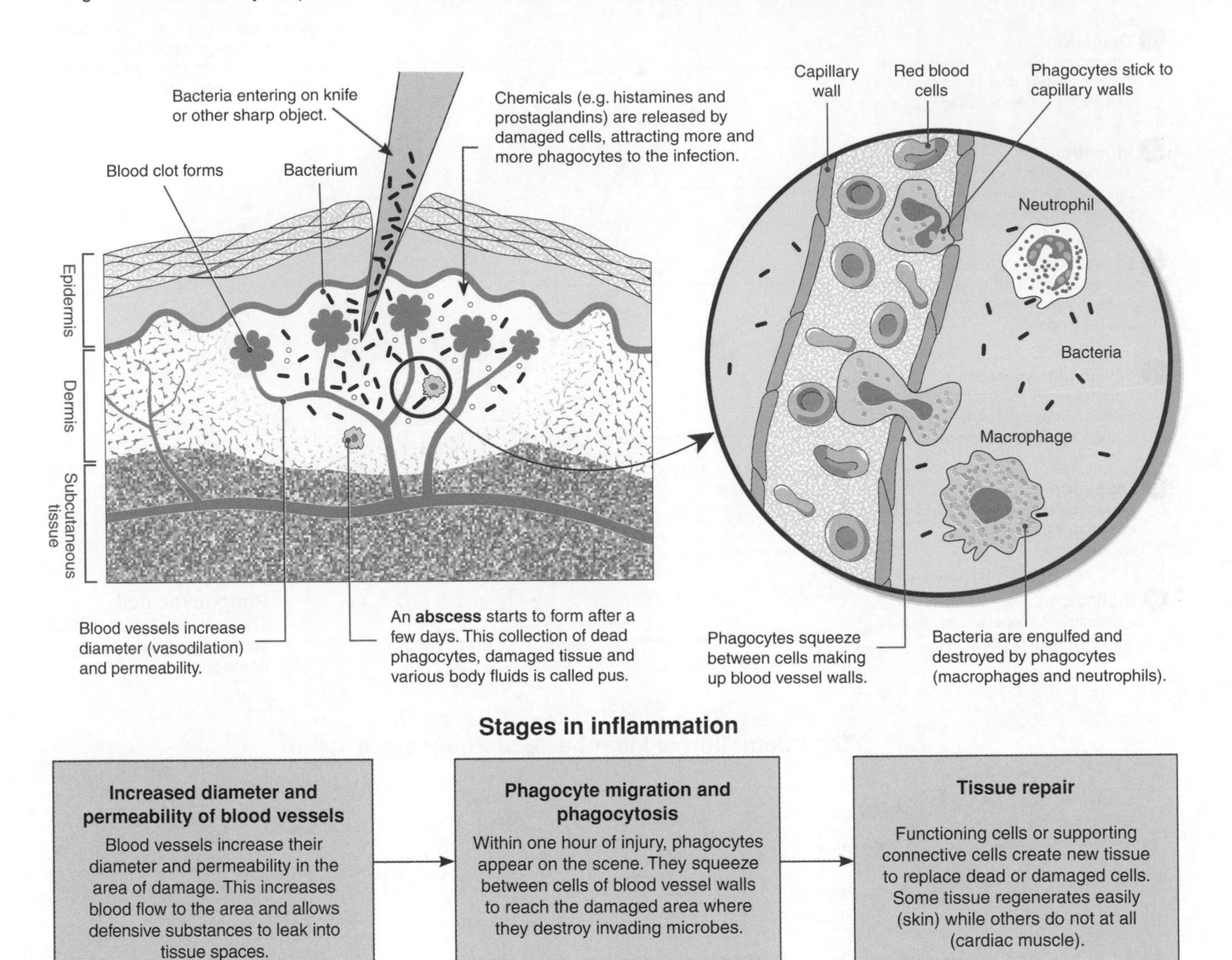

Stages in inflammation

Increased diameter and permeability of blood vessels	→	Phagocyte migration and phagocytosis	→	Tissue repair
Blood vessels increase their diameter and permeability in the area of damage. This increases blood flow to the area and allows defensive substances to leak into tissue spaces.		Within one hour of injury, phagocytes appear on the scene. They squeeze between cells of blood vessel walls to reach the damaged area where they destroy invading microbes.		Functioning cells or supporting connective cells create new tissue to replace dead or damaged cells. Some tissue regenerates easily (skin) while others do not at all (cardiac muscle).

1. Outline the three stages of inflammation and identify the beneficial role of each stage:

   (a) ______________________________

   ____

   (b) ______________________________

   ____

   (c) ______________________________

   ____

2. Describe two features of phagocytes important in the response to microbial invasion: ______________________________

   ____

3. State the role of histamines and prostaglandins in inflammation: ______________________________

4. Explain why pus forms at the site of infection: ______________________________

   ____

ISBN: 978-1-92717357-2

Fever

Fever is defined as an increase in body temperature above the normal range (36.2-37.2ºC). To a point, fever is beneficial, because it assists a number of the defense processes. The release of the protein **interleukin-1** helps to reset the thermostat of the body to a higher level and increases production of **T cells** (lymphocytes). High body temperature also intensifies the effect of **interferon** (an antiviral protein) and may inhibit the growth of some bacteria and viruses. Because high temperatures speed up the body's metabolic reactions, it may promote more rapid tissue repair. Fever also increases heart rate so that white blood cells are delivered to sites of infection more rapidly. Fevers of less than 40°C do not need treatment for hyperthermia, but excessive fever requires prompt attention (particularly in children). Death usually results if body temperature rises above 44.4 to 45.5°C.

1 Pathogen or toxin

The most frequent cause of fever is infection from bacteria (and their toxins) and viruses. A macrophage ingesting one of these will start the fever-causing process.

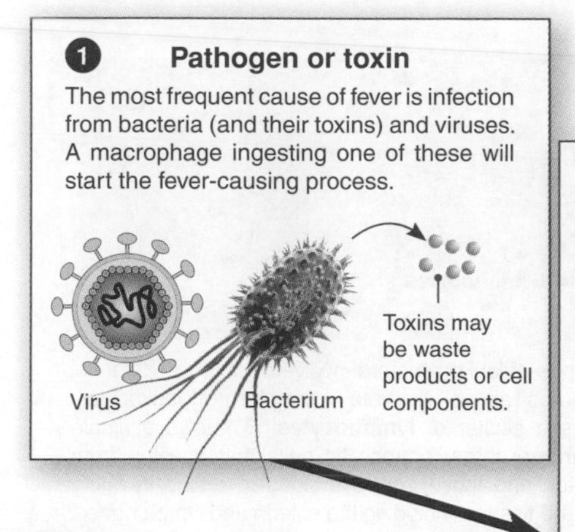

2 Macrophages respond

Macrophage ingests bacterium, destroying it and releasing endotoxins. The endotoxins induce the macrophage to produce a small protein called **interleukin-1**.

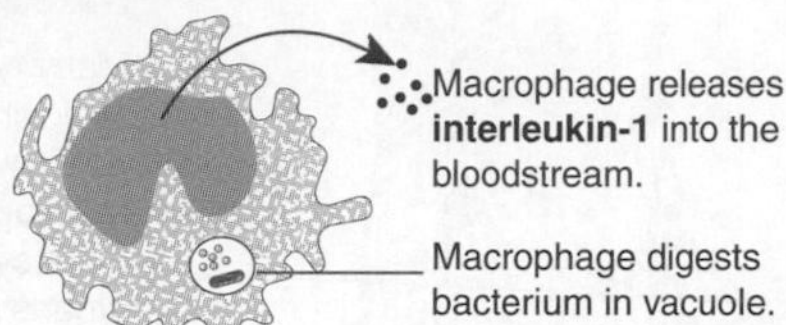

3 Thermostat is reset

Interleukin-1 circulates to the brain and induces the hypothalamus to produce more prostaglandins, which resets the 'thermostat' to a **higher temperature**, producing fever.

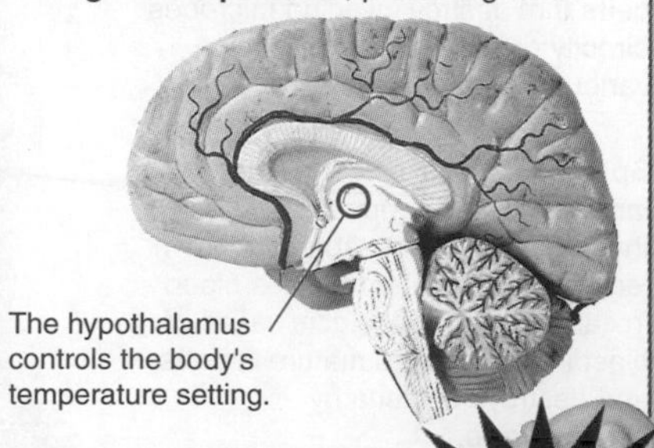

4 Fever onset

The body responds to a higher temperature set point by constricting blood vessels, increasing metabolic rate, and **shivering**. These responses raise body temperature beyond the normal range of 36.2 – 37.2 °C

5 Chill phase

Although body temperature is elevated above normal, the skin remains cold, and **shivering** increases. This condition, called a **chill**, is a definite sign that body temperature is rising. When the body reaches the setting of the thermostat, the chill disappears.

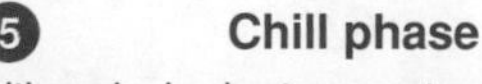

6 Crisis phase

Body temperature is maintained at the higher setting until the interleukin-1 has been eliminated. As the infection subsides, the **thermostat is reset to 37°C**. Heat losing mechanisms, such as sweating and vasodilation cause the person to feel warm. This **crisis phase** of the fever indicates that body temperature is falling.

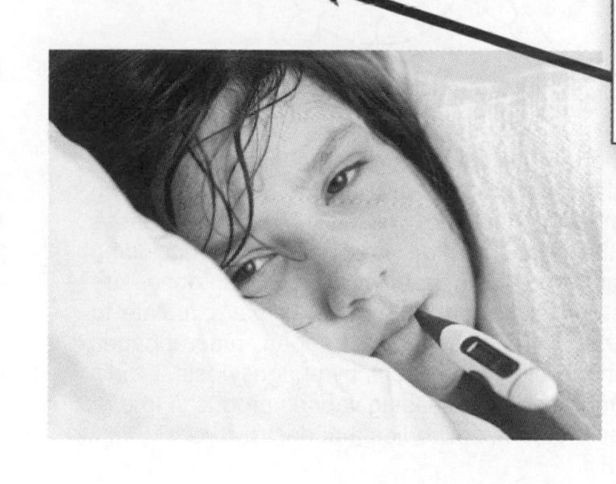

We are all familiar with the feeling of chill associated with a fever and the sweating that precedes the fever's resolution. Fevers of less than 40°C do not need treatment but high, prolonged fevers require prompt attention, because death usually results if the body temperature rises above 45.5°C.

1. Discuss the beneficial effects of fever on the body's ability to fight infections: ______

2. Summarize the key steps of how the body's thermostat is set at a higher level by infection: ______

ISBN: 978-1-92717357-2

Related activities: *The Body's Defenses*

The Lymphatic System

Fluid leaks out from capillaries and forms the tissue fluid, which is similar in composition to plasma but lacks large proteins. This fluid bathes the tissues, supplying them with nutrients and oxygen, and removing wastes. Some of the tissue fluid returns directly into the capillaries, but some drains back into the blood circulation through a network of lymph vessels. This fluid, called **lymph**, is similar to tissue fluid, but contains more leukocytes.

Apart from its circulatory role, the lymphatic system also has an important function in the immune response. Lymph nodes are the primary sites where pathogens and other foreign substances are destroyed. A lymph node that is fighting an infection becomes swollen and hard as the leukocytes reproduce rapidly to increase their numbers. The thymus, spleen, and bone marrow also contribute leukocytes to the lymphatic and circulatory systems.

Tonsils: Tonsils (and adenoids) comprise a collection of large lymphatic nodules at the back of the throat. They produce lymphocytes and antibodies and are well-placed to protect against invasion of pathogens.

Thymus gland: The thymus is a two-lobed organ located close to the heart. It is prominent in infants and diminishes after puberty to a fraction of its original size. Its role in immunity is to help produce **T cells** that destroy invading microbes directly or indirectly by producing various substances.

Spleen: The oval spleen is the largest mass of lymphatic tissue in the body, measuring about 12 cm in length. It stores and releases blood in case of demand (e.g. in cases of bleeding), produces mature **B cells**, and destroys bacteria by phagocytosis.

Bone marrow: Bone marrow produces red blood cells and many kinds of leukocytes: monocytes (and macrophages), neutrophils, eosinophils, basophils, and lymphocytes (B cells and T cells).

Lymphatic vessels: When tissue fluid is picked up by lymph capillaries, it is called **lymph**. The lymph is passed along lymphatic vessels to a series of lymph nodes. These vessels contain one-way valves that move the lymph in the direction of the heart until it is reintroduced to the blood at the subclavian veins.

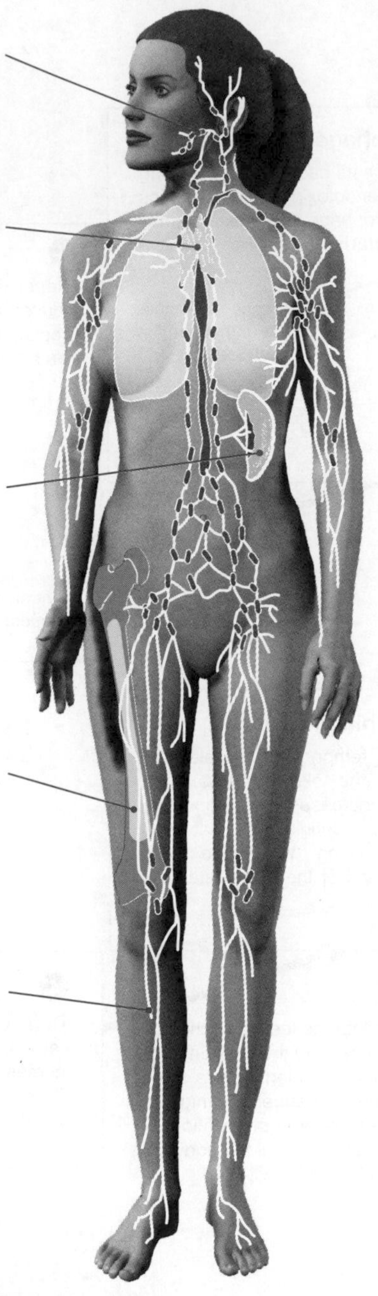

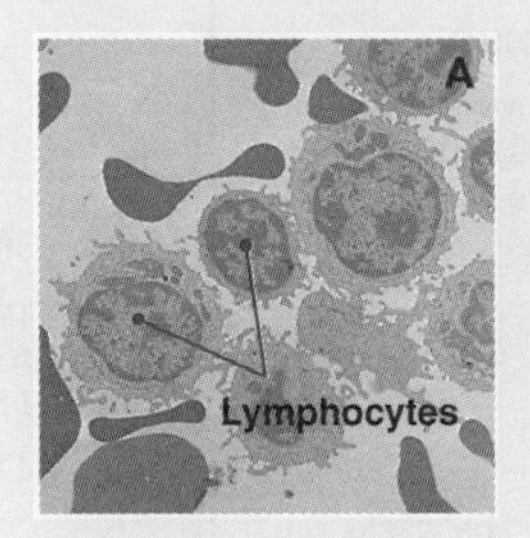

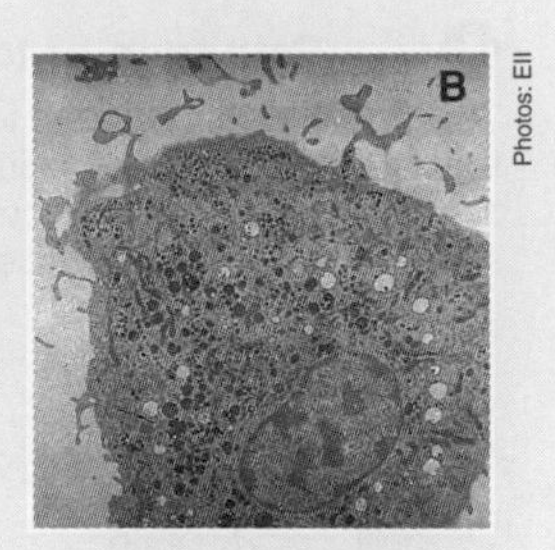

Many types of leukocytes are involved in internal defense. The photos above illustrate examples of leukocytes. **A** shows a cluster of **lymphocytes**. **B** shows a single **macrophage**: large, phagocytic cells that develop from monocytes and move from the blood to reside in many organs and tissues, including the spleen and lymph nodes.

Lymph node

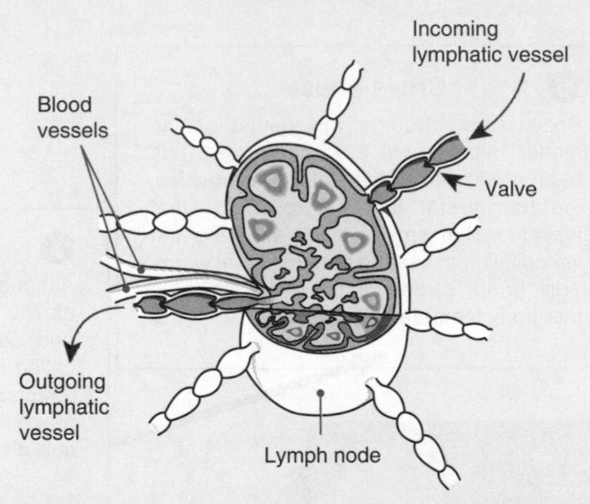

Lymph nodes are oval or bean-shaped structures, scattered throughout the body, usually in groups, along the length of lymphatic vessels. As lymph passes through the nodes, it filters foreign particles (including pathogens) by trapping them in fibers. Lymph nodes are also a "store" of **lymphocytes**, which may circulate to other parts of the body. Once trapped, macrophages destroy the foreign substances by phagocytosis. T cells may destroy them by releasing various products, and/or B cells may release antibodies that destroy them.

1. Briefly describe the composition of lymph: ______________________________

____

2. Discuss the various roles of lymph: ______________________________

____

____

3. Describe one role of each of the following in the lymphatic system:

   (a) Lymph nodes: ______________________________

   (b) Bone marrow: ______________________________

ISBN: 978-1-92717357-2

Related activities: *Formation of Tissue Fluid, The Immune System*

The Immune System

The efficient internal defense provided by the immune system is based on its ability to respond specifically against foreign substances and hold a memory of this response. There are two main components of the immune system: the humoral and the cell-mediated responses. They work separately and together to provide protection against disease. The **humoral immune response** is associated with the serum (the non-cellular part of the blood) and involves the action of **antibodies** secreted by **B-cell lymphocytes**. Antibodies are found in extracellular fluids including lymph, plasma, and mucus secretions. The humoral response protects the body against circulating viruses, and bacteria and their toxins. The **cell-mediated immune response** is associated with the production of specialized lymphocytes called **T-cells**. It is most effective against bacteria and viruses located within host cells, as well as against parasitic protozoa, fungi, and worms.

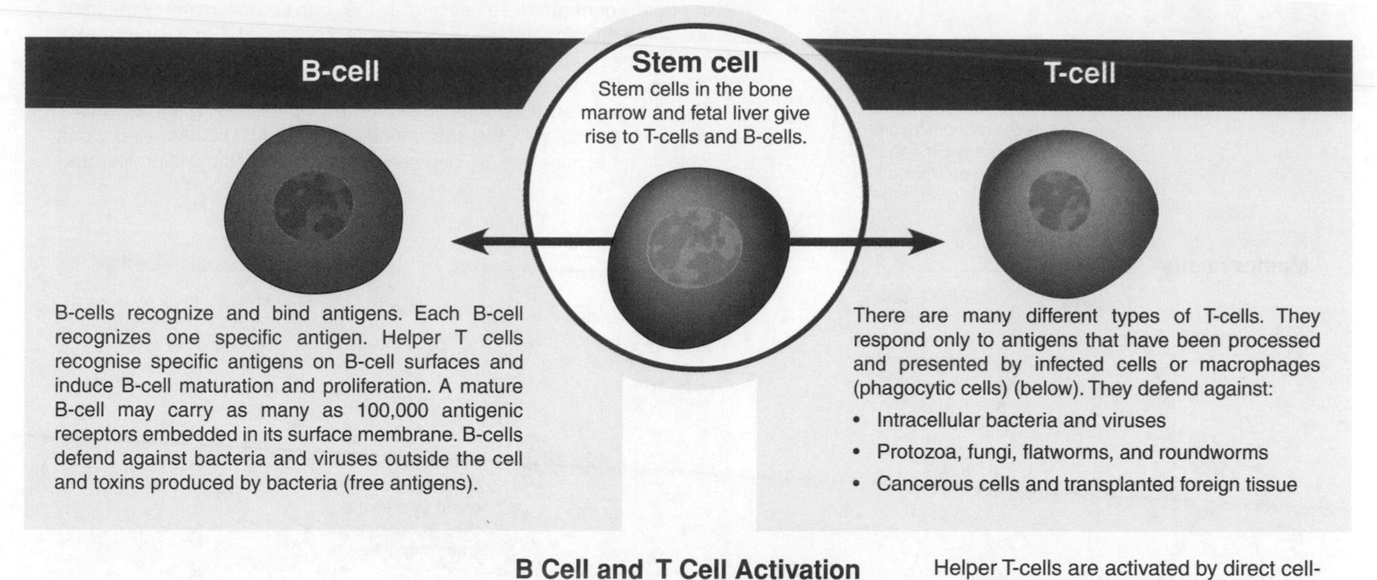

B Cell and T Cell Activation

Cytotoxic T-cell

Macrophage

Macrophage receptor

Other helper T-cells

Cytokines

Presented antigen

Helper T-cell

Antigen

Cytokines

B-cell

Helper T-cells are activated by direct cell-to-cell signaling and by signaling to nearby cells using **cytokines** from macrophages.

Macrophages ingest antigens, process them, and present them on the cell surface where they are recognized by helper T-cells. The helper T-cell binds to the antigen and to the macrophage receptor, which leads to activation of the helper T-cell.

The macrophage also produces and releases cytokines, which enhance T-cell activation. The activated T-cell then releases more cytokines which causes the proliferation of other helper T-cells (positive feedback) and helps to activate cytotoxic T-cells and antibody-producing B-cells.

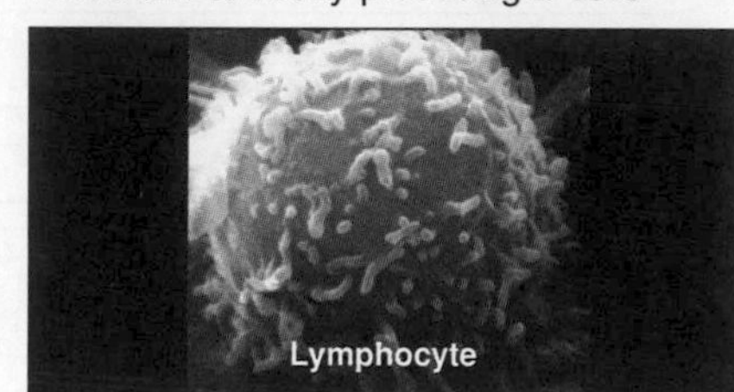

1. Describe the general action of the two major divisions in the immune system:

   (a) Humoral immune system: ____

   ____

   (b) Cell-mediated immune system: ____

   ____

2. Explain how an antigen causes the activation and proliferation of T-cells and B-cells: ____________________

____

____

____

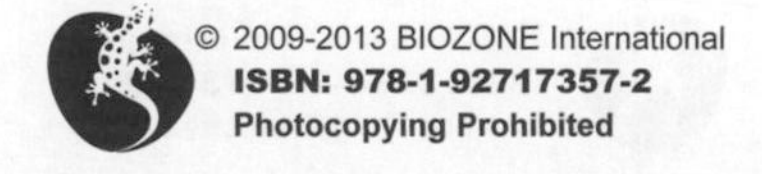

ISBN: 978-1-92717357-2

Clonal Selection

In 1955 Sir Frank Macfarlane Burnet proposed the **clonal selection theory** to explain how the immune system is able to respond to the large and unpredictable range of potential antigens in the environment. The diagram below describes clonal selection after antigen exposure for B-cells. In the same way, a T-cell stimulated by a specific antigen will multiply and develop into different types of T-cells. Clonal selection and differentiation of lymphocytes provide the basis for **immunological memory**.

Five (a-e) of the many B-cells generated during development. Each one can recognize only one specific antigen.

a b c d e

This B-cell encounters and binds an antigen. It is then stimulated to proliferate.

Clonal Selection Theory

Millions of B-cells form during development. Antigen recognition is randomly generated, so collectively they can recognize many antigens, including those that have never been encountered. Each B-cell makes antibodies corresponding to the specific antigenic receptor on its surface. The receptor reacts only to that specific antigen. When a B-cell encounters its antigen, it responds by proliferating and producing many clones all with the same kind of antibody. This is called clonal selection because the antigen selects the B-cells that will proliferate.

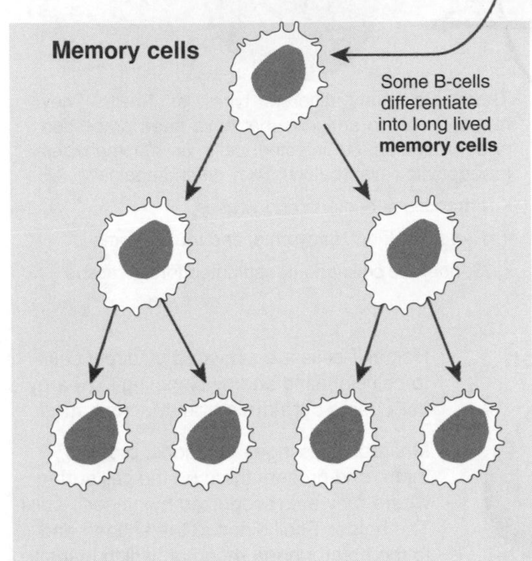

Plasma cells

Some B-cells differentiate into **plasma cells**

The antibody produced corresponds to the antigenic receptors on the cell surface.

Antibodies are secreted into the blood by plasma cells where they inactivate antigens.

Some B-cells differentiate into long lived **memory cells**. These are retained in the lymph nodes to provide future immunity (**immunological memory**). In the event of a second infection, B-memory cells react more quickly and vigorously than the initial B-cell reaction to the first infection.

Plasma cells secrete antibodies specific to the antigen that stimulated their development. Each plasma cell lives for only a few days, but can produce about 2000 antibody molecules per second. Note that during development, any B-cells that react to the body's own antigens are selectively destroyed in a process that leads to **self tolerance** (acceptance of the body's own tissues).

1. Describe how clonal selection results in the proliferation of one particular B-cell: ______________________

____

____

____

2. Describe the function of each of the following cells in the immune system response:

(a) Memory cells: ______________________

____

(b) Plasma cells: ______________________

____

3. (a) Explain the basis of **immunological memory**: ______________________

____

____

(b) Why is immunological memory important? ______________________

____

____

ISBN: 978-1-92717357-2

***Related activities:** The Immune System, Acquired Immunity*
***Weblinks:** The Immune Response, How Lymphocytes Produce Antibodies*

Antibodies

Antibodies and antigens play key roles in the response of the immune system. **Antigens** are foreign molecules that are able to bind to antibodies (or T-cell receptors) and provoke a specific immune response. Antigens include potentially damaging microbes and their toxins (see below) as well as substances such as pollen grains, blood cell surface molecules, and the surface proteins on transplanted tissues. **Antibodies** (also called **immunoglobulins**) are proteins that are made in response to antigens. They are secreted into the plasma where they circulate and can recognize, bind to, and help to destroy antigens. There are five classes of antibodies. Each plays a different role in the immune response (including destroying protozoan parasites, enhancing phagocytosis, protecting mucous surfaces, and neutralizing toxins and viruses). The human body can produce an estimated 100 million antibodies, recognizing many different antigens, including those it has never encountered. Each type of antibody is highly specific to only one particular antigen. The ability of the immune system to recognize and ignore the antigenic properties of its own tissues occurs early in development and is called **self-tolerance**. Exceptions occur when the immune system malfunctions and the body attacks its own tissues, causing an **autoimmune disorder**.

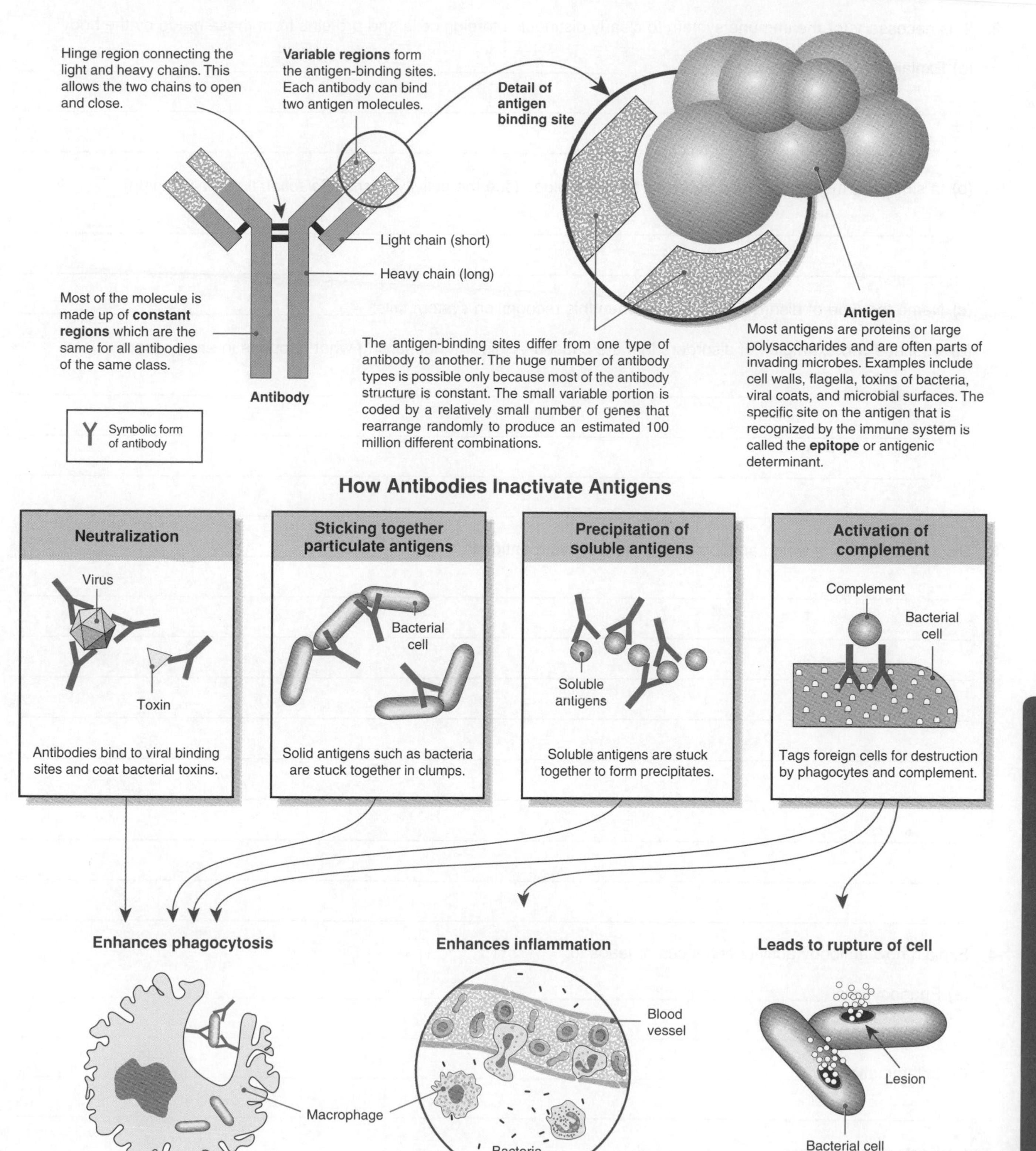

The Lymphatic System and Immunity

ISBN: 978-1-92717357-2

Related activities: *Clonal Selection, Acquired Immunity*
Weblinks: *The Humoral Response*

1. Distinguish between an **antibody** and an **antigen**:

2. It is necessary for the immune system to clearly distinguish foreign cells and proteins from those made by the body.

(a) Explain why this is the case:

(b) In simple terms, explain how **self tolerance** develops (see the activity *Clonal Selection* if you need help):

(c) Name the type of disorder that results when this recognition system fails:

(d) Describe two examples of disorders that are caused in this way, identifying what happens in each case:

3. Discuss the ways in which antibodies work to inactivate antigens:

4. Explain how antibody activity enhances or leads to:

(a) Phagocytosis:

(b) Inflammation:

(c) Bacterial cell lysis:

ISBN: 978-1-92717357-2

Acquired Immunity

We have natural or **innate resistance** to certain illnesses, including most diseases of other animal species. Immunity against microbes and foreign substances can also be developed during our lifetime. This is called **acquired immunity,** and it can be acquired either passively or actively. **Active immunity** develops when a person is exposed to foreign substances or to microorganisms (e.g. through infection) and the immune system responds. **Passive immunity** is acquired when antibodies are transferred from one person to another. Recipients do not make the antibodies themselves and the effect lasts only as long as the antibodies are present (usually several weeks or months). Either type of immunity may also be **naturally acquired** through natural exposure to microbes, or **artificially acquired** as a result of medical treatment.

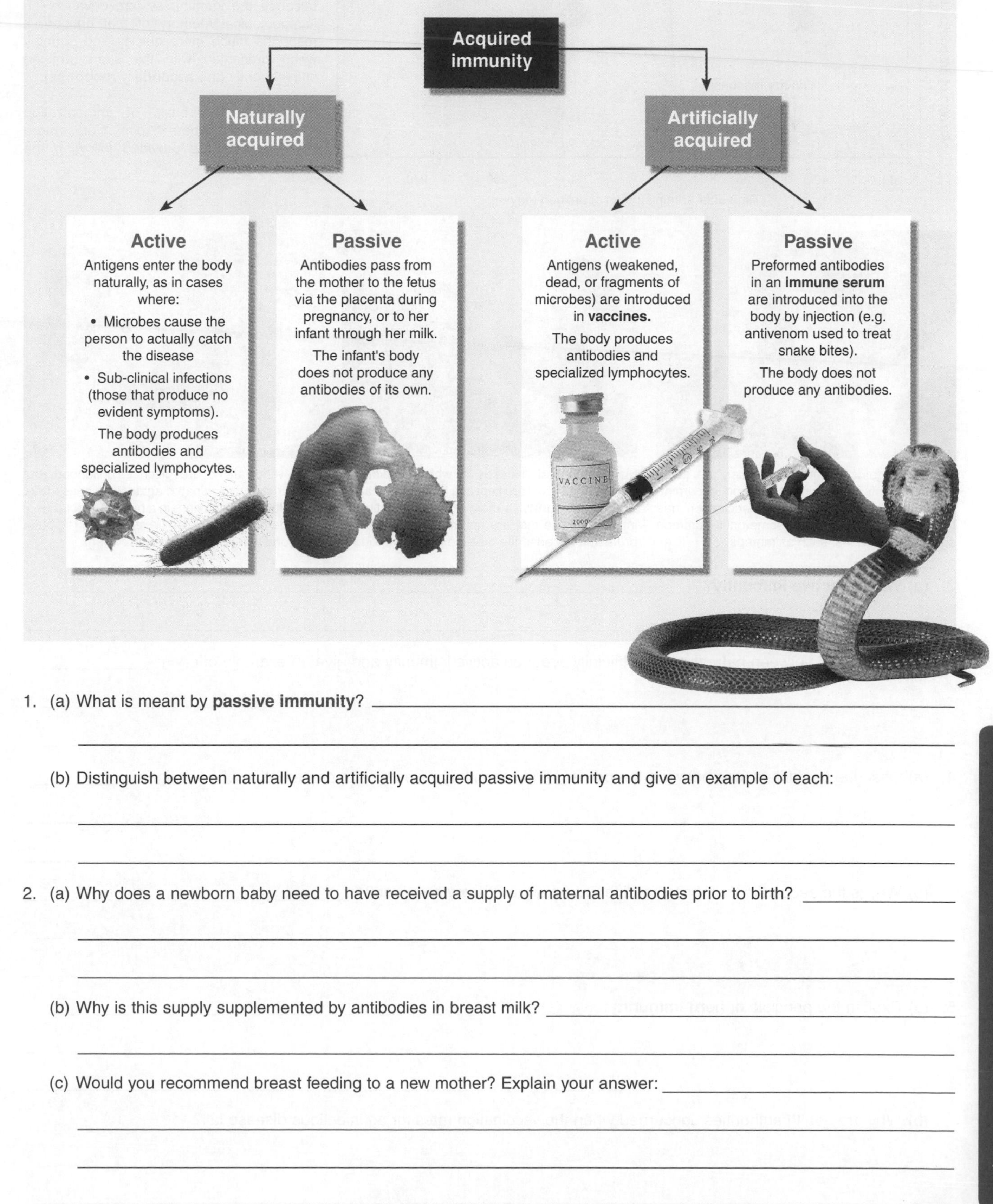

1. (a) What is meant by **passive immunity**? ______________________________

   (b) Distinguish between naturally and artificially acquired passive immunity and give an example of each:

   ____

2. (a) Why does a newborn baby need to have received a supply of maternal antibodies prior to birth? ______________________________

   (b) Why is this supply supplemented by antibodies in breast milk? ______________________________

   (c) Would you recommend breast feeding to a new mother? Explain your answer: ______________________________

ISBN: 978-1-92717357-2

Periodicals: Hard to swallow

Related activities: *Antibodies, Vaccines and Vaccination*

Primary and Secondary Responses to Antigens

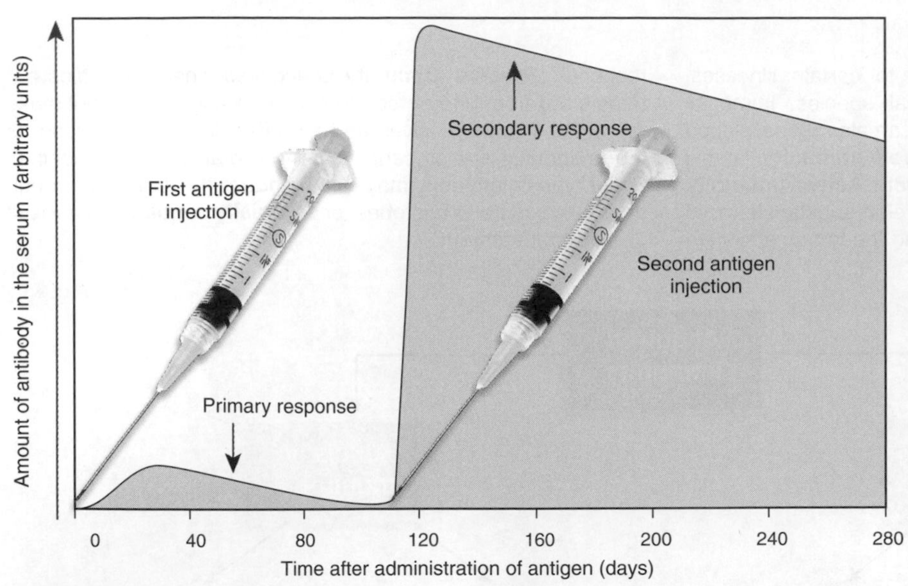

When B-cells encounter antigens and produce antibodies, the body develops **active immunity** against that antigen.

The initial response to antigenic stimulation, caused by the sudden increase in B-cell clones, is called the **primary response**. Antibody levels as a result of the primary response peak a few weeks after the response begins and then decline. However, because the immune system develops an immunological memory of that antigen, it responds much more quickly and strongly when presented with the same antigen subsequently (the **secondary response**).

This forms the basis of immunization programmes where one or more booster shots are provided following the initial vaccination.

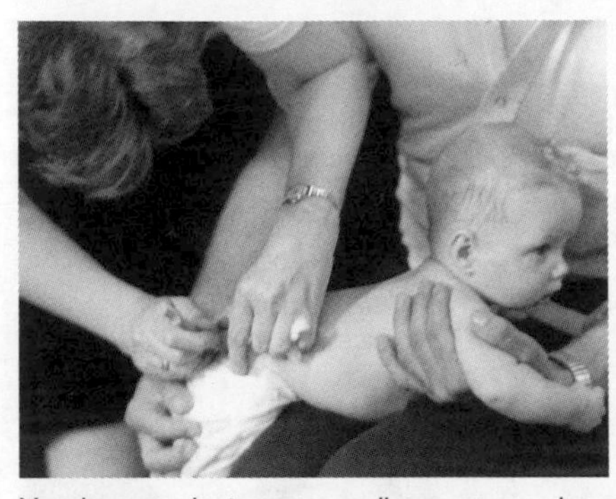

Vaccines against common diseases are given at various stages during childhood according to an immunization schedule. Vaccination has been behind the decline of some once-common childhood diseases, such as mumps.

Many childhood diseases for which vaccination programmes exist are kept at a low level because of **herd immunity**. If most of the population is immune, those that are not immunized may be protected because the disease is uncommon.

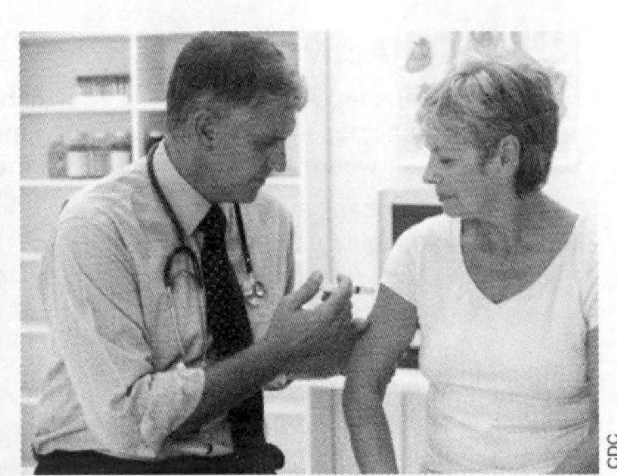

CDC

Most vaccinations are given in childhood, but adults may be vaccinated against a disease (e.g. TB, influenza) if they are in a high risk group (e.g. the elderly) or if they are travelling to a region in the world where a disease is prevalent.

3. (a) What is **active immunity**? ______________________________

(b) Distinguish between naturally and artificially acquired active immunity and give an example of each: ______________________________

4. (a) Describe two differences between the primary and secondary responses to presentation of an antigen: ______________________________

(b) Why is the secondary response so different from the primary response? ______________________________

5. (a) Explain the principle of **herd immunity**: ______________________________

(b) Why are health authorities concerned when the vaccination rates for an infectious disease fall?

____

ISBN: 978-1-92717357-2

Vaccines and Vaccination

A **vaccine** is a preparation of biological material that is deliberately introduced into the body to produce an immune response. A vaccine contains or resembles the components of a specific pathogen. The immune system produces antibodies to destroy the pathogen. The immune system remembers its response and will respond in the same way if it encounters the pathogen again. There are two basic types of vaccine: whole-agent vaccines and subunit vaccine. **Whole-agent vaccines** contain complete non-virulent microbes, either **inactivated** (killed), or alive but **attenuated** (weakened). Attenuated viruses make very effective vaccines and often provide life-long immunity without the need for booster immunizations. Inactivated viruses are less effective. **Subunit vaccines** contain only the parts of the pathogen that induce the immune response. They are safer than attenuated vaccines because they cannot reproduce in the recipient, and they have fewer side effects because they contain little or no extra material. The subunit vaccine loses its ability to cause disease while retaining its antigenic properties. Some of the most promising vaccines under development consist of naked DNA which is injected into the body and produces an antigenic protein. The safety of DNA vaccines is uncertain but they show promise against rapidly mutating viruses (e.g influenza and HIV).

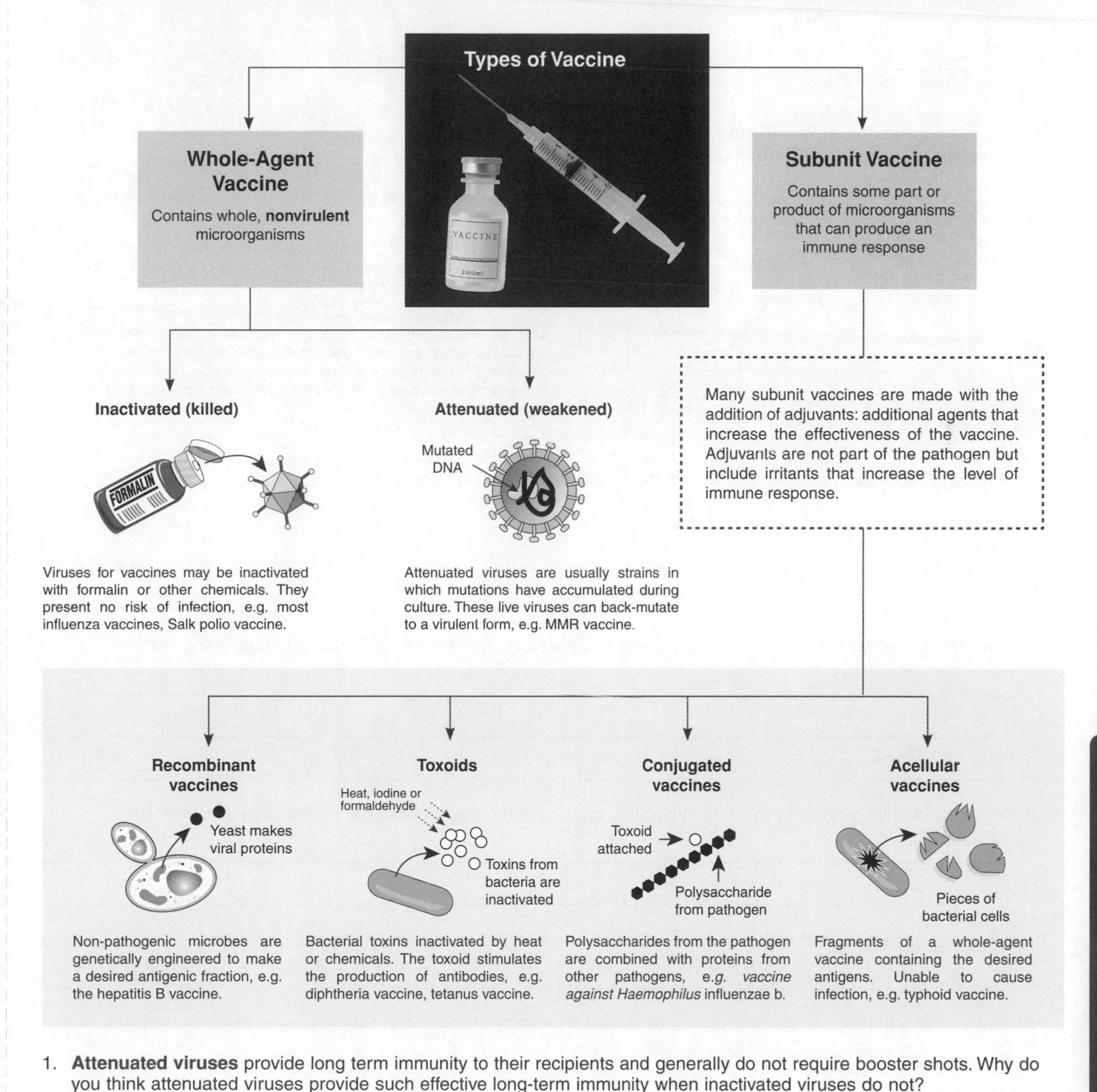

1. **Attenuated viruses** provide long term immunity to their recipients and generally do not require booster shots. Why do you think attenuated viruses provide such effective long-term immunity when inactivated viruses do not?

ISBN: 978-1-92717357-2

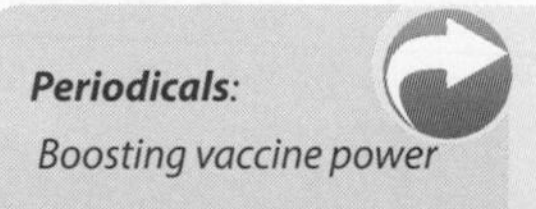

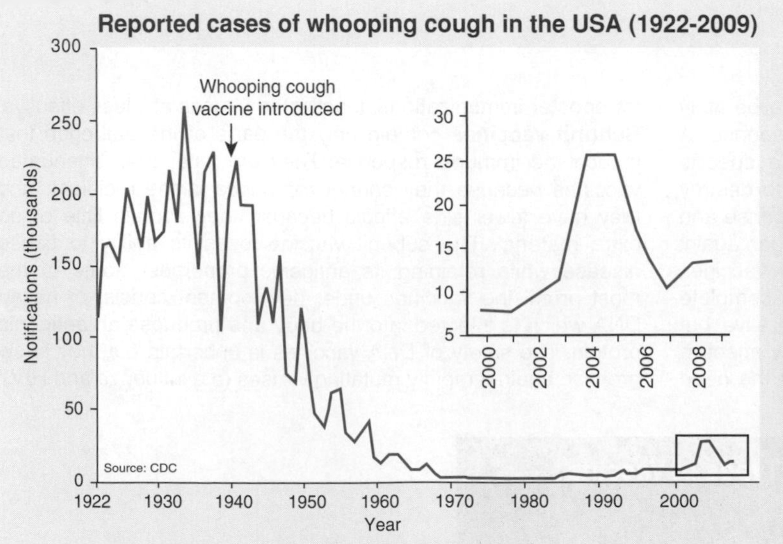

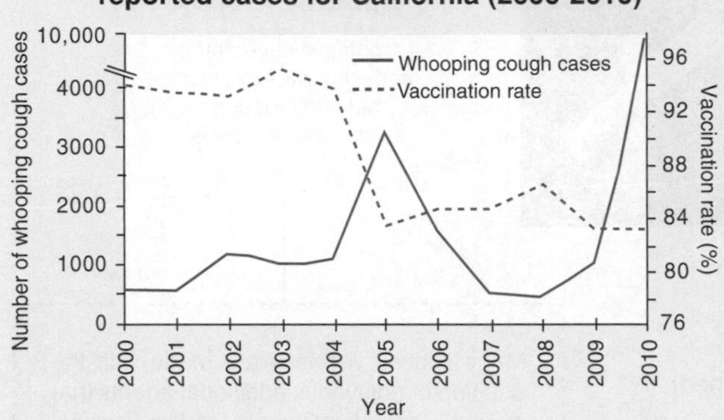

Case Study: Whooping Cough

Whooping cough is caused by the bacterium *Bordetella pertussis*, and infection may last for two to three months. It is characterized by painful coughing spasms, and a cough that sounds like a "whoop". Severe coughing fits may be followed by periods of vomiting. Inclusion of the whooping cough vaccine into the US immunization schedule in the 1940s has greatly reduced the incidence of the disease (left).

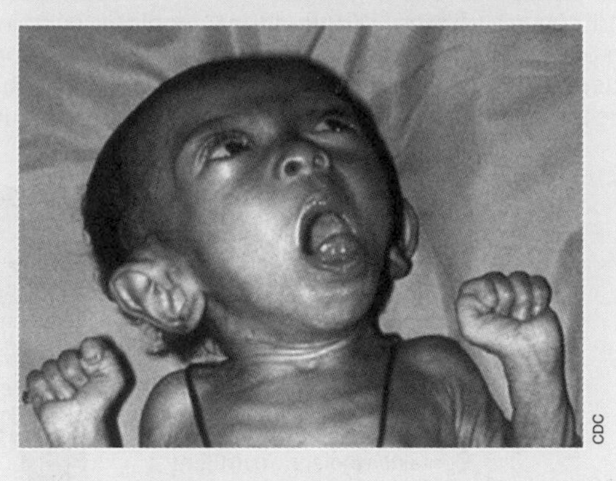

Above: Infants under six months of age are most at risk of developing complications or dying from whooping cough because they are too young to be fully protected by the vaccine. Ten infants died of whooping cough in California in 2010.

Left: In California, whooping cough vaccination rates have fallen amidst fears that it is responsible for certain health problems such as autism. As a result, rates of whooping cough have increased significantly since 2004. In 2010, over 9000 cases were reported, the highest level in 63 years.

2. How do high vaccination rates help to reduce the incidence of infectious disease?

3. (a) Describe the effect of introducing the whooping cough vaccine into the immunization schedule in the US:

(b) Why do you think whooping cough immunization rates have dropped significantly in California since 2004?

(c) What has been the effect of the lower immunization rates on the number of whooping cough cases?

(d) Suggest why the drop in immunization rates does not perfectly coincide with the increase in disease incidence:

4. Originally the whooping cough vaccine was a whole agent vaccine. In the 1990s it started being manufactured as an acellular vaccine. What advantages does an acellular vaccine have over a whole agent vaccine?

ISBN: 978-1-92717357-2

HIV and the Immune System

AIDS (Acquired Immune Deficiency Syndrome) first appeared in the news in 1981, with cases being reported in Los Angeles, in the US. By 1983, the pathogen causing the disease had been identified as a retrovirus that selectively infects **helper T cells**. The disease causes a massive deficiency in the immune system due to infection with **HIV** (human immunodeficiency virus). HIV is a **retrovirus** and is able to splice its genes into the host cell's chromosome. As yet, there is no cure or vaccine, and the disease has taken the form of a **pandemic**, spreading to all parts of the globe and killing more than a million people each year. HIV is transmitted in blood, vaginal secretions, semen, breast milk, and across the placenta. In developed countries, blood transfusions are no longer a likely source of infection because blood is tested for HIV antibodies. Historically, transmission of HIV in developed countries has been primarily through intravenous drug use and homosexual activity, but heterosexual transmission is increasing.

Capsid
Protein coat that protects the nucleic acids (RNA) within.

Viral envelope
A piece of the cell membrane budded off from the last human host cell.

Nucleic acid
Two identical strands of RNA contain the genetic blueprint for making more HIV viruses.

Reverse transcriptase
Two copies of this important enzyme convert the RNA into DNA once inside a host cell.

Surface proteins
These spikes allow HIV to attach to receptors on the host cells (T cells and macrophages).

The structure of HIV

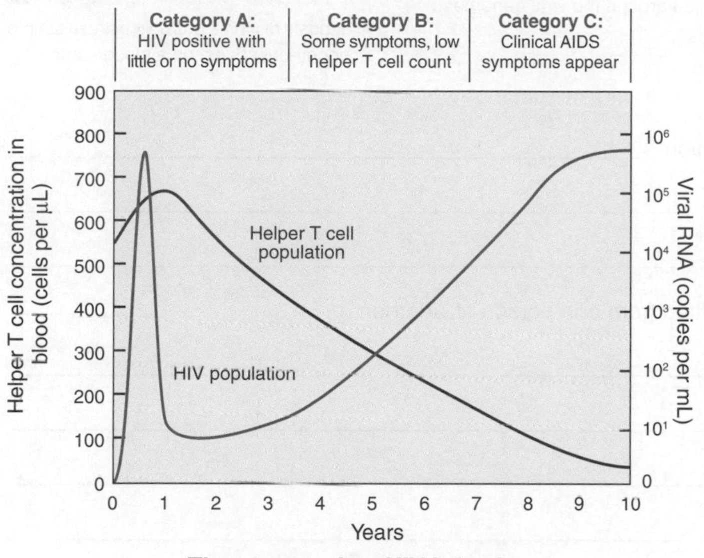

The stages of an HIV infection

AIDS is only the end stage of a HIV infection. Shortly after the initial infection, HIV antibodies appear in the blood. The progress of infection has three clinical categories (shown in the graph above).

AIDS: The end stage of an HIV infection

Individuals with human immunodeficiency virus (HIV) may have no symptoms, while medical examination may detect swollen lymph glands. Others may experience a short-lived illness when they first become infected (resembling infectious mononucleosis). The range of symptoms resulting from HIV infection is huge and not the result of the HIV infection directly. The symptoms arise from secondary infections that gain a foothold in the body due to the weakened immune system (due to the reduced number of T helper cells). These infections are from normally rare fungal, viral, and bacterial sources. Full blown AIDS can also feature some rare forms of cancer. Some symptoms are listed below:

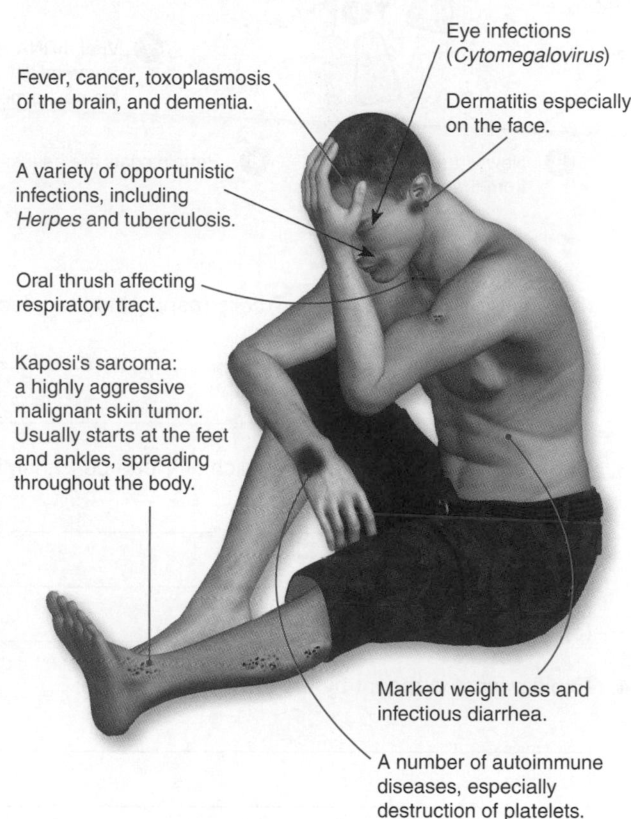

1. Explain why the HIV virus has such a devastating effect on the human body's ability to fight disease:

____

____

____

2. Consult the graph above showing the stages of HIV infection (remember, HIV infects and destroys helper T cells).

(a) Describe how the virus population changes with the progression of the disease: ____

____

____

ISBN: 978-1-92717357-2

Related activities: The Immune System
Weblinks: HIV Interactive Animation

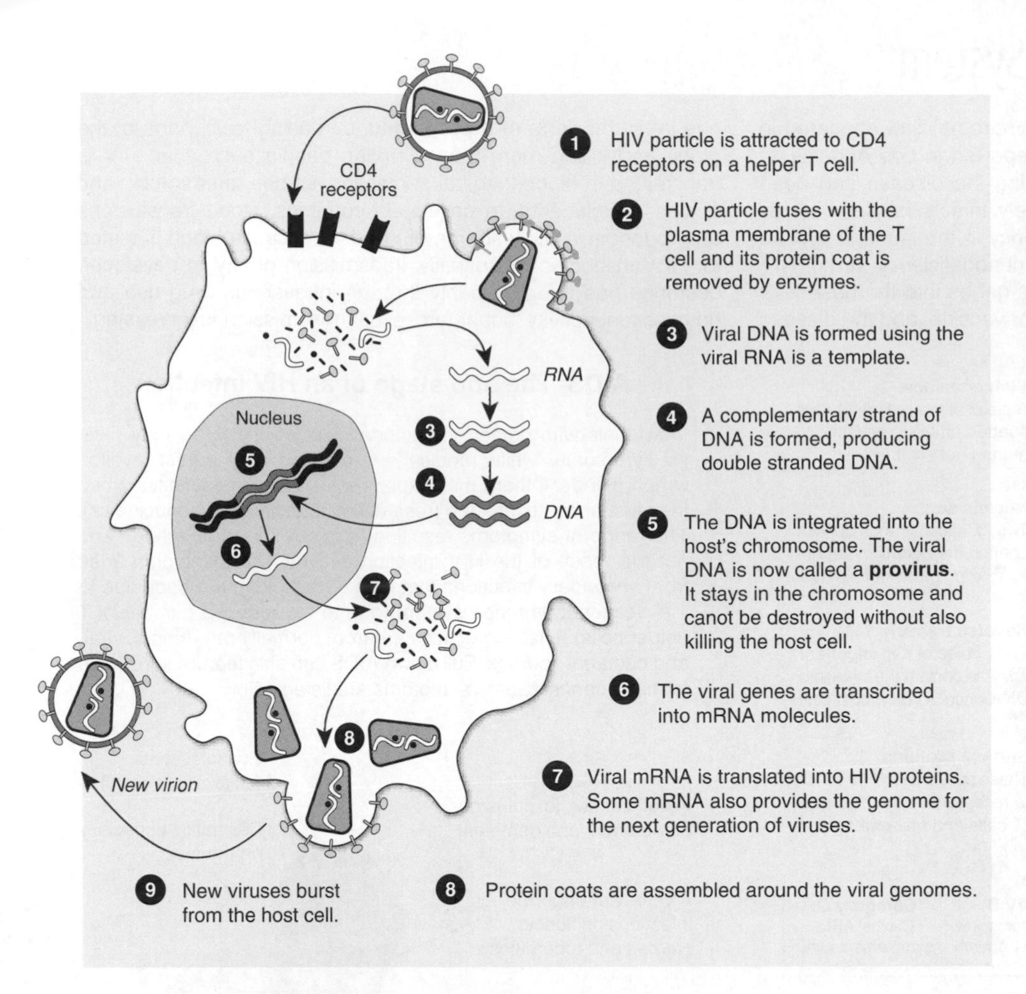

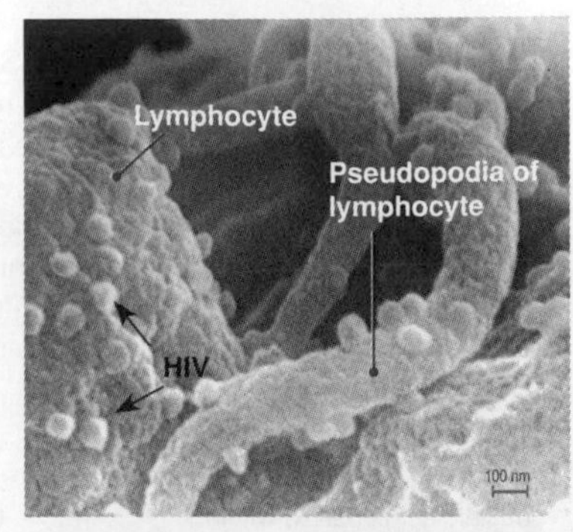

A SEM shows spherical HIV-1 virions on the surface of a human lymphocyte.

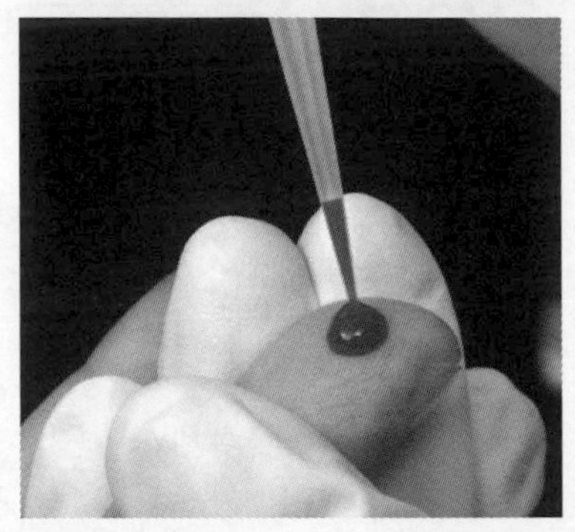

Diagnosis of HIV is possible using a simple antibody-based test on a blood sample.

(b) Describe how the helper T cells respond to the infection: ______________________

____

____

3. Name three common ways in which HIV can be transmitted from one person to another:

(a) ______________________

(b) ______________________

(c) ______________________

4. Explain what is meant by the term **HIV positive**: ______________________

____

____

5. In the years immediately following the discovery of the HIV pathogen, there was a sudden appearance of AIDS cases amongst **hemophiliacs** (people with an inherited blood disorder). State why this group was being infected with HIV:

____

____

6. Explain the significance of the formation of a provirus, both for the virus and for the human host: ______________________

____

____

____

____

____

ISBN: 978-1-92717357-2

Monoclonal Antibodies

A **monoclonal antibody** is an artificially produced antibody that binds to and neutralizes only one specific **antigen**. A monoclonal antibody binds an antigen in the same way that a normally produced antibody does. Monoclonal antibodies are used as diagnostic tools (e.g. detecting pregnancy) or to treat some types of cancer or autoimmune diseases. Therapeutic uses are still limited because the antibodies produced are from non-human cells and can cause side effects. In the future, production of monoclonal antibodies from human cells will probably result in fewer side effects. Monoclonal antibodies are produced in the laboratory by stimulating the production of B-cells in mice injected with the antigen. These B-cells produce an antibody against a specific antigen. Once isolated, they are made to fuse with immortal tumor cells, and they can be cultured indefinitely in a suitable growing medium (below). Monoclonal antibodies are useful for three reasons: they are all the same (i.e. clones), they can be produced in large quantities, and they are highly specific.

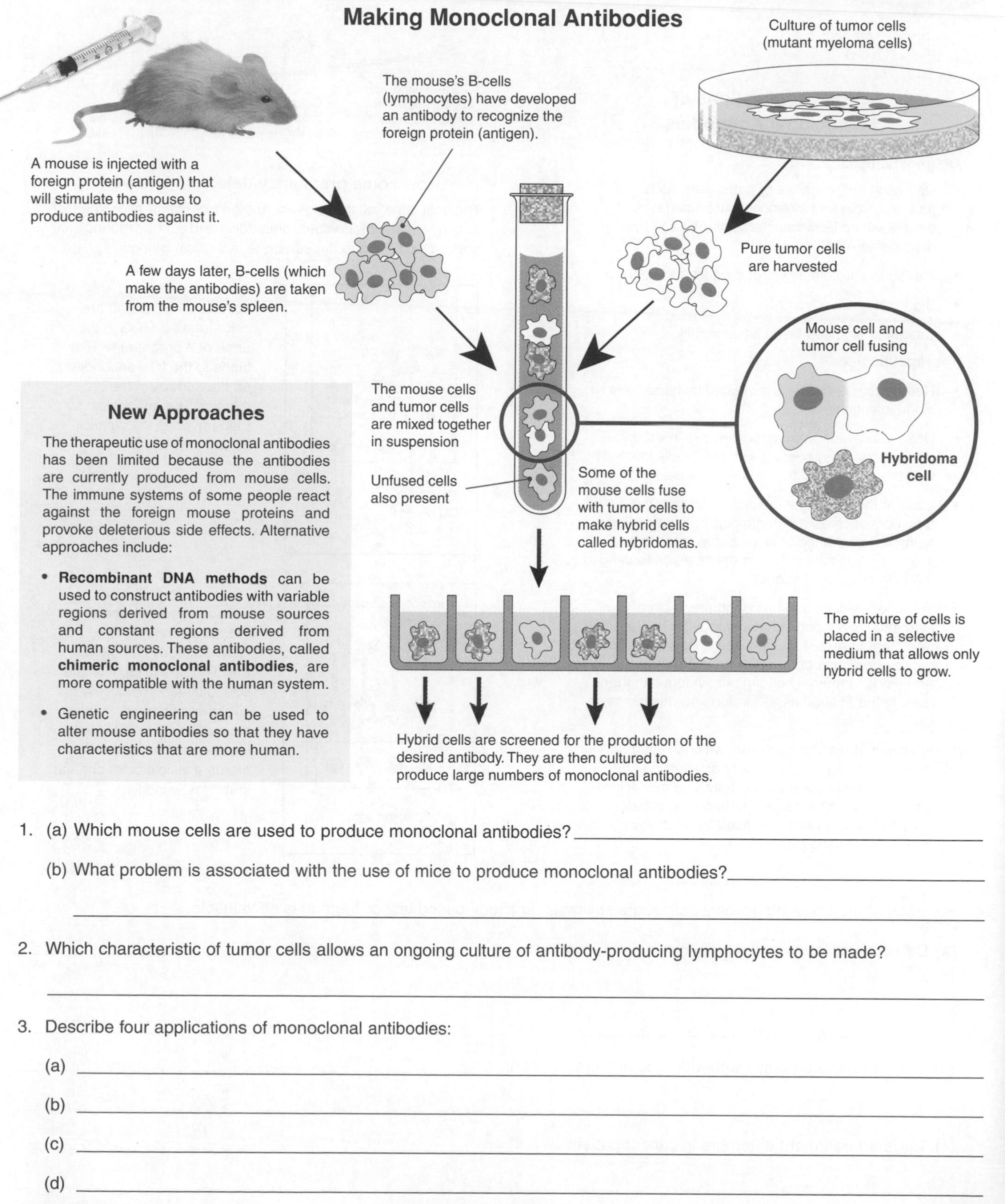

New Approaches

The therapeutic use of monoclonal antibodies has been limited because the antibodies are currently produced from mouse cells. The immune systems of some people react against the foreign mouse proteins and provoke deleterious side effects. Alternative approaches include:

- **Recombinant DNA methods** can be used to construct antibodies with variable regions derived from mouse sources and constant regions derived from human sources. These antibodies, called **chimeric monoclonal antibodies**, are more compatible with the human system.
- Genetic engineering can be used to alter mouse antibodies so that they have characteristics that are more human.

1. (a) Which mouse cells are used to produce monoclonal antibodies? ______

   (b) What problem is associated with the use of mice to produce monoclonal antibodies? ______

2. Which characteristic of tumor cells allows an ongoing culture of antibody-producing lymphocytes to be made?

3. Describe four applications of monoclonal antibodies:

   (a) ______

   (b) ______

   (c) ______

   (d) ______

ISBN: 978-1-92717357-2

Periodicals: *Monoclonals as medicines*

Related activities: *Antibodies*
Weblinks: *Monoclonal Antibody Production*

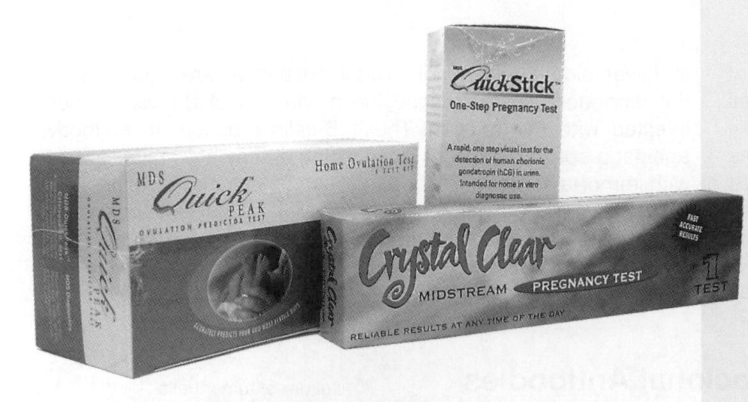

Detecting Pregnancy using Monoclonal Antibodies

When a woman becomes pregnant, a hormone called **human chorionic gonadotropin** (HCG) is released. HCG accumulates in the bloodstream and is excreted in the urine. Antibodies can be produced against HCG and used in simple test kits (below) to determine if a woman is pregnant. Monoclonal antibodies are also used in other home testing kits, such as those for detecting ovulation time (far left).

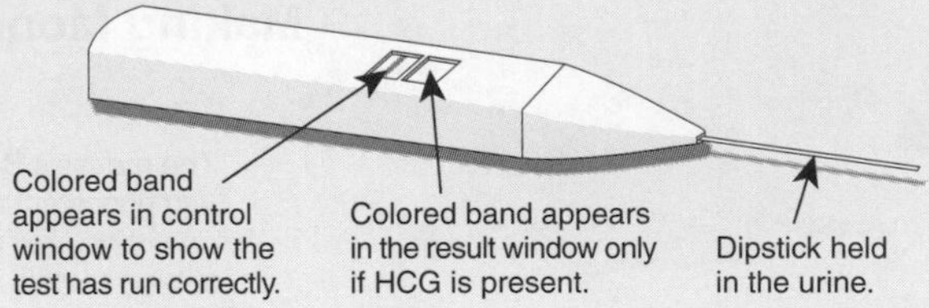

How home pregnancy detection kits work

The test area of the dipstick (below) contains two types of antibodies: free monoclonal antibodies and capture monoclonal antibodies, bound to the substrate in the test window.

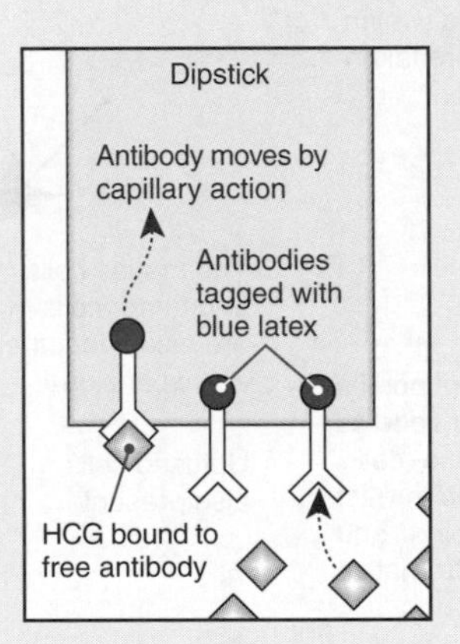

The free antibodies are specific for HCG and are color-labeled. HCG in the urine of a pregnant woman binds to the free antibodies on the surface of the dipstick. The antibodies then travel up the dipstick by capillary action.

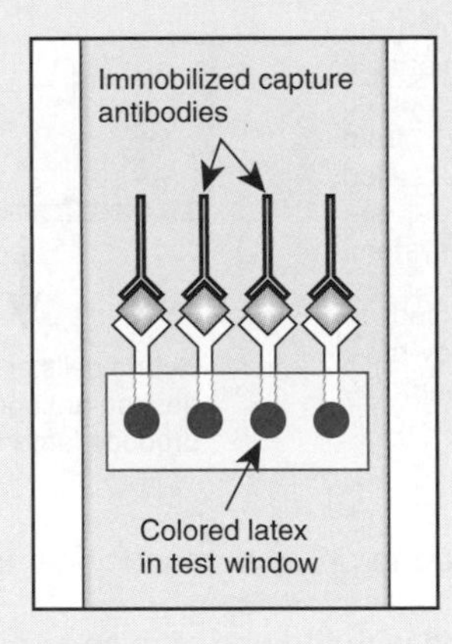

The capture antibodies are specific for the HCG-antibody complex. The HCG-antibody complexes traveling up the dipstick are bound by the immobilized capture antibodies, forming a sandwich. The color labeled antibodies then create a visible color change in the test window.

Other Applications of Monoclonal Antibodies

Diagnostic uses

- Detecting the presence of pathogens such as *Chlamydia* and streptococcal bacteria, distinguishing between *Herpesvirus* I and II, and diagnosing AIDS.
- Measuring protein, toxin, or drug levels in serum.
- Blood and tissue typing.
- Detection of antibiotic residues in milk.

Therapeutic uses

- Neutralizing endotoxins produced by bacteria in blood infections.
- Used to prevent organ rejection, e.g. in kidney transplants, by interfering with the T-cells involved with the rejection of transplanted tissue.
- Used in the treatment of some auto-immune disorders such as rheumatoid arthritis and allergic asthma. The monoclonal antibodies bind to and inactivate factors involved in the cascade leading to the inflammatory response.
- Immunodetection and immunotherapy of cancer. Herceptin is a monoclonal antibody for the targeted treatment of breast cancer. Herceptin recognizes receptor proteins on the outside of cancer cells and binds to them. The immune system can then identify the antibodies as foreign and destroy the cell.
- Inhibition of platelet clumping, which is used to prevent reclogging of coronary arteries in patients who have undergone angioplasty. The monoclonal antibodies bind to the receptors on the platelet surface that are normally linked by fibrinogen during the clotting process.

4. For each of the following applications, suggest why an antibody-based test or therapy is so valuable:

(a) Detection of toxins or bacteria in perishable foods: ______________________

____

____

(b) Detection of pregnancy without a doctor's prescription: ______________________

____

(c) Targeted treatment of tumors in cancer patients: ______________________

____

ISBN: 978-1-92717357-2

Herceptin: A Modern Monoclonal

Herceptin is the patented name of a **monoclonal antibody** for the targeted treatment of breast cancer. This drug (chemical name Trastuzumab) recognizes and is specific to the receptor proteins on the outside of cancer cells that are produced by the **proto-oncogene HER2**. The HER2 (**H**uman **E**pidermal growth factor **R**eceptor **2**) gene codes for cell surface proteins that signal to the cell when it should divide. Cancerous cells contain 20-30% more of the HER2 gene than normal cells and this causes **over-expression** of HER2, and large amounts of HER2 protein.

The over-expression causes the cell to divide more often than normal, producing a tumor. Cancerous cells are designated **HER2+** indicating receptor protein over-expression. The immune system fails to destroy these cells because they are not recognized as being abnormal. Herceptin's role is to recognize and bind to the HER2 protein on the surface of the cancerous cell. The immune system can then identify the antibodies as foreign and destroy the cell. The antibody also has the effect of blocking the cell's signalling pathway and thus stops the cell from dividing.

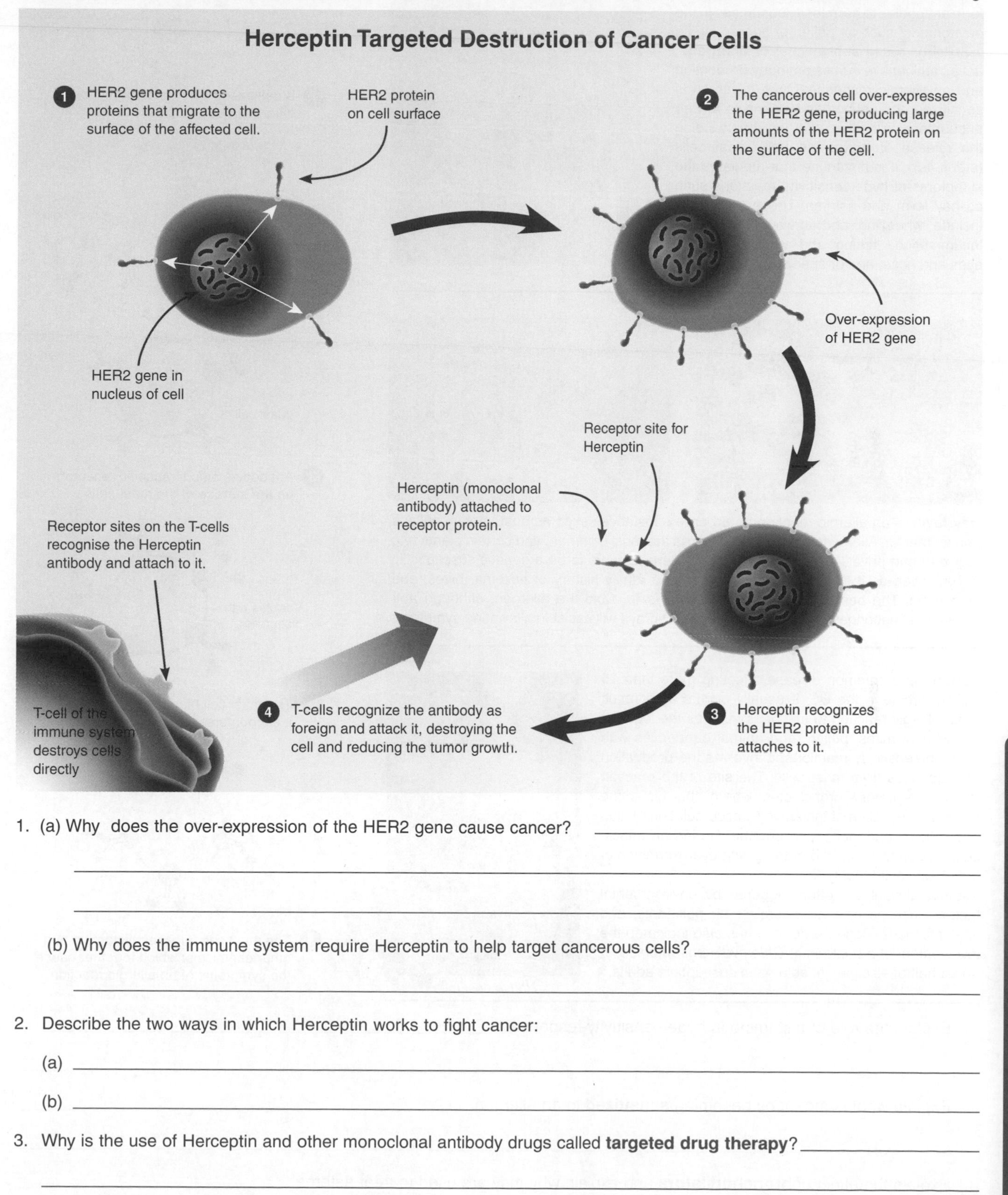

1. (a) Why does the over-expression of the HER2 gene cause cancer? ______

   (b) Why does the immune system require Herceptin to help target cancerous cells? ______

2. Describe the two ways in which Herceptin works to fight cancer:

   (a) ______

   (b) ______

3. Why is the use of Herceptin and other monoclonal antibody drugs called **targeted drug therapy**? ______

The Lymphatic System and Immunity

ISBN: 978-1-92717357-2

Related activities: *Monoclonal Antibodies*
Weblinks: *How Herceptin Works*

Allergies and Hypersensitivity

Sometimes the immune system may overreact, or react to the wrong substances instead of responding appropriately. This is termed **hypersensitivity** and the immunological response leads to tissue damage rather than immunity. Hypersensitivity reactions occur after a person has been **sensitized** to an antigen. In some cases, this causes only localized discomfort, as in the case of hayfever. More generalized reactions (such as anaphylaxis from insect venom or drug injections), or localized reactions that affect essential body systems (such as asthma), can cause death through asphyxiation and/or circulatory shock.

Hypersensitivity

A person becomes **sensitized** when they form antibodies to harmless substances in the environment such as pollen or spores (steps 1-2 right). These substances, or **allergens**, act as antigens to induce antibody production and an allergic response. Once a person is sensitized, the antibodies respond to further encounters with the allergen by causing the release of **histamine** from mast cells (steps 4-5). It is histamine that mediates the symptoms of hypersensitivity reactions such as hay fever and asthma. These symptoms include wheezing and airway constriction, inflammation, itching and watering of the eyes and nose, and/or sneezing.

Eyewire

EII

Hay fever is an allergic reaction to airborne substances such as dust, molds, pollens, and animal fur. Allergy to wind-borne pollen is the most common, and certain plants (e.g. ragweed and privet) are highly allergenic. There appears to be a genetic susceptibility to hay fever, as it is common in people with a family history of eczema, hives, and/ or asthma. The best treatment for hay fever is to avoid the allergen, although anti-histamines, decongestants, and steroid nasal sprays will assist in alleviating symptoms.

Asthma is a common disease affecting more than 15 million people in the US. It usually occurs as a result of an allergic reaction to allergens such as the feces of house dust mites, pollen, and animal dander. As with all hypersensitivity reactions, it involves the production of histamines from mast cells. The site of the reaction is the respiratory bronchioles where the histamine causes constriction of the airways, accumulation of fluid and mucus, and inability to breathe. During an attack, sufferers show labored breathing with overexpansion of the chest cavity (right).

Asthma attacks are often triggered by environmental factors such as cold air, exercise, air pollutants, and viral infections. Recent evidence has also indicated the involvement of a bacterium: *Chlamydia pneumoniae*, in about half of all cases of asthma in susceptible adults.

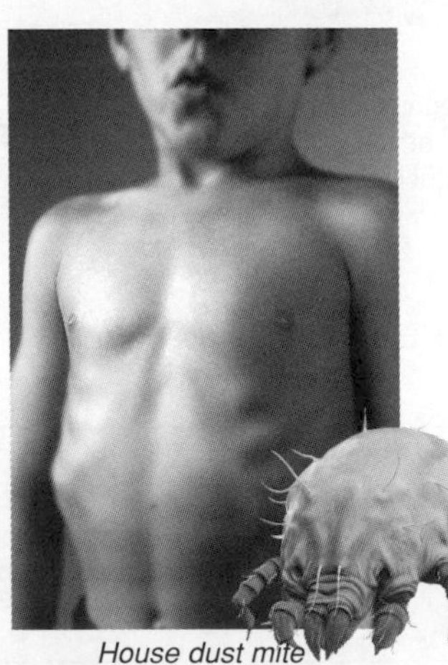

House dust mite

The Basis of Hypersensitivity

B cell

1. B cell encounters the allergen and differentiates into plasma cells

Plasma cell

Antibodies

2. The plasma cell produces antibodies

Mast cell

3. Antibodies bind to specific receptors on the surface of the mast cells

Vesicles with histamine

4. The mast cell binds the allergen when it encounters it again.

5. The mast cell releases histamine and other chemicals, which together cause the symptoms of an allergic reaction.

1. Explain the role of histamine in hypersensitivity responses: ______

2. Explain what is meant by becoming **sensitized** to an allergen: ______

3. Explain the effect of **bronchodilators** and explain why they are used to treat asthma: ______

ISBN: 978-1-92717357-2

Related activities: The Immune System, Respiratory Diseases

Organ and Tissue Transplants

Transplant surgery replaces a damaged organ or tissue with a healthy, living one from a **donor**. The donor is usually someone who is brain dead (their breathing and heartbeat are maintained artificially), but transplant material (e.g. bone marrow, blood, kidney) can also be taken from a living donor. Two major problems associated with organ transplants are the lack of donors and **tissue rejection**. Recall that the genetically determined MHC antigens on the surface of all cells are responsible for the immune system's recognition of the body's own tissues. Cells from donor tissue will have different MHC antigens to those of the recipient, and the donor tissue will be recognized as foreign and attacked (rejected). A number of factors have been involved in increasing the success of transplants in recent years. These include better **tissue-typing** and more effective **immunosuppressant drugs**, both of which decrease the MHC response. There have also been marked improvements in surgical techniques and methods for organ preservation and transport. More recently, tissue engineering and stem cell technology have been used to create **bioartificial organs** (organs created from donor cells grown on an artificial scaffold). This technology has been limited to simple, hollow structures (e.g. windpipe and bladder) but there have already been several successful transplants.

Organ Transplants

Currently, there are five organs that are routinely transplanted (below). In addition to organs, whole hand transplants and partial or whole face transplants are now possible.

Face: Facial reconstructions began initially with reconnection of a patient's own facial components damaged in accidents. More recently, medical techniques have developed so that partial reconstructions (usually of nose and mouth) are possible using facial material from a dead donor. In 2008, a French medical team completed the world's first full face transplant. These types of transplants require careful connection of blood vessels, skin, muscles, and bone, tendons, and other connective tissues. Performing face transplants also involves addressing a number of ethical concerns.

Heart (H): These transplants are carried out after a patient suffers heart failure due to heart attack, viral infections of the heart, or congenital, irreparable defects.

Lungs (Ls): Replacement of organs damaged by cystic fibrosis or emphysema. Typically, lungs are transplanted together, but single lung transplants and heart-lung transplants are also possible.

Hands: In 1999 the first successful hand transplant was performed on Matthew Scott in the USA. A year and a half after the operation, he could sense temperature, pressure and pain, and could write, turn the pages of a newspaper, tie shoelaces and throw a baseball. Many successful hand transplants have been undertaken since.

Liver (Li): Substitute for a liver destroyed by cirrhosis, congenital defects, or hepatitis.

Pancreas (P): Restores insulin production in Type I diabetics (caused by autoimmune destruction of the insulin producing cells of the pancreas).

Kidneys (K): Used in cases of renal failure, diabetes, high blood pressure, inherited illnesses, and infection. The failing kidneys are usually left *in situ* and the transplant (Kt) is placed in a location different from the original kidney.

Tissue Transplants

A large number of tissues are currently used in transplant procedures. An estimated 200 patients can potentially benefit from the organs and tissues donated from a single body.

Cornea: Transplants can restore impaired vision.

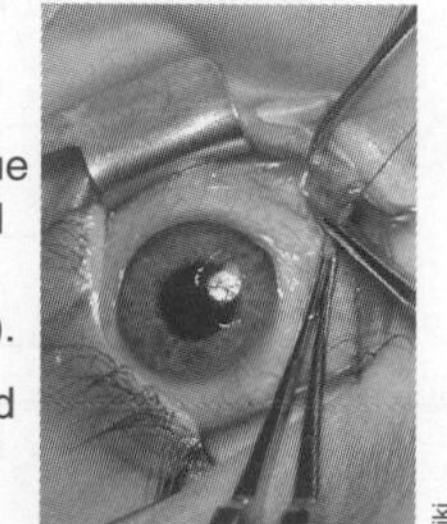
Wiki

Dental powder: This tissue is prepared to help rebuild defects in the mandible (which supports the teeth).

Jaw: The mandible is used in facial reconstruction.

Ear bones: The three bones of the inner ear can be transplanted to improve some forms of deafness.

Pericardium: The pericardium surrounding the heart is made of tough tissue that can be used to cover the brain after surgery. Transplants of the brain coverings themselves are no longer performed because of the risk of transmitting prion infections.

Blood and blood vessels: Blood transfusions are transplants of blood tissue. Blood vessels, mostly veins, can be transplanted to reroute blood around blockages, such as this atherosclerotic plaque (right) in the body.

Plaque

Wiki

Hip joints: Joints can be reconstructed by transplanting the head of the femur.

Bone marrow: Marrow is extracted from living donors and used to help people with a wide variety of illnesses, such as leukemia.

Bones: Long bones of the arms and legs can be used in limb reconstruction; ribs can be used for spinal fusions and facial repair.

Cartilage and ligaments: Orthopedic surgeons use these materials to rebuild ankle, knee, hip, elbow and shoulder joints.

Skin: Skin can be used as a temporary covering for burn injuries until the patient's own skin grows back.

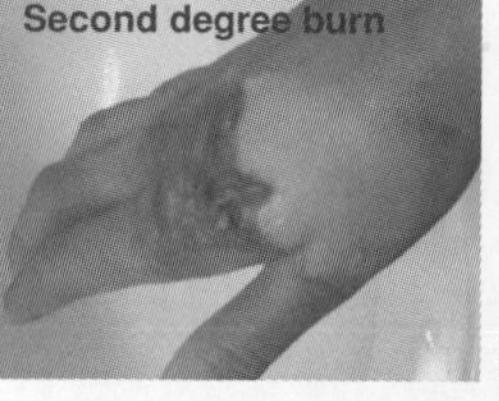

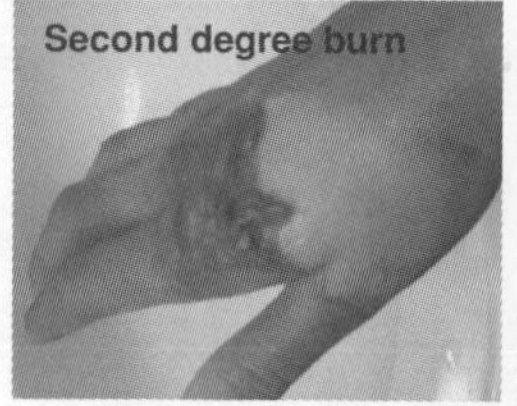
Wiki

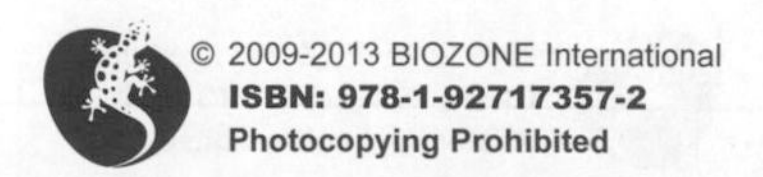

ISBN: 978-1-92717357-2

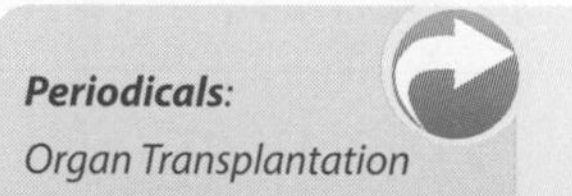

Commonly Performed Transplants

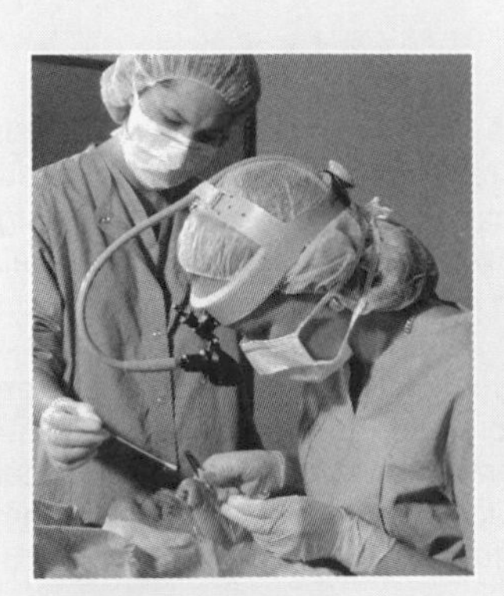

Corneal transplants can be used to restore sight in patients with damaged vision. The cornea naturally has a poor blood supply so rejection is less of a problem than with some other tissues

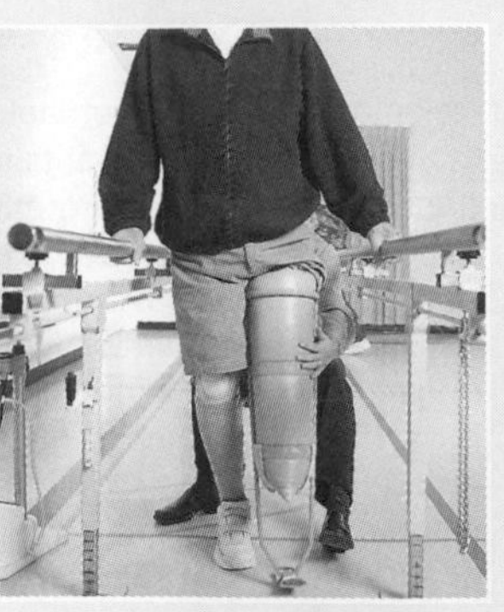

For many amputees, being fitted with an artificial limb is the first step towards mobility. In the future, such prostheses may be replaced with limb transplants, in much the same way as current hand transplants.

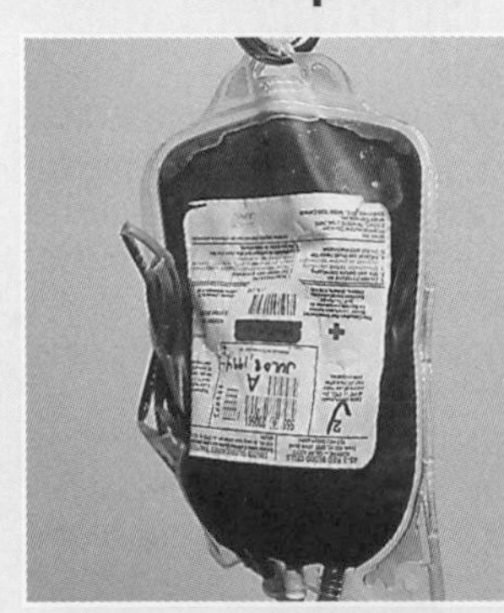

Transplants of whole blood, blood plasma, platelets, and other blood components are crucial to many medical procedures. The donor blood is carefully typed to ensure compatibility with the recipient.

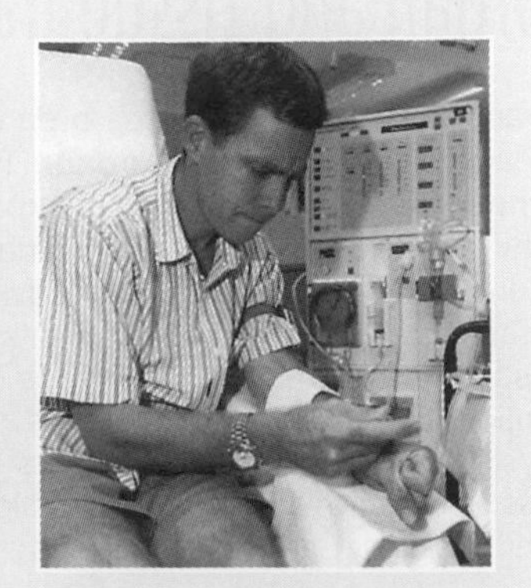

Many patients with kidney failure rely on regular dialysis in order to function. This is expensive and inconvenient, and carries health risks. Such patients are usually waiting for a kidney transplant.

1. Describe three major technical advances that have improved the success rate of organ transplantation:

   (a) ______________________________

   (b) ______________________________

   (c) ______________________________

2. (a) Explain the basis for organ and tissue rejection: ______________________________

   (b) Explain the role of **tissue typing** and **immunosuppressant drugs** in reducing or preventing this response:

   (c) A major side-effect of using immunosuppressant drugs is an increased susceptibility to infections. Explain why:

3. Discuss the ethical issues associated with organ and tissue transplants. Consider costs, benefits, source of tissue, and criteria for choosing recipients. If required, debate the issue, or develop your arguments as a separate report:

ISBN: 978-1-92717357-2

Stem Cell Technology

Stem cells are undifferentiated cells able to give rise to a number of other specialized cell types. Their two important properties, **self renewal** and **potency**, make them valuable as a source of cell lines for research and therapy. Two types of stem cells are important in medicine and research. **Embryonic stems cells**, which are found in the early embryo, and **adult stem cells**, which occur in some tissues of adults and children. Stem cell research is still at an early stage and much has to be learned about the environments that cells require in order to differentiate into specific cell types. However, the potential applications include tissue engineering and cell therapy to replace diseased or damaged cells with new ones grown in culture.

Embryonic Stem Cells

Embryonic stem cells (ESCs) are pluripotent stem cells from the inner cell mass of blastocysts (below). Blastocysts are embryos of about 50-150 cells that are about five days old. Most ESCs come from IVF embryos, which have been donated for research. When grown *in vitro*, ESCs retain their pluripotency for many cell divisions, provided they are not stimulated to differentiate.

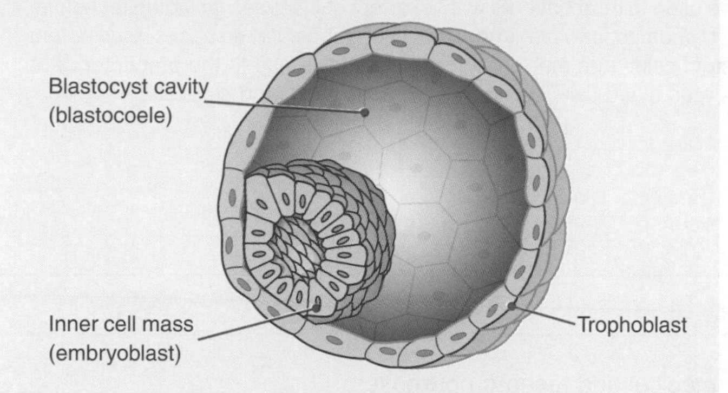

Stem Cell Properties

Self renewal: The ability to divide many times while maintaining an unspecialized state.

Potency: The ability to differentiate into specialized cells.

The Potency of Stem Cells

Totipotent: Stem cells that can differentiate into all the cells in an organism. In humans, only the zygote and its first few divisions are totipotent.

Pluripotent: Stem cells that can become any cells of the body, except extra-embryonic cells, such as the placenta. Embryonic stem cells are pluripotent.

Multipotent: Stem cells that give rise a limited number of cell types, usually related to their tissue of origin (e.g. hematopoietic stem cells give rise to all blood cell types).

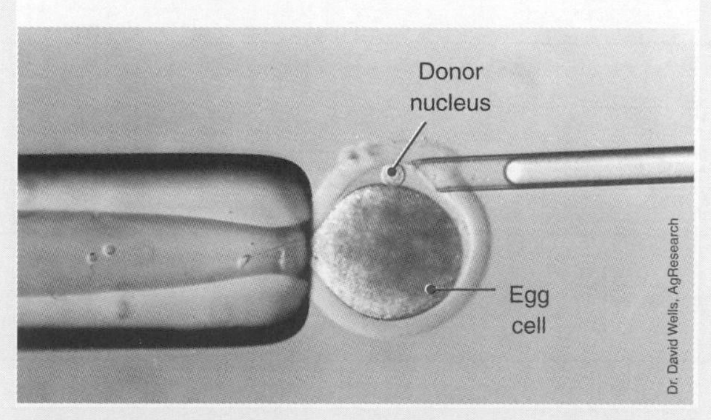

Histocompatibility refers to the compatibility of cells and tissues between different individuals. If donor material is poorly matched to the recipient, the recipient's immune system rejects the donor cells. Stem cell cloning (also called **therapeutic cloning**) provides a way around this problem. Stem cell cloning produces cells that have been derived from the recipient, and are therefore histocompatible. Transplantation success rates are much improved and immunosuppressant drugs are no longer required.

Embryonic Stem Cell Cloning

When ESCs are provided with appropriate growth conditions, they will differentiate into specialized cell types. Scientists can control this process by manipulating the *in vitro* culture conditions to produce cells of a particular type for a particular purpose (e.g. heart cells to replaced damaged heart tissue).

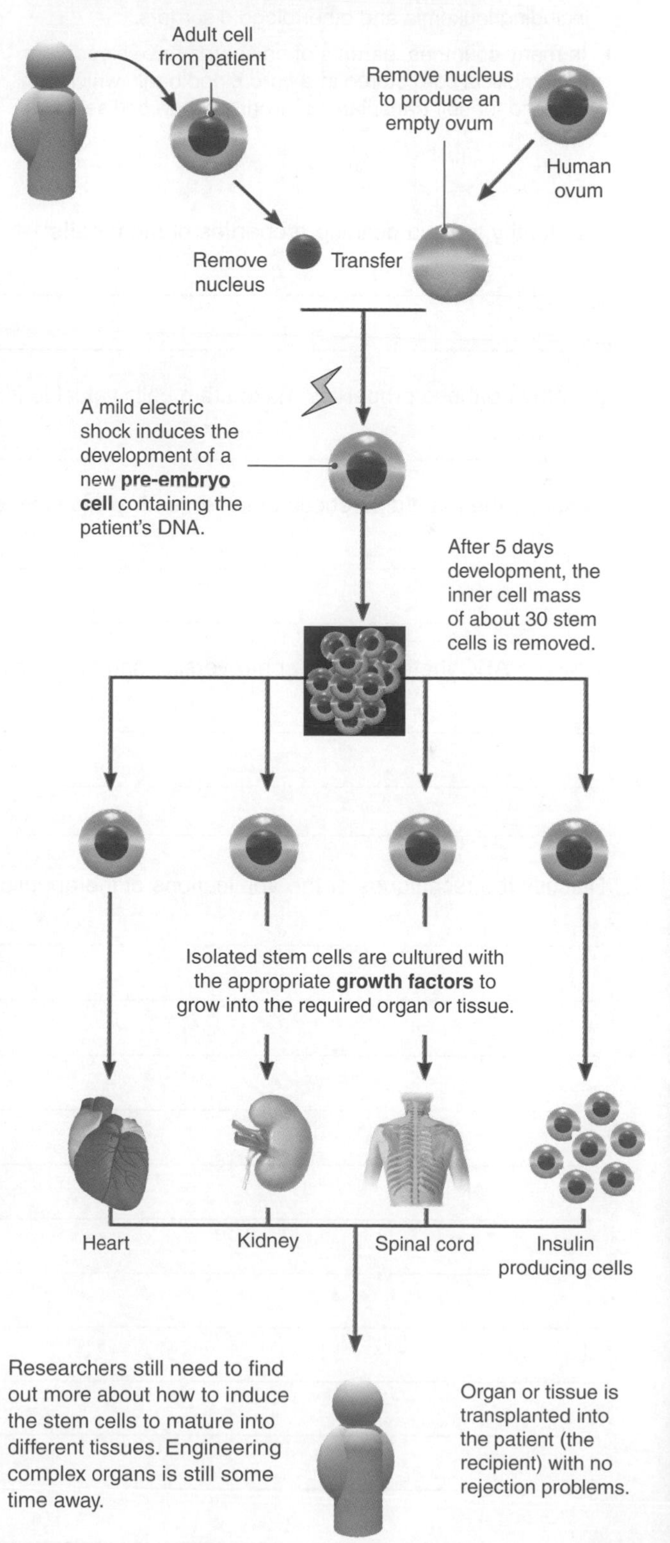

ISBN: 978-1-92717357-2

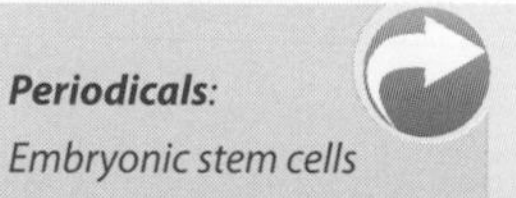

Adult Stem Cells

- **Adult stem cells** (ASC) are undifferentiated cells found in several types of tissues in adults and children (e.g. brain, bone marrow, skin, and liver), and in umbilical cord blood. ASCs are also called somatic stem cells.
- ASCs are **multipotent**, meaning they can only differentiate into a limited number of cell types, usually related to the tissue of origin.
- The function of ASCs in the body is to replace dying cells and repair damaged tissue.
- There are fewer ethical issues associated with using ASCs because they are derived from adult tissues, whereas ESCs are harvested from human embryos.
- ASCs are already used to treat a number of diseases including leukemia and other blood disorders.
- In many countries, parents of newborns have blood from the umbilical cord stored in a cord blood bank, which can be used for self transplants to treat a range of diseases.

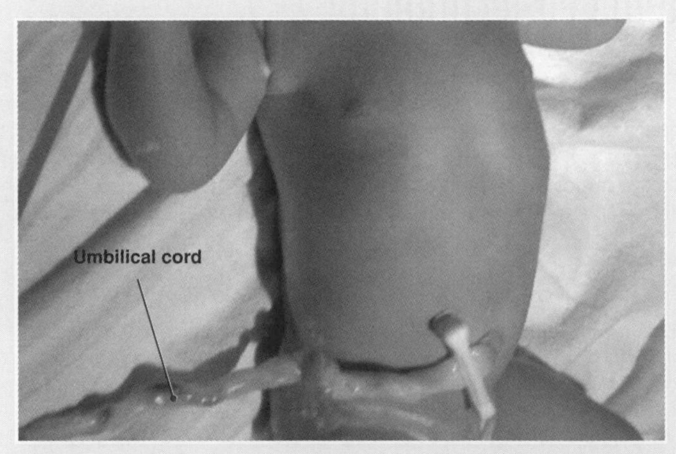

Cells obtained from umbilical cord blood (above) or bone marrow could be used to treat patients with a variety of diseases including leukemia, lymphomas, anemia, and a range of congenital diseases. Multipotent stem cells from marrow or cord blood give rise to the precursor cells for red blood cells, all white blood cell types, and platelets.

1. (a) Identify the two defining properties of **stem cells**:

   (b) Why do these properties make stem cells valuable for medical and research purposes:

2. Describe the main differences between embryonic stem cells (ESC) and adult stem cells (ASC):

3. Why are ASC therapies less controversial than therapies using ESC?

4. Discuss the techniques or the applications of therapeutic stem cell cloning (including ethical issues where relevant):

ISBN: 978-1-92717357-2

Gene Therapy for Immune Dysfunction

Gene therapy involves correcting genetic disorders by replacing a defective or missing gene with a normally functioning version. Gene therapies vary in their technical detail, but all are based around the same technique. Normal (non-faulty) DNA containing the correct copy of the gene is inserted into a vector, which transfers the DNA into the patient's cells in a process called **transfection**. Viruses and liposomes are commonly used vectors. The vector is introduced into a sample of the patient's cells and these are cultured to **amplify the gene**. The cultured cells are then transferred back to the patient, and the corrected gene begins to be expressed. The treatment of somatic cells or stem cells may be therapeutic, but the changes are not inherited. Modification of gametes (germ-line therapy) would enable genetic changes to be inherited. Gene therapy has had limited success because transfection of targeted cells is usually inefficient and the side effects can be severe or even fatal.

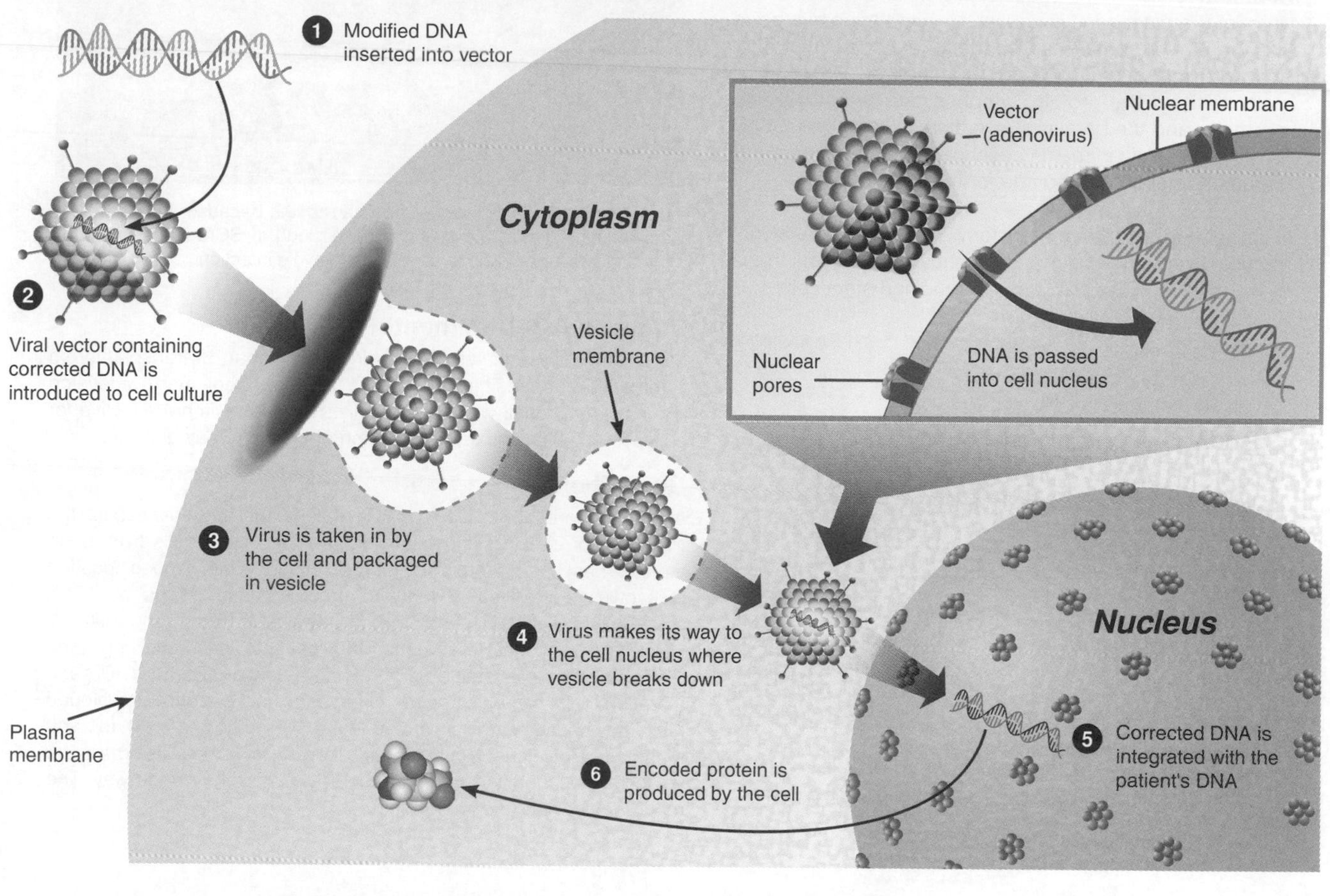

1. (a) Describe the general principle of **gene therapy**: ______

   (b) Describe the medical areas where gene therapy might be used: ______

2. Explain the significance of transfecting **germ-line cells** rather than **somatic** (body) cells: ______

3. What do you think is the purpose of **amplifying the gene** prior to transferring the cultured cells back to the patient: ______

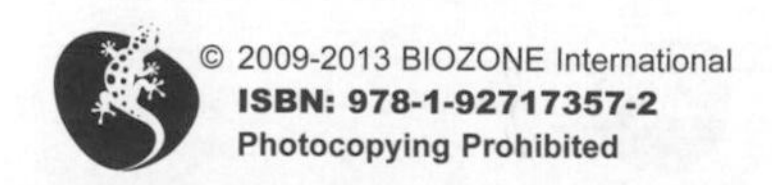

ISBN: 978-1-92717357-2

Periodicals: *Genes can come true, Gene therapy,are we there yet?*

Related activities: *Stem Cell Technology*
Weblinks: *Gene Therapy, Gene Therapy Primer*

Treating SCID Using Gene Therapy

What is SCID?

SCID (severe combined immunodeficiency) describes a group of inherited disorders affecting production of B- and T-cells. Children born with SCID have no specific immune system, and are susceptible to life-threatening infections. Without early diagnosis and treatment, they usually die before their second birthday.

X-linked SCID is the most common and most severe form of the disease. It is caused by a mutation to a gene on the X chromosome. The gene encodes a protein that forms part of a receptor complex on white blood cells. Without the protein, lymphocytes do not develop normally, and the body is susceptible to infections. SCID is more common in males because they lack a second X chromosome to compensate for the defective one.

The second most common form of SCID (ADA-SCID) results from an enzyme deficiency. Both forms have been treated with transplants of modified stem cells.

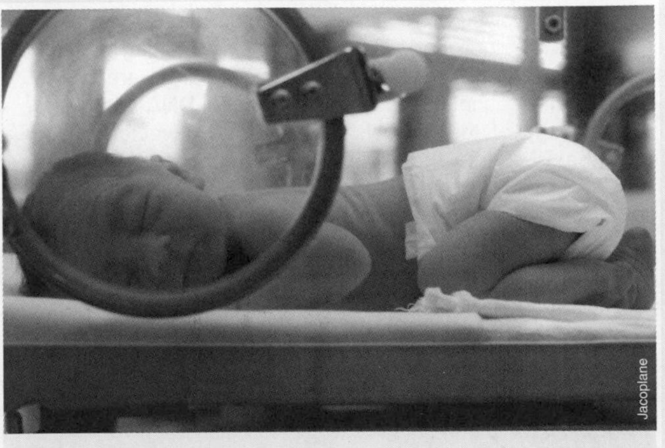

Detecting SCID is difficult in the first months because the mother's antibodies still circulate in the infant's blood. If SCID is suspected, patients are kept in sterile conditions to avoid infection.

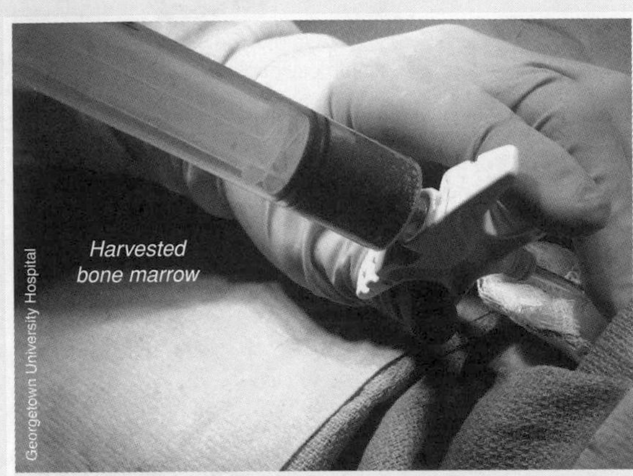

Bone marrow, which contains stem cells, is extracted from the patient. The corrected gene is inserted into the stem cells, and the stem cells are returned to the patient.

Conventional Treatment

SCID is most commonly treated using a bone marrow transplant from a compatible donor. The stem cells in the donor bone marrow develop into new, functional, white blood cells. However, it is not effective for all types of SCID, and tissue rejection or infection may develop.

Gene Therapy

Although there have been very few gene therapy successes, the treatment of SCID using genetically modified stem cells from bone marrow has been promising. Only a single gene, whose location is known, is involved in the two most common forms of SCID, so SCID is an ideal candidate for treatment with gene therapy. Unlike conventional treatments, the patients are receiving their own cells, so there is little risk of tissue rejection. Early SCID gene therapy trials were stopped when several patients developed leukemia because the corrected gene was inserted next to (and switched on) a gene regulating cell growth. New trials using a safer vector and a modified gene promoter are currently underway. The results look promising to date.

4. What is SCID? ______________________________

____

5. (a) How are bone marrow transplants used to treat SCID? ______________________________

____

____

(b) What are some of the disadvantages of bone marrow transplants? ______________________________

____

____

(c) What are some of the advantages of treating SCID with gene therapy? ______________________________

____

____

6. Discuss the potential risks associated with gene therapy: ______________________________

____

____

____

ISBN: 978-1-92717357-2

KEY TERMS: Mix and Match

INSTRUCTIONS: Test your vocabulary by matching each term to its definition, as identified by its preceding letter code.

active immunity ....

allergies ....

antibody ....

antigen ....

artificially acquired immunity ....

B-cells (B lymphocytes) ....

cell-mediated immunity ....

clonal selection ....

fever ....

HIV ....

humoral immunity ....

hypersensitivity ....

immune system ....

inflammatory response ....

lymphatic system ....

major histocompatibility complex (MHC) ....

monoclonal antibodies ....

naturally acquired immunity ....

non-specific resistance ....

passive immunity ....

phagocytes ....

primary response ....

secondary response ....

specific resistance ....

T-cells (T lymphocytes) ....

vaccination ....

A Immune response involving the activation of macrophages, specific T-cells, and cytokines against antigens.

B Immunity acquired through the development of antibodies via encountering a pathogen or by obtaining antibodies from the mother via the placenta or breast milk.

C The response of the body to a second (or subsequent) exposure to an antigen the has already been encountered. The response if often quicker and stronger than the initial response.

D Cellular part of an immune response. These respond only to antigenic fragments that have been processed and presented by infected cells or macrophages.

E The undesirable over-reaction of a sensitized person's immune system to a perceived antigen. It includes allergic reactions and autoimmune diseases.

F An immune response that is associated with the non-cellular part of the blood and involves the action of antibodies secreted by B-cell lymphocytes.

G A defensive response to damage caused by physical agents, microbial infections or chemical agents. The inflammation process involves pain, redness, heat and swelling.

H Human immunodeficiency virus. A retrovirus which is able to splice its genes into the host cell's chromosomes.

I A protein produced by the body in response to a specific antigen and aimed at targeting and destroying it.

J Abnormal reactions of the immune system that occur in response to otherwise harmless substances.

K The delivery of antigenic material (the vaccine) to produce immunity to a disease.

L Artificially produced antibodies designed to recognize and bind a specific antigen.

M Immunity acquired by medical treatment. Includes vaccinations and the introduction of an immune serum that contains preformed antibodies.

N Immunity that is induced in the host itself by the antigen, and is long lasting.

O Immunity gained by the receipt of ready-made antibodies. Immunity can be gained naturally (from mother to fetus) or artificially.

P The body's initial defense against any foreign material. It includes physical barriers, antimicrobial substances, inflammation, fever, and phagocytosis.

Q A type of lymphocyte that makes antibodies against specific antigens. The cell type responsible for the humoral immune response.

R Part of the circulatory system, distinct from the cardiovascular system, with functions including internal defense, and transport and removal of tissue fluid.

S A theory for how B- cells and T-cells are selected to target specific antigens invading the body.

T The structures and processes in the body that provide defense against disease.

U A set of molecules displayed on cell surfaces that are responsible for lymphocyte recognition and antigen presentation.

V White blood cells that destroys microbes by digesting them.

W A foreign molecule that stimulates an immune response in the body.

X The targeted response of the immune system to specific, identified antigens.

Y The initial response of the immune system to exposure to an antigen.

Z An increase in body temperature above the normal range (36.2-37.2ºC).

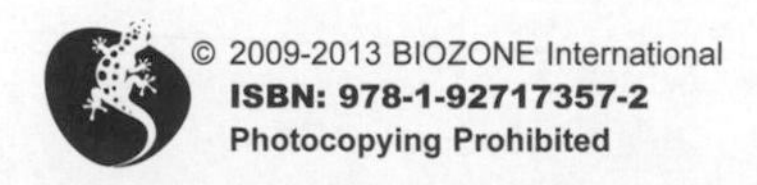

ISBN: 978-1-92717357-2

Interacting Systems

The Respiratory System

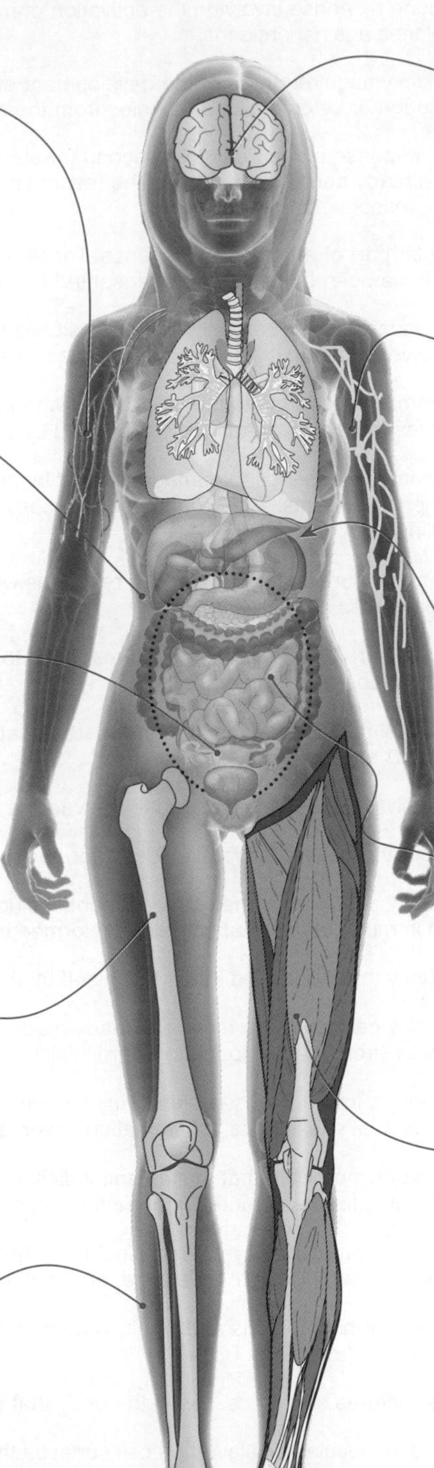

Cardiovascular system

- Blood transports respiratory gases.
- The carbonate-bicarbonate system in blood contributes to blood buffering.
- Blood proteins (e.g. immunoglobulins) also contribute to blood buffering.

Nervous system

- Control centers in the medulla and the pons regulate the rate and depth of breathing.
- Sensory feedback to the respiratory control centers is provided by stretch receptors in the bronchioles and chemoreceptors in the aorta and carotid arteries.

Urinary system

- Kidneys dispose of the waste products of respiratory metabolism (other than CO_2 which is breathed out).

Lymphatic system and immunity

- Immune system provides general protection against pathogens; specifically the tonsils protect against pharyngeal and upper respiratory tract infections.
- Lymphatic system helps to maintain blood volume required for efficient transport of respiratory gases.

Endocrine system

- Epinephrine (from the adrenal medulla atop the kidneys) acts as a bronchodilator.
- Testosterone promotes enlargement of the larynx at puberty in males.
- Cortisol has role in lung maturation and a number of hormones directly or indirectly influence breathing rates.

Reproductive system

- Pregnancy has significant effects on breathing mediated largely through sex hormones. The enlarged uterus (its position at full term shown by dotted line) pushes abdominal organs up and outwards and compromises the functioning of the diaphragm.

Digestive system

- Digestive system provides the nutrients required by the respiratory system.

Skeletal system

- Bones enclose and protect the lungs and bronchi from damage.
- Expansion and elastic recoil of the ribcage produce the volume changes necessary for inhalation/exhalation (breathing).

Muscular system

- Diaphragm and intercostal muscles (with the ribcage) produce the volume changes necessary for breathing.
- Regular aerobic exercise improves the efficiency of the respiratory system.

Integumentary system

- Skin forms a surface barrier protecting the organs of the respiratory system.

General Functions and Effects on all Systems

The respiratory system provides an interface for gas exchange with the external environment. It is ultimately responsible for providing all the cells of the body with oxygen and disposing of waste carbon dioxide produced as a result of cellular respiration. These respiratory gases are transported in the blood.

Disease

Symptoms of disease

- Chest pain
- Excessive mucus production
- Coughing, sneezing
- Difficulty breathing, cyanosis

Infectious respiratory diseases

- Bacterial pneumonia
- Pulmonary tuberculosis
- Influenza

Non-infectious respiratory diseases

- Asthma
- Fibrosis (scarring)
- Smoking-related diseases
- Inherited diseases (e.g. CF)

Medicine & Technology

Diagnosis of disorders

- Chest X-ray and CT scans
- Pulmonary function tests
- Sputum cultures and biopsy
- DNA tests and screening

Preventing and treating diseases of the respiratory system

- Drug therapies (e.g. antibiotics)
- Vaccination
- Surgery (e.g. transplants)
- Radiotherapy
- Gene or cell therapy (e.g. CF)
- Behavior modification

CDC

- Asthma
- Lung cancer
- Chronic bronchitis
- Emphysema
- Asbestosis

- Stem cell therapy
- Gene therapy
- X rays
- Vaccination

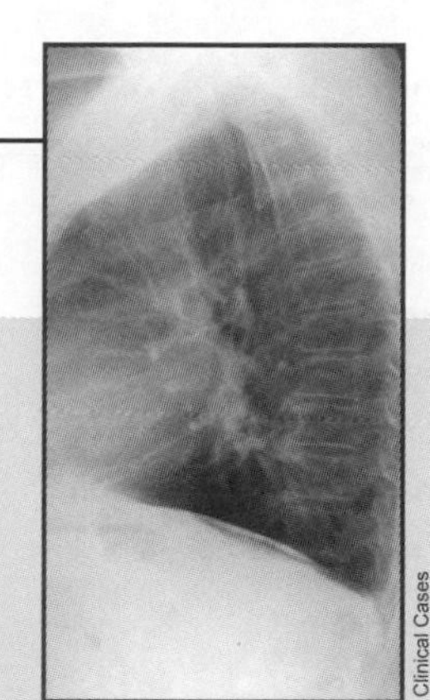

Clinical Cases

Taking a Breath

The Respiratory System

The respiratory system can be affected by disease and undergoes changes associated with training and aging.

Medical technologies can be used to diagnose and treat respiratory disorders. Exercise and lifestyle management can prevent some respiratory diseases.

- Lung cancer
- Chronic bronchitis
- Emphysema

- VO_2max
- Ventilation efficiency
- Ventilation rhythm

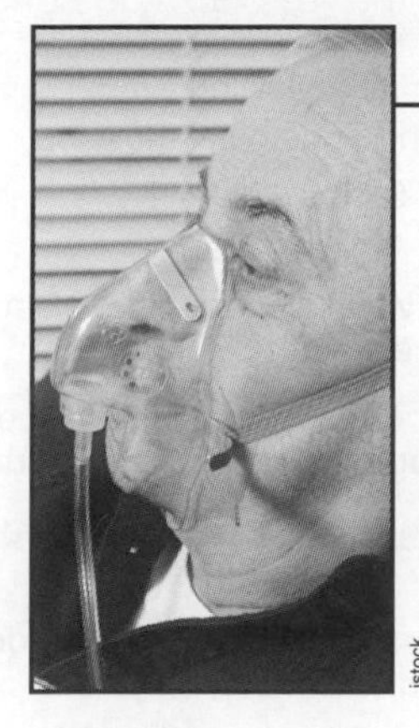

istock

Exercise can delay or reverse some age-related changes to respiratory function

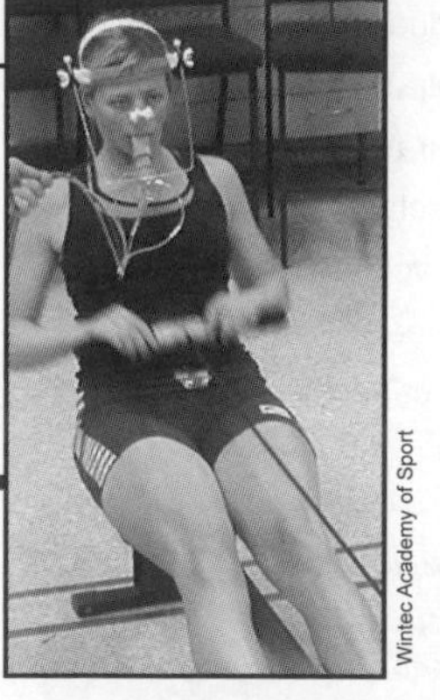

Wintec Academy of Sport

Aging and the respiratory system

- Decline in respiratory capacity
- Decline in aerobic capacity (VO_2max)
- Increased incidence of chronic respiratory disease

Effects of exercise on the respiratory system

- Increased rate and depth of breathing
- Increased aerobic capacity (VO_2max)
- Increased respiratory efficiency
- Improved oxygen loading -unloading
- Better diaphragmatic performance

The Effects of Aging

Exercise

The Respiratory System

Key terms

alveolar-capillary membrane
alveoli (sing. alveolus)
asthma
breathing
breathing rate
bronchi
bronchioles
chronic bronchitis
emphysema
expiration (=exhalation)
gas exchange system
hemoglobin
inspiration (=inhalation)
intercostal muscles
intrapleural pressure
lungs
myoglobin
obstructive pulmonary disease
oxygen-hemoglobin dissociation curve
partial pressure
pulmonary ventilation rate (PV)
residual volume
respiratory center
respiratory pigment
surfactant
tidal volume
trachea
vital capacity

Periodicals:
Listings for this chapter are on page 280

Weblinks:
www.thebiozone.com/weblink/AnaPhy-3572.html

BIOZONE APP:
Student Review Series
The Respiratory System

Key concepts

- Respiratory gases are exchanged with the environment across gas exchange membranes in the alveoli of the lungs.
- The specific features of the gas exchange system maximize exchange rates.
- Hemoglobin transports oxygen in the blood and delivers it to the tissues.
- The respiratory center responds to changes in oxygen demand by changing the breathing rate.
- Lung function can be measured using spirometry.

Learning Objectives

☐ 1. Use the **KEY TERMS** to compile a glossary for this topic.

The Gas Exchange System

pages 194-195, 198-202

☐ 2. Distinguish between **cellular respiration** and **gas exchange**. Identify the **respiratory gases** and explain how they are exchanged across gas exchange surfaces.

☐ 3. Describe the location and function of the gas exchange surfaces and related structures, including: **trachea**, **bronchi**, **bronchioles**, **lungs**, **alveoli** (and alveolar epithelium). Explain the features contributing to efficient gas exchange.

☐ 4. Describe the gross structure of the lungs and their relationship to the thoracic wall. Include reference to the visceral and parietal pleura, pleural fluid, and pleural space.

☐ 5. Describe the distribution and function of the cartilage, ciliated epithelium, goblet cells, smooth muscle, and elastic fibers in the gas exchange system.

☐ 6. Describe the features of the alveoli, explaining their role in maximizing gas exchange rates. Draw a diagram of an alveolus to show its relationship with the blood vessels in the lung tissue. Annotate the diagram to show the movement of O_2 and CO_2.

☐ 7. Describe the functionally important features of the **alveolar-capillary membrane** (gas exchange membrane). Comment on the role of the alveolar macrophages.

Breathing and Gas Transport

pages 198-206, 211-212

☐ 8. Explain **breathing** (lung ventilation), including reference to the following:
- The role of the diaphragm, **intercostal muscles**, and the abdominal muscles.
- The role of negative **intrapleural pressure** in preventing lung collapse.
- The difference between quiet and forced breathing.

☐ 9. Explain **spirometry** is used to measure lung function and ventilation rates. Explain how lung volumes and ventilation rates are calculated and expressed.

☐ 10. Describe the role of **hemoglobin** the transport and delivery of oxygen to the tissues. Describe oxygen transport in relation to the **oxygen-hemoglobin dissociation curve**.

☐ 11. Describe the effect of pH (CO_2 level) on the oxygen-hemoglobin dissociation curve (the Bohr effect) and explain its significance.

☐ 12. Explain how CO_2 is transported, including the role of the following: carbonic anhydrase, chloride shift, plasma proteins, and hydrogen-carbonate (bicarbonate) ions.

☐ 13. Explain how basic rhythm of breathing is controlled through the **respiratory center** in the medulla and its output via the phrenic nerve and the intercostal nerves.

☐ 14. Explain how the breathing rate responds to changes in oxygen demand, e.g. during exercise. Include reference to the activity of the carotid and aortic chemoreceptors.

☐ 15. Describe the short and long term effects of high altitude and **training**.

Respiratory Disease

pages 197, 207-210

☐ 16. Explain how pathogens and environmental factors contribute to lung diseases with specific, recognizable symptoms.

☐ 17. Distinguish between restrictive and **obstructive pulmonary diseases** and describe examples. Outline the link between tobacco smoking and respiratory disease.

Living With Chronic Lung Disease

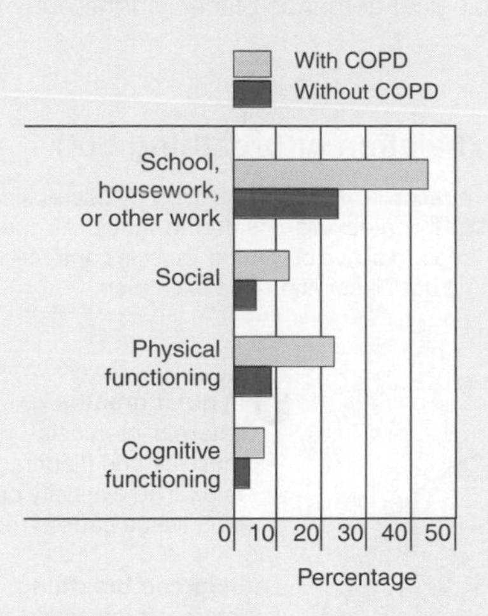

Cognitive, physical, social and activity-related limitations are more common among people with chronic obstructive pulmonary disease.

The Impact of COPD in the US

Chronic obstructive pulmonary disease (COPD) is a serious lung disease which makes it difficult for people to breathe because their airways are partially blocked or because the alveoli of the lungs lose their elasticity or become damaged. COPD includes **chronic bronchitis** and **emphysema**, and affects about 12 million people in the US. Most of those affected are over the age of 40 and smoking is the cause in the vast majority of cases. This relationship is clear; people who have never smoked rarely develop COPD. The symptoms of COPD and asthma are similar, but COPD causes permanent damage to the airways, and so symptoms are chronic (persistent) and treatment is limited.

COPD severely limits the capacity of sufferers to carry out even a normal daily level of activity. A survey by the American Lung Association of hundreds of people living with COPD found that nearly half became short of breath while washing, dressing, or doing light housework (left). Over 25% reported difficulty in breathing while sitting or lying still. Lack of oxygen also places those with COPD at high risk of heart failure. As the disease becomes more severe, sufferers usually require long-term oxygen therapy, in which they are more or less permanently attached to an oxygen supply.

COPD is estimated to cost the US $32 billion dollars each year, $14 billion of which are indirect costs, such as lost working days. A 'flare-up' of COPD, during which the symptoms worsen, is one of the commonest reasons for admission to hospital and the disease places a substantial burden on health services. COPD is the only major disease with an increasing death rate in the US, rising 16%, and is the third leading cause of death. At least 120,000 people die each year from the end stages of COPD, but the actual number may be higher as COPD is often present in patients who die from heart failure and stroke. Many of these people have several years of ill health before they die. Being able to breathe is something we don't often think about. What must it be like to struggle for each breath, every minute of every day for years?

A Personal Story

Deborah Ripley's message from her mother Jenny (used with permission)

"Fear, anxiety, depression, and carbon monoxide are ruining whatever life my mother has left. I posted this portrait of my Mum on the photo website Flickr because she wants to send a warning to anyone who's still smoking. I've just returned from visiting her in a nursing home where she's virtually shackled to the bed. Getting up to go to the bathroom practically kills her. She was admitted to hospital after a bout of pneumonia, which required intensive antibiotic therapy and left her hardly able to breathe. She has moderate dementia caused by a series of mini-strokes, which is aggravated by the pneumonia. She has no recollection of who has visited her or when, so consequently thinks she's alone most of the time, which is upsetting and disturbing for her.

This is all caused by damage to her brain and lungs as a result of 65 years of smoking. In those moments when she is lucid, she asks me who she can warn that this could happen to them. She said 'if people could see me lying here like this it would put them off...' None of her other known blood relatives suffered this sort of decline in their old age and, as far as I know, none of them smoked".

Thankfully, Jenny's pneumonia has since subsided and her COPD is being well managed. However, constant vigilance is important because flare-ups are common with COPD and recovery from lung infections is difficult when breathing is already compromised.

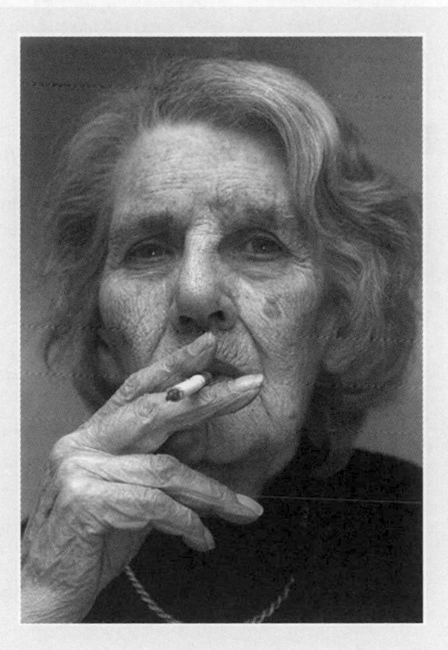

Used with permission ©deborahripley.com

1. Describe the economic impact of smoking-related diseases, such as emphysema and chronic bronchitis:

2. Discuss the personal costs of a smoking-related disease and comment on the value of personal testimonials such as those from Deborah's mother:

ISBN: 978-1-92717357-2

Breathing

Breathing (ventilation) provides a continual supply of fresh air to the lungs and helps to maintain a large diffusion gradient for **respiratory gases** (oxygen and carbon dioxide) across the gas exchange surface. Oxygen must be delivered regularly to supply the needs of respiring cells. Similarly, carbon dioxide, which is produced as a result of cellular metabolism, must be quickly eliminated. Adequate lung ventilation is essential to these exchanges. The cardiovascular system participates by transporting respiratory gases to and from the cells of the body. The volume of gases exchanged during breathing varies according to the physiological demands placed on the body (e.g. by exercise).

Inspiration (inhalation or breathing in)

During quiet breathing, inspiration is achieved by increasing the space (therefore decreasing the pressure) inside the lungs (and alveoli). Air then flows into the alveoli in response to the decreased pressure. Inspiration is always an active process involving muscle contraction.

Expiration (exhalation or breathing out)

During quiet breathing, expiration is achieved passively by decreasing the space (thus increasing the pressure) inside the lungs. Air then flows passively out of the lungs. In active breathing, muscle contraction is involved in bringing about both inspiration and expiration.

Intercostal muscles

Rib

Intrapleural cavity

1a External intercostal muscles contract causing the ribcage to expand and move up.

1b Diaphragm contracts and moves down.

2 Thoracic volume increases, lungs expand, and the pressure inside the lungs decreases.

3 Air flows into the lung alveoli in response to the pressure gradient.

Diaphragm contracts and moves down

Diaphragm relaxes and moves up

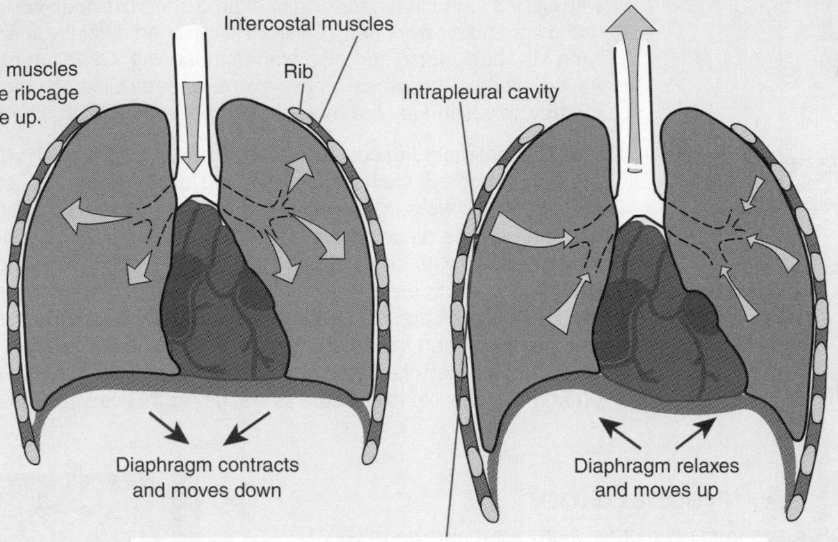

1 In **quiet breathing**, external intercostal muscles and diaphragm relax. The elasticity of the lung tissue causes recoil.

In **forced breathing**, the internal intercostals and abdominal muscles also contract to increase the force of the expiration

2 Thoracic volume decreases and the pressure inside the lungs increases.

3 Air flows passively out of the lung alveoli in response to the pressure gradient.

The intrapleural pressure (the pressure inside the intrapleural cavity) is always negative and relatively large, whereas alveolar (intrapulmonary) pressure moves from slightly positive to slightly negative as a person breathes. This allows the alveolar pressure to exert force and stop the lungs collapsing.

1. Explain the purpose of breathing: ______

2. (a) Describe the sequence of events involved in quiet breathing: ______

   (b) Explain the essential difference between this and the situation during heavy exercise or forced breathing:

3. Identify what other gas is lost from the body in addition to carbon dioxide: ______

4. Explain the role of the elasticity of the lung tissue in normal, quiet breathing: ______

5. Under normal circumstances, the intrapleural pressure is always negative:

   (a) What would happen to the intrapleural pressure if someone was punctured through the chest wall (a pneumothorax)?

   (b) What would then happen to the lung on that side? ______

Related activities: *Measuring Lung Function, Gas Transport in Humans*

Weblinks: *Respiratory Basics, Intrapleural Pressure Changes During Breathing*

ISBN: 978-1-92717357-2

The Respiratory System

The **respiratory system** (or gas exchange system) includes all the structures associated with exchanging **respiratory gases** with the environment. The paired lungs are located within the thorax and are connected to the outside air by way of a system of tubular passageways: the trachea, bronchi, and bronchioles. Ciliated, mucus-secreting epithelium lines this system of tubules, trapping and removing dust and pathogens before they reach the gas exchange surface. Each lung is divided into several lobes, each receiving its own bronchus. Each bronchus divides many times, and ends in the respiratory bronchioles and their **alveoli** (air sacs). The alveoli provide a very large surface area (around 70 m^2) for the exchange of respiratory gases by diffusion between the alveoli and the blood in the capillaries. This exchange occurs across the respiratory (alveolar-capillary) membrane.

Structures of the Respiratory System

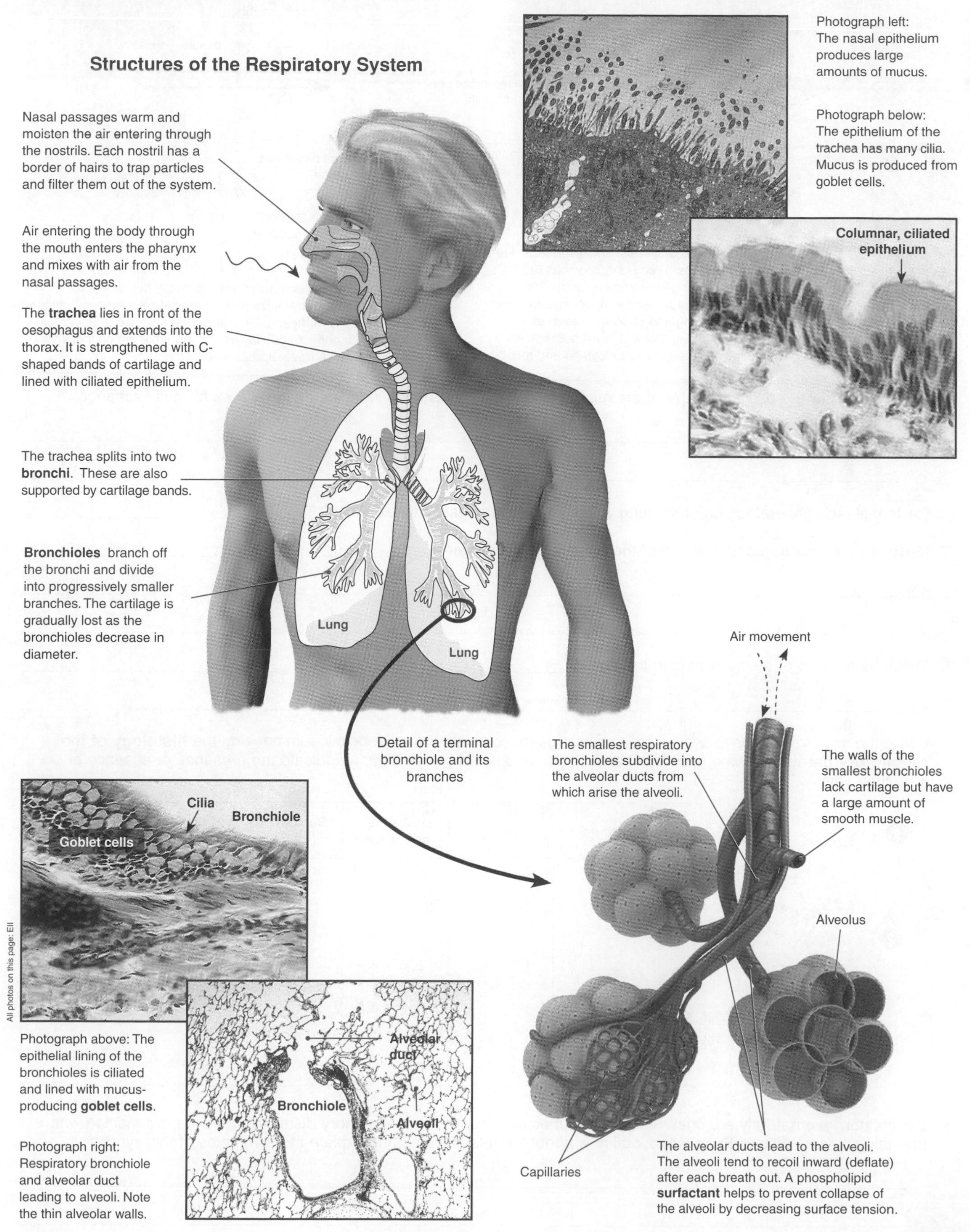

Photograph left: The nasal epithelium produces large amounts of mucus.

Photograph below: The epithelium of the trachea has many cilia. Mucus is produced from goblet cells.

Photograph above: The epithelial lining of the bronchioles is ciliated and lined with mucus-producing **goblet cells**.

Photograph right: Respiratory bronchiole and alveolar duct leading to alveoli. Note the thin alveolar walls.

ISBN: 978-1-92717357-2

Periodicals: *Gas exchange in the lungs*

Related activities: *Gas Transport in Humans*
Weblinks: *Respiratory Basics, Interactive Lungs*

Cross Section Through An Alveolus

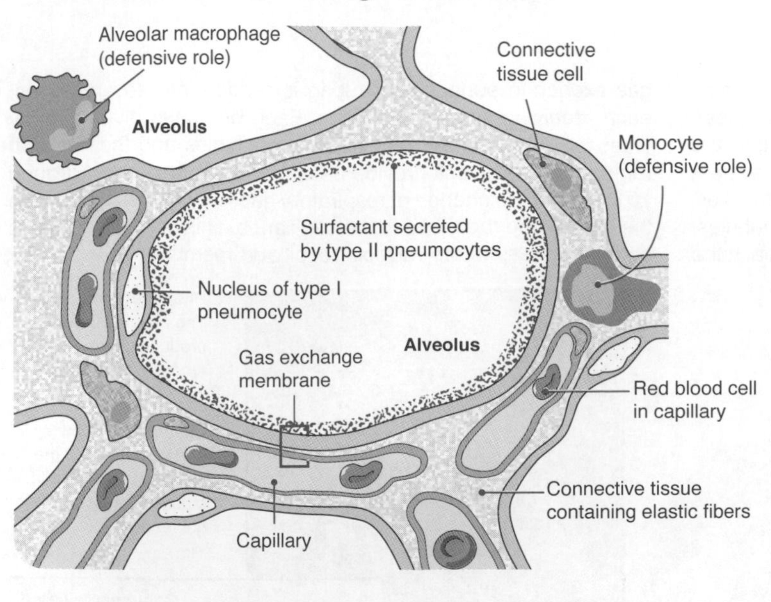

The alveoli are very close to the blood-filled capillaries. The alveolus is lined with alveolar epithelial cells or **pneumocytes**. Type I pneumocytes (90-95% of aveolar cells) contribute to the gas exchange membrane (right). Type II pneumocytes secrete a **surfactant**, which decreases surface tension within the alveoli and prevents them from collapsing and sticking to each other. Macrophages and monocytes defend the lung tissue against pathogens. Elastic connective tissue gives the alveoli their ability to expand and recoil.

The Respiratory Membrane

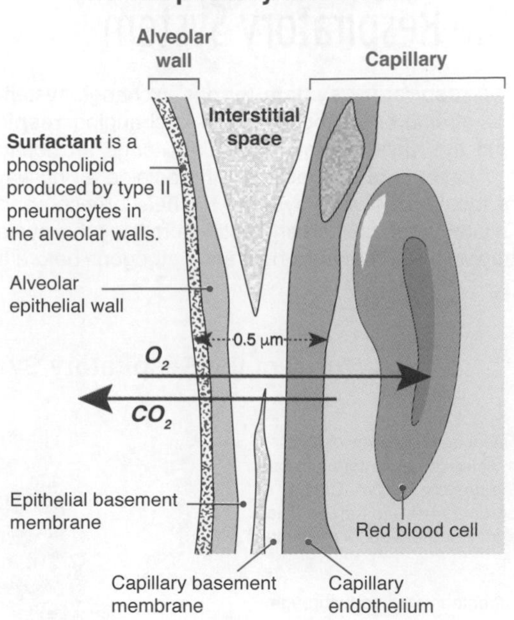

The respiratory (or gas exchange) membrane is the term for the layered junction between the alveolar epithelial cells (pneumocytes), the endothelial cells of the capillary, and their associated basement membranes (thin, collagenous layers that underlie the epithelial tissues). Gases move freely across this membrane.

1. (a) Explain how the basic structure of the human respiratory system provides such a large area for gas exchange:

   ____

   ____

   (b) Identify the general region of the lung where exchange of gases takes place: ____________________

2. Describe the structure and purpose of the respiratory membrane: ____________________

   ____

   ____

3. Describe the role of the surfactant in the alveoli: ____________________

   ____

4. Using the information above and on the previous page, complete the table below summarizing the **histology of the respiratory pathway**. Name each numbered region and use a tick or cross to indicate the presence or absence of particular tissues.

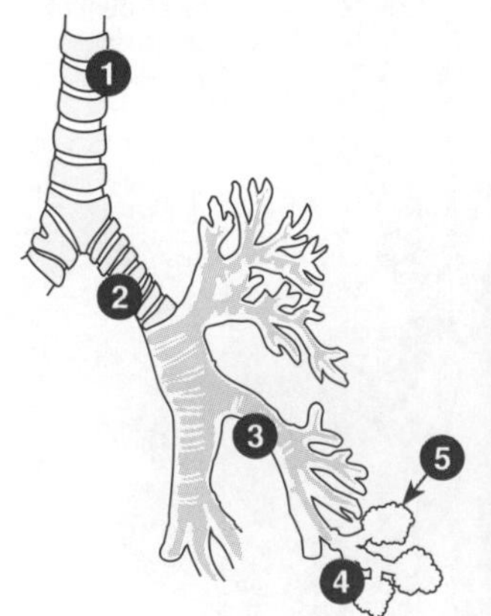

	Region	Cartilage	Ciliated epithelium	Goblet cells (mucus)	Smooth muscle	Connective tissue
1						✓
2						
3		gradually lost				
4	Alveolar duct		✗	✗		
5					very little	

5. Babies born prematurely are often deficient in surfactant. This causes respiratory distress syndrome; a condition where breathing is very difficult. From what you know about the role of surfactant, explain the symptoms of this syndrome:

   ____

   ____

ISBN: 978-1-92717357-2

Measuring Lung Function

Changes in lung volume can be measured using a technique called **spirometry**. Total adult lung capacity varies between 4 and 6 liters (L or dm^3) and is greater in males. The **vital capacity**, which describes the volume exhaled after a maximum inspiration, is somewhat less than this because of the residual volume of air that remains in the lungs even after expiration. The exchange between fresh air and the residual volume is a slow process and the composition of gases in the lungs remains relatively constant. Once measured, the tidal volume can be used to calculate the **pulmonary ventilation rate** or PV, which describes the amount of air exchanged with the environment per minute. Measures of respiratory capacity provide one way in which a reduction in lung function can be assessed (for example, as might occur as result of disease or an obstructive lung disorder such asthma).

Determining changes in lung volume using spirometry

The apparatus used to measure the amount of air exchanged during breathing and the rate of breathing is a spirometer (also called a respirometer). A simple spirometer consists of a weighted drum, containing oxygen or air, inverted over a chamber of water (to provide a stable temperature). A tube connects the air-filled chamber with the subject's mouth, and soda lime in the system absorbs the carbon dioxide breathed out. Oxygen consumption is detected by the amount of fluid displacement in a manometer connected to the container. Breathing results in a trace called a spirogram, from which lung volumes can be measured directly.

During inspiration
Air is removed from the chamber, the drum sinks, and an upward deflection is recorded on the paper on the rotating drum.

During expiration
Air is added to the chamber, the drum rises, and a downward deflection is recorded.

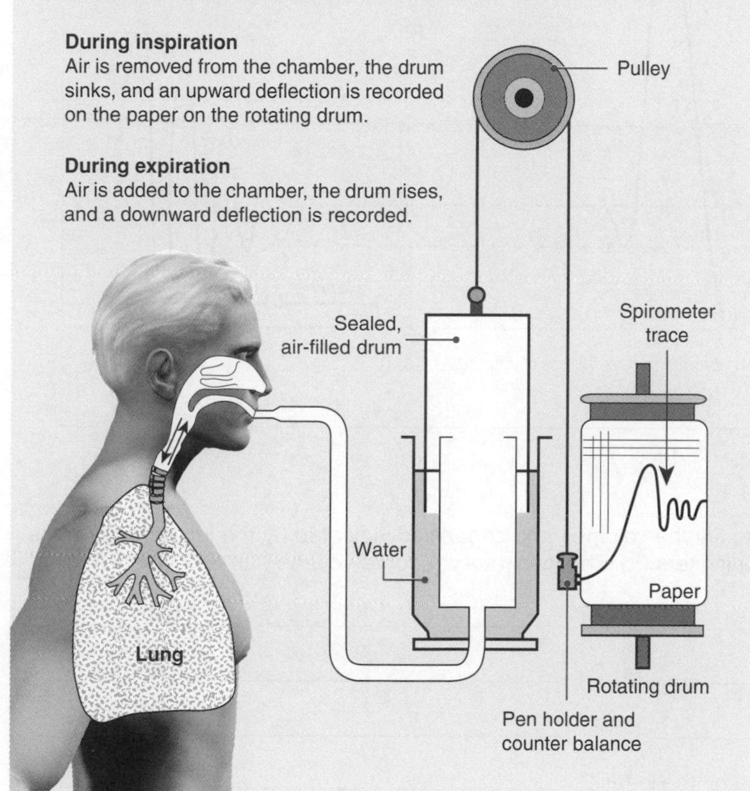

Lung Volumes and Capacities

The air in the lungs can be divided into volumes. Lung capacities are combinations of volumes.

DESCRIPTION OF VOLUME	Vol (L)
Tidal volume (TV) Volume of air breathed in and out in a single breath	**0.5**
Inspiratory reserve volume (IRV) Volume breathed in by a maximum inspiration at the end of a normal inspiration	**3.3**
Expiratory reserve volume (ERV) Volume breathed out by a maximum effort at the end of a normal expiration	**1.0**
Residual volume (RV) Volume of air remaining in the lungs at the end of a maximum expiration	**1.2**
DESCRIPTION OF CAPACITY	
Inspiratory capacity (IC) = TV + IRV Volume breathed in by a maximum inspiration at the end of a normal expiration	**3.8**
Vital capacity (VC) = IRV + TV + ERV Volume that can be exhaled after a maximum inspiration.	**4.8**
Total lung capacity (TLC) = VC + RV The total volume of the lungs. Only a fraction of TLC is used in normal breathing	**6.0**

PRIMARY INDICATORS OF LUNG FUNCTION

Forced expiratory volume in 1 second (FEV_1)
The volume of air that is maximally exhaled in the first second of exhalation.

Forced vital capacity (FVC)
The total volume of air that can be forcibly exhaled after a maximum inspiration.

1. Describe how each of the following might be expected to influence values for lung volumes and capacities obtained using spirometry:

   (a) Height: ____________________

   (b) Gender: ____________________

   (c) Age: ____________________

2. A percentage decline in FEV_1 and FVC (to <80% of normal) are indicators of impaired lung function, e.g in asthma:

   (a) Explain why a forced volume is a more useful indicator of lung function than tidal volume:

   ____

   (b) Asthma is treated with drugs to relax the airways. Suggest how spirometry could be used during asthma treatment:

   ____

ISBN: 978-1-92717357-2

Related activities: The Human Respiratory System
Weblinks: Respiratory Basics, Spirometry

Respiratory gas	Approximate percentages of O_2 and CO_2		
	Inhaled air	Air in lungs	Exhaled air
O_2	21.0	13.8	16.4
CO_2	0.04	5.5	3.6

Above: The percentages of respiratory gases in air (by volume) during normal breathing. The percentage volume of oxygen in the alveolar air (in the lung) is lower than that in the exhaled air because of the influence of the **dead air volume** (the air in the spaces of the nose, throat, larynx, trachea and bronchi). This air (about 30% of the air inhaled) is unavailable for gas exchange.

Left: During exercise, the breathing rate, tidal volume, and PV increase up to a maximum (as indicated below).

Spirogram for a male during quiet and forced breathing, and during exercise

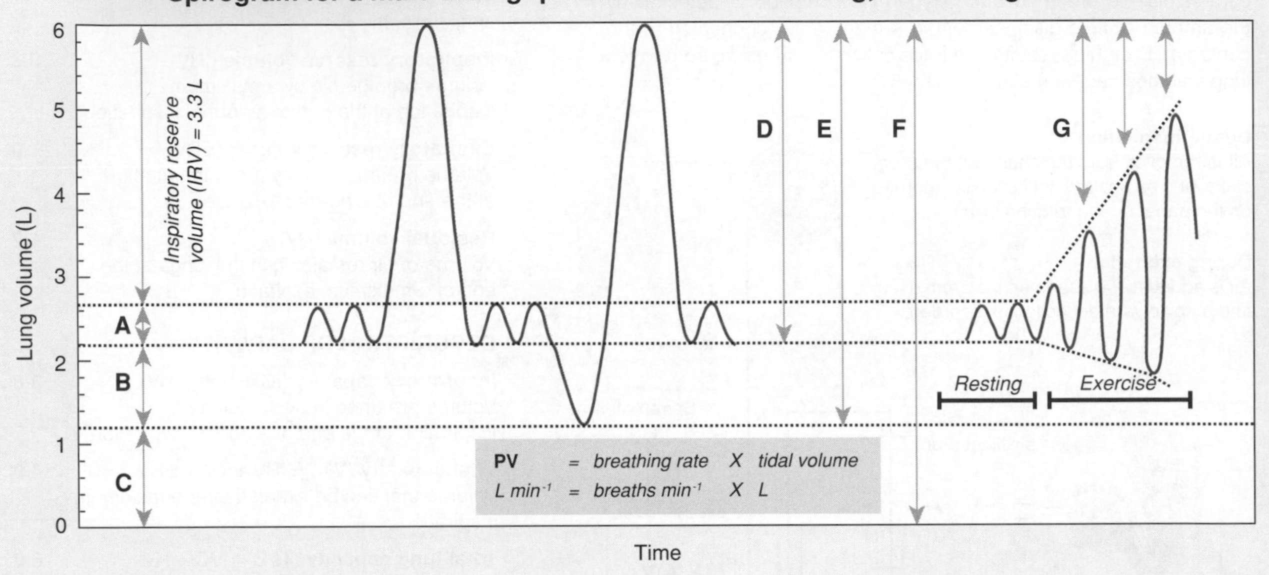

3. Using the definitions given on the previous page, identify the volumes and capacities indicated by the letters **A-F** on the spirogram above. For each, indicate the volume (vol) in liters (L). The inspiratory reserve volume has been identified:

   (a) **A**: ______ Vol: ______ (d) **D**: ______ Vol: ______

   (b) **B**: ______ Vol: ______ (e) **E**: ______ Vol: ______

   (c) **C**: ______ Vol: ______ (f) **F**: ______ Vol: ______

4. Explain what is happening in the sequence indicated by the letter **G**: ______

5. Calculate PV when breathing rate is 15 breaths per minute and tidal volume is 0.4 L: ______

6. (a) Describe what would happen to PV during strenuous exercise: ______

   (b) Explain how this is achieved: ______

7. The table above gives approximate percentages for respiratory gases during breathing. Study the data and then:

   (a) Calculate the difference in CO_2 between inhaled and exhaled air: ______

   (b) Explain where this 'extra' CO_2 comes from: ______

   (c) Explain why the dead air volume raises the oxygen content of exhaled air above that in the lungs: ______

ISBN: 978-1-92717357-2

Gas Transport

The transport of respiratory gases around the body is the role of the blood and its respiratory pigments. Oxygen is transported throughout the body chemically bound to the respiratory pigment **hemoglobin** inside the red blood cells. In the muscles, oxygen from hemoglobin is transferred to and retained by **myoglobin**, a molecule that is chemically similar to hemoglobin except that it consists of only one heme-globin unit. Myoglobin has a greater affinity for oxygen than hemoglobin and acts as an oxygen store within muscles, releasing the oxygen during periods of prolonged or extreme muscular activity. If the myoglobin store is exhausted, the muscles are forced into oxygen debt and must respire anaerobically. The waste product of this, lactic acid, accumulates in the muscle and is transported (as lactate) to the liver where it is metabolized under aerobic conditions.

Gas Exchange and Transport

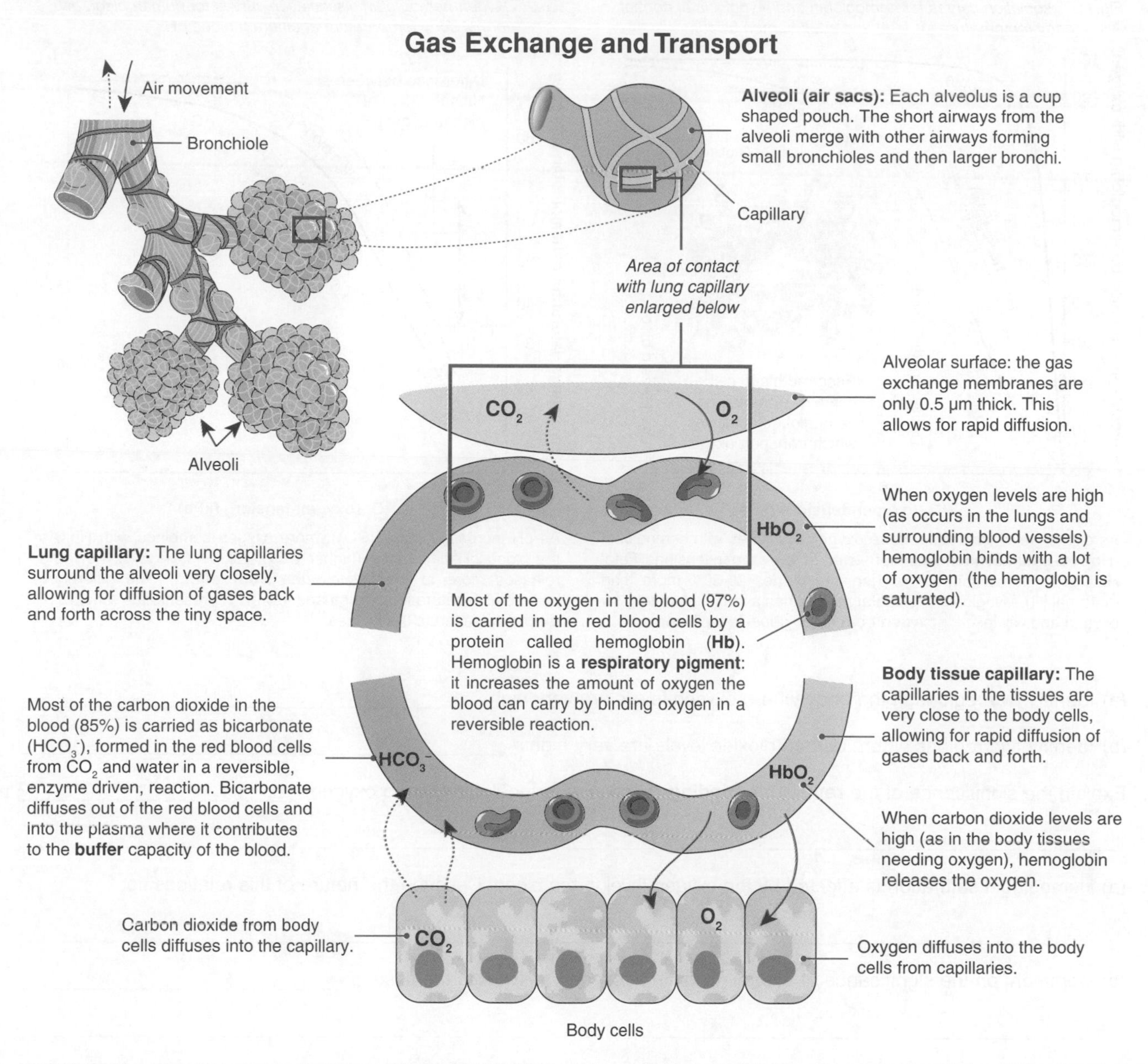

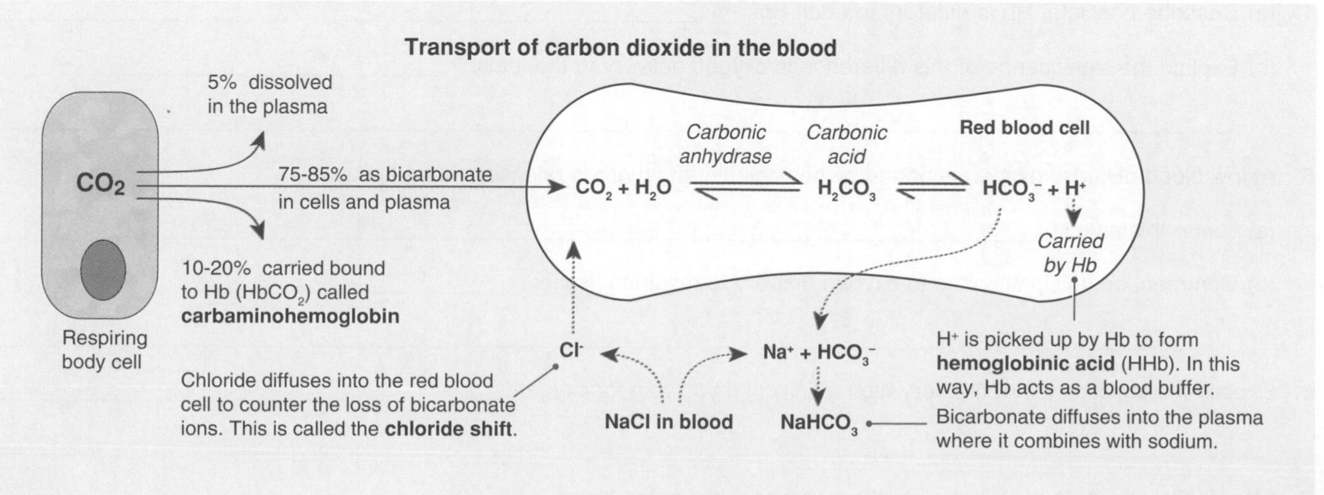

ISBN: 978-1-92717357-2

Related activities: *The Human Respiratory System, Respiratory Pigments*
Weblinks: *Gas transport*

Oxygen does not easily dissolve in blood, but is carried in chemical combination with hemoglobin (Hb) in red blood cells. The most important factor determining how much oxygen is carried by Hb is the level of oxygen in the blood. The greater the oxygen tension, the more oxygen will combine with Hb. This relationship can be illustrated with an oxygen-hemoglobin dissociation curve as shown below (Fig. 1). In the lung capillaries, (high O_2), a lot of oxygen is picked up and bound by Hb. In the tissues, (low O_2), oxygen is released. In skeletal muscle, myoglobin picks up oxygen from hemoglobin and therefore serves as an oxygen store when oxygen tensions begin to fall. The release of oxygen is enhanced by the **Bohr effect** (Fig. 2).

Respiratory Pigments and the Transport of Oxygen

Fig.1: Dissociation curves for hemoglobin and myoglobin at normal body temperature for fetal and adult human blood.

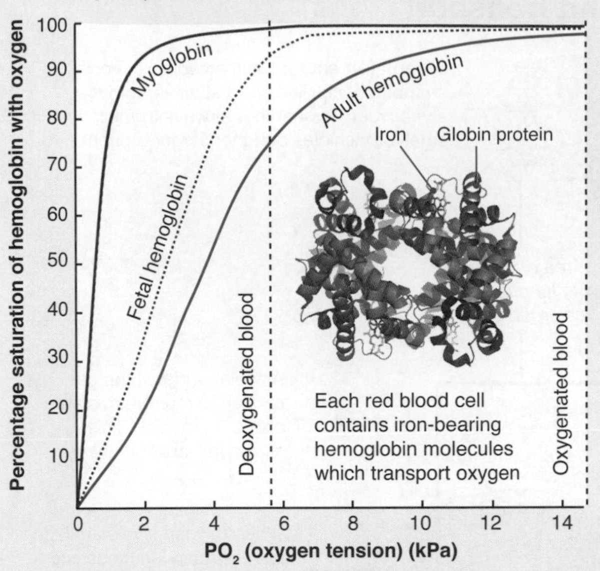

As oxygen level increases, more oxygen combines with hemoglobin (Hb). Hb saturation remains high, even at low oxygen tensions. Fetal Hb has a high affinity for oxygen and carries 20-30% more than maternal Hb. Myoglobin in skeletal muscle has a very high affinity for oxygen and will take up oxygen from hemoglobin in the blood.

Fig.2: Oxygen-hemoglobin dissociation curves for human blood at normal body temperature at different blood pH.

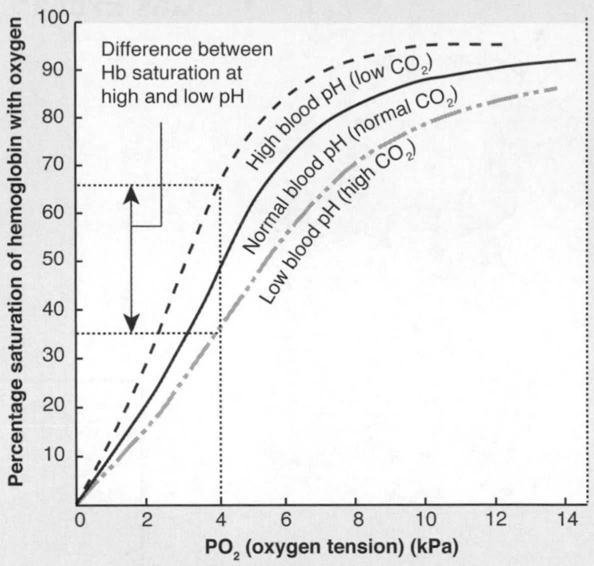

As pH increases (lower CO_2), more oxygen combines with Hb. As the blood pH decreases (higher CO_2), Hb binds less oxygen and releases more to the tissues (the **Bohr effect**). The difference between Hb saturation at high and low pH represents the amount of oxygen released to the tissues.

1. (a) Identify two regions in the body where oxygen levels are very high: __________

   (b) Identify two regions where carbon dioxide levels are very high: __________

2. Explain the significance of the **reversible binding** reaction of hemoglobin (Hb) to oxygen: __________

3. (a) Hemoglobin saturation is affected by the oxygen level in the blood. Describe the nature of this relationship:

   (b) Comment on the significance of this relationship to oxygen delivery to the tissues: __________

4. (a) Describe how fetal Hb is different to adult Hb: __________

   (b) Explain the significance of this difference to oxygen delivery to the fetus: __________

5. At low blood pH, less oxygen is bound by hemoglobin and more is released to the tissues:

   (a) Name this effect: __________

   (b) Comment on its significance to oxygen delivery to respiring tissue: __________

6. Explain the significance of the very high affinity of myoglobin for oxygen: __________

7. Identify the two main contributors to the buffer capacity of the blood: __________

ISBN: 978-1-92717357-2

Control of Breathing

The basic rhythm of breathing is controlled by the **respiratory center**, a cluster of neurons located in the medulla oblongata. This rhythm is adjusted in response to the physical and chemical changes that occur when we carry out different activities. Although the control of breathing is involuntary, we can exert some degree of conscious control over it. The diagram below illustrates these controls.

The Respiratory Center and the Control of Breathing

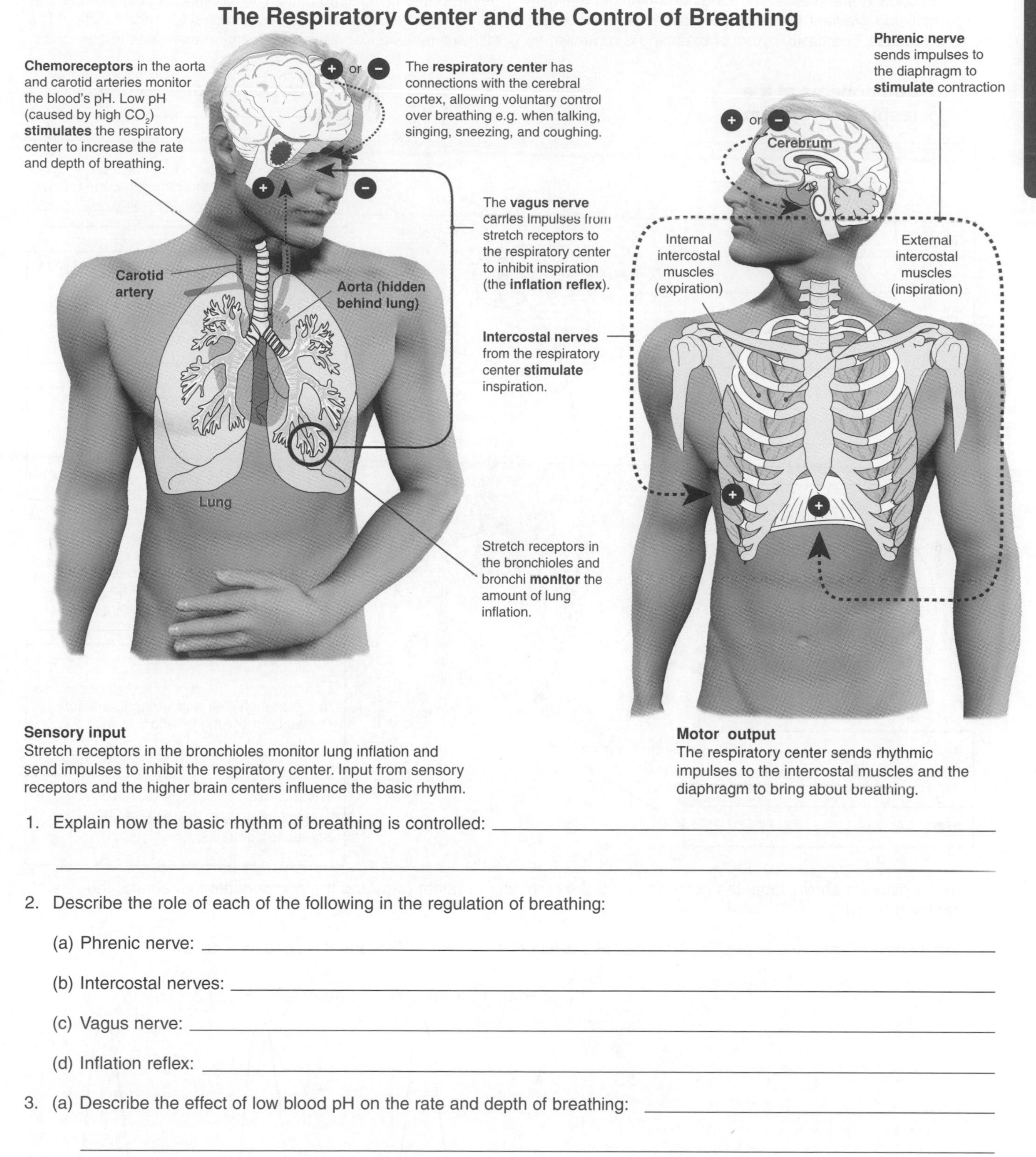

Sensory input
Stretch receptors in the bronchioles monitor lung inflation and send impulses to inhibit the respiratory center. Input from sensory receptors and the higher brain centers influence the basic rhythm.

Motor output
The respiratory center sends rhythmic impulses to the intercostal muscles and the diaphragm to bring about breathing.

1. Explain how the basic rhythm of breathing is controlled: ______

2. Describe the role of each of the following in the regulation of breathing:

   (a) Phrenic nerve: ______

   (b) Intercostal nerves: ______

   (c) Vagus nerve: ______

   (d) Inflation reflex: ______

3. (a) Describe the effect of low blood pH on the rate and depth of breathing: ______

   (b) Explain how this effect is mediated: ______

   (c) Suggest why blood pH is a good mechanism by which to regulate breathing rate: ______

ISBN: 978-1-92717357-2

Related activities: *Gas Transport in Humans*

Review of Lung Function

The respiratory system in humans (and other air breathing vertebrates) includes the lungs and the system of tubes through which the air reaches them. Breathing (ventilation) provides a continual supply of fresh air to the lungs and helps to maintain a large diffusion gradient for respiratory gases across the gas exchange surface. The basic rhythm of breathing is controlled by the respiratory center in the medulla of the hindbrain. The volume of gases exchanged during breathing varies according to the physiological demands placed on the body. These changes can be measured using spirometry. The following activity summarizes the key features of respiratory system structure and function. The stimulus material can be found in earlier exercises in this topic.

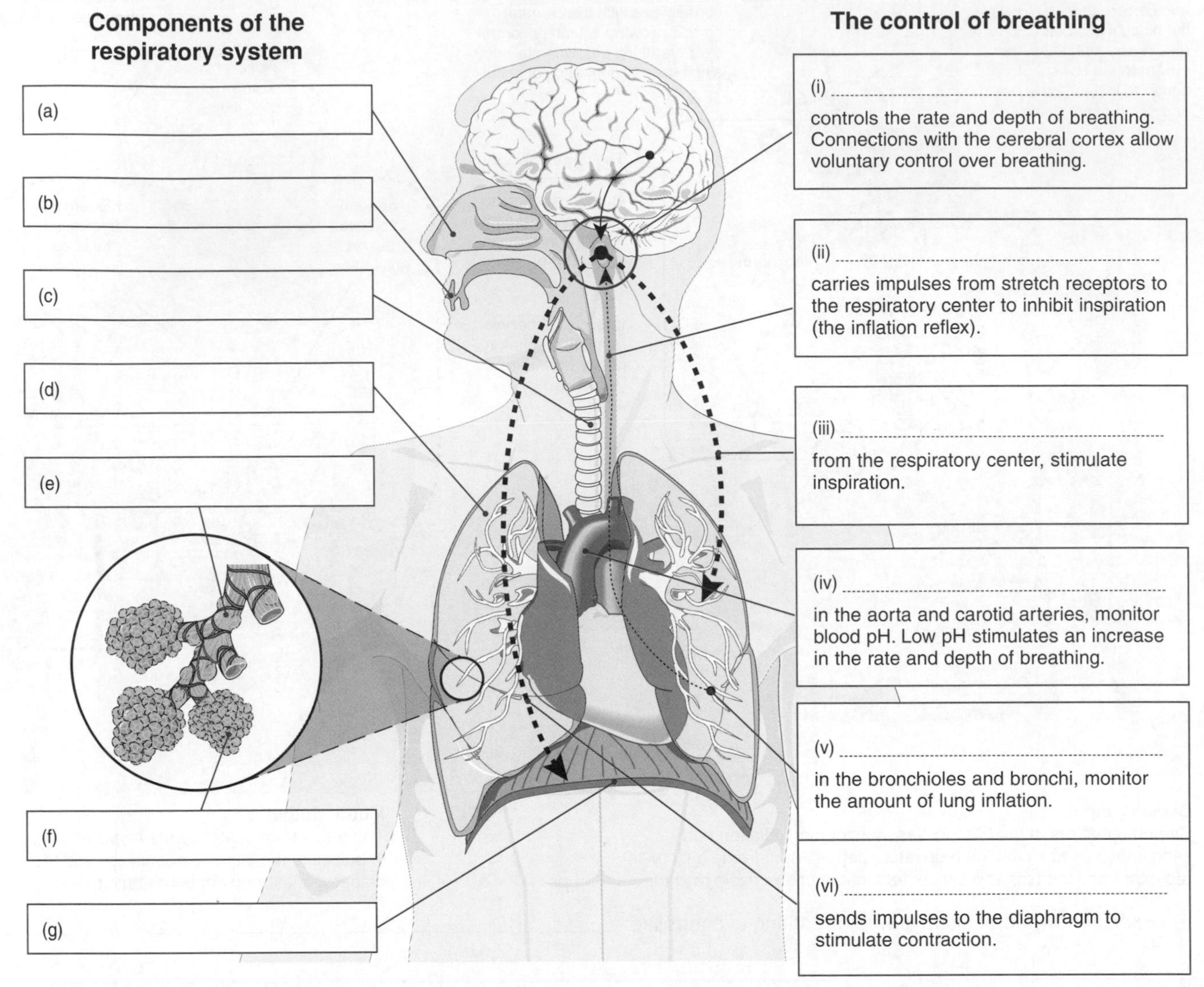

1. On the diagram above, label the components of the respiratory system (a-g) and the components that control the rate of breathing (i - vi).

2. Identify the volumes and capacities indicated by the letters A - E on the diagram of a spirogram below.

A = ...

B = ...

C = ...

D = ...

F = ...

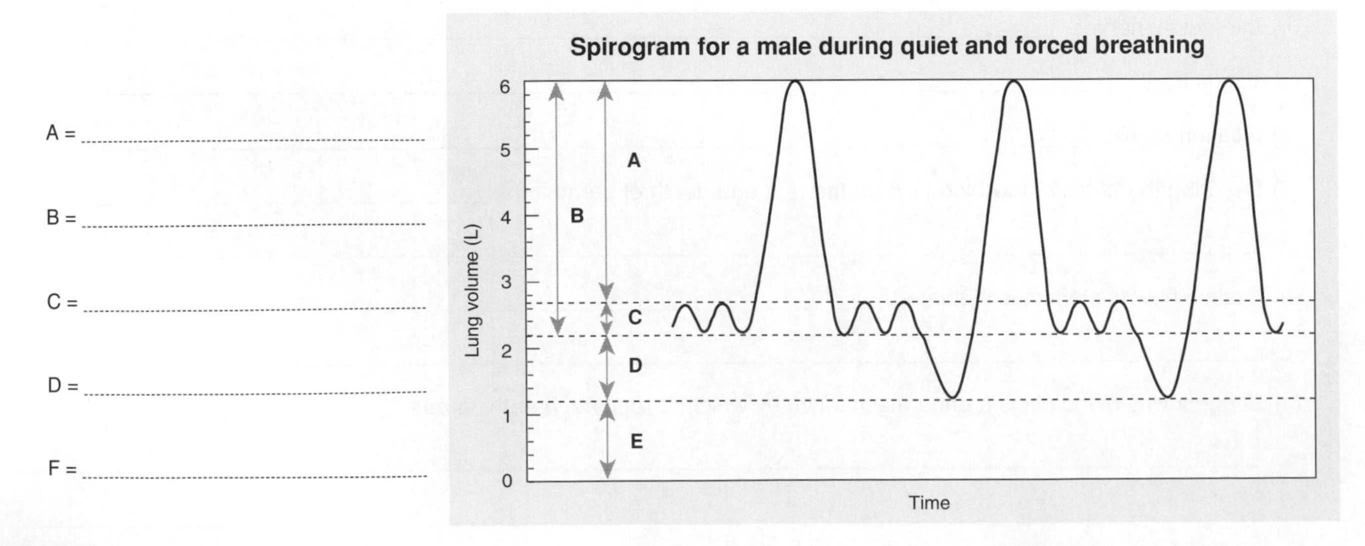

ISBN: 978-1-92717357-2

***Related activities:** Breathing in Humans, The Human Respiratory System, Measuring Lung Function, Control of Breathing*

Respiratory Diseases

Respiratory diseases are diseases of the gas exchange system, including diseases of the lung, bronchial tubes, trachea, and upper respiratory tract. Respiratory diseases include mild and self-limiting diseases such as the common cold, to life-threatening infections such as tuberculosis. One in six people in the US is affected by some form of chronic lung disease, the most common being asthma and chronic obstructive pulmonary disease (including emphysema and chronic bronchitis). Non-infectious respiratory diseases are categorized according to whether they prevent air reaching the alveoli (**obstructive**) or whether they affect the gas exchange tissue itself (**restrictive**).

Such diseases have different causes and different symptoms (below) but all are characterized by difficulty in breathing and the end result is similar in that gas exchange rates are too low to meet metabolic requirements. Non-infectious respiratory diseases are strongly correlated with certain behaviors and are made worse by exposure to air pollutants. Obstructive diseases, such as emphysema, are associated with an inflammatory response of the lung to noxious particles or gases, most commonly tobacco smoke. In contrast, scarring (**fibrosis**) of the lung tissue underlies restrictive lung diseases such as **asbestosis** and **silicosis**. Such diseases are often called occupational lung diseases.

Chronic bronchitis
Excess mucus blocks airway, leading to inflammation and infection

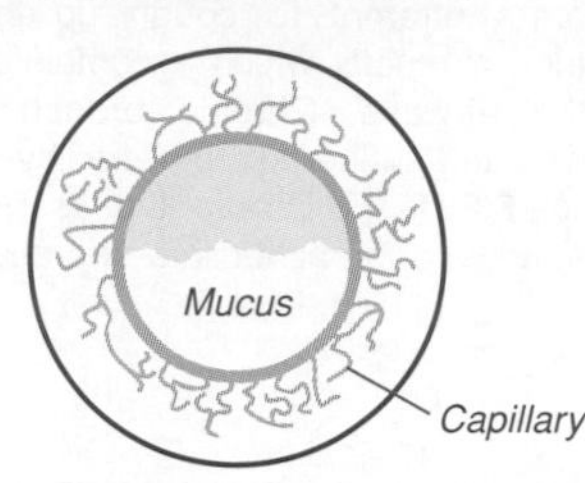

Asthma
Thickening of bronchiole wall and muscle hypertrophy. Bronchioles narrow.

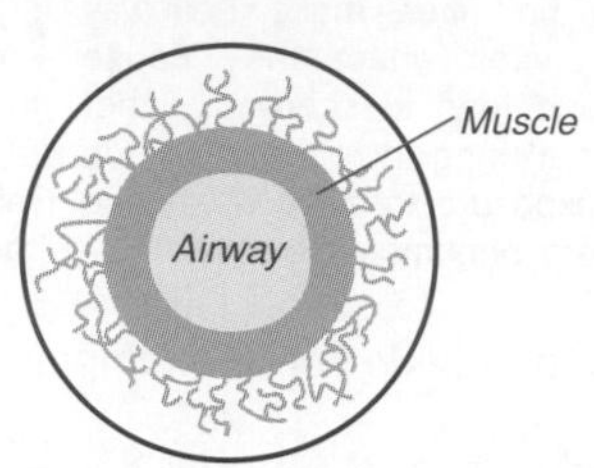

Emphysema
Destruction of capillaries and structures supporting the small airways and lung tissue

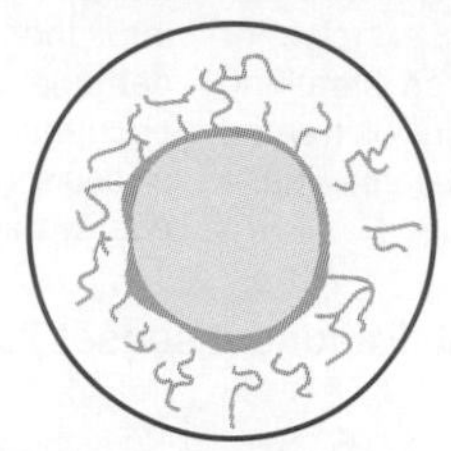

Cross sections through a bronchiole with various types of obstructive lung disease

A peak flow meter is a small, hand-held device used to monitor a person's ability to breathe out air. It measures the airflow through the bronchi and thus the degree of obstruction in the airways.

Obstructive lung disease – passage blockage –

In obstructive lung diseases, a blockage prevents the air getting to the gas exchange surface.

The flow of air may be obstructed because of constriction of the airways (as in **asthma**), excess mucus secretion (as in chronic **bronchitis**), or because of reduced lung elasticity, which causes alveoli and small airways to collapse (as in **emphysema**). Shortness of breath is a symptom in all cases and chronic bronchitis is also associated with a persistent cough.

Chronic bronchitis and emphysema often occur together and are commonly associated with cigarette smoking, but can also occur with chronic exposure to air pollution.

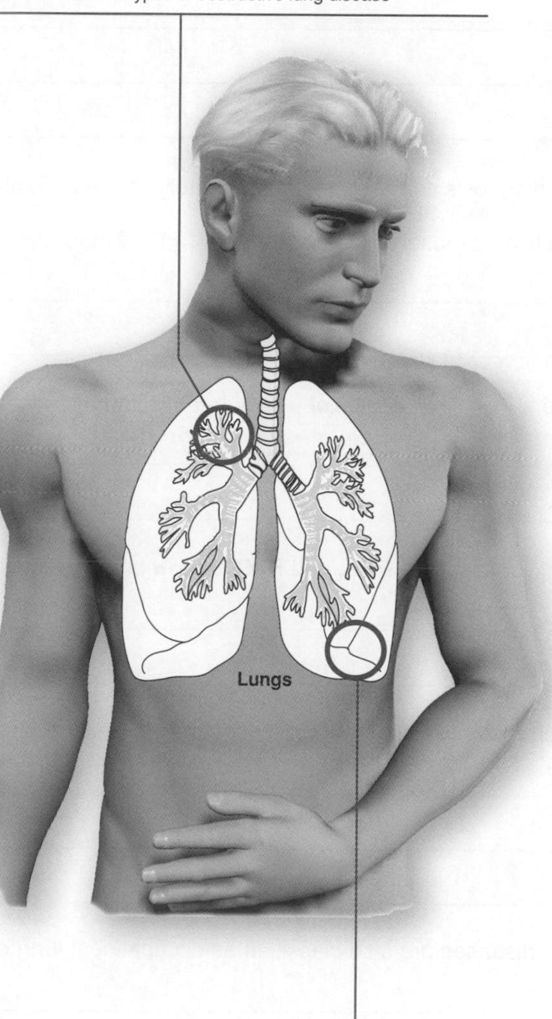

Scarring (fibrosis) makes the lung tissue stiffer and prevents adequate gas exchange

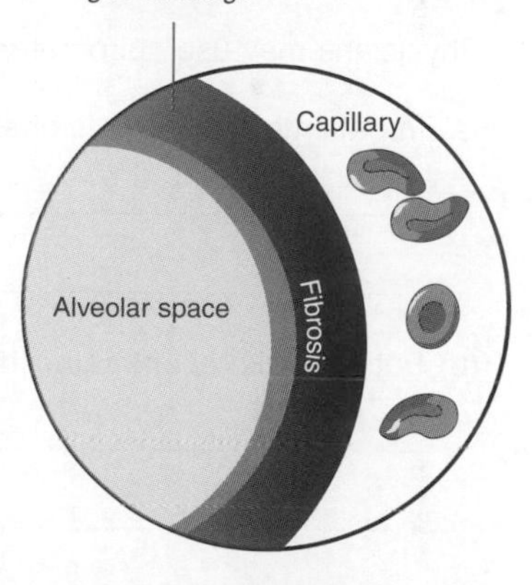

Restrictive lung disease – scarring –

Restrictive lung diseases are characterized by scarring or **fibrosis** within the gas exchange tissue of the lung (above). As a result of the scarring, the lung tissue becomes stiffer and more difficult to expand, leading to shortness of breath.

Restrictive lung diseases are usually the result of exposure to inhaled substances (especially dusts) in the environment, including **inorganic dusts** such as silica, asbestos, or coal dust, and **organic dusts**, such as those from bird droppings or moldy hay. Like most respiratory diseases, the symptoms are exacerbated by poor air quality (such as occurs in smoggy cities).

USGS

SEM of asbestos fibers. Asbestos has different toxicity depending on the type. Some types are very friable, releasing fibers into the air, where they can be easily inhaled.

ISBN: 978-1-92717357-2

Related activities: *Measuring Lung Function, Allergies and Hypersensitivity, Smoking and the Lungs* ***Weblinks:*** *Asthma, Emphysema*

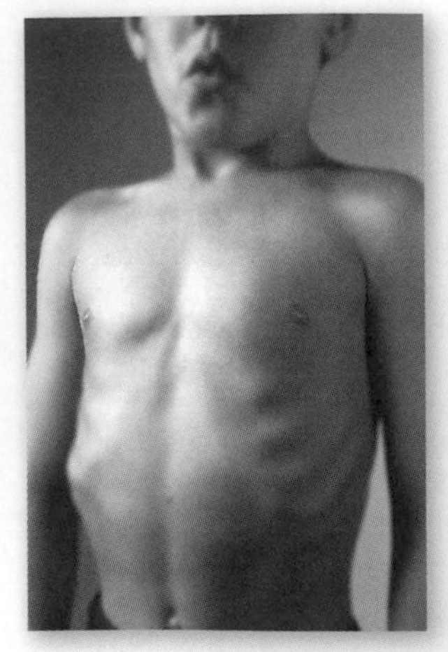

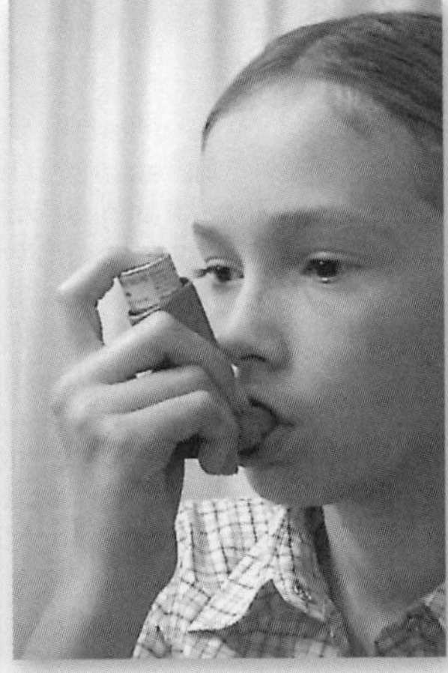

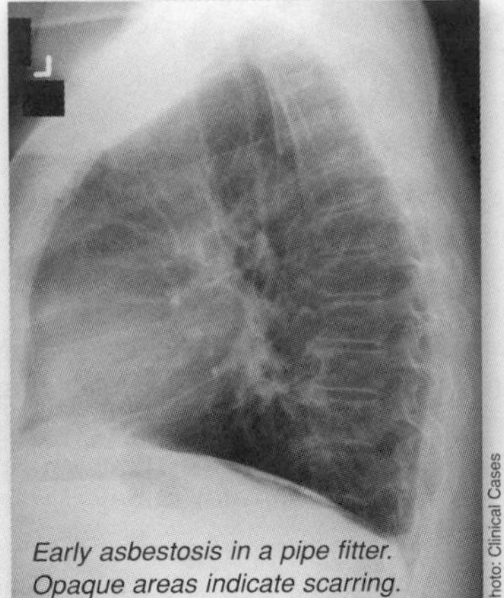

Early asbestosis in a pipe fitter. Opaque areas indicate scarring.

Photo: Clinical Cases

Asthma is a common disease affecting millions of people worldwide (20 million in the US alone). Asthma is the result of a hypersensitive reaction to allergens such as house dust or pollen, but attacks can be triggered by environmental factors such as cold air, exercise, or air pollutants. During an attack, sufferers show labored breathing with overexpansion of the chest cavity (above left). Asthma is treated with drugs that help to expand the airways (bronchodilators). These are usually delivered via a nebulizer or inhaler (above).

Asbestosis is a restrictive lung disease caused by breathing in asbestos fibers. The tiny fibers make their way into the alveoli where they cause damage and lead to scarring. Other occupational lung diseases include silicosis (exposure to silica dust) and coal workers' pneumoconiosis.

Chronic bronchitis is accompanied by a persistent, productive cough, where sufferers attempt to cough up the sputum or mucus which accumulates in the airways. Chronic bronchitis is indicated using **spirometry** by a reduced FEV_1/FVC ratio that is not reversed with bronchodilator therapy.

1. Distinguish between obstructive and restrictive lung diseases, and provide some examples:

2. Physicians may use spirometry to diagnosis certain types of respiratory disease. Explain the following typical results:

   (a) In patients with chronic obstructive pulmonary disease, the FEV_1 / FVC ratio declines (to <70% of normal):

   (b) Patients with asthma also have a FEV_1 / FVC ratio of <70%, but this improves following use of bronchodilators:

   (c) In patients with restrictive lung disease, both FEV_1 and FVC are low but the FEV_1 / FVC ratio is normal to high:

3. Describe the mechanisms by which restrictive lung diseases reduce lung function and describe an example:

4. Suggest why many restrictive lung diseases are also classified as occupational lung diseases:

5. Describe the role of histamine in the occurrence of an asthma attack:

ISBN: 978-1-92717357-2

Smoking and the Lungs

Tobacco smoking has been accepted as a major health hazard only relatively recently in historical terms, despite its practice in Western countries for more than 400 years, and much longer elsewhere. Cigarettes became popular at the end of World War I because they were cheap, convenient, and easier to smoke than pipes and cigars. They remain popular for the further reason that they are more addictive than other forms of tobacco. The milder smoke can be more readily inhaled, allowing **nicotine** (a powerful addictive poison) to be quickly absorbed into the bloodstream. **Lung cancer** is the most widely known and most harmful effect of smoking; 98% of cases are associated with cigarette smoking. Symptoms include chest pain, breathlessness, and coughing up blood. Tobacco smoking is also directly associated with coronary artery disease, emphysema, chronic bronchitis, peripheral vascular disease, and stroke. The damaging components of cigarette smoke include tar, carbon monoxide, nitrogen dioxide, and nitric oxide. Many of these harmful chemicals occur in greater concentrations in sidestream smoke (as occurs as a result of **passive smoking**) than in mainstream smoke (inhaled) due to the presence of a filter in the cigarette.

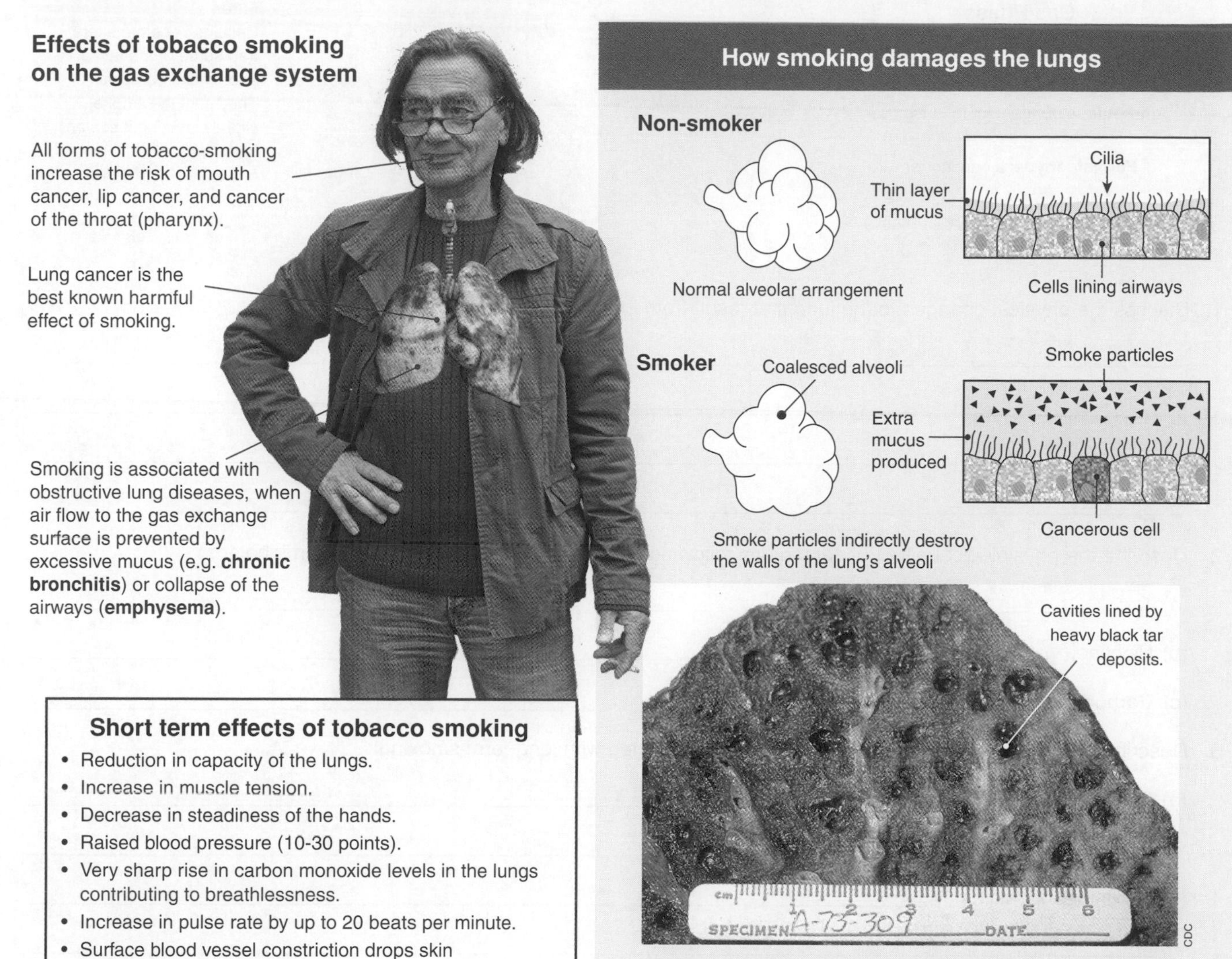

Gross pathology of lung tissue from a patient with emphysema. Tobacco tar deposits can be seen. Tar contains at least 17 known carcinogens.

Short term effects of tobacco smoking

- Reduction in capacity of the lungs.
- Increase in muscle tension.
- Decrease in steadiness of the hands.
- Raised blood pressure (10-30 points).
- Very sharp rise in carbon monoxide levels in the lungs contributing to breathlessness.
- Increase in pulse rate by up to 20 beats per minute.
- Surface blood vessel constriction drops skin temperature by up to 5°C.
- Dulling of appetite and sense of smell and taste.

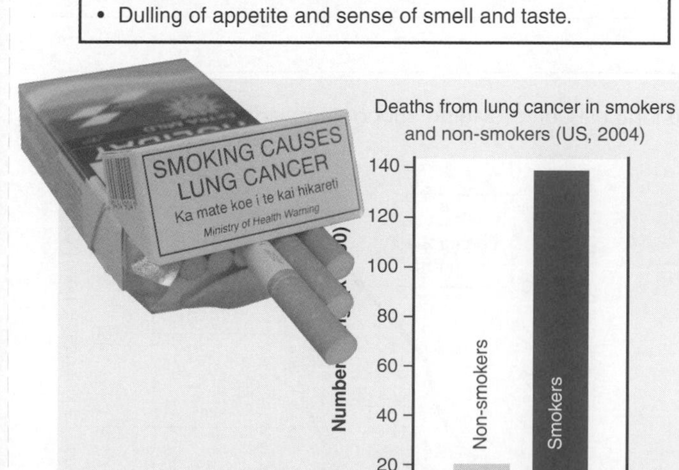

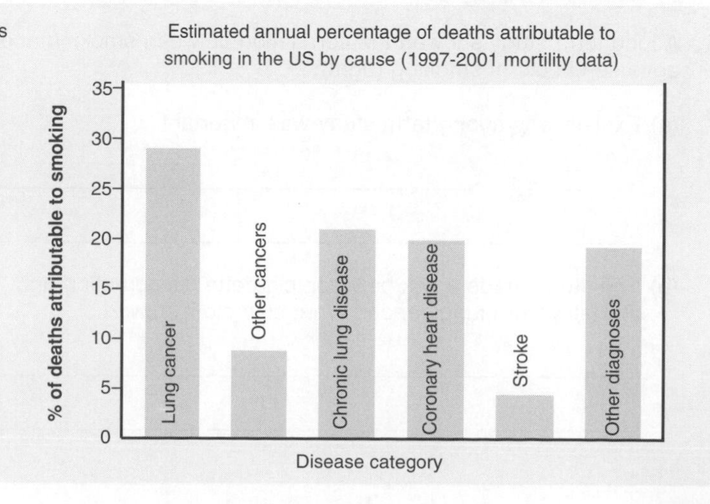

ISBN: 978-1-92717357-2

Related activities: *Gas Transport In Humans, Cardiovascular Disease*
Weblinks: *CDC: The Health Consequences of Smoking*

Components of Cigarette Smoke

Particulate Phase

Nicotine: a highly addictive alkaloid

Tar: composed of many chemicals

Benzene: carcinogenic hydrocarbon

Gas Phase

Carbon monoxide: a poisonous gas

Ammonia: a pungent, colourless gas

Formaldehyde: a carcinogen

Hydrogen cyanide: a highly poisonous gas

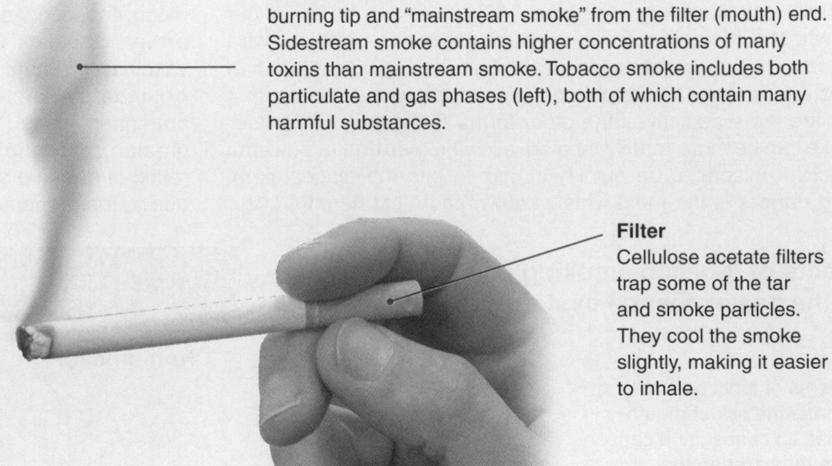

Tobacco smoke is made up of "sidestream smoke" from the burning tip and "mainstream smoke" from the filter (mouth) end. Sidestream smoke contains higher concentrations of many toxins than mainstream smoke. Tobacco smoke includes both particulate and gas phases (left), both of which contain many harmful substances.

1. Discuss the physical changes to the lung that result from long-term smoking:

2. Describe the physiological effect of each of the following constituents of tobacco smoke when inhaled:

   (a) Tar:

   (b) Nicotine:

   (c) Carbon monoxide:

3. Describe the symptoms of the following diseases associated with long-term smoking:

   (a) Emphysema:

   (b) Chronic bronchitis:

   (c) Lung cancer:

4. A long term study showed the correlation between smoking and lung cancer, providing supporting evidence for the adverse effects of smoking (right):

   (a) Explain why a long term study was important:

   (b) The study made a link between cigarette consumption and mortality from lung cancer. What else did it show?

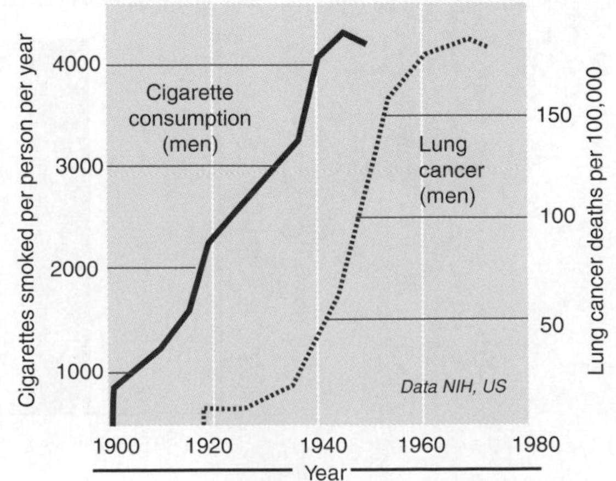

ISBN: 978-1-92717357-2

Responding to Exercise

Physical exercise places greater demands on the abilities of the body to maintain a steady state. Extra heat generated during exercise must be dissipated, oxygen demands increase, and there are more waste products produced. The body has an immediate response to exercise but also, over time, responds to the stress of repeated exercise (**training**) by adapting and improving its capacity for exercise and the efficiency with which it performs. This concept is illustrated below. Training causes tissue damage and depletes energy stores, but the body responds by repairing the damage, replenishing energy stores, and adjusting its responses in order to minimize the impact of exercise in the future. Maintaining homeostasis during exercise is principally the job of the circulatory and respiratory systems, although the skin, kidneys, and liver are also important.

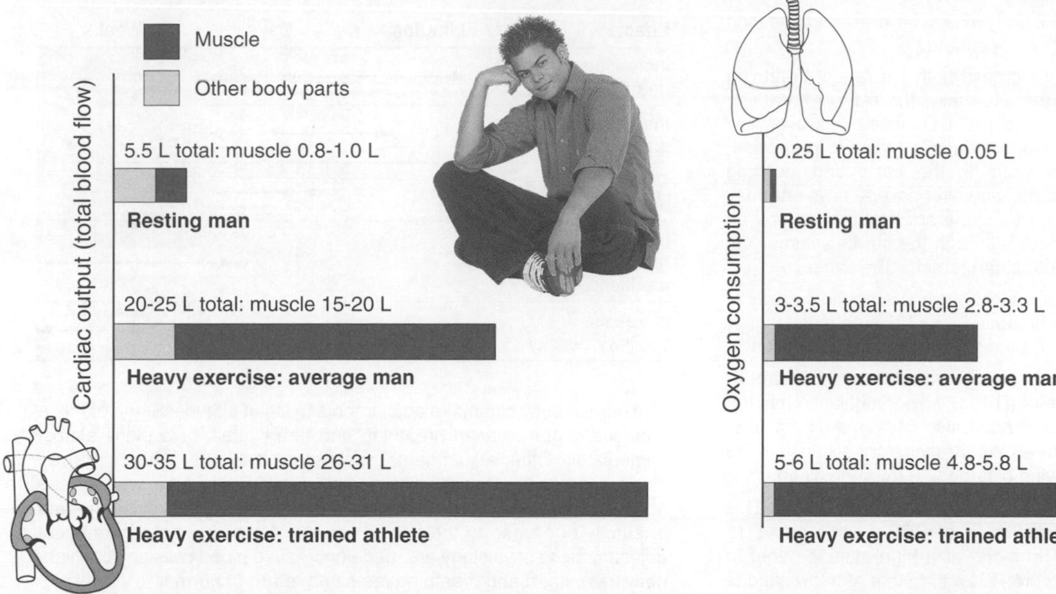

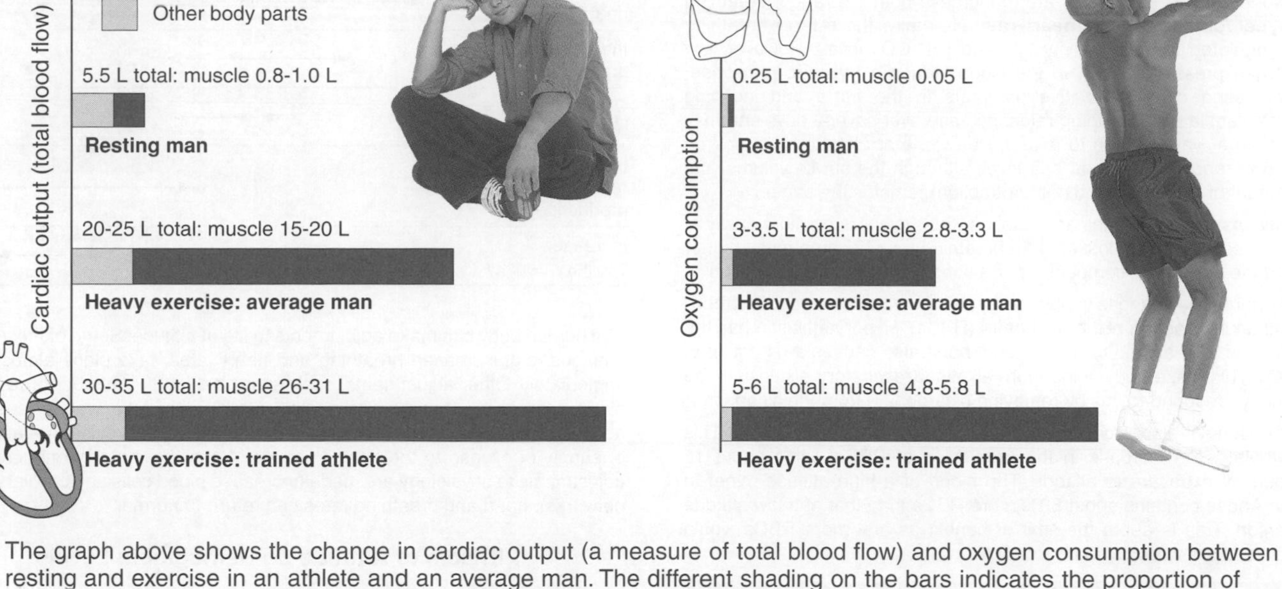

1. The graph above shows the change in cardiac output (a measure of total blood flow) and oxygen consumption between resting and exercise in an athlete and an average man. The different shading on the bars indicates the proportion of oxygen or blood flow in skeletal muscle compared to other body parts.

   (a) Describe what happens to the output of the heart (total blood flow) during heavy exercise: ______________________

   (b) Explain why this is the case: ______________________

   (c) List the organ(s) and tissues responsible for adjusting blood flow during exercise: ______________________

2. (a) Describe what happens to oxygen consumption during heavy exercise: ______________________

   (b) Explain why this is the case: ______________________

   (c) Explain the change in the proportion of oxygen consumed by the muscles during exercise: ______________________

3. Explain the difference in oxygen consumption and blood flow between a trained athlete and an average man:

____

ISBN: 978-1-92717357-2

Related activities: *Extrinsic Control of Heart Rate, Exercise and Blood Flow*

The Effects of High Altitude

Air pressure decreases with altitude so the pressure (therefore amount) of oxygen in the air also decreases. Many of the physiological effects of high altitude arise from the low oxygen pressure, not the low air pressure *per se*. Sudden exposure to an altitude of 2000 m causes breathlessness and above 7000 m most people would become unconscious. Humans and other animals can make both short and long term physiological adjustments to altitude. These changes are referred to **acclimation**, and are different to the evolutionary adaptations of high altitude populations, which are inherited.

Physiological Adjustments to Altitude

When exposed to high altitude, the body makes several short and long term physiological adjustments to compensate.

Two immediate responses are to increase both the rate of breathing (**hyperventilation**) and **heart rate**. Normally, the rate of breathing is regulated by a sensitivity to blood pH (CO_2 level). However, low oxygen pressures (pO_2) in the blood induce a **hypoxic response**, stimulating oxygen-sensitive receptors in the aorta and inducing hyperventilation. Breathing rates increase over several days and may remain elevated for up to a year. Heart rate at altitude increases up to 50% above the rate at sea level, although the stroke volume (the amount of blood pumped per contraction) remains the same.

Fluid loss also becomes an issue, as water evaporates more easily at low pressure. Water losses from breathing thus become much higher and the body must compensate by taking in more fluids.

Longer terms changes in physiology include **acid-base readjustment** and an increase in **red blood cells** (RBCs). Hyperventilation has the effect of increasing O_2 in the blood, but it also causes a reduction of CO_2. This make body fluids more alkaline (respiratory alkalosis). The kidneys respond to this by removing bicarbonate from the blood.

The kidneys also produce the hormone **erythropoietin** (**EPO**). This stimulates an increase in the production of RBCs within about 15 hours of exposure to altitude. The blood of a high altitude miner in the Andes contains about 38% more RBCs than that of a low altitude person. This is about the limit of benefit as any more RBCs would cause blood viscosity to be too high.

Ascending to altitude (4500 m+) too rapidly results in mountain sickness. The symptoms include breathlessness and nausea. Continuing to ascend with mountain sickness can result in fatal accumulation of fluid on the lungs and brain.

Time Scales of Physiological Changes at Altitude

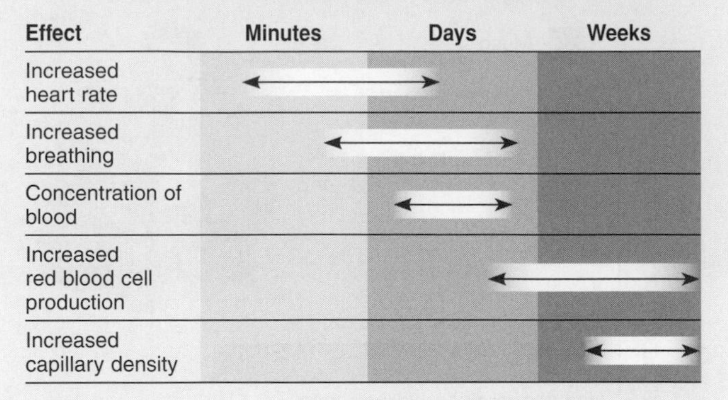

The human body can make adjustments to life at altitude. Some of these changes, e.g. increased breathing and heart rates, take place almost immediately. Other adjustments, such as increasing the number of red blood cells and associated hemoglobin level, may take weeks (see above and below). These responses are all aimed at improving the rate of supply of oxygen to the body's tissues. When the more permanent adjustments to physiology are made (increased blood cells and capillary networks), heart and breathing rates can return to normal.

Effects of Altitude on Hemoglobin Levels

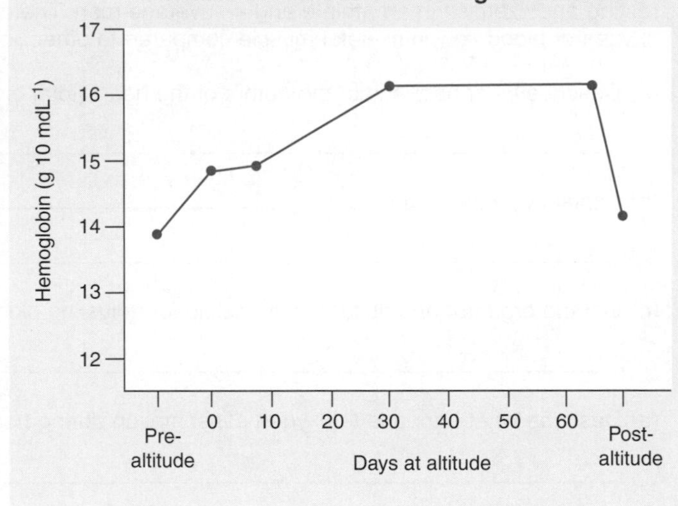

1. (a) Describe the initial effects of high altitude on the body in people who are not acclimated: ______________________

   (b) Name the general term given to describe these effects: ______________________

2. (a) Name one short term and long term physiological adjustment that humans make to high altitude: ______________________

   (b) Explain how these adjustments help to increase the amount of oxygen the body receives: ______________________

Related activities: *Control of Breathing*

Periodicals: *Humans with Altitude*

ISBN: 978-1-92717357-2

KEY TERMS: Mix and Match

INSTRUCTIONS: Test your vocabulary by matching each term to its definition, as identified by its preceding letter code.

- alveoli (sing. alveolus) ....
- asthma ....
- breathing ....
- breathing rate ....
- bronchi ....
- bronchioles ....
- chronic bronchitis ....
- emphysema ....
- expiration (=exhalation) ....
- gas exchange system ....
- hemoglobin ....
- inspiration (=inhalation) ....
- intercostal muscles ....
- intrapleural pressure ....
- lungs ....
- myoglobin ....
- obstructive pulmonary disease ....
- oxygen-hemoglobin dissociation curve ....
- pulmonary ventilation rate (PV) ....
- residual volume ....
- respiratory center ....
- respiratory pigment ....
- surfactant ....
- tidal volume ....
- trachea ....
- vital capacity ....

A A specialised system to improve the efficiency of oxygen and carbon dioxide exchanges in multicellular organisms.

B The volume of air breathed in and out in a single breath.

C A progressive obstructive lung disease resulting in reduced elasticity of the lung tissue. Alveoli and small airways collapse preventing air reaching the gas exchange surface.

D Internal gas exchange structures found in air-breathing vertebrates.

E A phospholipid produced by cells in the alveolar walls, responsible for reducing surface tension.

F The muscles between the ribs that assist in the expansion and contraction of the chest cavity during breathing,

G The pressure within the pleural cavity. It is always negative and relatively large relative to the pressure inside the alveoli.

H Microscopic structures in the lungs of air-breathing vertebrates that form the terminus of the bronchioles. The site of gas exchange.

I The act of breathing out or removing air from the lungs.

J A respiratory disease where the airways become narrow restricting the flow of air to the lungs. Often results from hypersensitive to allergens and environmental factors.

K Inflammation of the bronchi in the lungs, where the flow of air is obstructed because of excess mucus secretion, it is accompanied by a persistent productive cough.

L A substance carried in blood that is able to bind oxygen for transport to cells.

M The rate at which a cycle of inspiration and expiration occurs (usually per minute).

N Heme-containing protein found in muscle that binds oxygen.

O Large air tubes that branch from the trachea to enter the lungs in vertebrates.

P The maximum amount of air a person can expel from the lungs after a maximum inspiration.

Q Tube that conveys air from the mouth to the bronchi. Also known as the windpipe.

R Volume of air remaining in the lungs at the end of a maximum expiration.

S A large iron-containing protein, which transports oxygen in the blood.

T A region in the medulla oblongata that controls the basic rhythm of breathing.

U The volume of air exchanged between the lungs and the environment per minute.

V The act of inhaling and exhaling air into and out of the lungs.

W A graph showing how readily hemoglobin acquires and releases oxygen molecules at various partial pressures of oxygen.

X The act of breathing in or filling the lungs with air.

Y Small air tubes in the lungs of vertebrates that divide from the bronchi and become progressively smaller.

Z Disease of the lungs that include chronic bronchitis and emphysema, most often together. Symptoms include difficulty in breathing.

ISBN: 978-1-92717357-2

Interacting Systems

The Digestive System

Respiratory system

- Respiratory system provides O_2 to the organs of the digestive system and disposes of CO_2 produced by cellular respiration.

Cardiovascular system

- Digestive system absorbs iron required for synthesis of hemoglobin and water for maintenance of blood volume.
- Hepatic portal system transports nutrient-rich blood from substantial parts of the gastrointestinal tract to the liver. Ultimately the cardiovascular system distributes nutrients throughout the body.
- The liver produces angiotensinogen, a precursor of the protein angiotensin, which is involved in the system regulating blood pressure and fluid volume.
- Blood distributes hormones of the digestive tract.

Endocrine system

- The liver removes hormones from circulation and prevents their continued activity.
- Pancreas contains endocrine cells that produce hormones for regulating blood sugar.
- Local hormones (e.g. gastrin from the stomach, cholecystokinin and secretin from the intestinal mucosa) help to regulate digestive function, including secretion of digestive juices and gut motility.

Skeletal system

- Digestive system absorbs calcium needed for bone maintenance, growth, and repair.
- Skeletal system protects some of the digestive organs from major damage.
- Bone acts as a storage depot for some nutrients (e.g. calcium).

Integumentary system

- Digestive system provides fats for insulation in dermal and subcutaneous tissues.
- Skin provides external covering to protect the digestive organs.
- The skin synthesizes a precursor to vitamin D, which is required for absorption of calcium from the gut.

Nervous system

- The feeding center of the hypothalamus stimulates hunger. The satiety center suppresses the feeding center's activity after eating.
- Autonomic NS activity regulates much of gut function. Generally, parasympathetic stimulation increases and sympathetic stimulation decreases gut activity.
- There are reflex and voluntary controls over defecation.

Lymphatic system and immunity

- The lymphatic vessels of the small intestine (the lacteals) drain fat-laden lymph from the gut to the liver.
- Acidic gastric secretions destroy pathogens (non-specific defense).
- Lymphoid tissues in the gut mesenteries and intestinal wall house macrophages and leukocytes that protect against infection.

Urinary system

- Kidneys excrete toxins and the breakdown products of hormones which have been metabolized by the liver.
- Final activation of vitamin D, which is involved in calcium and phosphorus metabolism, occurs in the kidneys.

Reproductive system

- The digestive system provides nutrients required both for normal growth and repair, and the extra nutrition required to support pregnancy and lactation.

Muscular system

- Liver removes and metabolizes lactic acid produced by intense muscular activity.
- Calcium absorbed in the gut as part of the diet is required for muscle contraction.
- Activity of skeletal muscles increases the motility of the gastrointestinal tract, aiding passage of food through the gut.

General Functions and Effects on all Systems

The digestive system is responsible for the physical and chemical digestion and absorption of ingested food. Ultimately, it provides the nutrients required by all body systems for energy metabolism, growth, repair, and maintenance of tissues. Some nutrients may be stored (e.g. in bone, liver, and adipose tissue).

Disease

Symptoms of disease
- Pain (moderate to severe)
- Bleeding or change in bowel function
- Gastric reflux, nausea or vomiting
- Nutritional deficiencies

Infectious diseases of the digestive system
- Cholera
- Viral hepatitis
- Bacterial food poisoning
- Viral gastroenteritis

Non-infectious disorders of the digestive system
- Bowel cancer
- Appendicitis
- Inflammatory bowel diseases
- Food allergies or intolerance
- Cirrhosis of the liver

Medicine & Technology

Diagnosis of disorders
- Endoscopy and colonoscopy
- Gastrointestinal biopsy
- MRI scans and barium enema X-ray
- Blood tests

Preventing and treating diseases of the digestive system
- Drug therapies
- Surgery
- Radiotherapy
- Dietary management
- Behavior modification

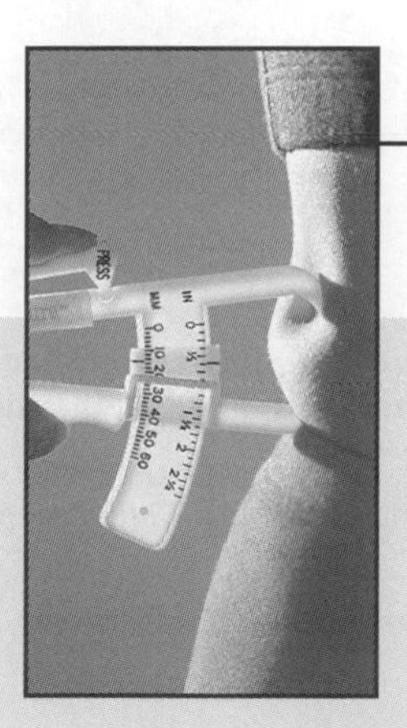

- Appendicitis
- Lactose intolerance
- Celiac disease
- Salmonellosis
- Cholera

- Endoscopy
- MRI scanning
- Appendectomy
- Diet for health

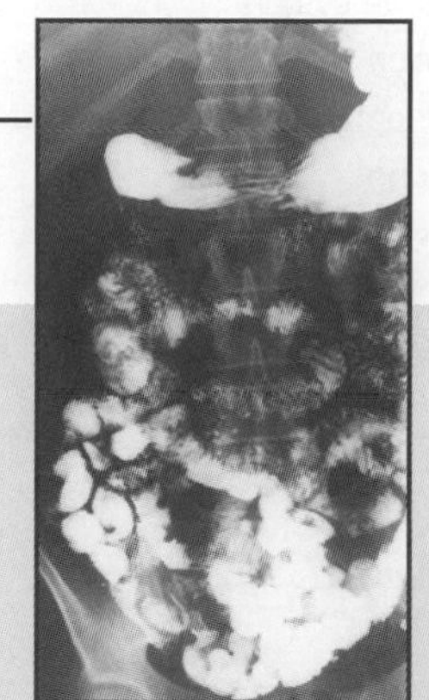

Eating to Live

The Digestive System

The digestive system provides for the energy and nutritional needs of all the body's systems.

While the digestive system is fairly robust against degenerative changes, gastrointestinal disorders are common. Gut function is improved by moderate exercise

- Constipation
- Gastric emptying
- Bowel cancer

- GI blood flow
- Sports nutrition
- Carbo-loading
- Nutrition & recovery

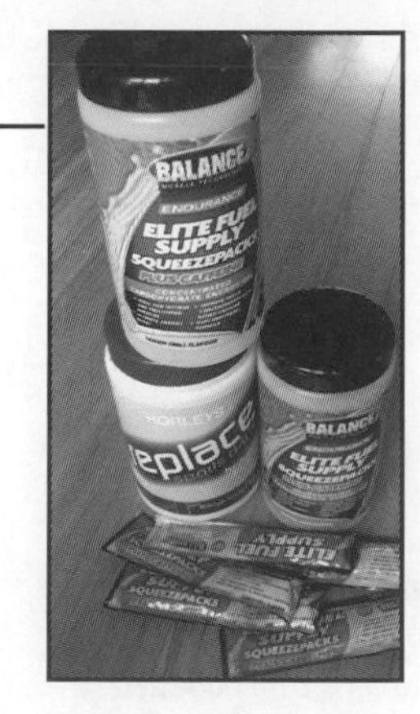

Effects of aging on the digestive system
- Increased risk of bowel cancers
- Slower passage of food, constipation
- Fibrosis of some organs (pancreas)
- Decline in gastric emptying rate
- Reduced gastric capacity

The Effects of Aging

Effects of exercise on the digestive system
- Reduced blood flow to gut (short term)
- Decreased intestinal transit time
- Improved digestive function (long term)
- GI upset in highly trained athletes

Exercise

The Digestive System

Key concepts

- The digestive system takes in and processes food, making it available for assimilation.
- Food moves through the gut by persistalsis.
- Food is broken down by enzymes secreted by the organs of the digestive tract.
- Nutrients are absorbed across the gut wall into the blood. The liver is important in nutrient processing.
- Nervous and hormonal signals regulate digestion.
- Dietary and digestive disorders may be associated with inadequate nutrient intake and infection.

Key terms

absorption
anus
assimilation
diet
digestion
digestive enzymes
digestive system
egestion
esophagus
exocrine
feces
gall bladder
ingestion
intestinal villus
large intestine
liver
malnutrition
mouth
oral rehydration solution
pancreas
peristalsis
pharynx
saliva
small intestine
stomach
swallowing
tooth
type 2 diabetes mellitus

Learning Objectives

☐ 1. Use the **KEY TERMS** to compile a glossary for this topic.

The Alimentary Canal

pages 218-222

☐ 2. Describe the overall function of the **digestive system**. Recognize the stages in processing food: **ingestion**, **digestion**, **absorption**, and **egestion**.

☐ 3. Annotate a diagram of the human digestive system, including the **mouth**, **esophagus**, **stomach**, **small intestine**, **large intestine**, **anus**, **liver**, **pancreas**, and **gall bladder**.

☐ 4. Describe the structure and function of the mouth, **pharynx**, and esophagus. Describe deciduous and permanent **dentition**, and the basic anatomy of a **tooth**. Describe the composition and function of **saliva**.

☐ 5. Explain how food is moved through the digestive tract by **peristalsis**. Describe the mechanism of **swallowing**.

☐ 6. Describe the structure and function of the stomach.

☐ 7. Describe the structure and function of the small intestine (duodenum, jejunum, ileum).

☐ 8. Describe the **exocrine** role of the pancreas.

☐ 9. Identify the main **digestive enzymes**, their substrates and their end products.

The Liver

pages 225-227

☐ 10. Describe the structure and histology of the liver. Describe the liver's role in digestion.

☐ 11. Describe the central role of the liver in metabolism.

Absorption and Transport

pages 221-224

☐ 12. Distinguish between **absorption** and **assimilation**. Identify where each of the following is absorbed: water, small molecules (alcohol, glucose), breakdown products of carbohydrate, protein, and fat digestion.

☐ 13. Describe the structure of an **intestinal villus** and relate the structure to its role in absorption. Explain the ultrastructure of an epithelial cell of a villus.

☐ 14. Describe nutrient absorption in the ileum. Explain the role of micelles and chylomicrons in lipid absorption and transport.

☐ 15. List the materials that are not absorbed and explain the role of the large intestine and rectum in **feces** formation and egestion. Identify defecation as a reflex activity.

☐ 16. Outline the control of digestive secretions by nerves and hormones. Outline the role of the appetite control center in the brain in regulating food intake.

Diet, Nutrition, and Disease

pages 128, 217, 229-234

☐ 17. Explain what is meant by a balanced **diet**. Discuss the consequences of **malnutrition**.

☐ 18. Recognize that **type 2 diabetes mellitus** can be partly managed through diet.

☐ 19. Describe how cholera infection can affect normal ion and water transport in the digestive tract. Explain why treatment with an **oral rehydration solution** is important.

Periodicals:
Listings for this chapter are on page 280

Weblinks:
www.thebiozone.com/weblink/AnaPhy-3572.html

BIOZONE APP:
Student Review Series
The Digestive System

A Balanced Diet

Nutrients are required for metabolism, tissue growth and repair, and as an energy source. A **diet** refers to the quantity and nature of the food eaten. A **balanced diet** is one that provides all the nutrients and energy to maintain good health. Conversely poor nutrition (**malnutrition**) may cause ill-health or **deficiency diseases**. While not all foods contain all the representative nutrients, we can obtain the required balance of different nutrients by eating a wide variety of foods. In a recent overhaul of previous dietary recommendations, the health benefits of monounsaturated fats (such as olive and canola oils), fish oils, and whole grains have been recognized, and people are being urged to reduce their consumption of highly processed foods and saturated (rather than total) fat. Those on diets that restrict certain food groups (e.g. vegans) must take care to balance their intake of foods to ensure an adequate supply of protein and other nutrients (e.g. iron and B vitamins). Dietary information, including **Recommended Daily Amounts** (RDAs) for energy and nutrients, is provided to consumers through the food labeling. Such information helps individuals to assess their nutrient and energy intake and adjust their diet accordingly.

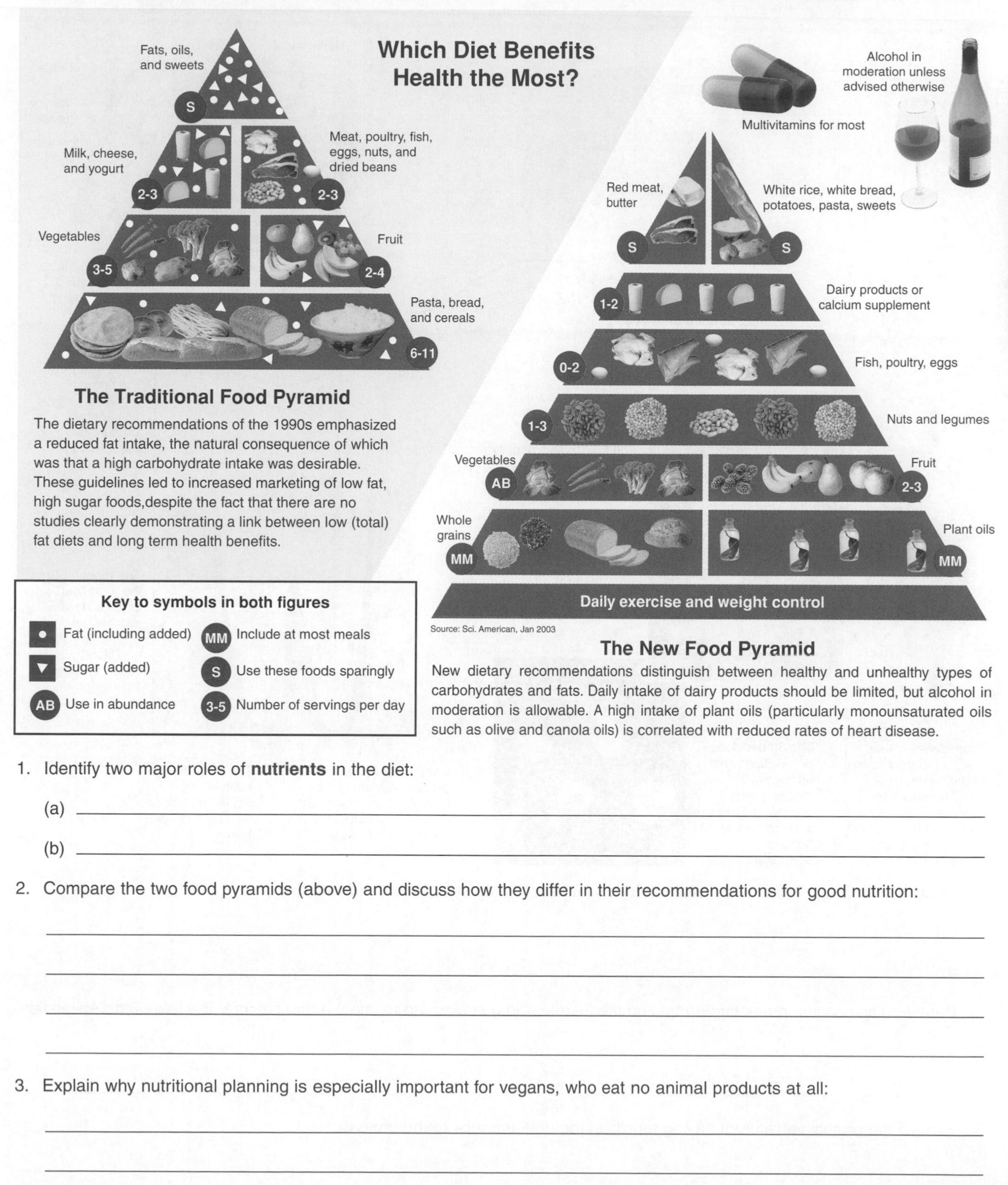

The dietary recommendations of the 1990s emphasized a reduced fat intake, the natural consequence of which was that a high carbohydrate intake was desirable. These guidelines led to increased marketing of low fat, high sugar foods,despite the fact that there are no studies clearly demonstrating a link between low (total) fat diets and long term health benefits.

Key to symbols in both figures

- ● Fat (including added)
- ▼ Sugar (added)
- AB Use in abundance
- MM Include at most meals
- S Use these foods sparingly
- 3-5 Number of servings per day

The New Food Pyramid

New dietary recommendations distinguish between healthy and unhealthy types of carbohydrates and fats. Daily intake of dairy products should be limited, but alcohol in moderation is allowable. A high intake of plant oils (particularly monounsaturated oils such as olive and canola oils) is correlated with reduced rates of heart disease.

1. Identify two major roles of **nutrients** in the diet:

   (a) ______________________________

   (b) ______________________________

2. Compare the two food pyramids (above) and discuss how they differ in their recommendations for good nutrition:

3. Explain why nutritional planning is especially important for vegans, who eat no animal products at all:

ISBN: 978-1-92717357-2

Related activities: *Deficiency Diseases, Malnutrition and Obesity*

The Mouth and Pharynx

The mouth (**oral cavity**) is the first part of the gut where food is **ingested**. It is formed by the cheeks, hard and soft palate, and tongue and leads to the **pharynx** (the first part of the throat). The contains the teeth, which are hard structures, specialized for chewing food (**mastication**). The tongue moves food around and the teeth and the salivary glands, which produce saliva, begin digestion. The oral cavity is divided into quadrants and the number of teeth in each quadrant given by a **dental formula**. There are 32 adult (permanent) teeth, organized as shown below left. The basic structure of a tooth is described in the inset below.

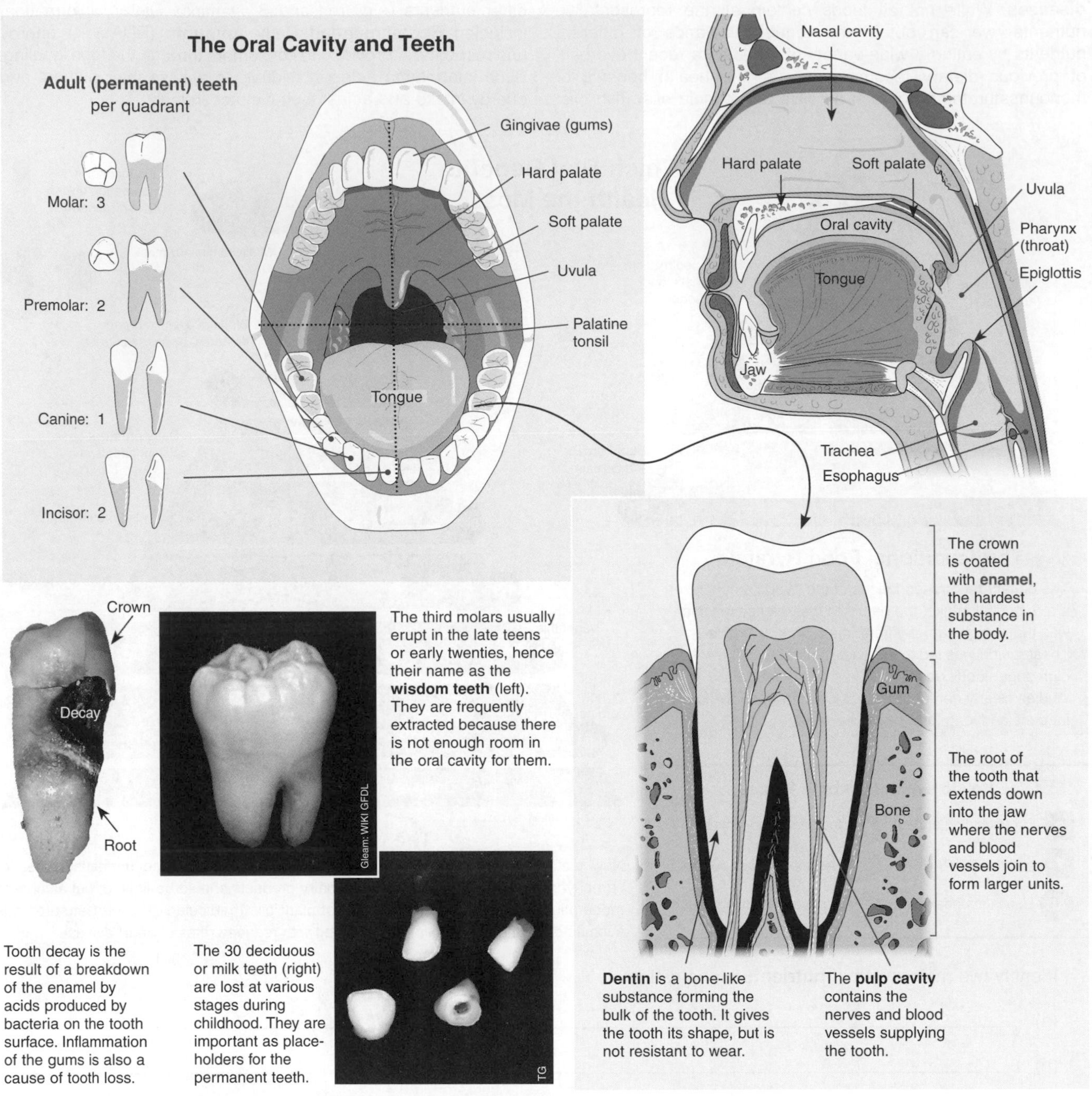

The crown is coated with **enamel**, the hardest substance in the body.

The root of the tooth that extends down into the jaw where the nerves and blood vessels join to form larger units.

Dentin is a bone-like substance forming the bulk of the tooth. It gives the tooth its shape, but is not resistant to wear.

The **pulp cavity** contains the nerves and blood vessels supplying the tooth.

The third molars usually erupt in the late teens or early twenties, hence their name as the **wisdom teeth** (left). They are frequently extracted because there is not enough room in the oral cavity for them.

Tooth decay is the result of a breakdown of the enamel by acids produced by bacteria on the tooth surface. Inflammation of the gums is also a cause of tooth loss.

The 30 deciduous or milk teeth (right) are lost at various stages during childhood. They are important as place-holders for the permanent teeth.

1. Describe two major roles of the **oral cavity** and its associated structures in digestion:

   (a) ____

   (b) ____

2. Based on its position projecting up behind the tongue and guarding the tracheal entrance, infer the role of the epiglottis:

   ____

   ____

3. Explain the protective value of having tonsils at the oral entrance to the pharynx: ____

   ____

   ____

ISBN: 978-1-92717357-2

Moving Food Through the Digestive Tract

Food is moved through the digestive tract by a series of wave-like, smooth muscle contractions called **peristalsis**. Contraction superior to the mass of food (**bolus**) forces it along the digestive tract, while relaxation in the segment inferior to the bolus allows it to move forward. The gut wall has two layers of muscle which enable this movement. Inner circular muscles contract to narrow the tube, and outer longitudinal muscles contract to widen and shorten the tube. When one set of muscles contracts, the other relaxes. Digestive **sphincters** are circular muscular valves that close an orifice or passage in the digestive tract. They regulate the movement of food and prevent back-flow of the bolus.

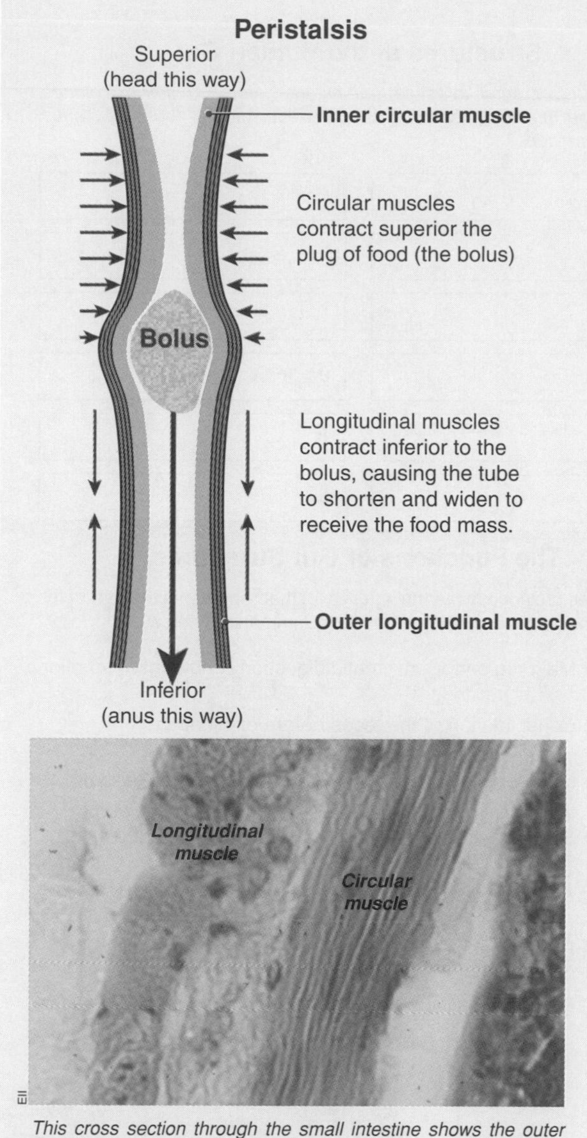

This cross section through the small intestine shows the outer longitudinal and inner circular muscles involved in peristalsis. In a cross sectional view, the longitudinal muscles have a circular appearance because they are viewed end on, while the circular muscle appear as long stands.

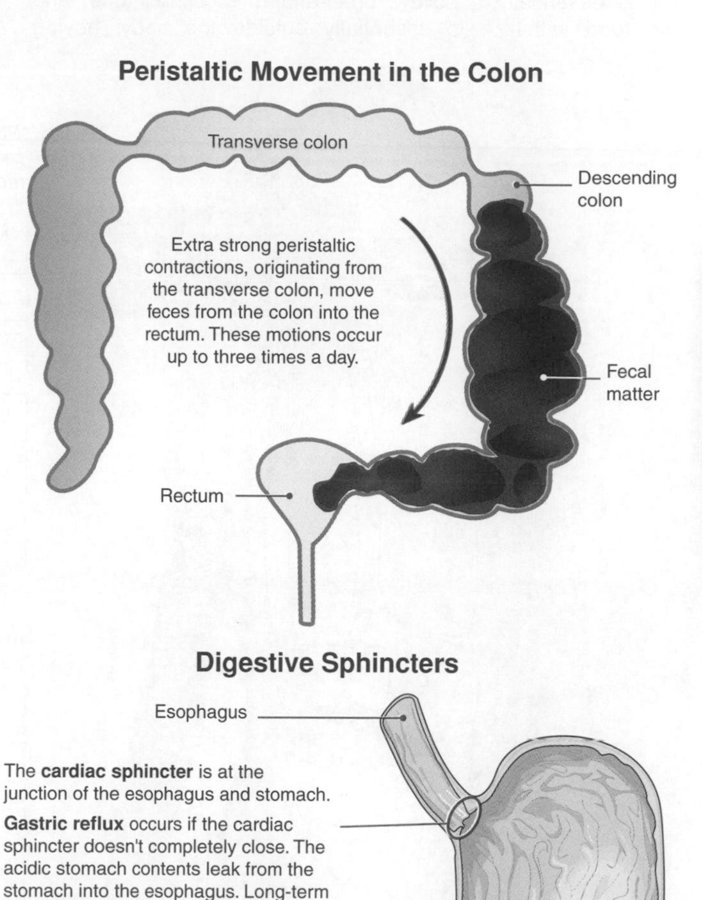

The Digestive System

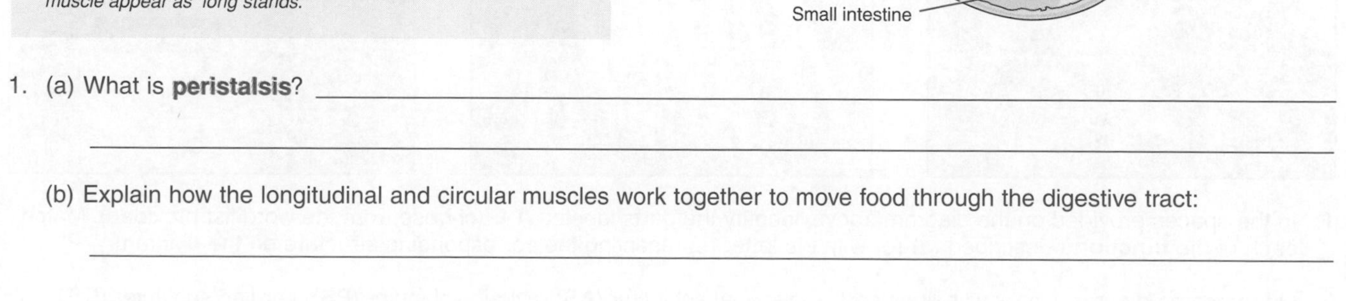

1. (a) What is **peristalsis**? ______________________________

   (b) Explain how the longitudinal and circular muscles work together to move food through the digestive tract:

   ____

2. (a) What is a **sphincter**? ______________________________

   (b) What is the function of a sphincter in the digestive system? ______________________________

ISBN: 978-1-92717357-2

Related activities: *The Digestive Tract*
Weblinks: *Peristalsis Animation, Peristalsis and Gastric Emptying*

The Digestive Tract

An adult consumes an estimated metric tonne of food a year. Food provides the source of the energy required to maintain **metabolism**. The digestive tract or **gut** is regionally specialized to maximize the efficiency of physical and chemical breakdown (**digestion**), **absorption**, and **elimination**. The gut is essentially a hollow, open-ended, muscular tube, and the food within it is essentially outside the body, having contact only with the cells lining the tract. Food is physically moved through the gut tube by waves of muscular contraction, and subjected to chemical breakdown by enzymes contained within digestive secretions. The products of this breakdown are then absorbed across the gut wall. A number of organs are associated with the gut along its length and contribute, through their secretions, to the digestive process at various stages.

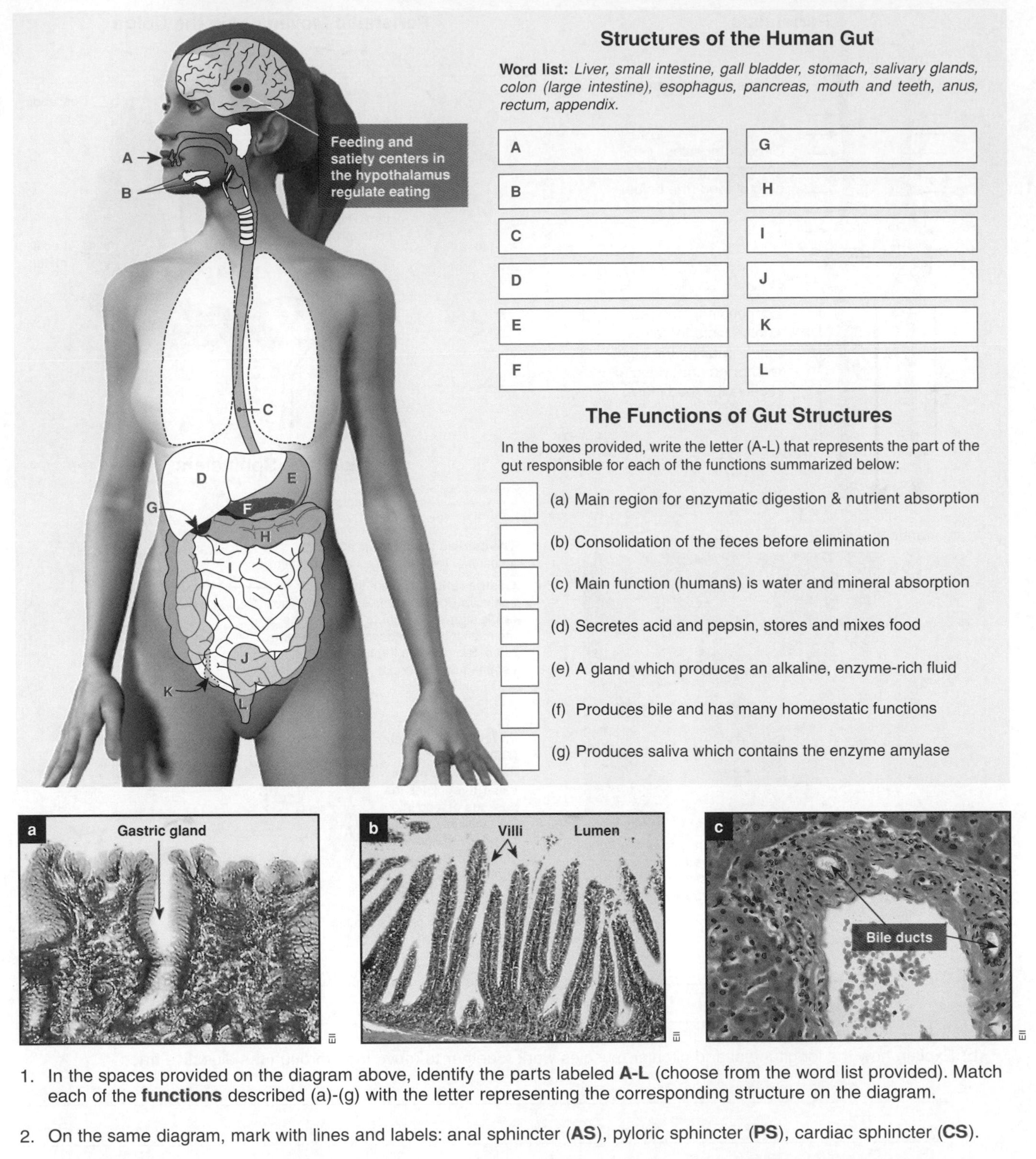

1. In the spaces provided on the diagram above, identify the parts labeled **A-L** (choose from the word list provided). Match each of the **functions** described (a)-(g) with the letter representing the corresponding structure on the diagram.

2. On the same diagram, mark with lines and labels: anal sphincter (**AS**), pyloric sphincter (**PS**), cardiac sphincter (**CS**).

3. Identify the region of the gut illustrated by the photographs (**a**)-(**c**) above. For each one, explain the identifying features:

   (a) ______________________________

   (b) ______________________________

   (c) ______________________________

Related activities: *Absorption and Transport*
Weblinks: *Digestion Animation, Interactive Digestion Quiz*

Periodicals:
The anatomy of digestion
The pancreas & pancreatitis

ISBN: 978-1-92717357-2

The Stomach, Duodenum, and Pancreas

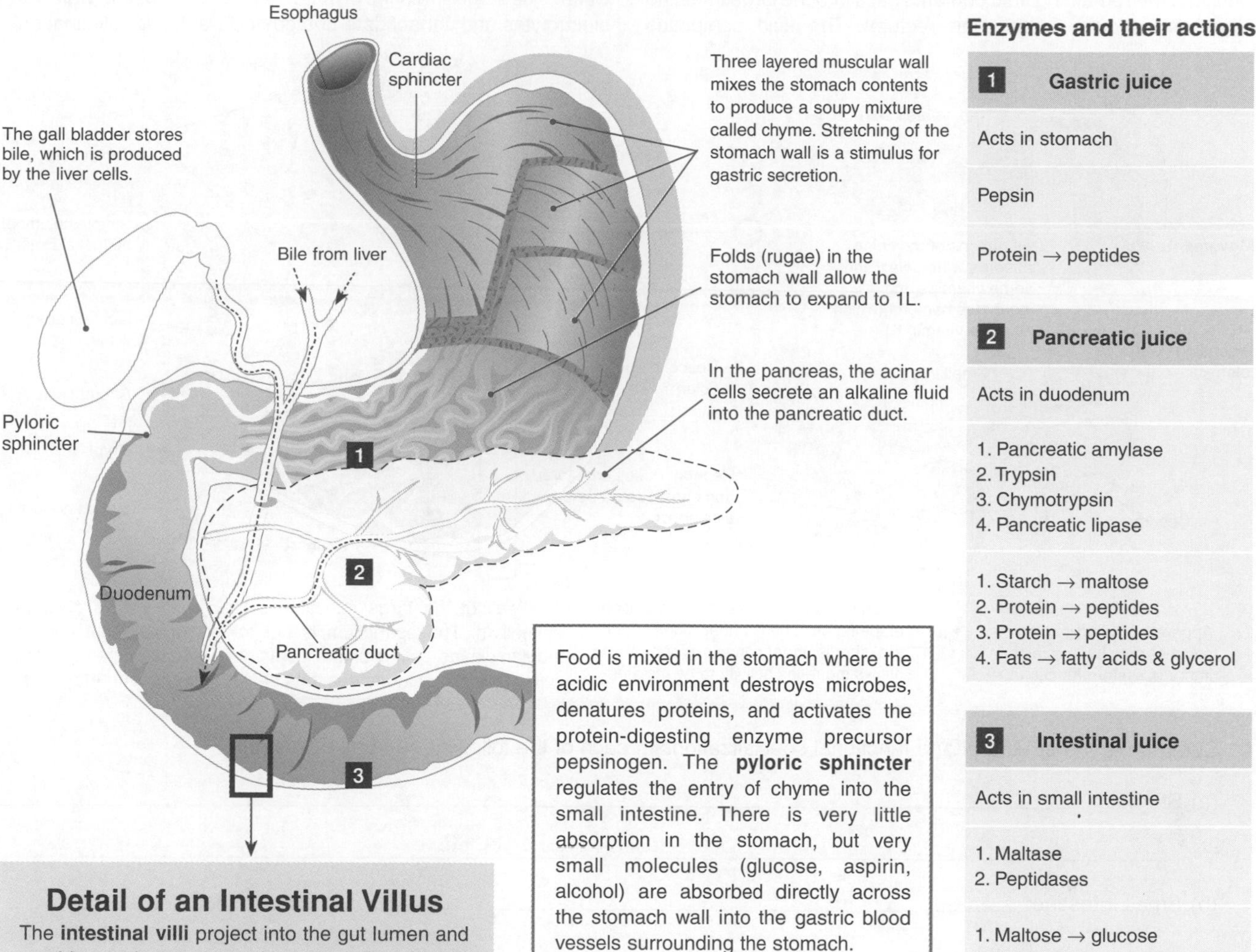

Food is mixed in the stomach where the acidic environment destroys microbes, denatures proteins, and activates the protein-digesting enzyme precursor pepsinogen. The **pyloric sphincter** regulates the entry of chyme into the small intestine. There is very little absorption in the stomach, but very small molecules (glucose, aspirin, alcohol) are absorbed directly across the stomach wall into the gastric blood vessels surrounding the stomach.

Enzymes and their actions

1 Gastric juice
Acts in stomach
Pepsin
Protein → peptides

2 Pancreatic juice
Acts in duodenum
1. Pancreatic amylase 2. Trypsin 3. Chymotrypsin 4. Pancreatic lipase
1. Starch → maltose 2. Protein → peptides 3. Protein → peptides 4. Fats → fatty acids & glycerol

3 Intestinal juice
Acts in small intestine
1. Maltase 2. Peptidases
1. Maltose → glucose 2. Polypeptides → amino acids

Detail of an Intestinal Villus

The **intestinal villi** project into the gut lumen and provide an immense surface area for nutrient absorption. The villi are lined with **epithelial cells** and each has a brush border of many **microvilli** which further increase the surface area.

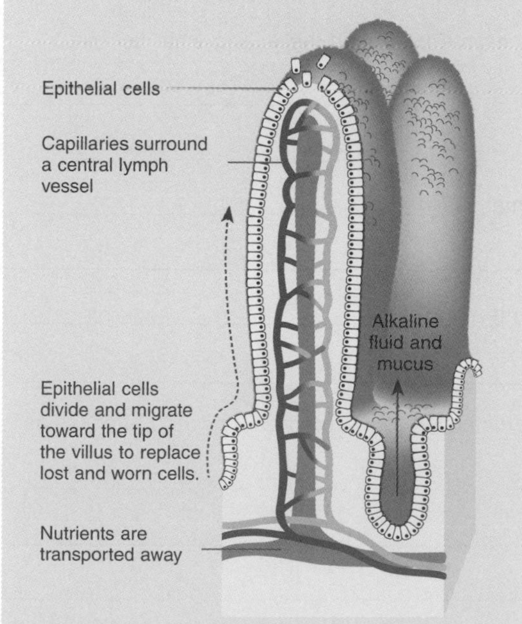

Enzymes bound to the surfaces of the epithelial cells break down peptides and carbohydrate molecules. The breakdown products are then absorbed into the underlying blood and lymph vessels. Tubular exocrine glands and goblet cells secrete alkaline fluid and mucus into the lumen.

Detail of a Gastric Gland (Stomach Wall)

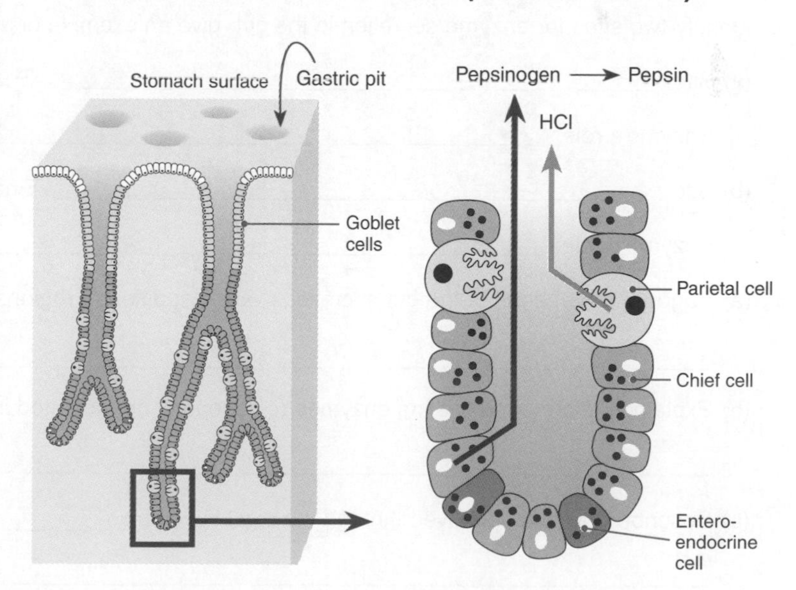

Gastric secretions are produced by **gastric glands**, which pit the lining of the stomach. Chief cells in the gland secrete pepsinogen, a precursor of the enzyme pepsin. Parietal cells produce hydrochloric acid, which activates the pepsinogen. Goblet cells at the neck of the gastric gland secrete mucus to protect the stomach mucosa from the acid. Enteroendocrine cells in the gastric gland secrete the hormone gastrin which acts on the stomach to increase gastric secretion.

ISBN: 978-1-92717357-2

The Large Intestine

After most of the nutrients have been absorbed in the small intestine, the remaining fluid contents pass into the large intestine (appendix, cecum, colon, and rectum). The fluid comprises undigested material, bacteria, dead cells, mucus, bile, ions, and water. The large intestine's main role is to reabsorb water and electrolytes and consolidate undigested wastes for elimination.

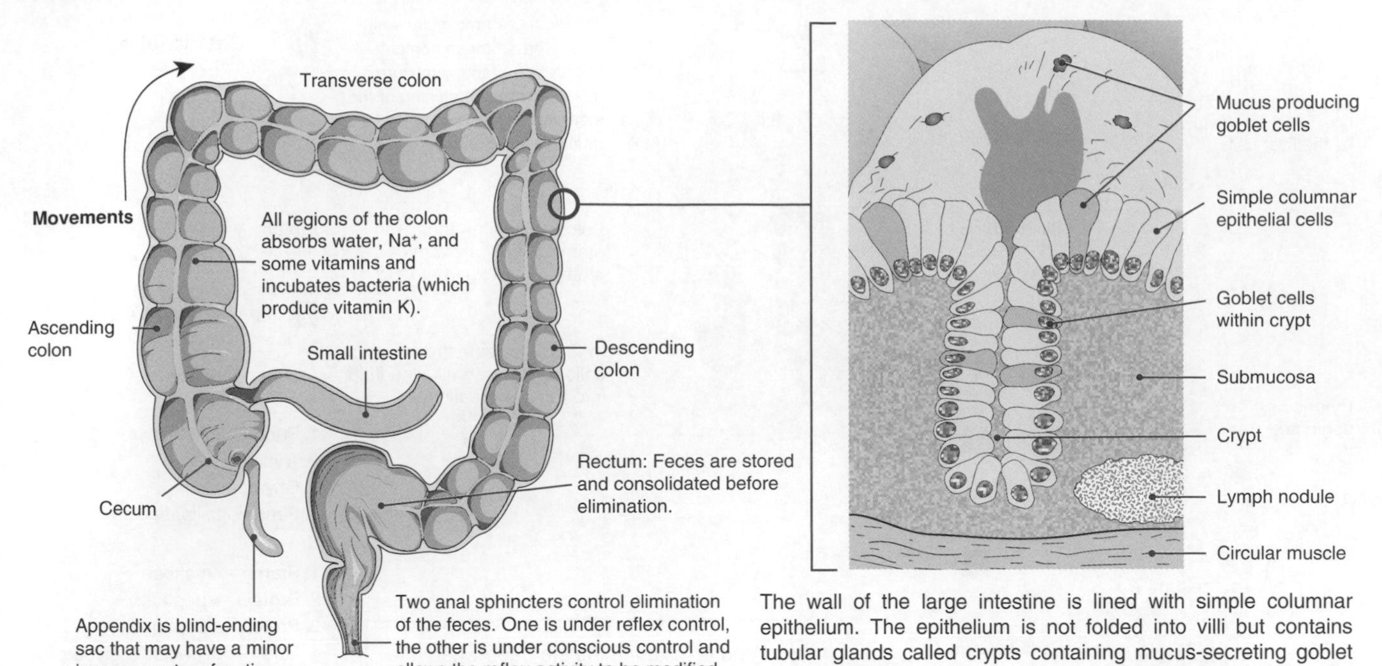

The wall of the large intestine is lined with simple columnar epithelium. The epithelium is not folded into villi but contains tubular glands called crypts containing mucus-secreting goblet cells. The mucus lubricates the colon and helps form the feces.

4. Summarize the structural and functional specializations in each of the following regions of the gut:

   (a) Stomach: ___

   (b) Small intestine: ___

   (c) Large intestine: ___

5. Identify two sites for enzyme secretion in the gut, give an example of an enzyme produced there, and state its role:

   (a) Site: ___ Enzyme: ___

   Enzyme's role: ___

   (b) Site: ___ Enzyme: ___

   Enzyme's role: ___

6. (a) Suggest why the pH of the gut secretions varies at different regions in the gut: ___

   (b) Explain why protein-digesting enzymes (e.g. pepsin) are secreted in an inactive form and then activated after release:

7. (a) Describe how food is moved through the digestive tract: ___

   (b) Explain how the passage of food through the tract is regulated: ___

8. (a) Predict the consequence of food moving too rapidly through the gut: ___

   (b) Predict the consequence of food moving too slowly through the gut: ___

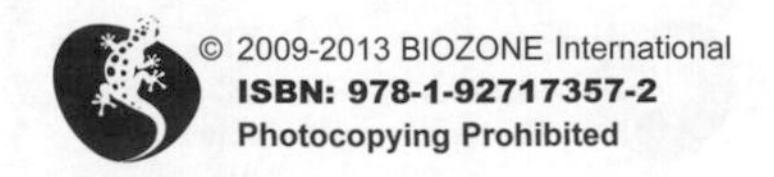

ISBN: 978-1-92717357-2

Absorption and Transport

All the chemical and physical processes of digestion from the mouth to the small intestine are aimed at the breakdown of food molecules into forms that can pass through intestinal lining into the underlying blood and lymph vessels. These breakdown products include monosaccharides, amino acids, fatty acids, glycerol, and glycerides. Passage of these molecules from the gut into the blood or lymph is called **absorption**. After absorption, nutrients are transported directly or indirectly to the liver for storage or processing. Some of the features of nutrient absorption and transport are shown below. For simplicity, all nutrients are shown in the lumen of the intestine, even though some nutrients are digested on the epithelial cell surfaces.

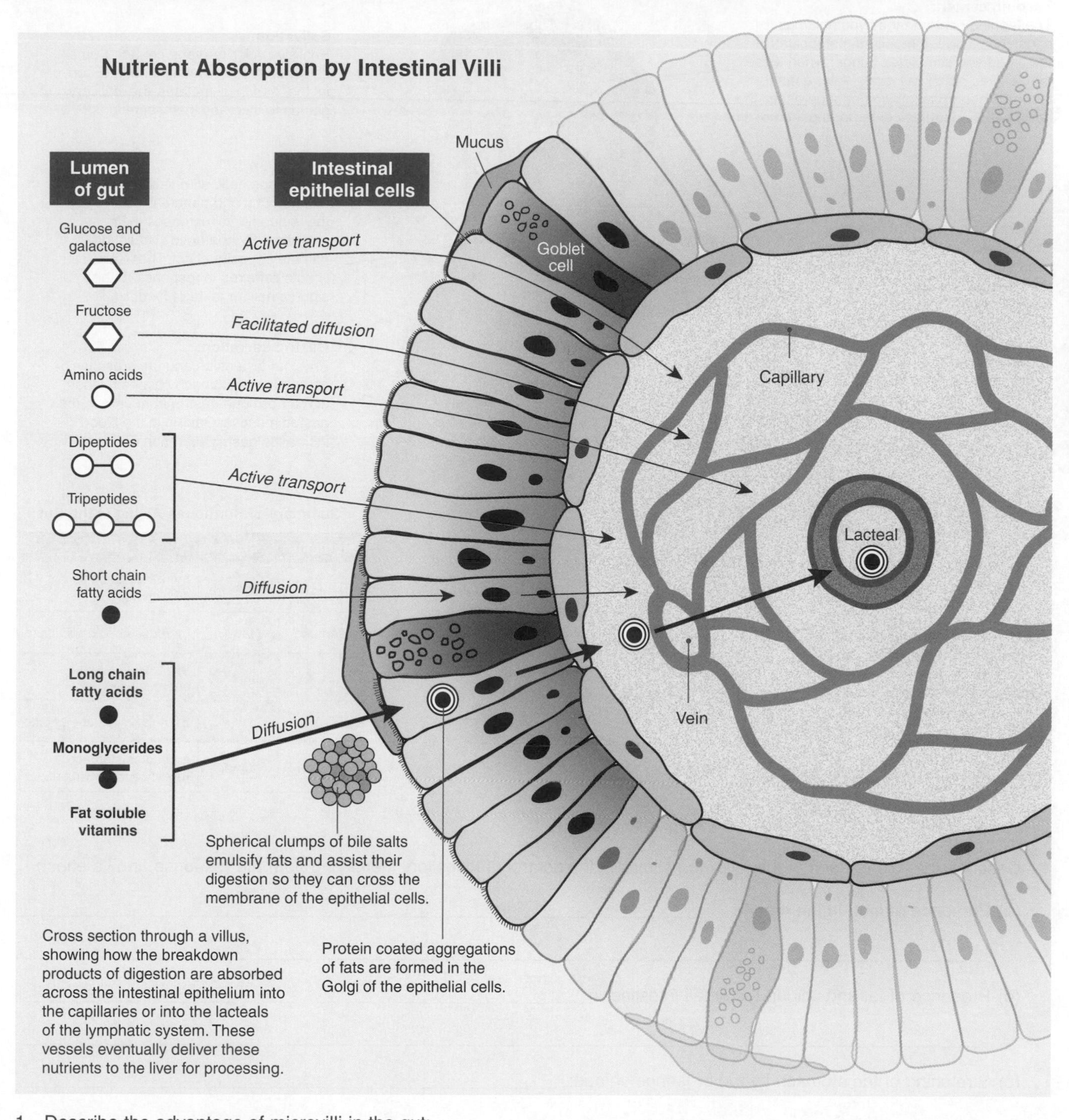

Cross section through a villus, showing how the breakdown products of digestion are absorbed across the intestinal epithelium into the capillaries or into the lacteals of the lymphatic system. These vessels eventually deliver these nutrients to the liver for processing.

1. Describe the advantage of microvilli in the gut: ______

______

2. Describe how each of the following nutrients is absorbed by the intestinal villi in mammals:

(a) Glucose and galactose: ______

(b) Fructose: ______

(c) Amino acids: ______

(d) Dipeptides: ______

(e) Tripeptides: ______

(f) Short chain fatty acids: ______

(g) Monoglycerides: ______

(h) Fat soluble vitamins: ______

ISBN: 978-1-92717357-2

Related activities: *The Digestive Tract, The Liver*
Weblinks: *Digestion Animation, Fatty Acid Metabolism*

The Control of Digestion

The majority of digestive juices are secreted only when there is food in the gut and both nervous and hormonal mechanisms are involved in coordinating and regulating this activity appropriately. The digestive system is innervated by branches of the **autonomic nervous system**. Hormonal regulation is achieved through the activity of several hormones, which are released into the bloodstream in response to nervous or chemical stimuli and influence the activity of gut and associated organs.

Feeding center:
The feeding center in the hypothalamus is constantly active. It monitors metabolites in the blood and stimulates hunger when these metabolites reach low levels. After a meal, the neighboring satiety center suppresses the activity of the feeding center for a period of time.

Salivation:
Entirely under nervous control. Some saliva is secreted continuously. Food in the mouth stimulates the salivary glands to increase their secretions.

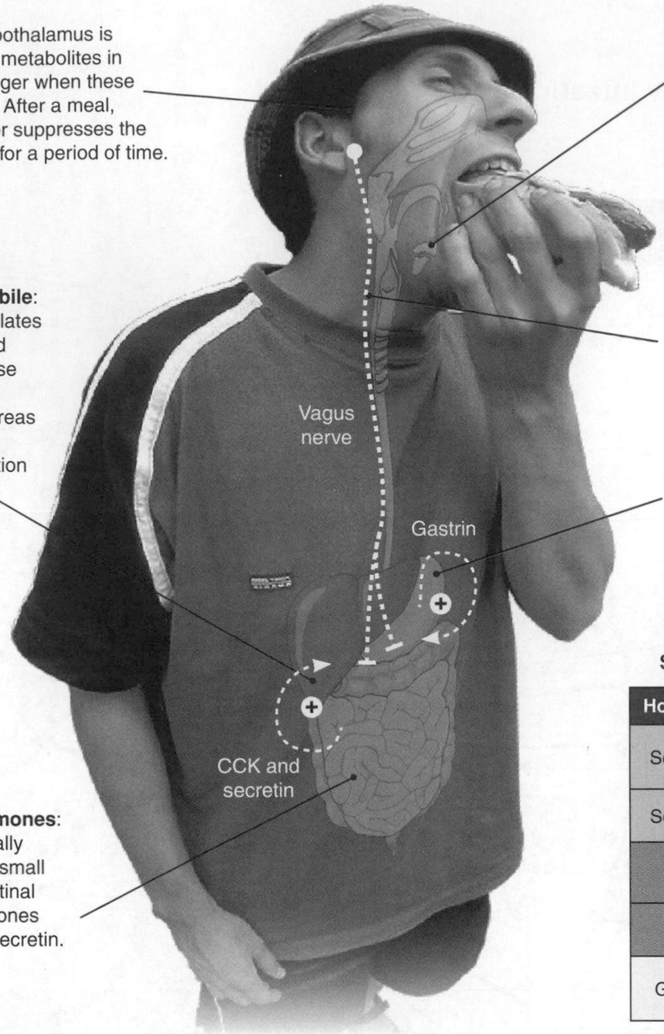

Parasympathetic stimulation of the stomach and pancreas via the vagus nerve increases their secretion. Sympathetic stimulation has the opposite effect. These are simple **reflexes** in response to the sight, smell, or taste of food.

Pancreatic secretions and bile:
Cholecystokinin (CCK) stimulates secretion of enzyme-rich fluid from the pancreas and release of bile from the gall bladder. Secretin stimulates the pancreas to increase its secretion of alkaline fluid and the production of bile from the liver cells.

Gastric secretion:
Physical distension and the presence of food in the stomach causes release of the hormone gastrin from cells in the gastric mucosa. Gastrin in the blood increases gastric secretion and motility.

Intestinal secretion of hormones:
The entry of **chyme** (especially fat and gastric acid) into the small intestine stimulates the intestinal mucosa to secrete the hormones cholecystokinin (CCK) and secretin.

Summary of Hormones Acting in the Gut

Hormone	Organ	Effect
Secretin	Pancreas	Increases secretion of alkaline fluid
Secretin	Liver	Increases bile production
CCK	Pancreas	Increases enzyme secretion
CCK	Liver	Stimulates release of bile
Gastrin	Stomach	Increases stomach motility and secretion

1. Describe the role of each of the following stimuli in the control of digestion, identifying both the response and its effect:

   (a) Presence of food in the mouth: __________

   (b) Presence of fat and acid in the small intestine: __________

   (c) Stretching of the stomach by the presence of food: __________

2. Outline the role of the vagus nerve in regulating digestive activity: __________

3. Discuss the role of nerves and hormones in controlling digestion: __________

Related activities: The Digestive Tract, the Liver

Weblinks: Three Phases of Gastric Secretion, Role of CCK

Periodicals:
Alimentary thinking

ISBN: 978-1-92717357-2

The Liver

The liver the largest homeostatic organ. It is located just below the diaphragm and makes up 3-5% of body weight. It performs a vast number of functions including production of bile, storage and processing of nutrients, and detoxification of poisons and metabolic wastes. The liver receives a dual blood supply from the hepatic portal vein and hepatic arteries, and up to 20% of the total blood volume flows through it at any one time. This rich vascularization makes it the central organ for regulating activities associated with the blood and circulatory system. In spite of its many functions, the liver tissue and the liver cells themselves are structurally relatively simple. Features of liver structure and function are outlined below.

Role of the Liver in Digestion and Nutrient Processing

- Liver cells produce (secrete) bile, important in emulsifying fats in digestion.
- Metabolizes amino acids, fats, and carbohydrates.
- Synthesizes glucose from non-carbohydrate sources when glycogen stores are exhausted (gluconeogenesis).
- Stores iron, copper, and some vitamins (A, D, E, K, B_{12}).
- Synthesizes cholesterol from acetyl coenzyme A.
- Converts unwanted amino acids to urea (ornithine cycle).
- Manufactures heparin and plasma proteins (e.g. albumin).

Other Homeostatic Functions of the Liver

- Detoxifies poisons or turns them into less harmful forms.
- Some liver cells phagocytose worn-out blood cells.

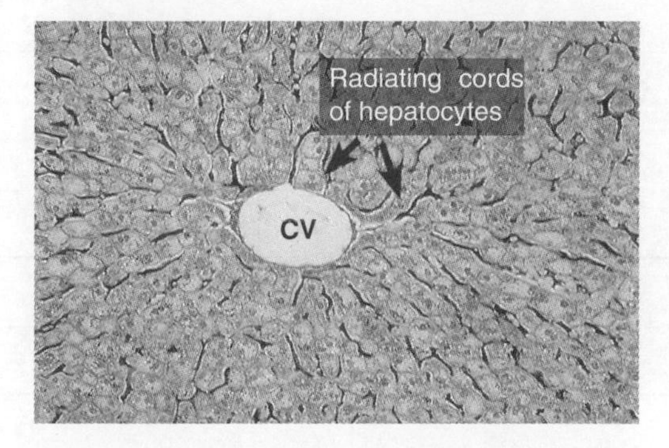

The photograph above shows the histology of a **liver lobule**, with the central vein, cords of hepatocytes (liver cells), and sinusoids (dark spaces).

The Gross Structure of the Liver

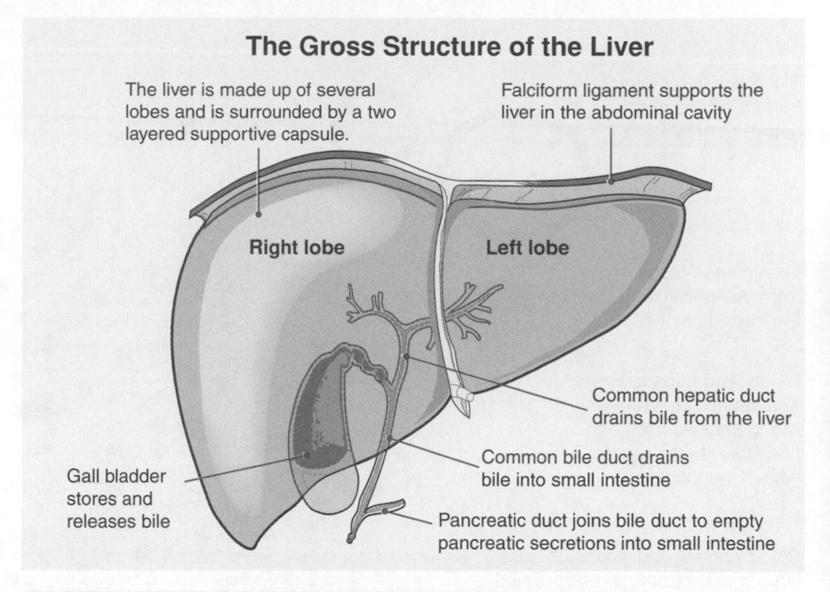

The Internal Structure of the Liver

Schematic of the arrangement of lobules and portal triad in the liver

Bile ductule
Branch of hepatic portal vein
Branch of hepatic artery
Portal triad (tract)
Bile flow
Blood flow towards the central vein
Central vein
Central vein
Fibrous connective tissue capsule surrounds lobules
Lobule

The liver is divided into functional units called lobules with rows of liver cells arranged around a **central vein**. Branches of the hepatic artery, hepatic portal vein, and bile duct lie between the lobules, forming a **portal triad** (tract).

1. What cells produce bile? ______________________

2. Identify the components of a portal triad: ______________________

3. (a) Name one vascular function of the liver: ______________________

   (b) Name two metabolic functions of the liver: ______________________

   ____

   (c) Name one digestive function of the liver: ______________________

   (d) Name one excretory function of the liver: ______________________

   (e) Name on storage function of the liver: ______________________

ISBN: 978-1-92717357-2

Periodicals:
Metabolic powerhouse

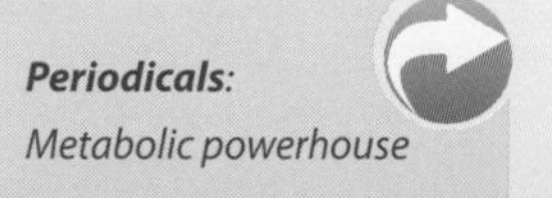

***Related activities**: The Histology of the Liver*

The Histology of the Liver

The functional repeating unit of the liver is the **lobule**, which is made up of tightly packed rows (cords) of liver cells radiating from a central vein and surrounded by small blood vessels called sinusoids. Branches of the hepatic artery and the hepatic portal vein supply the lobules. This highly vascular structure is a reflection of the liver's important role as dynamic blood reservoir, able to both store and release blood as required. More than half of the 10-20% of the total blood volume normally in the liver resides in the sinusoids. Sinusoids are similar to capillaries but have a more porous endothelium. The increased permeability of the sinusoids allows small and medium-sized proteins, such as albumin, to readily enter and leave the bloodstream.

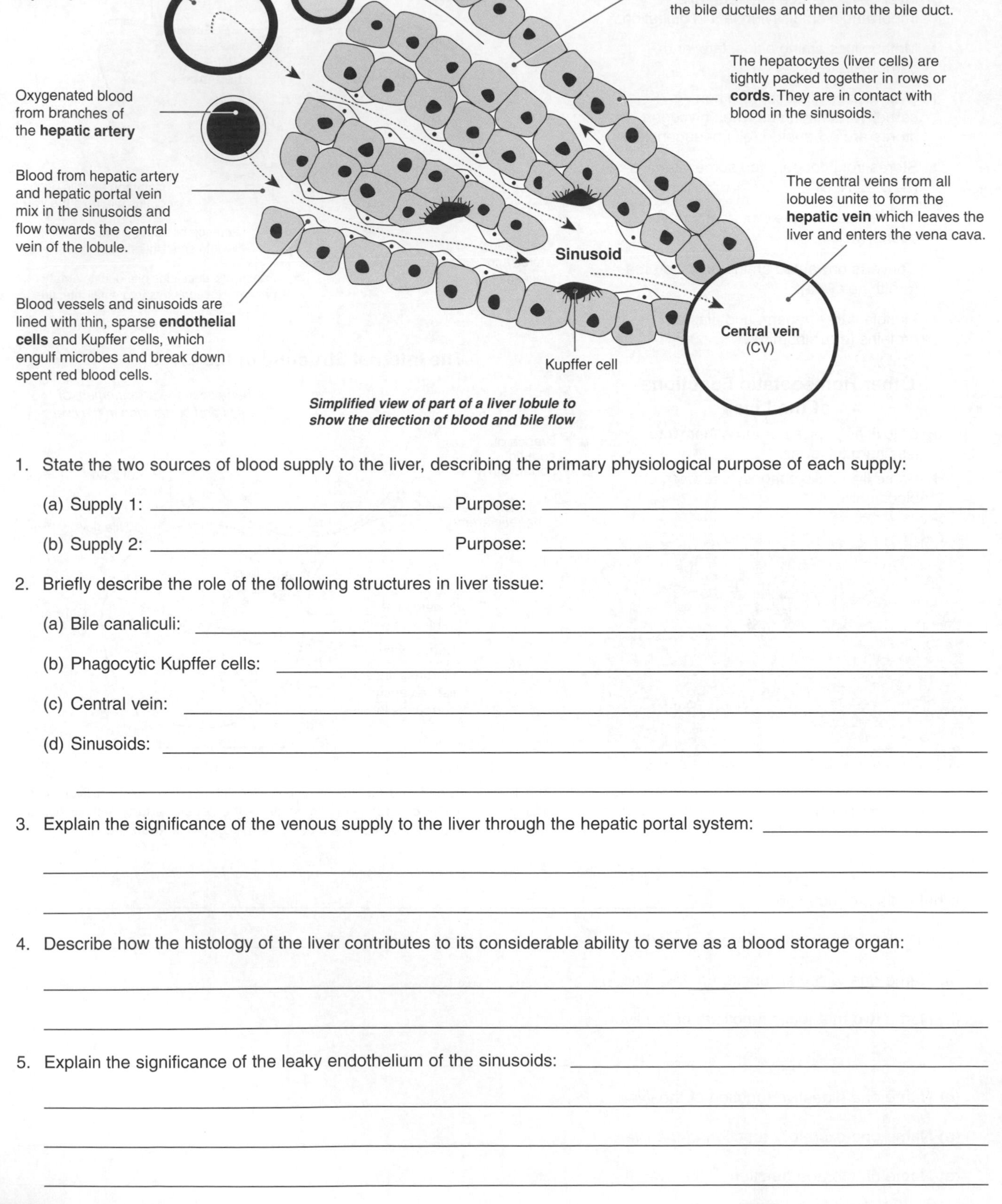

Simplified view of part of a liver lobule to show the direction of blood and bile flow

1. State the two sources of blood supply to the liver, describing the primary physiological purpose of each supply:

   (a) Supply 1: ______________________ Purpose: ______________________

   (b) Supply 2: ______________________ Purpose: ______________________

2. Briefly describe the role of the following structures in liver tissue:

   (a) Bile canaliculi: ______________________

   (b) Phagocytic Kupffer cells: ______________________

   (c) Central vein: ______________________

   (d) Sinusoids: ______________________

3. Explain the significance of the venous supply to the liver through the hepatic portal system: ______________________

4. Describe how the histology of the liver contributes to its considerable ability to serve as a blood storage organ:

5. Explain the significance of the leaky endothelium of the sinusoids: ______________________

ISBN: 978-1-92717357-2

Related activities: *The Liver*

Protein Metabolism in the Liver

The liver has a crucial role in the metabolism of proteins and the storage and detoxification of hormones and ingested or absorbed poisons (including alcohol). The most critical aspects of protein metabolism occurring in the liver are deamination and transamination of amino acids, removal of ammonia from the body by synthesis of urea, and synthesis of non-essential amino acids. Hepatocytes are responsible for synthesis of most of the plasma proteins, including albumins, globulins, and blood clotting proteins. Urea formation via the ornithine cycle occurs primarily in the liver. The urea is formed from ammonia and carbon dioxide by condensation with the amino acid ornithine, which is recycled through a series of enzyme-controlled steps.

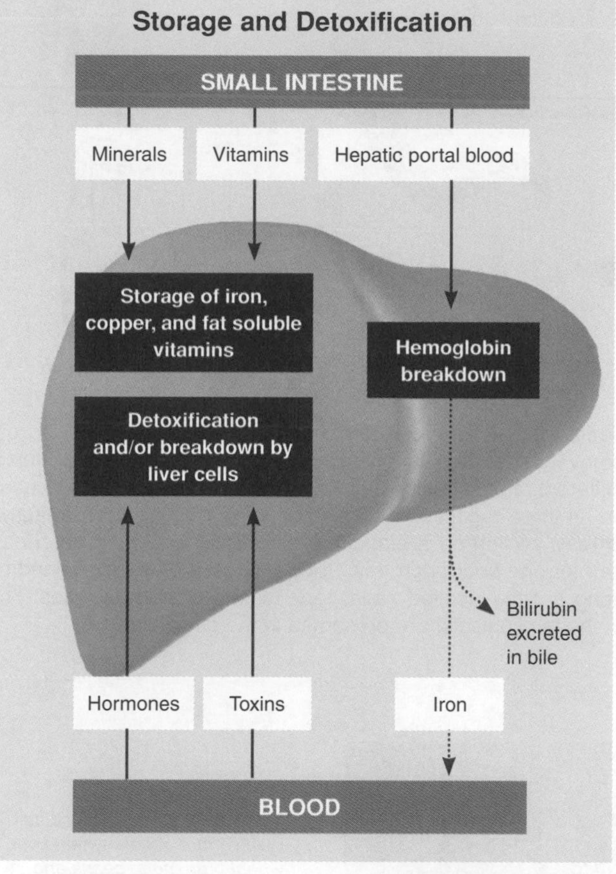

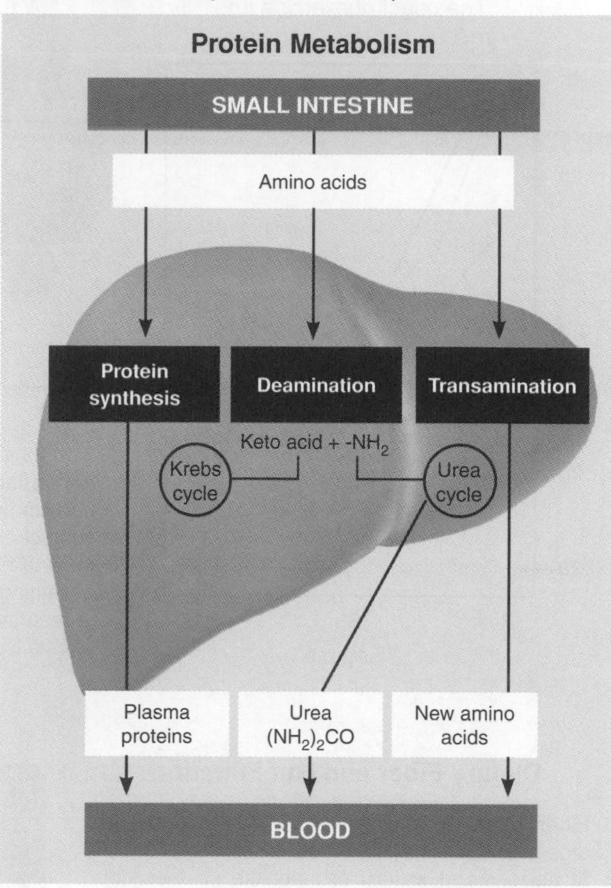

1. Explain three aspects of protein metabolism in the liver:

(a) ______________________________

(b) ______________________________

(c) ______________________________

2. Identify the waste products arising from deamination of amino acids and describe their fate:

____

3. An X-linked disorder of the ornithine cycle results in sufferers lacking the enzyme to convert ornithine to citrulline. Suggest what the symptoms and the prognosis might be:

____

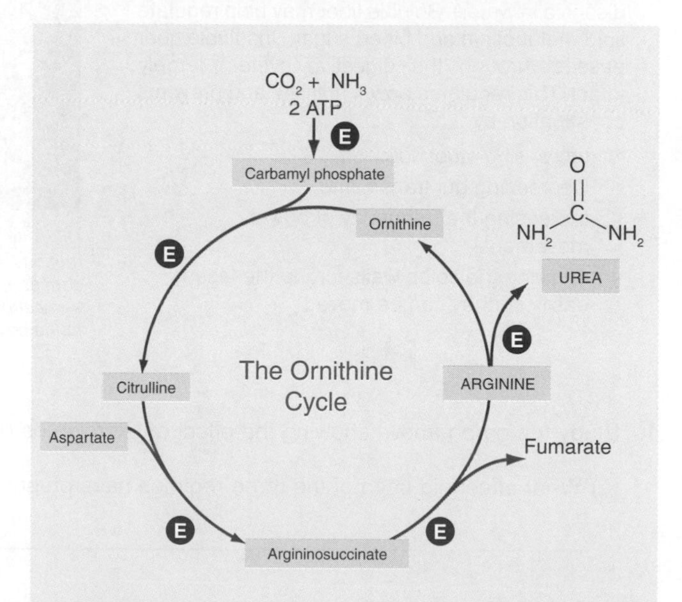

Ammonia (NH_3), the product of protein metabolism, is toxic in even small amounts and must be removed. It is converted to the less toxic urea via the ornithine cycle and is excreted from the body by the kidneys. The liver contains a system of carrier molecules and enzymes (E) which quickly convert the ammonia (and CO_2) into urea. One turn of the cycle consumes two molecules of ammonia (one comes from aspartate) and one molecule of CO_2, creates one molecule of urea, and regenerates a molecule of ornithine.

ISBN: 978-1-92717357-2

Periodicals: *The liver in health and disease, Urea*

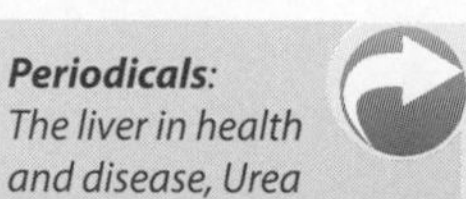

Related activities: *Carbohydrate Metabolism in the Liver*

Exercise, Fiber, and Gut Function

Exercise has both short and long term effects on the function of the digestive system. Regular light to moderate exercise has several advantages including strengthening the muscles which support the digestive system and stimulating intestinal contractions. However, people undertaking strenuous and prolonged exercise can suffer from digestive disorders such as vomiting, diarrhea, and reflux. Exercise and dietary composition influence the time taken for food to move through the digestive tract.

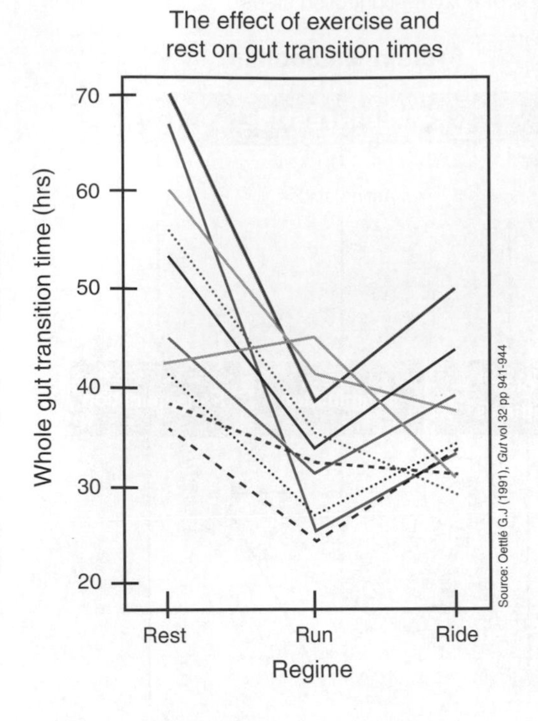

Rest
Gut transit time: 51.2 hrs

Run
Gut transit time: 34 hrs

Ride
Gut transit time: 36.6 hrs

Ten subjects were involved in an study to determine if exercise influenced gut transit time. The study was divided into three one week periods, and each subject participated in a different regime each week. Over the experimental period, each subject completed all three regimes. The regimes were running on a treadmill for one hour each day, cycling on a stationary bicycle for one hour each day, or resting in a chair for one hour each day. Gut transit time was determined by measuring how long a radio-labelled pellet took to be egested as feces. The results for each of the ten subjects, are presented in the graph (left).

Dietary Fiber and Gut Function

Fiber refers to plant material that cannot be digested. Fiber is classified as either soluble (it dissolves in water), or insoluble (it does not dissolve in water). Soluble fiber may help regulate lipid metabolism and blood sugar. Insoluble fiber passes through the digestive system largely intact. This regulates bowel activity and prevents constipation by:

- Increasing stool bulk
- Decreasing gut transit time
- Increasing the frequency of bowel movements
- Allowing the colon walls to grip the feces easily so they can be moved

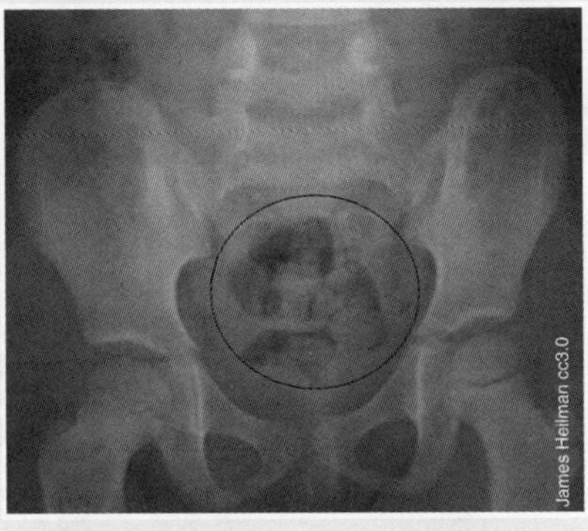

Difficulty or infrequency in passing bowel motions is called constipation. It has many causes including a lack of dietary fiber. The x-ray above shows hard and compacted feces in the colon a patient with constipation.

Most plant-based foods contain dietary fiber. Whole-wheat flour, wheat bran, nuts, beans and vegetables, are good sources of insoluble fiber.

1. Study the graph above showing the effect of exercise on gut transit times.

   (a) What effect did each of the three regimes have on gut transit times? ______

   (b) Explain how exercise may have influenced gut transition times: ______

2. Why is dietary fiber important in maintaining healthy gut function? ______

ISBN: 978-1-92717357-2

Related activities: A Balanced Diet, The Digestive Tract

Malnutrition and Obesity

Malnutrition is the term for nutritional disorders resulting from not having enough food or not enough of the right food. In economically developed areas of the world, most (but not all) forms of malnutrition are the result of poorly balanced nutrient intakes rather than a lack of food *per se*. Amongst the most common of these is **obesity**, defined as BMI values in excess of 30 (see below). Although some genetic and hormonal causes are known, obesity is commonly the result of excessive energy intake, usually associated with a highly processed diet, high in fat and sugar. In addition, incidental physical activity is declining: we drive more, use labor-saving machines, and exercise less. Obesity is a risk factor in a number of chronic diseases, including hypertension, cardiovascular disease, and type 2 diabetes. Obesity in developed countries is more common in poorly educated, lower socio-economic groups than amongst the wealthy, who often have more options in terms of food choices.

Obesity and Malnutrition

In adults, the level of obesity is determined by reference to the Body Mass Index (BMI). A score of 30+ on the BMI indicates mild obesity, while those with severe or morbid obesity have BMIs of 40+. Child obesity is based on BMI-for-age, and is assessed in relation to the weight of other children of a similar age and gender. Central or abdominal obesity refers to excessive fat around the abdomen and is now classified as an independent risk factor for some serious diseases. While the explanation for excessive body fat is simple (energy in exceeds energy out), a complex of biological and socio-economic factors are implicated in creating the problems of modern obesity.

Obesity Prevalence

In 2010, an estimated 35% of adult Americans were obese and a similar number were overweight. The prevalence of obesity (BMI > 30) continues to be a health concern for adults, children, and adolescents in the US (and other developed countries), as indicated by self reported BMI assessment data below.

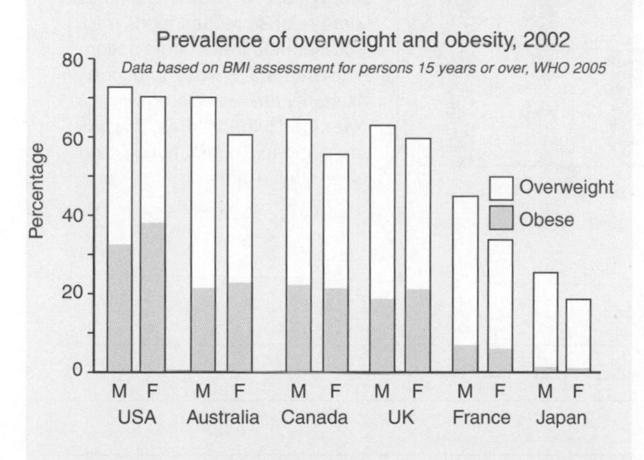

Recent data show clear differences in the percentage of overweight and obese adults between Australia, Canada, the USA, and the UK (which are similar) and Japan and France, where obesity is much less common. These differences are attributable largely to customary dietary differences between these nations.

Health Effects of Obesity

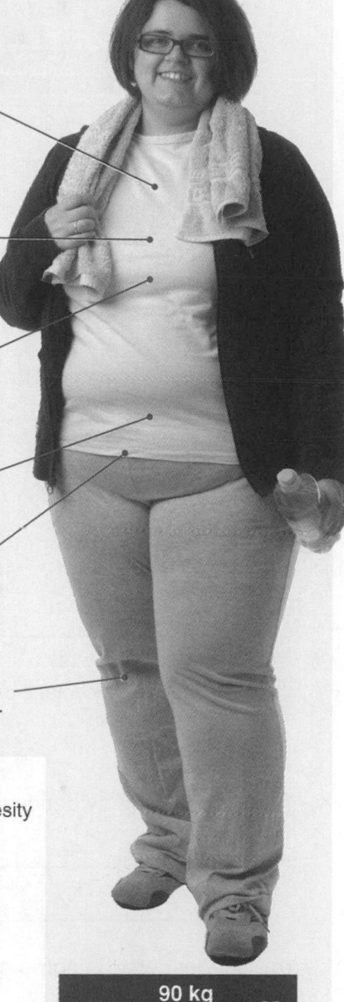

Obesity more than doubles the risk of hypertension and stroke.

Obesity is a major independent risk factor for cardiovascular disease because it is associated with an increased prevalence of cardiovascular risk factors, including **type 2 diabetes** and **high blood lipids**.

The heaviness of the chest wall and a higher-than-normal oxygen requirement in obese people restricts normal physical activity and increases respiratory problems.

Obesity is associated with high bile cholesterol levels, gallstones and gall bladder disease.

Obesity is clearly associated with higher risk of certain types of cancers, including rectal, colon and breast cancer. Cancer survival rates are also lower among obese patients.

Obesity in pre-menopausal women is associated with irregular menstrual cycles and infertility.

Obese people are at higher risk of osteoarthritis in their weight-bearing joints.

$$\text{BMI} = \frac{90 \text{ kg}}{(1.68)^2} = 32$$

Body Mass Index

A common method for assessing obesity is the **body mass index** (BMI).

$$\text{BMI} = \frac{\text{weight of body (in kg)}}{\text{height (in metres) squared}}$$

A BMI of:
- 17 to 20 = underweight
- 20 to 25 = normal weight
- 25 to 30 = overweight
- over 30 = obesity

1. (a) Explain why obesity is regarded as a form of malnutrition: ______________________

   (b) Describe the two basic energy factors that determine how a person's weight will change: ______________________

2. Using the BMI, calculate the minimum and maximum weight at which a 1.85 m tall man would be considered:

   (a) Overweight: ______________ (b) Obese: ______________

3. BMI is routinely used to assess healthy weight. Explain why BMI might sometimes not be a reliable in this respect:

   ____

ISBN: 978-1-92717357-2

Related activities: *Deficiency Diseases, Cardiovascular Disease, Type 2 Diabetes*

The Problem of the Malnourished Obese

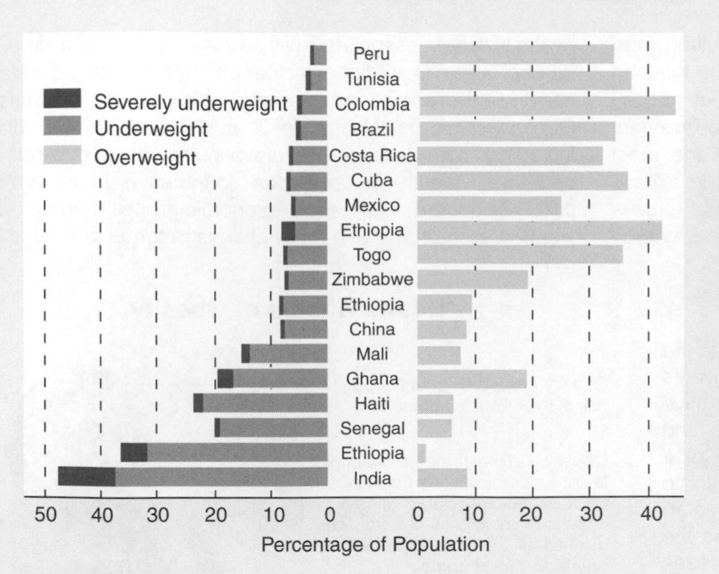

Increasingly, obesity is a problem for the world's poor, who often rely on nutritionally poor carbohydrate staples and are deficient in the micronutrients required for good health. A 1999 United Nations study found obesity to be growing rapidly in all developing countries, even in countries where hunger exists, such as China. The well-intentioned policies of some countries can compound the problem. For example, the Egyptian Food Subsidy Programme, which reduced the relative prices of energy-dense, nutrient-poor food items, is one of the major factors contributing to the emergence of obese and micronutrient deficient mothers in the country.

Left: Weight profiles in selected developing nations. WHO 1997, via FAO

Below: Percentage obesity in relation to income and education. American Journal of Clinical Nutrition, 2004

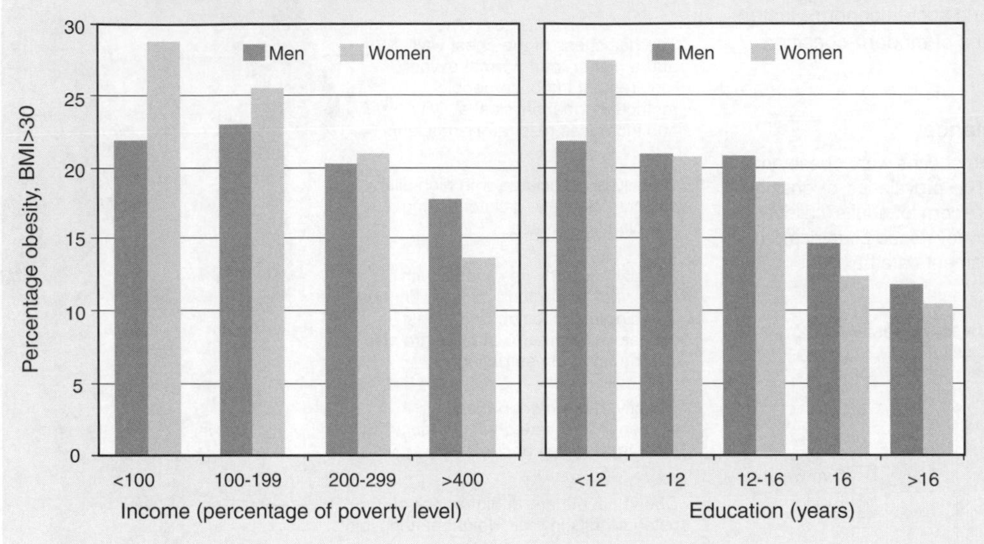

In both developed and developing countries, many disparities in health are linked to inequalities in education and income. The highest rates of obesity occur amongst the most poorly educated, impoverished sectors of the population. In addition, there is an inverse relation between energy density (MJ per kg) and energy cost ($ per MJ), such that energy-dense refined foods may represent the lowest-cost option to consumers. Poverty and food insecurity are associated with lower food expenditures, low fruit and vegetable consumption, and lower-quality diets.

4. (a) Describe the evidence linking obesity to economic hardship: ______

(b) Suggest why, in some countries, obesity is associated with a lack of education: ______

(c) Explain how these factors could also explain the rising incidence of obesity in developing nations: ______

5. Discuss the factors that contribute to micronutrient deficiencies among the obese: ______

ISBN: 978-1-92717357-2

Deficiency Diseases

A deficiency disease can occur when there is an inadequate intake of a specific nutrient. Vitamin and mineral deficiencies are the most common but, in developing countries, protein and energy deficiencies also cause disease. Specific vitamin and mineral deficiencies are associated with specific disorders, e.g. scurvy (vitamin C), rickets (vitamin D), visual defects (vitamin A), osteoporosis (calcium), or anemia (iron). In developed countries, deficiency diseases were once limited to people with very restricted diets, intestinal disorders, or drug and alcohol problems. However, some deficiencies, e.g. iron or vitamin D deficiencies, are relatively more common. This is mainly because diets often lack the nutrients in the quantities required to maintain good health and normal development. Children and pregnant women are particularly susceptible to nutritional deficiencies because of their specific metabolic requirements.

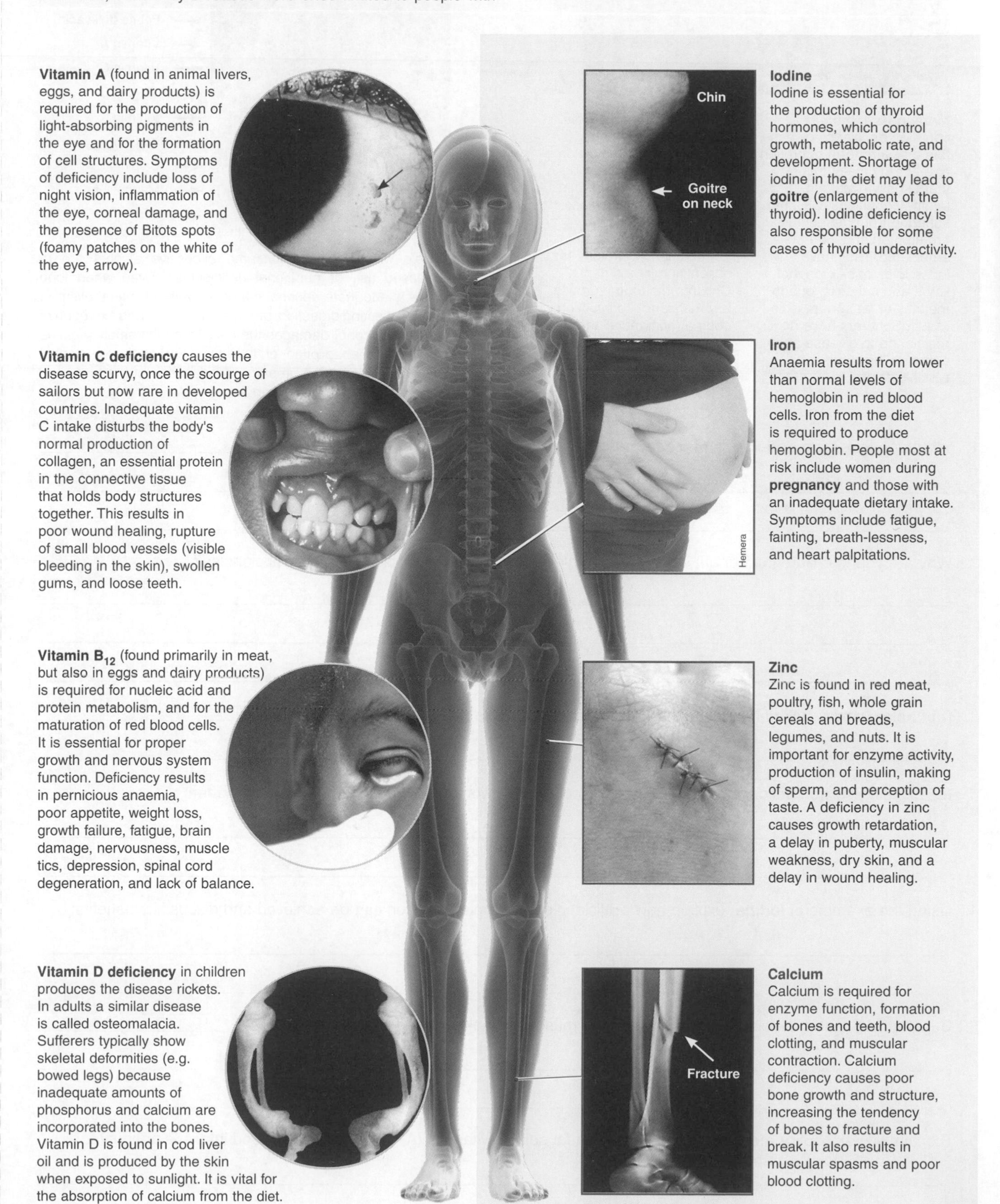

Vitamin A (found in animal livers, eggs, and dairy products) is required for the production of light-absorbing pigments in the eye and for the formation of cell structures. Symptoms of deficiency include loss of night vision, inflammation of the eye, corneal damage, and the presence of Bitots spots (foamy patches on the white of the eye, arrow).

Vitamin C deficiency causes the disease scurvy, once the scourge of sailors but now rare in developed countries. Inadequate vitamin C intake disturbs the body's normal production of collagen, an essential protein in the connective tissue that holds body structures together. This results in poor wound healing, rupture of small blood vessels (visible bleeding in the skin), swollen gums, and loose teeth.

Vitamin B_{12} (found primarily in meat, but also in eggs and dairy products) is required for nucleic acid and protein metabolism, and for the maturation of red blood cells. It is essential for proper growth and nervous system function. Deficiency results in pernicious anaemia, poor appetite, weight loss, growth failure, fatigue, brain damage, nervousness, muscle tics, depression, spinal cord degeneration, and lack of balance.

Vitamin D deficiency in children produces the disease rickets. In adults a similar disease is called osteomalacia. Sufferers typically show skeletal deformities (e.g. bowed legs) because inadequate amounts of phosphorus and calcium are incorporated into the bones. Vitamin D is found in cod liver oil and is produced by the skin when exposed to sunlight. It is vital for the absorption of calcium from the diet.

Iodine
Iodine is essential for the production of thyroid hormones, which control growth, metabolic rate, and development. Shortage of iodine in the diet may lead to **goitre** (enlargement of the thyroid). Iodine deficiency is also responsible for some cases of thyroid underactivity.

Iron
Anaemia results from lower than normal levels of hemoglobin in red blood cells. Iron from the diet is required to produce hemoglobin. People most at risk include women during **pregnancy** and those with an inadequate dietary intake. Symptoms include fatigue, fainting, breath-lessness, and heart palpitations.

Zinc
Zinc is found in red meat, poultry, fish, whole grain cereals and breads, legumes, and nuts. It is important for enzyme activity, production of insulin, making of sperm, and perception of taste. A deficiency in zinc causes growth retardation, a delay in puberty, muscular weakness, dry skin, and a delay in wound healing.

Calcium
Calcium is required for enzyme function, formation of bones and teeth, blood clotting, and muscular contraction. Calcium deficiency causes poor bone growth and structure, increasing the tendency of bones to fracture and break. It also results in muscular spasms and poor blood clotting.

All photos CDC unless indicated otherwise

ISBN: 978-1-92717357-2

Related activities: *A Balanced Diet, Malnutrition and Obesity*

Protein and Energy Deficiencies

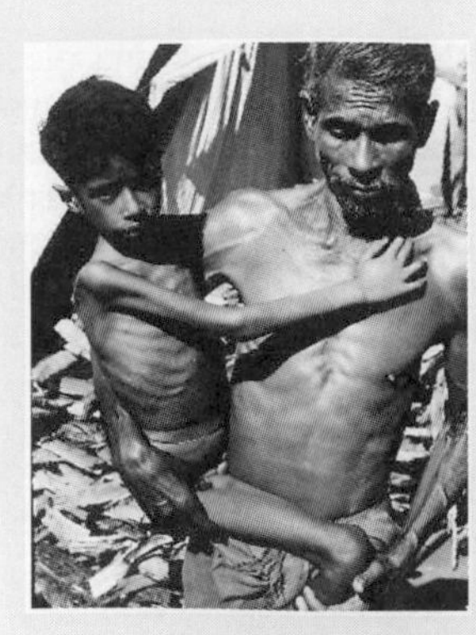

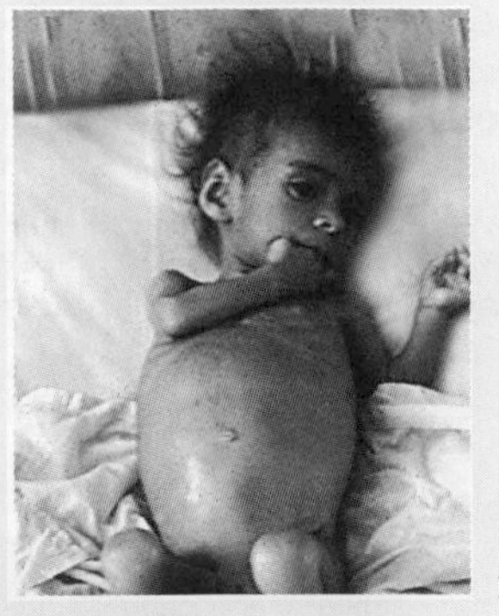

J. Armstrong

Marasmus is the most common deficiency disease. It is a severe form of protein and energy malnutrition that usually occurs in famine conditions. Children suffering from marasmus are stunted and extremely emaciated. They have loose folds of skin on the limbs and buttocks, due to the loss of fat and muscle tissue. Sufferers have no resistance to disease and common infections are typically fatal.

Kwashiorkor is a severe type of protein-energy deficiency in young children (1-3 years old), occurring mainly in poor rural areas in the tropics. Kwashiorkor occurs when a child is weaned on to a diet that is low in calories, protein, and essential micronutrients. Children have poor growth, low resistance to infection, edema (accumulation of fluid in the tissues), and are inactive, apathetic and weak.

Alcohol Abuse and Nutritional Deficiency

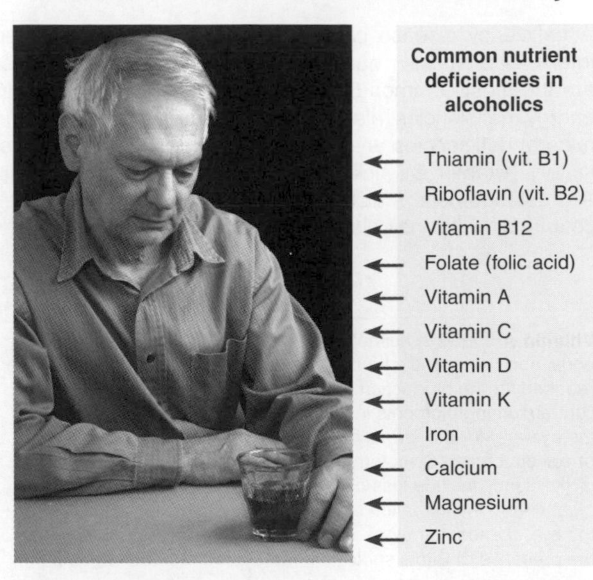

People who regularly consume excessive alcohol are at increased risk of nutritional deficiencies. Even when food intake is adequate, alcohol interferes with the metabolism of food by affecting digestion, storage, utilization, and excretion of nutrients. Alcohol damages the cells lining the small intestine and impairs absorption of nutrients. For example, alcohol inhibits fat absorption, impairs the digestion of proteins, and interferes with glucose metabolism.

1. What does the term **deficiency disease** mean? ______________________

2. Why are young children, pregnant women, and athletes the most susceptible to dietary deficiencies?

3. (a) Explain why a lack of iron leads to the symptoms of anemia (fatigue and breathlessness): ______________________

   (b) Explain why iron deficiency is relatively more common in women of child-bearing age than in men: ______________________

4. Using the example of **iodine**, explain how artificial dietary supplementation can be achieved and discuss its benefits:

5. Suggest why a zinc deficiency is associated with muscular weakness and a delay in puberty: ______________________

6. Explain why alcoholics are likely to be deficient in fat soluble vitamins (A, D, K) even when food intake is adequate:

ISBN: 978-1-92717357-2

Infection and Gut Function

Cholera is an acute intestinal infection caused by the bacterium ***Vibrio cholerae***. It results in copious watery diarrhoea and is responsible for 100,000-120,000 deaths a year. Cholera is spread primarily by the consumption of contaminated drinking water or food and in the developing world, the incidence of the disease is an indication of levels of sanitation. Cholera can be prevented by hygienic disposal of human faeces, provision of an adequate supply of safe drinking water, safe food handling and preparation (e.g. preventing contamination of food and water), and effective general hygiene (e.g. hand washing with soap). Once contracted, the only treatment for cholera is the administration of **oral rehydration solutions (ORS)** to prevent dehydration or death. In severe cases the rehydration solution is administered intravenously, and the patient may be prescribed antibiotics to reduce the infection time. With prompt and appropriate ORS treatment, the fatality rate from cholera infection is less than 1%.

Cholera and Diarrhoea

The cholera bacterium (below) produces an enterotoxin, which binds to membrane receptors on the small intestine, opening the ion channels and increasing permeability of the mucosal epithelium to chloride ions. Water follows the salt across the membrane by osmosis, resulting in copious, painless, watery diarrhoea that can lead to severe dehydration, electrolyte imbalance, kidney failure, and death within hours if untreated. Treating the diarrhoea by drinking water alone is ineffective for two reasons. During bouts of diarrhoea the large intestine is losing rather than absorbing water, and secondly, electrolyte loss is not addressed.

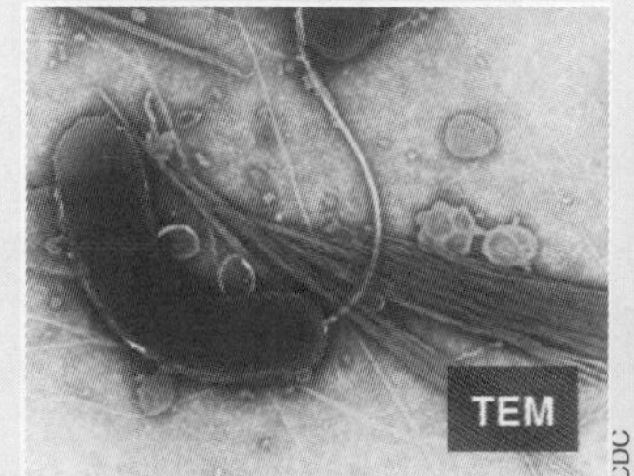

Dartmouth Electron Microscope Facility

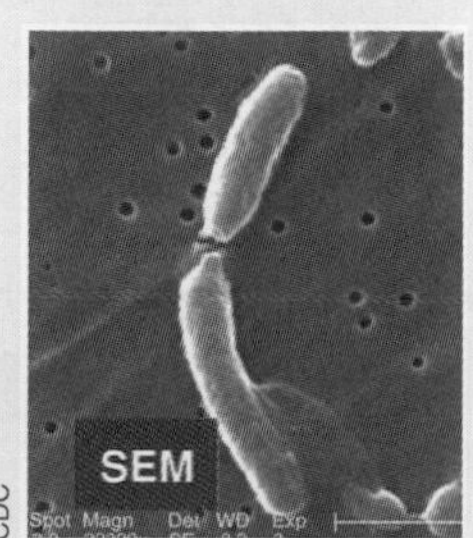

CDC

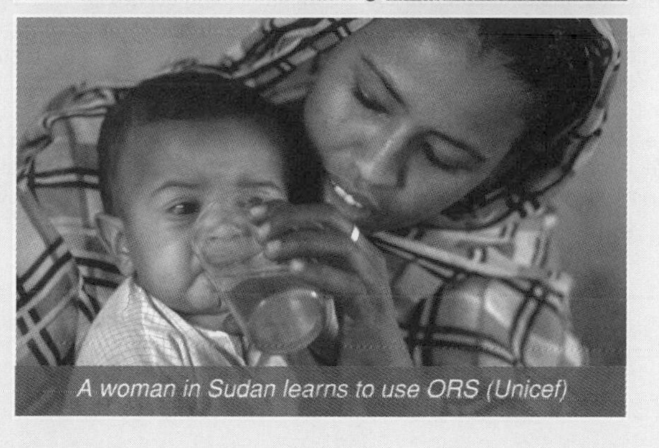

A woman in Sudan learns to use ORS (Unicef)

Roger LeMoyne UNICEF (www.unicef.org.nz)

Development of Oral Rehydration Solutions

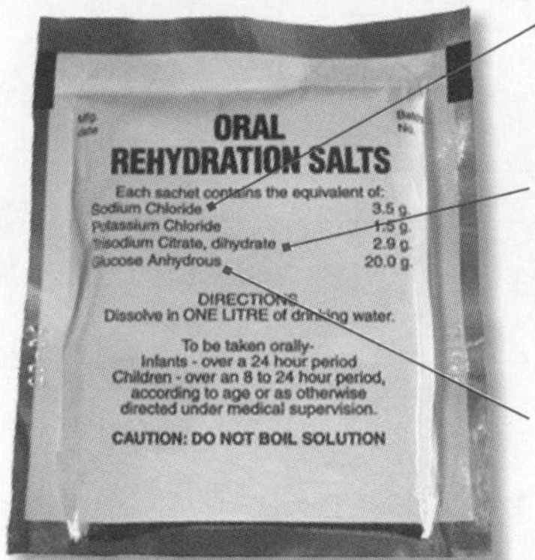

Modern **oral rehydration solutions** (ORS) are a simple, inexpensive product and can be administered with no medical training. They contain water and salts in ratios designed to replenish fluids and electrolytes. Carbohydrates, such as glucose or sucrose, are added to enhance electrolyte absorption in the intestinal tract. Sugars may actually increase the rate of diarrhoea at first, but they are essential for the cotransport of sodium into the intestinal cells. Many scientific disciplines have been involved in developing modern ORS. Key discoveries include:

- **1950s**: Physiologists first noted that glucose and sodium were transported together across the intestinal epithelium.
- **1960s**: The first ORS formulations were developed to treat severe **diarrhoea**. In addition to electrolytes, they also contained glucose, which had been proven to increase water reabsorption.
- Scientists discovered that the cholera **enterotoxin** was responsible for diarrhoea by interfering with cAMP activity and G-proteins.
- **Current**: The development of low osmolarity solutions which use alternative carbohydrate sources such as rice, instead of sugars, to minimize diarrhoeal effect.

1. Which pathogen causes cholera? ______________________

2. Why is the severe diarrhoea caused by cholera infection so dangerous if not treated quickly? ______________________

____

____

3. Why are ORS more effective in treating the symptoms of cholera than water alone? ______________________

____

____

4. (a) Why might glucose make diarrhoea worse? ______________________

(b) Why is glucose included in ORS, despite the fact that it can make the diarrhoea worse? ______________________

____

(c) What is the risk to the treatment outcome during this initial period? ______________________

____

ISBN: 978-1-92717357-2

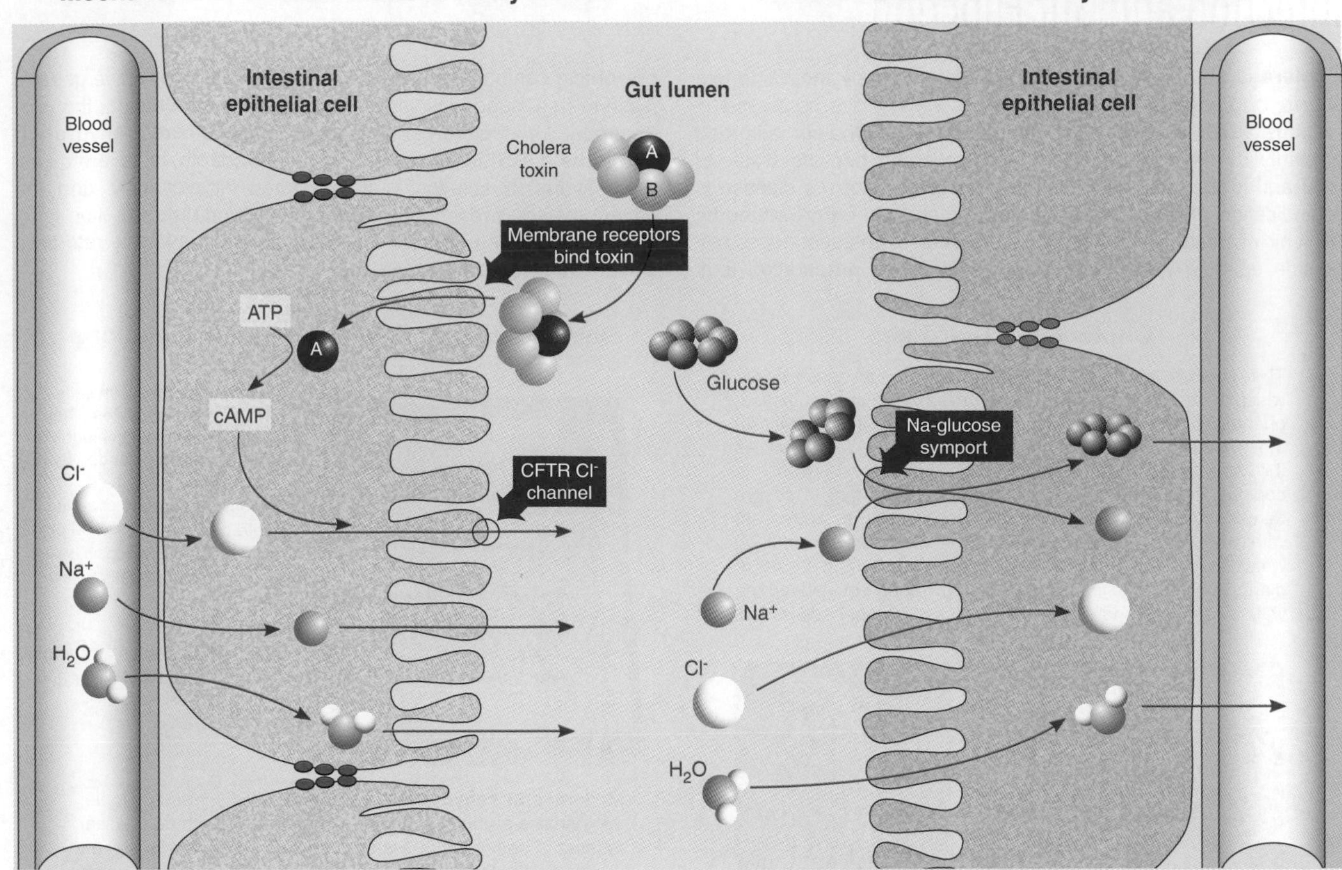

Cholera toxin contains two subunits (A and B), which bind to receptors on the cell membrane. Subunit A enters the cell and activates pathways that cause the production of cAMP from ATP. cAMP opens the Cystic Fibrosis Transmembrane conductance Regulator (CFTR) channel, causing the loss of chloride ions (Cl^-) into the gut lumen. This results in a negative charge within the lumen. Sodium follows down its electrochemical gradient, while water follows down its osmotic gradient. These are replaced from the blood, causing a drop in blood volume.

Glucose and sodium enter the gut lumen in high concentrations and are transported into the cell by the sodium-glucose symport (ion pump). The influx of sodium causes a positive charge within the cell and results in Cl^- moving down its electrochemical gradient and back into the cell. Water follows down its osmotic gradient. Water can then reenter the blood to restore blood volume (rehydration). This effect can occur rapidly, but oral rehydration must continue until the patient's immune system is able to eliminate the cholera bacteria.

5. How do ORS cause water to reenter the cells and blood? ______________________

____

____

____

6. Why was it important to understand ion transport in the intestine when devising a treatment for cholera?

____

____

____

7. How does the cholera toxin cause a loss of water into the gut lumen? ______________________

____

____

____

8. Discuss some of the ethical issues associated with trialing new ORS formulations on humans: ______________________

____

____

____

ISBN: 978-1-92717357-2

KEY TERMS: Mix and Match

INSTRUCTIONS: Test your vocabulary by matching each term to its definition, as identified by its preceding letter code.

Term		Definition
absorption	A	The muscular tube through which food passes from the mouth to the stomach.
anus	B	The body opening through where food enters the digestive tract. Includes the cheeks, hard and soft palates, tongue, lips and teeth.
assimilation	C	The body's largest internal organ. Has important roles in digestion and homeostasis.
diet	D	Wave-like smooth muscle contractions that move food through the digestive tract.
digestion	E	Removal of undigested food (in feces) from the gut.
digestive enzymes	F	A secretion into a duct (as opposed as into the blood directly).
digestive system	G	A large organ that has both digestive and endocrine functions. It is both an exocrine and endocrine gland.
egestion	H	A watery secretion in the mouth produced by the salivary glands. It has roles in food digestion and lubrication.
esophagus	I	Finger-like projections lining the surface of the intestine which increase the surface area for absorption.
exocrine	J	The process of making food pass from the mouth, to the pharynx, and into the esophagus. Also called deglutition.
feces	K	The types of food and drink a person regularly consumes.
gall bladder	L	Process by which the products of digestion move across the gut lining into the blood or lymph.
ingestion	M	The external opening of the rectum through which feces leave the body.
intestinal villus	N	Large muscular digestive organ located between the esophagus and small intestine. It secretes protein digesting enzymes and strong acids to aid digestion of food.
large intestine	O	The waste products of digestion which are discharged from the digestive tract through the anus.
liver	P	Globular proteins that break down specific types of food molecules.
malnutrition	Q	A disease where the body does not respond correctly to insulin, it results in elevated blood glucose levels.
mouth	R	The upper part of the gut comprising of the duodenum, jejunum and ileum. The main site of absorption of food.
oral rehydration solution	S	The area situated below the mouth and nasal cavity, and above the esophagus. Also called the throat.
pancreas	T	The absorption and conversion or utilization by a cell, of simple substances into more complicated ones.
peristalsis	U	Hard enamel-coated structures in the jaws, used for biting and chewing food.
pharynx	V	Small organ located near the liver which aids in digestion of fats by adding bile.
saliva	W	The body system where ingestion, digestion, absorption, and elimination of food occurs.
small intestine	X	The taking in of water or food into the body (by drinking or eating).
stomach	Y	The physical and chemical breakdown of food.
swallowing	Z	An oral dehydration treatment consisting of a solution of salt, sugars, and minerals.
tooth	AA	The lower part of the gut comprising of the appendix, cecum and colon.
type 2 diabetes mellitus	BB	Insufficient, excessive or imbalanced consumption of nutrients.

The Digestive System

ISBN: 978-1-92717357-2

Interacting Systems

The Urinary System

Respiratory system

- Respiratory system provides O_2 to the urinary system and disposes of CO_2 produced by cellular respiration.
- An enzyme in the cells of the lung capillaries converts angiotensin I to angiotensin II (involved in regulation of glomerular filtration).

Nervous system

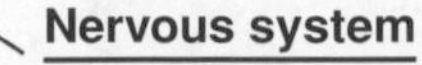

- Renal regulation of the Na^+, K^+, and Ca^{2+} content in the extracellular fluid is necessary for nerve function.
- Micturition (urination) is controlled by voluntary and reflex nervous activity.
- Sympathetic NS activity triggers the renin-angiotensin system for regulation of blood volume.

Cardiovascular system

- Regulation of salt and water balance in the kidney is important in regulation of blood pressure.
- Regulation of blood composition of Na^+, K^+, and Ca^{2+} helps maintain heart function.
- Arterial blood pressure is the driving force for glomerular filtration.
- Heart muscle cells secrete a peptide (ANP) in response to high blood pressure. ANP results in greater excretion of Na^+ and water from the kidney.
- Blood distributes the hormones that influence renal function (e.g. ADH and aldosterone).

Lymphatic system and immunity

- The lymphatic vessels return leaked fluid to the general circulation and help to maintain the blood volume/pressure required for kidney function.
- The immune system protects the urinary organs from infection and cancer.

Endocrine system

- Kidneys produce erythropoietin, a hormone which promotes the formation of red blood cells in the bone marrow.
- Regulation of salt and water balance by the kidneys maintains the blood volume necessary for hormone transport.
- ADH, ANP, aldosterone and other hormones interact to regulate reabsorption of water and electrolytes in the kidney.

Digestive system

- The final activation of vitamin D occurs in the kidneys. Bioactive vitamin D is required for calcium absorption in the gut.
- The liver synthesizes most of the body's urea, which is then excreted via the kidneys.
- The digestive system provides nutrients for maintenance and health of the urinary organs.

Reproductive system

- Urinogenital systems are closely aligned. Urethra has an excretory function in both sexes and a reproductive function in males, for the passage of semen.

Skeletal system

- Bones of the ribcage provide some protection for the kidneys.
- Erythropoietin from the kidneys promotes the formation of red blood cells in the bone marrow.

Muscular system

- Renal regulation of the Na^+, K^+, and Ca^{2+} content in the extracellular fluid is necessary for muscle function.
- Muscles of the pelvic floor and external urethral sphincter are involved in voluntary control of micturition.
- Creatinine, which is a break-down product of creatine phosphate in muscle, is produced at a fairly constant rate by the body and must be excreted by the kidneys.

Integumentary system

- The skin provides an external protective barrier.
- Final activation of vitamin D (synthesized in the skin) occurs in the kidneys.
- Skin is a site of water loss.

General Functions and Effects on all Systems

The urinary system (kidneys and associated structures) is responsible for disposing of nitrogenous wastes toxins, and metabolic breakdown products. The kidneys maintain the fluid, electrolytes, and acid-base balance of the body fluids, which is essential for the proper functioning of all body systems.

Disease

Symptoms of disease

- Pain (moderate to severe)
- Abnormal urine composition or volume
- Abnormal electrolyte balance
- Abnormal fluid balance

Diseases and disorders of the urinary system

- Kidney stones
- Hereditary disorders
- Nephrotic syndrome
- Congenital diseases (malformations)
- Bladder and kidney cancer
- Chronic kidney disease (CKD)
- Incontinence

Medicine & Technology

Diagnosis of disorders

- MRI scans
- Kidney biopsy
- Urine tests
- Blood tests

Prevention of urinary system disorders

- Control of hypertension
- Control of diabetes
- Control of weight
- Behavior to control risk of UTIs

Treatment of urinary system disorders

- Drug therapy (e.g. antibiotics)
- Physical therapy (e.g for pelvic floor)
- Surgery (e.g. removal of kidney stones)
- Transplant (e.g. of donor kidney)
- Renal dialysis

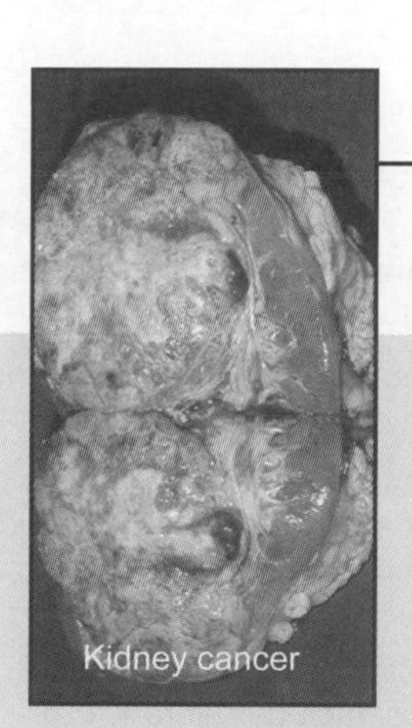
Kidney cancer

- Kidney stones
- CKD & renal failure
- Polycystic kidney disease

- MRI scans
- Urine analysis
- Kidney transplants
- Renal dialysis

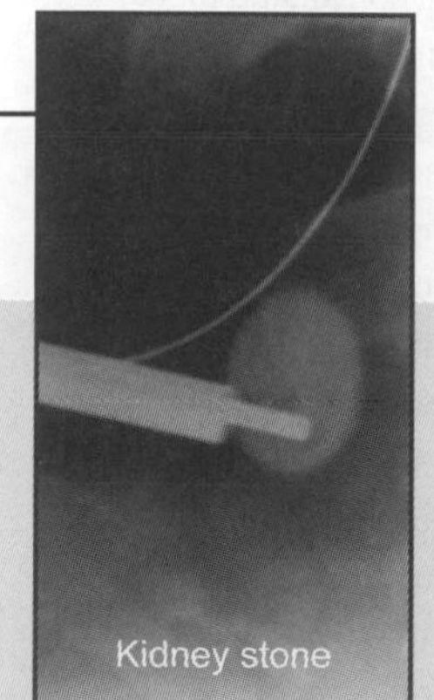
Kidney stone

Waste Removal

The Urinary System

The urinary system had a primary role in excretion of nitrogenous wastes, and in fluid and electrolyte balance.

Degenerative changes in kidney and bladder function can be severe and debilitating. Renal dialysis and transplants are options for sufferers of kidney disease.

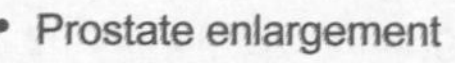

- Prostate enlargement
- Poor renal function
- Incontinence

- Effects of exercise on health
- Creatine metabolism
- Dehydration

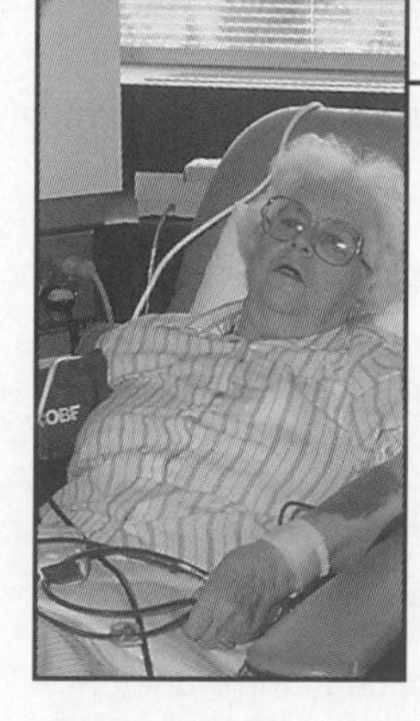

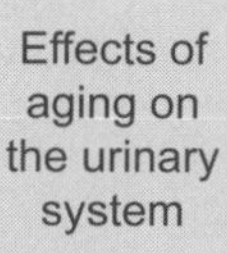

Effects of aging on the urinary system

- Lower number of functional nephrons
- Reduction in glomerular filtration rate
- Reduced response to ADH
- Prostate enlargement (males)
- Loss of bladder/sphincter muscle tone

Effects of exercise on the urinary system

- Lowered blood pressure, therefore...
- Reduced risk of chronic kidney disease
- Increased rates of creatinine excretion
- (Rarely) exercise-induced renal failure (dehydration or electrolyte imbalance)

The Effects of Aging

Exercise

The Urinary **System**

Key concepts

- The urinary system is the primary system for the excretion of nitrogenous and other wastes.
- The composition of the body's fluids is regulated through feedback mechanisms involving the blood, and respiratory and urinary systems.
- Urine production in the kidney is the result of ultrafiltration, secretion, and reabsorption.
- Urine analysis can assist in detecting disease.
- Renal failure can be addressed through dialysis or kidney transplant.

Key terms

aldosterone
antidiuretic hormone (ADH)
bladder
Bowman's capsule
collecting duct
cortex (of kidney)
distal convoluted tubule
electrolyte
excretion
extracellular fluid
glomerulus
intracellular fluid
kidney
kidney transplant
loop of Henle
medulla (of kidney)
micturition (= urination)
nephron
osmoreceptors
proximal convoluted tubule
renal corpuscle
renal dialysis
renal failure
renal pelvis
ultrafiltration
urethra
ureter
urinalysis
urine

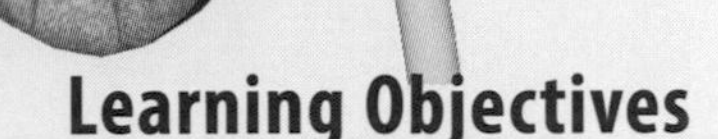

Learning Objectives

☐ 1. Use the **KEY TERMS** to compile a glossary for this topic.

The Urinary System

pages 239-245

☐ 2. On a diagram of the **urinary system**, identify **kidneys**, **ureters**, renal blood vessels, **bladder**, **urethra**. Describe the role of the **kidneys** in regulating body fluids.

☐ 3. Describe the gross structure of the kidney including the **cortex**, **medulla**, and **renal pelvis**. Interpret features of kidney histology as seen with a light microscope.

☐ 4. Describe how **nephrons** are arranged in the kidney. Annotate a diagram to show the structure of a nephron and its relationship to the surrounding blood vessels. Include reference to **glomerulus**, **Bowman's capsule**, **proximal** and **distal convoluted tubules**, **loop of Henle**, and **collecting duct**.

☐ 5. On your annotated diagram, outline each of the following to describe how the nephron produces **urine** and regulates electrolyte balance. Include reference to:
(a) **Ultrafiltration** in the **glomerulus**.
(b) The ultrastructure of the **renal corpuscle** (glomerlulus + Bowman's capsule).
(c) Selective secretion and reabsorption of water and solutes in the **proximal convoluted tubule** and **distal convoluted tubule**.
(d) The role of the **loop of Henle** and the counter-current multiplier system in creating and maintaining the ionic (salt) gradient in the kidney.
(e) The role of the ionic gradient in the kidney in osmotic withdrawal of water in the collecting duct (and concentration or urine).

☐ 6. Describe **micturition** (urination or voiding), including reference to the involvement of urethral sphincters and voluntary and involuntary controls over micturition.

Fluid and Electrolyte Balance

pages 247-248

☐ 7. Describe the main fluid compartments in the body: the **intracellular fluid** and the **extracellular fluid**. Recognize that homeostasis is maintained through constant adjustments of these fluid compartments.

☐ 8. Explain how electrolyte balance and volume are regulated. Include reference to:
(a) The role of **osmoreceptors** in the hypothalamus
(b) Release of antidiuretic hormone (**ADH**) and its action on the kidney.
(c) The regulation of ADH output.
(d) The role of **aldosterone** in promoting sodium reabsorption in the kidney.
(e) The regulation of aldosterone release through the renin-angiotensin system.

☐ 9. Recognize the requirement for maintenance of blood pH at pH 7.35-7.45. Describe how the acid-base balance of the body fluids is maintained through bicarbonate buffer system in the blood, respiratory system controls, and renal mechanisms.

Diseases of the Urinary System

pages 246, 249-250

☐ 10. Describe the characteristics of healthy urine and describe the use of urine analysis (**urinalysis**) in the diagnosis of renal disease.

☐ 11. Describe the causes of **renal failure** and discuss options for treatment, including **kidney transplant** and **renal dialysis**. Explain the advantages and problems associated with these treatment options.

Periodicals:
Listings for this chapter are on page 280

Weblinks:
www.thebiozone.com/weblink/AnaPhy-3572.html

***BIOZONE** APP:*
Student Review Series
The Urinary System

Waste Products in Humans

In humans, a number of organs are involved in the excretion of the waste products of metabolism: mainly the kidneys, lungs, skin, and gut. The liver is a particularly important organ in the initial treatment of waste products, particularly the breakdown of hemoglobin and the formation of urea from ammonia. Excretion should not be confused with the elimination or egestion of undigested and unabsorbed food material from the gut. Note that the breakdown products of hemoglobin (blood pigment) are excreted in bile and pass out with the feces, but they are not the result of digestion.

CO_2
Water

Lungs
Excretion of carbon dioxide (CO_2) with some loss of water.

Skin
Excretion of water, CO_2, hormones, salts and ions, and small amounts of urea as sweat.

Liver
Produces urea from ammonia in the urea cycle. Breakdown of hemoglobin in the liver produces the bile pigments e.g. bilirubin.

Gut
Excretion of bile pigments in the feces. Also loses water, salts, and carbon dioxide.

Bladder
Storage of urine before it is expelled to the outside.

All cells
All the cells that make up the body carry out cellular respiration; they break down glucose to release energy and produce the waste products, carbon dioxide and water.

Excretion in Humans

In humans, the kidney is the main organ of excretion, although the skin, gut, and lungs also play important roles. As well as ridding the body of nitrogenous wastes, the kidneys are also able to excrete many unwanted poisons and drugs that are taken in from the environment. Usually these are ingested with food or drink, or inhaled. As long as these are not present in toxic amounts, they can usually be slowly eliminated from the body.

Kidney
Filtration of the blood to remove urea. Unwanted ions, particularly hydrogen (H^+) and potassium (K^+), and some hormones are also excreted by the kidneys. Some poisons and drugs (e.g. penicillin) are also excreted by active secretion into the urine. Water is lost in excreting these substances and extra water may be excreted if necessary.

Substance	Origin*	Organ(s) of excretion
Carbon dioxide		
Water		
Bile pigments		
Urea		
Ions (K^+, HCO_3^-, H^+)		
Hormones		
Poisons		
Drugs		

*Origin refers to from where in the body each substance originates

1. Complete the table above summarizing the origin of excretory products and the main organ(s) of excretion for each.
2. Explain the role of the liver in excretion, even though it is not primarily an organ of excretion: ____________________
3. Tests for pregnancy are sensitive to an excreted substance in the urine. Suggest what type of substance this might be: ____________________
4. In people suffering renal failure, the kidneys cease to produce filtrate. Based on your knowledge of the central role of the kidneys in fluid and electrolyte balance, as well as nitrogen excretion, describe the typical symptoms of kidney failure: ____________________

ISBN: 978-1-92717357-2

Periodicals:
Urea

Related activities: *Acid-Base Balance, Kidney Dialysis*

Water Budget in Humans

We cannot live without water for more than about 100 hours and adequate water is a requirement for physiological function and health. Body water content varies between individuals and through life, from above about 90% of total weight as a fetus to 74% as an infant, 60% as a child, and around 50-59% in adults, depending on gender and age. Gender differences (males usually have a higher water content than females) are the result of differing fat levels. Water intake and output are highly variable but closely matched to less than 0.1% over an extended period. Typical values for water gains and losses, as well as daily water transfers are given below. Men need more water than women due to their higher (on average) fat-free mass and energy expenditure. Infants and young children need more water in proportion to their body weight as they cannot concentrate their urine as efficiently as adults. They also have a greater surface area relative to weight, so water losses from the skin are greater.

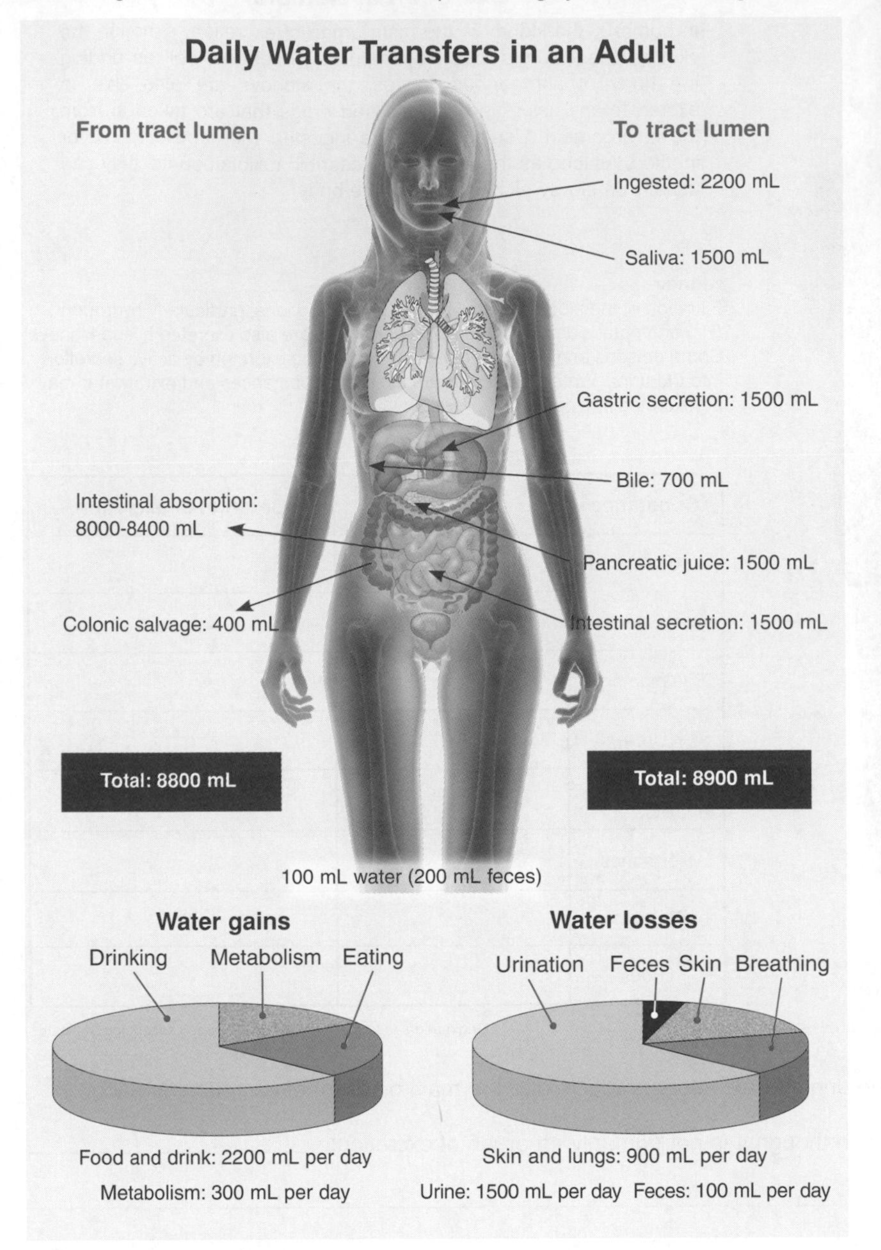

About 63% of our daily requirement for water is met through drinking fluids, 25% is obtained from food, and the remaining 12% comes from metabolism (the oxidation of glucose to ATP, CO_2, and water).

Typically, we lose 60% of body water through urination, 36% through the skin and lungs, and 4% in feces. Losses through the skin and from the lungs (breathing) average about 900 mL per day or more during heavy exercise. These are called **insensible losses**.

1. Explain how metabolism provides water for the body's activities: ______________________________

____

2. Describe four common causes of physiological dehydration:

(a) ______________________ (c) ______________________

(b) ______________________ (d) ______________________

3. Some recent sports events have received media coverage because athletes have collapsed after excessive water intakes. This condition, called **hyponatremia** or water intoxication, causes nausea, confusion, diminished reflex activity, stupor, and eventually coma. From what you know of fluid and electrolyte balances in the body, explain these symptoms:

____

____

Related activities: *Control of Urine Output, Fluid and Electrolyte Balance*

ISBN: 978-1-92717357-2

The Urinary System

The urinary system consists of the kidneys and bladder, and their associated blood vessels and ducts. The **kidneys** have a plentiful blood supply from the renal artery. The blood plasma is filtered by the **kidney nephrons** to form urine. Urine is produced continuously, passing along the **ureters** to the **bladder**, a hollow muscular organ lined with smooth muscle and stretchable epithelium. Each day the kidneys filter about 180 L of plasma. Most of this is reabsorbed, leaving a daily urine output of about 1 L. By adjusting the composition of the fluid excreted, the kidneys help to maintain the body's internal chemical balance. Human kidneys are very efficient, producing a urine that is concentrated to varying degrees depending on requirements.

Urinary System

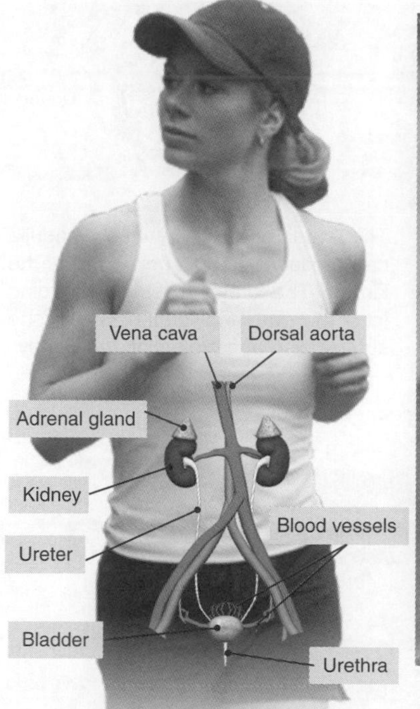

Kidneys *in-situ* (Rat)

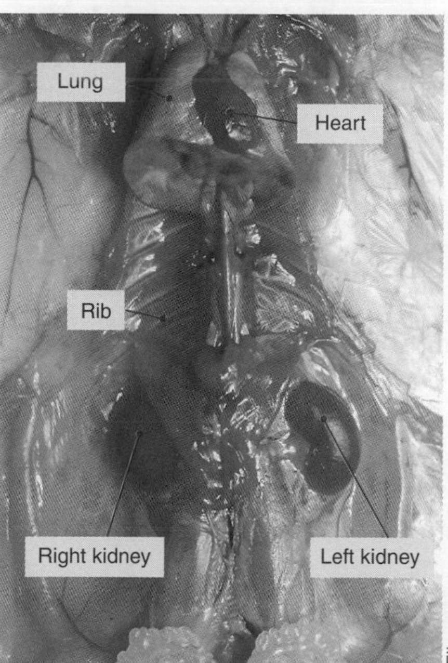

Sagittal Section of Kidney (Pig)

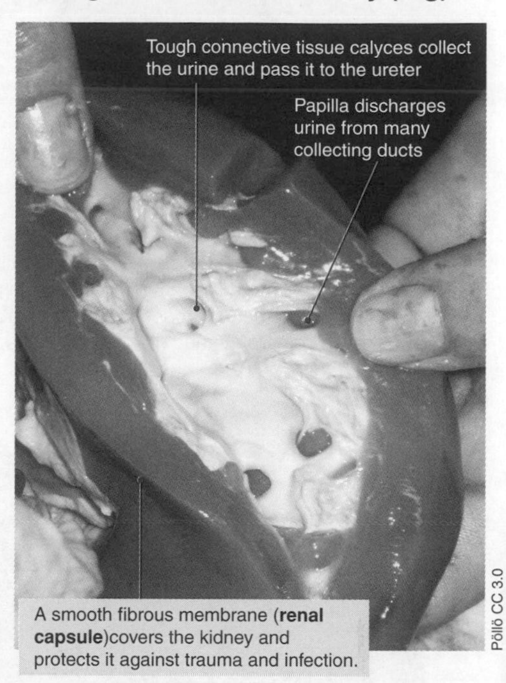

The kidneys of humans (above), rats (dissection, above center), and many other mammals (e.g. pig above right) are distinctive, bean shaped organs that lie at the back of the abdominal cavity to either side of the spine. The kidneys lie outside the peritoneum of the abdominal cavity (**retoperitoneal**) and are partly protected by the lower ribs (see kidneys *in-situ* above center).

Human kidneys are ~100-120 mm long and 25 mm thick. A cut through in a sagittal plane (see photo above right), reveals numerous tough connective tissue calyces. These collect the urine from the papillae where it is discharged and drain it into the ureter.

The Kidneys and Their Blood Supply

Vena cava returns blood to the heart

Dorsal aorta supplies oxygenated blood to the body.

Kidney produces urine and regulates blood volume.

Adrenal glands are associated with, but not part of, the urinary system.

Renal vein returns blood from the kidney to the venous circulation.

Ureters carry urine to the bladder.

Renal artery carries blood from the aorta to the kidney.

1. State the function of each of the following components of the urinary system:

(a) Kidney: ______

(b) Ureters: ______

(c) Bladder: ______

(d) Urethra: ______

(e) Renal artery: ______

(f) Renal vein: ______

(g) Renal capsule: ______

2. Calculate the percentage of the plasma reabsorbed by the kidneys: ______

ISBN: 978-1-92717357-2

***Related activities**: The Physiology of the Kidney, Control of Urine Output*

Internal Structure of the Human Kidney

Nephrons are arranged with all the collecting ducts pointing towards the renal pelvis.

Outer cortex contains the renal corpuscles and convoluted tubules.

Inner medulla is organized into pyramids.

Each pyramid ends in a papilla or opening.

Urine enters the **calyces**.

Urine collects in a space near the ureter called the renal pelvis, before leaving the kidney via the ureter.

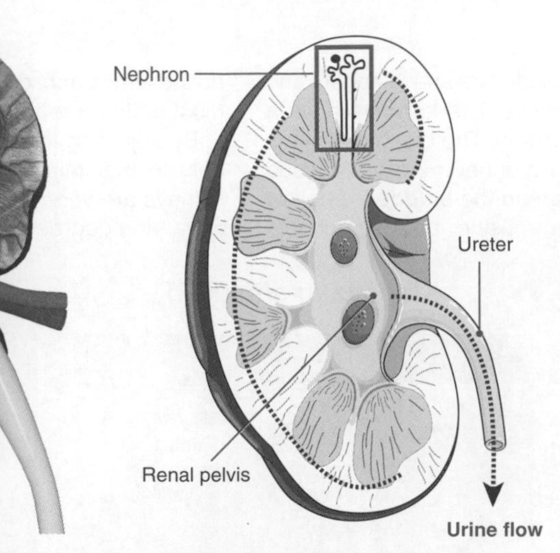

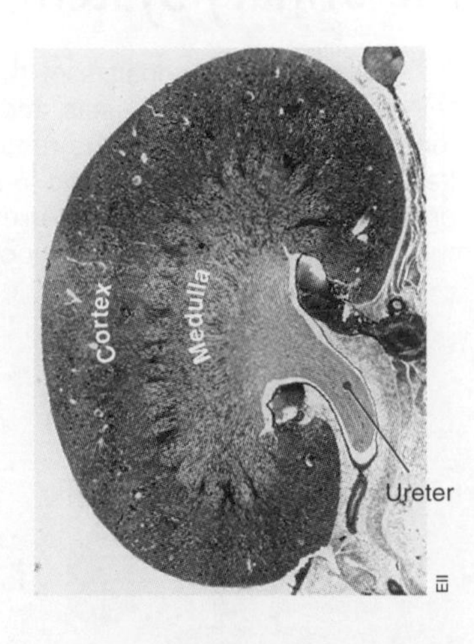

The functional units of the kidney are selective filter elements called **nephrons**. Each kidney contains more than 1 million nephrons and they are precisely aligned so that urine is concentrated as it flows towards the ureter (model and diagram above). The alignment gives the kidney tissue a striated (striped) appearance and makes it possible to accommodate all the filtering units needed.

The outer cortex and inner medulla can be seen in a low power LM of the kidney. The ureter is seen extending into the fat and connective tissue surrounding and protecting the kidney.

The Bladder

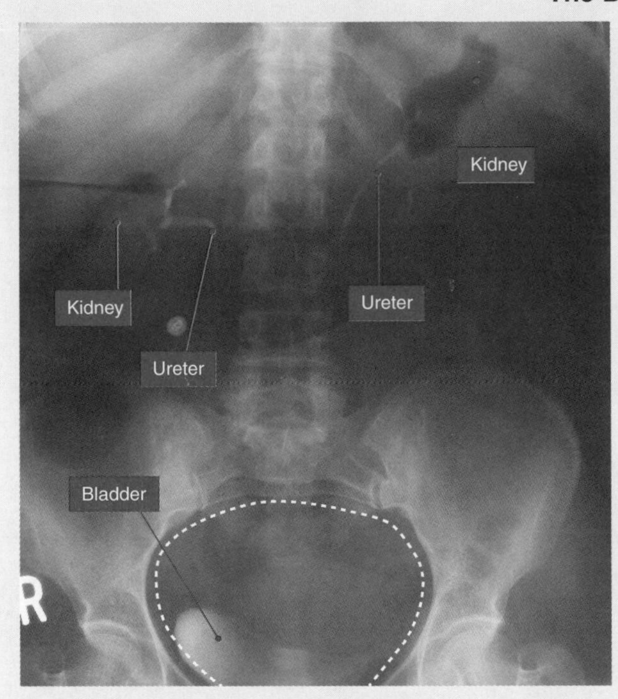

The bladder is a hollow stretchable organ, which stores the urine before it leaves the body via the urethra. In this X-ray, it is empty and resembles a deflated balloon. The dotted line shows where it would sit if full.

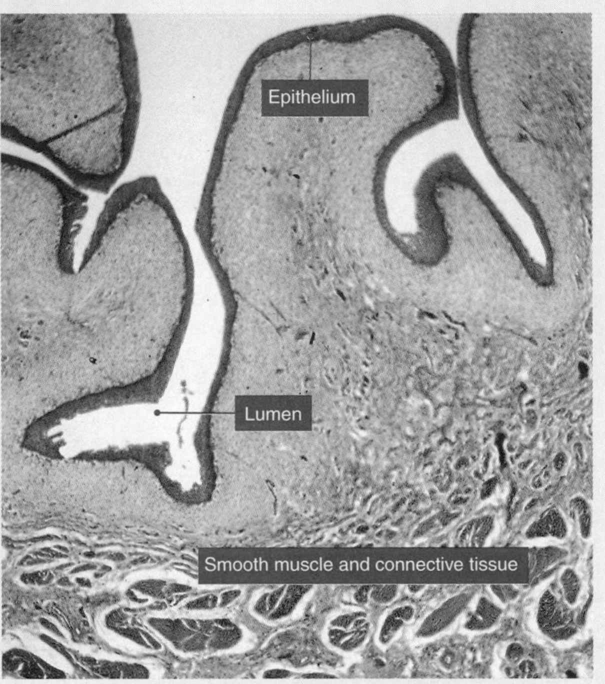

The bladder is lined with **transitional epithelium**. This type of epithelium is layered, or **stratified**, so it can be stretched without the outer cells breaking apart from each other. This image shows the bladder in a deflated state.

3. (a) What is a nephron? ______________________________

   (b) What is its role in excretion? ______________________________

   ____

4. (a) Where would you find transitional epithelium in the urinary system: ______________________________

   (b) Why do you find this type of epithelium here? ______________________________

5. In adults, the opening of the urethra is regulated by a voluntary sphincter muscle. What is the purpose of this sphincter?

   ____

   ____

ISBN: 978-1-92717357-2

The Physiology of the Kidney

The kidney **nephron**, is a selective filter element, comprising a **renal corpuscle** and its associated tubules and ducts. **Ultrafiltration**, i.e. forcing fluid and dissolved substances through a membrane by pressure, occurs in the first part of the nephron, across the membranes of the capillaries and the glomerular capsule. The passage of water and solutes into the nephron and the formation of the glomerular filtrate depends on the pressure of the blood entering the afferent arteriole (below). If it increases, filtration rate increases. When it falls, glomerular filtration rate also falls. This process is so precisely regulated that, in spite of fluctuations in arteriolar pressure, glomerular filtration rate per day stays constant. After formation of the initial filtrate, the **urine** is modified through secretion and tubular reabsorption according to physiological needs at the time.

Nephron Structure

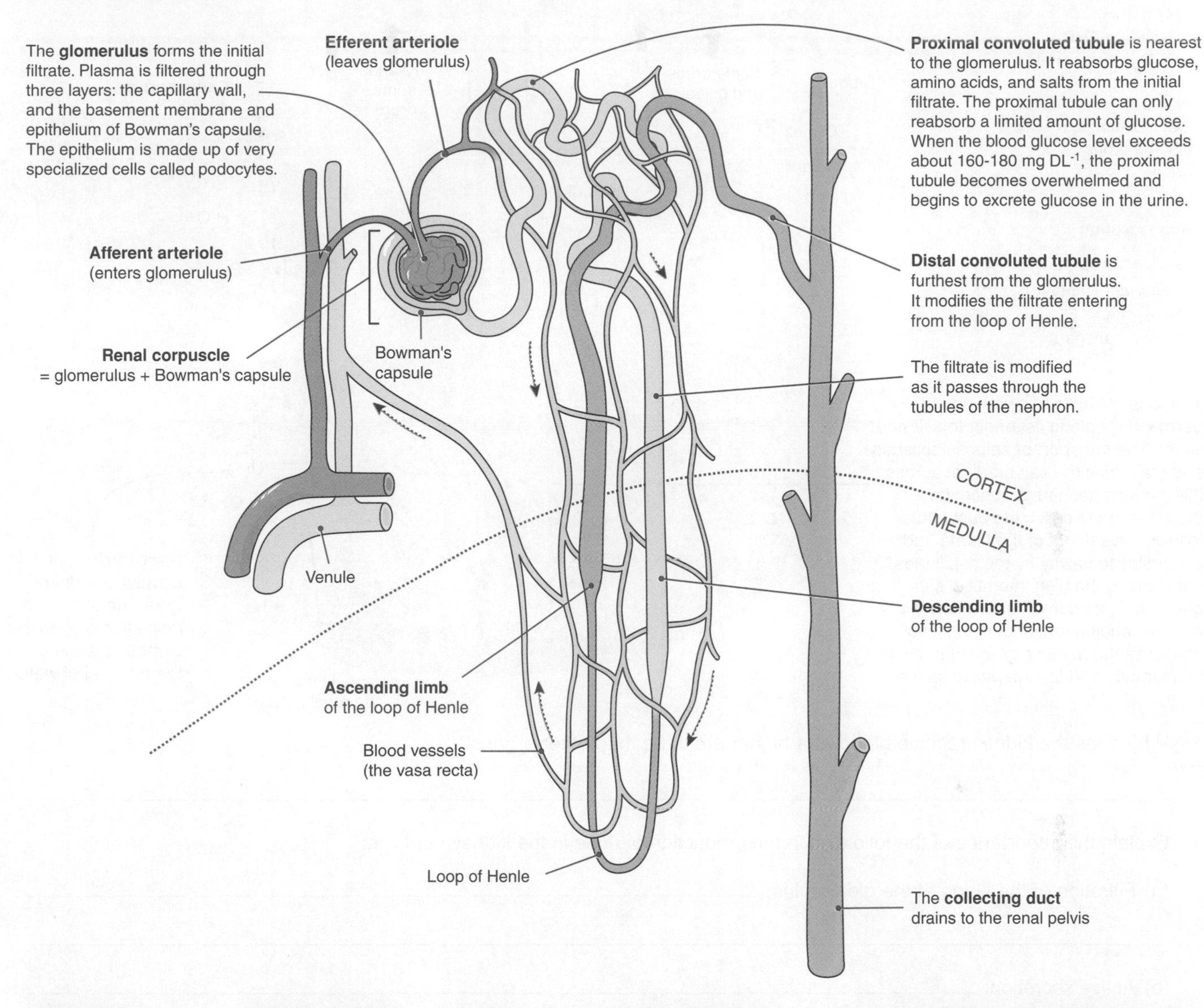

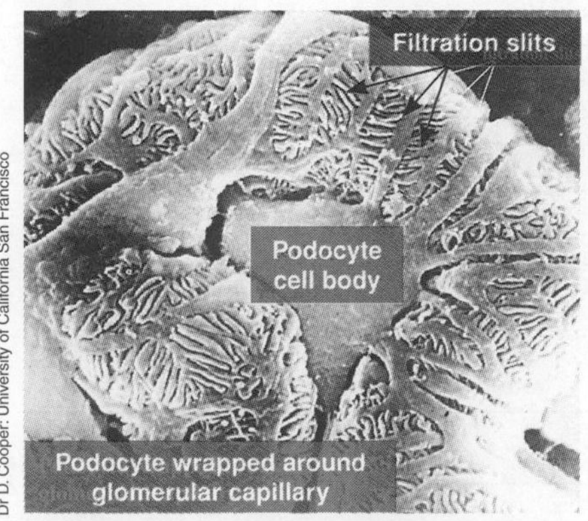

The epithelium of Bowman's capsule is made up of specialized cells called **podocytes**. The finger-like cellular processes of the podocytes wrap around the capillaries of the glomerulus, and the plasma filtrate passes through the filtration slits between them.

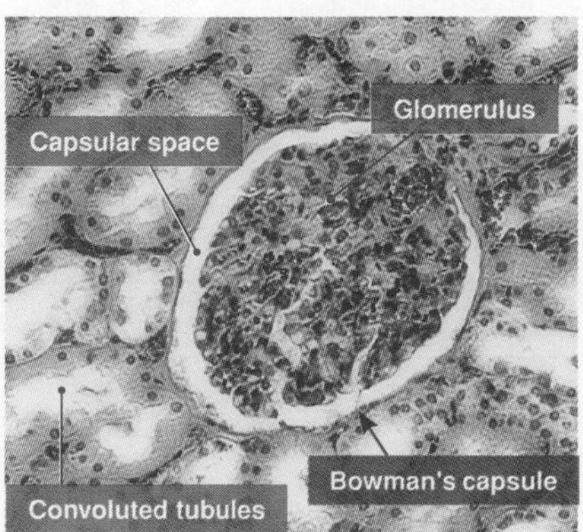

Bowman's capsule is a double walled cup, lying in the cortex of the kidney. It encloses a dense capillary network called the **glomerulus**. The capsule and its enclosed glomerulus form a **renal corpuscle**. In this section, the convoluted tubules can be seen surrounding the renal corpuscle.

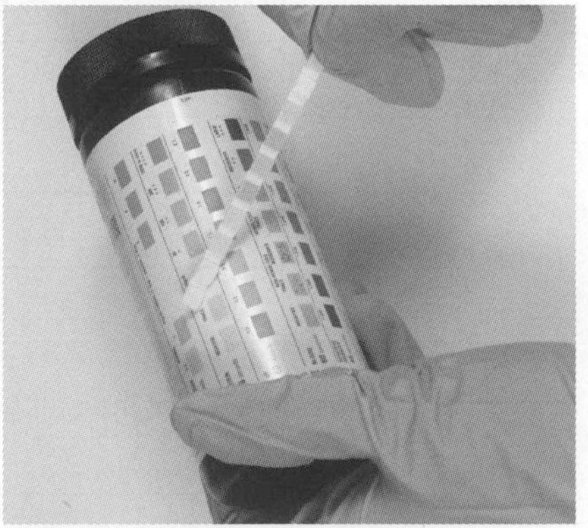

Dipstick urinalysis is commonly used to detect metabolic errors. Less than 0.1% of glucose filtered by the glomerulus normally appears in urine. The presence of glucose in the urine is usually due to untreated diabetes mellitus, which is characterized by high blood glucose levels.

ISBN: 978-1-92717357-2

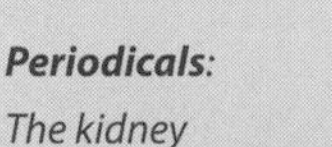

Periodicals: *The kidney*

Weblinks: *Kidney Vascular System, Interactive Kidney Quiz, The Juxtaglomerular Apparatus*

Urine formation begins by **ultrafiltration** of the blood, as fluid is forced through the capillaries of the glomerulus, forming a filtrate similar to blood but lacking cells and proteins. The filtrate is then modified by **secretion** and **reabsorption** to add or remove substances (e.g. ions). The processes involved in urine formation are summarized below. The loop of Henle acts as a **countercurrent multiplier**, establishing and increasing the salt gradient through the medullary region. This is possible because the descending loop is freely permeable to water but the ascending loop is not.

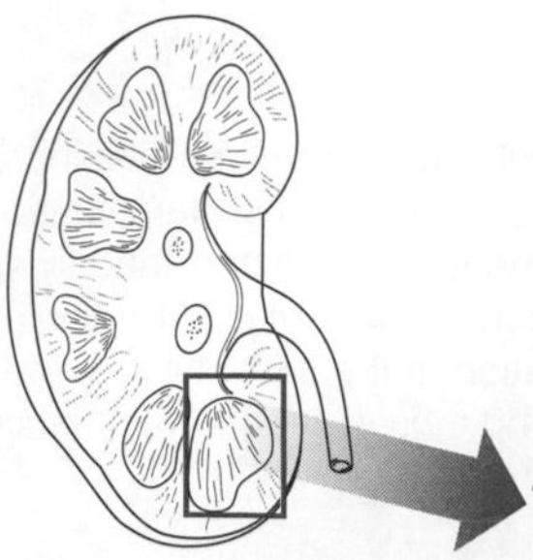

Filtrate

- H_2O
- Salts (e.g. NaCl)
- HCO_3^- (bicarbonate)
- H^+
- Urea
- Glucose, amino acids
- Some drugs

Reabsorption

- Active transport
- Passive transport
- Secretion (active transport)

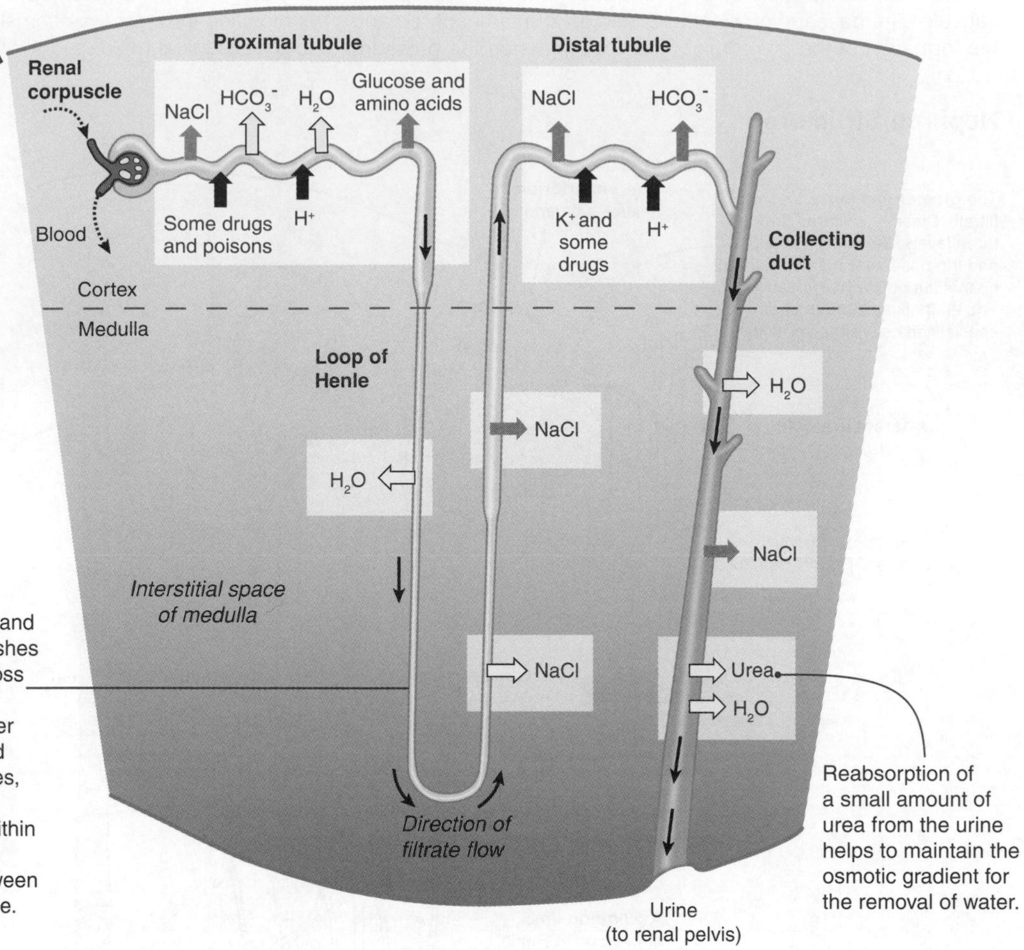

The loop of Henle has varying permeability along its length to salt and water. The transport of salts establishes and maintains the salt gradient across the medulla needed to concentrate the urine in the collecting duct. Water follows the salt out of the filtrate and is transported away by the capillaries, maintaining the high interstitial salt gradient. The countercurrent flow within the descending and ascending limb multiplies the osmotic gradient between tubular fluid and the interstitial space.

Reabsorption of a small amount of urea from the urine helps to maintain the osmotic gradient for the removal of water.

1. Why does the kidney receive blood at a higher pressure than other organs? ______________________________

2. Explain the importance of the following in the production of urine in the kidney nephron:

   (a) Filtration of the blood at the glomerulus: ______________________________

   (b) Active secretion: ______________________________

   (c) Reabsorption: ______________________________

   (d) Osmosis: ______________________________

3. (a) What is the purpose of the salt gradient in the kidney? ______________________________

   (b) How is this salt gradient produced? ______________________________

ISBN: 978-1-92717357-2

Control of Urine Output

Variations in salt and water intake, and in the environmental conditions to which we are exposed, contribute to fluctuations in blood volume and composition. The primary role of the kidneys is to regulate blood volume and composition (including the removal of nitrogenous wastes), so that homeostasis is maintained. This is achieved through varying the volume and composition of the urine. Two hormones, **antidiuretic hormone** (ADH) and **aldosterone**, are involved in the process.

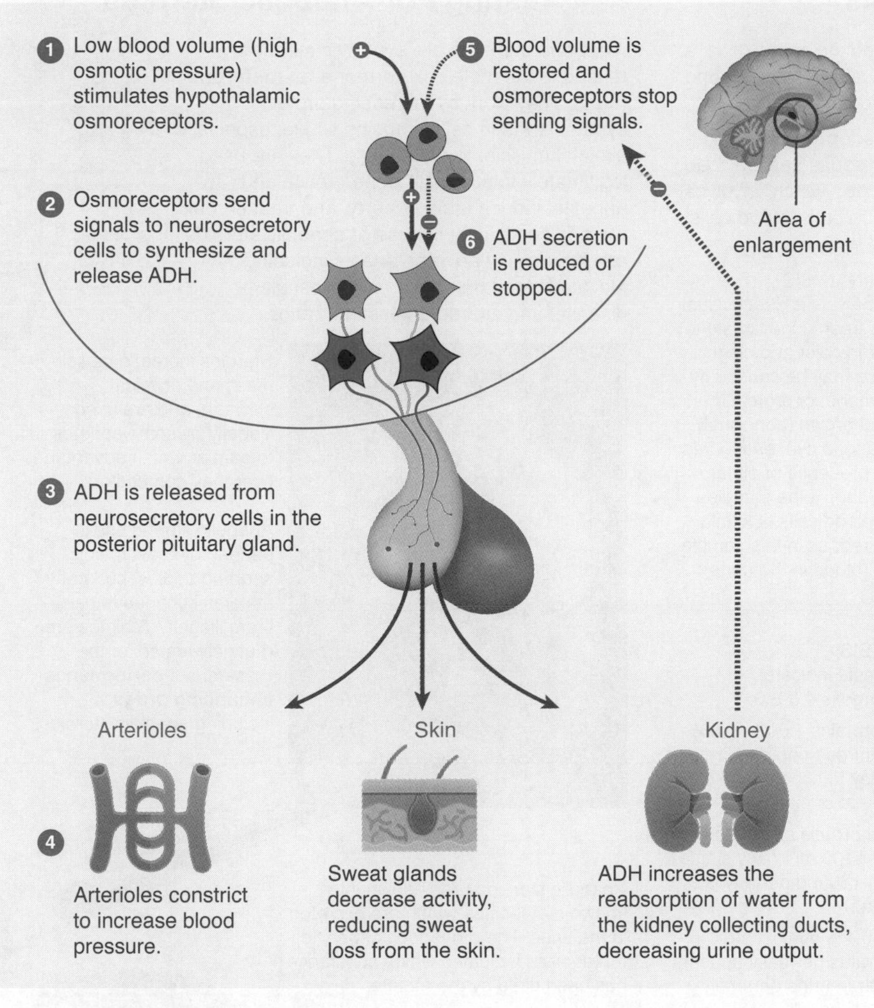

Osmoreceptors in the **hypothalamus** of the brain respond to changes in blood volume. A blood volume stimulates the synthesis and secretion of the hormone ADH (antidiuretic hormone), which is released from the posterior pituitary into the blood. ADH increases the permeability of the kidney collecting duct to water so that more water is reabsorbed and urine volume decreases. A second hormone, aldosterone, helps by increasing sodium reabsorption.

Factors causing ADH release

- Low blood volume
  - = More negative water potential
  - = High blood sodium levels
  - = Low fluid intake
- Nicotine and morphine

Factors inhibiting ADH release

- High blood volume
  - = Less negative water potential
  - = Low blood sodium levels
- High fluid intake
- Alcohol consumption

Factors causing the release of aldosterone

Low blood volumes also stimulate secretion of aldosterone from the adrenal cortex. This is mediated through a complex pathway involving osmoreceptors near the kidney glomeruli and the hormone renin from the kidney.

1. (a) **Diabetes insipidus** is a disease caused by a lack of ADH. Based on what you know of the role of ADH in regulating urine volumes, describe the symptoms of this disease:

   (b) Suggest how this disorder might be treated:

2. Explain why alcohol consumption (especially to excess) causes dehydration and thirst:

3. Explain how negative feedback mechanisms operate to regulate blood volume and urine output:

4. **Diuretics** are drugs that increase urine volume. Many work by inhibiting the active transport of sodium and chloride in the nephron. Explain how this would lead to an increase in urine volume:

ISBN: 978-1-92717357-2

***Related activities**: Water Budget in a Human, The Physiology of the Kidney, The Endocrine System, Hormones of the Pituitary*

Urine Analysis

Urine is the liquid waste product of the body. It contains water, electrolytes, and other waste metabolites which are filtered out of the blood by the kidneys. **Urine analysis** (urinalysis) is used as a medical diagnostic tool for a wide range of metabolic disorders. In addition, urine analysis can be used to detect the presence of illicit (non-prescription) drugs and for diagnosing pregnancy.

Diagnostic Urinalysis

A urinalysis (UA) is an array of tests performed on urine. It is a common method of medical diagnosis, as most tests are quick and easy to perform, non-invasive, and well understood diagnostically.

A typical urinalysis usually includes a **macroscopic analysis**, a **dipstick chemical analysis**, in which the test results can be read as color changes, and a **microscopic analysis**, which involves centrifugation of the sample and examination for crystals, blood cells, or microbial contamination.

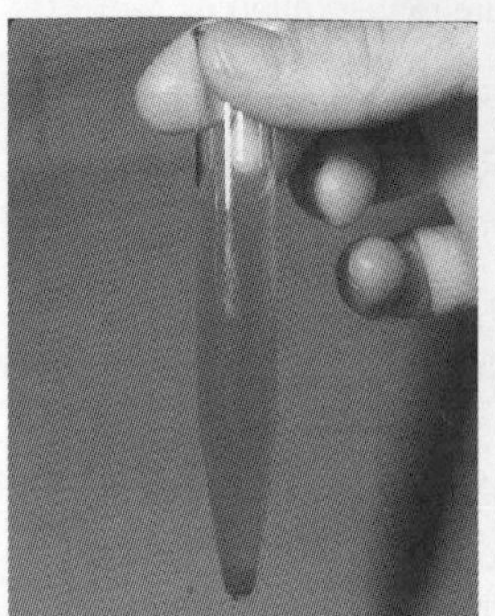

MACROSCOPIC URINALYSIS
The first part of a urinalysis is direct visual observation. Normal, fresh urine is pale to dark yellow or amber in color and clear. Turbidity or cloudiness may be caused by excessive cellular material or protein in the urine. A red or red-brown (abnormal) color could be from a food dye, eating fresh beets, a drug, or the presence of either hemoglobin or myoglobin. If the sample contained many red blood cells, it would be cloudy as well as red, as in this sample indicating hematuria (blood in the urine).

DIPSTICK URINALYSIS
Commonly dipstick tests indicate:
Urine pH: normal range is 4.5-8.0.

Specific gravity: Normal is 1.002 - 1.035 Specific gravity measures urine density, or the ability of the kidney to concentrate or dilute the urine over that of plasma.

Protein: Normal total protein excretion does not exceed 10 mg per 100 ml in any single specimen. More than 150 mg per day is defined as proteinuria.

Glucose: Less than 0.1% of glucose filtered by the glomerulus normally appears in urine. Excess sugar in urine generally indicates diabetes mellitus.

Ketones: Ketones in the urine result from diabetic ketosis or some other form of calorie deprivation (starvation).

Nitrite: Nitrites indicate that bacteria may be present in significant numbers.

Leukocyte esterase: A positive leukocyte esterase test results from the presence of whole or lysed white blood cells (indicating infection).

Testing For Anabolic Steroids

Anabolic steroids are synthetic steroids related to the male sex hormone **testosterone** (right). They work by increasing protein synthesis within cells, causing tissue, especially skeletal muscle, to build mass. They are used legitimately to stimulate bone growth and appetite, induce male puberty, and treat chronic wasting conditions. Misuse of anabolic steroids can have many adverse effects including elevated blood pressure, cardiovascular disease, and altered cholesterol ratios.

Steroids increase muscle mass and physical strength, and are used illegally by some athletes to gain an unfair advantage over their competitors.

Anabolic steroid use is banned by most major sporting bodies, but many athletes continue to use them illegally. Athletes are routinely tested for the presence of **performance enhancing drugs**, including anabolic steroids.

Anabolic steroids break down into known metabolites which are excreted in the urine. The presence of specific metabolites indicates which substance has been used by the athlete.

Some steroid metabolites stay in the urine for weeks or months after being taken, while others are eliminated quite rapidly. Athletes using anabolic steroids can escape detection by stopping use of the drugs prior to competition. This allows the body time to break down and eliminate the components, and the drug use goes undetected.

1. Explain why urinalysis is a frequently used diagnostic technique for many common disorders: ____________________

____

2. Explain why the pH of normal urine (4.5-8.0) is much more variable than the pH of the blood (pH 7.35-7.45):

____

3. Identify what each of the following might indicate in a urine sample:

(a) Cloudy, red color: ____________________ (b) Positive leukocyte esterase test: ____________________

4. Explain why athletes exploiting illegal drugs might withhold them for a period before competition: ____________________

____

____

ISBN: 978-1-92717357-2

Related activities: *Blood, Acid-Base Balance, The Hormones of Pregnancy*

Fluid and Electrolyte Balance

The body's fluid and electrolyte balance is critical to metabolic function. Water makes up around 60% of the body and is found within two main fluid compartments. The **intracellular fluid** makes up 60-65% of the water in the body and is found within the body's cells. The **extracellular fluid** makes up the rest of the body's water and can be divided into **intravascular fluid** (mostly blood) and the **extravascular fluid** (interstitial fluid around the cells). Electrolytes in the body fluids are responsible for maintaining osmotic gradients and permitting ion exchanges. For example, in the blood plasma, electrolytes help to maintain the blood volume by keeping water moving into the capillaries. When electrolyte (mostly Na^+) levels fall, water moves out of the capillaries and into the tissues. This causes blood volume and pressure to fall and plasma to thicken. Two hormones are involved in regulating blood volume: **ADH**, which promotes water reabsorption in the kidney collecting ducts, and **aldosterone**. Aldosterone promotes sodium reabsorption in the kidney tubules and the most important mechanism for regulating its release is the **renin-angiotensin system (RAS)**. The RAS is mediated by the **juxtaglomerular (JG) apparatus** in the renal tubules.

The Renin-Angiotensin System

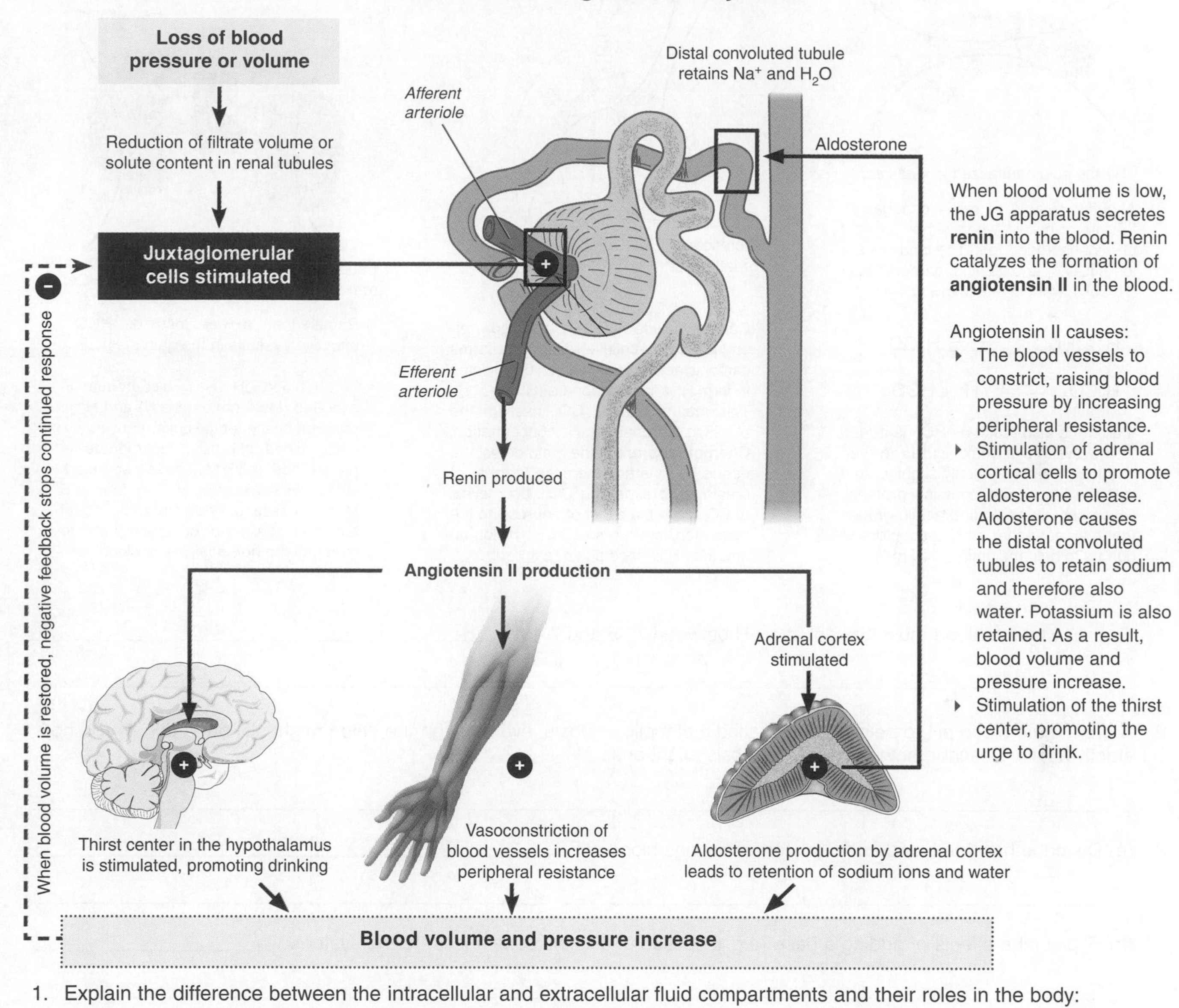

When blood volume is low, the JG apparatus secretes **renin** into the blood. Renin catalyzes the formation of **angiotensin II** in the blood.

Angiotensin II causes:

- The blood vessels to constrict, raising blood pressure by increasing peripheral resistance.
- Stimulation of adrenal cortical cells to promote aldosterone release. Aldosterone causes the distal convoluted tubules to retain sodium and therefore also water. Potassium is also retained. As a result, blood volume and pressure increase.
- Stimulation of the thirst center, promoting the urge to drink.

1. Explain the difference between the intracellular and extracellular fluid compartments and their roles in the body:

2. (a) Describe two situations that could cause a fall in blood volume:

   (b) Explain how the renin-angiotensin system responds to this loss of blood volume:

ISBN: 978-1-92717357-2

Related activities: Control of Urine Output, Kidney Transplants
Weblinks: The Juxtaglomerular Apparatus

Acid-Base Balance

The pH of the body's fluids must be maintained within a very narrow range (pH 7.35-7.45). The products of metabolic activity are generally acidic and could alter pH considerably without a buffer system to counteract pH changes. The carbonic acid-bicarbonate buffer works throughout the body to maintain the pH of blood plasma close to 7.40. The body maintains the buffer by eliminating either the acid (carbonic acid) or the base (bicarbonate ions). The blood buffers, the lungs, and the kidneys represent the three defense systems against disturbances of pH homeostasis. Changes in carbonic acid concentration can be effected within seconds through increased or decreased respiration. The renal system, although acting more slowly, can permanently eliminate metabolic acids and regulate the levels of alkaline substances, controlling pH by either excreting or retaining bicarbonate ions.

The Blood Buffer System

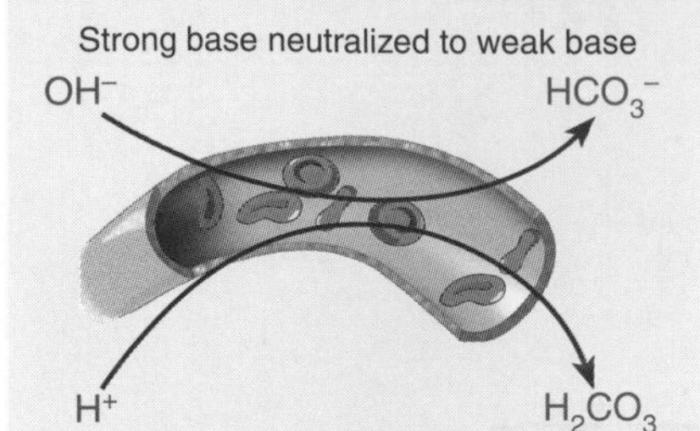

A buffer is able to resist changes to the pH of a fluid when either an acid or base is added to it. The bicarbonate ion (HCO_3^-) and its acid, carbonic acid (H_2CO_3), work in the following way:

$$H^+ + HCO_3^- \rightleftharpoons H_2CO_3$$

$$H_2CO_3 \rightleftharpoons H^+ + HCO_3^-$$

If a strong acid (such as HCl) is added to the system a weak acid is formed and thus the pH falls only slightly. Note that the blood also contains proteins, which contain basic and acidic groups that may act either as H^+ acceptors or donors to help maintain blood pH.

The Respiratory System

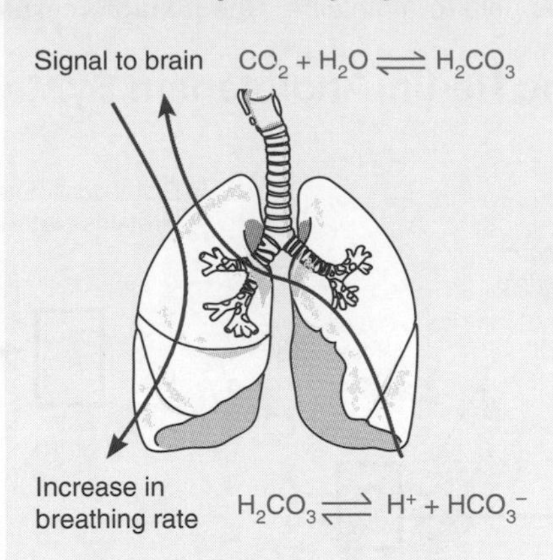

Carbon dioxide (CO_2) in the blood, an end-product of cellular respiration, forms carbonic acid (H_2CO_3) which dissociates to form H^+ and bicarbonate (HCO_3^-). This means that as CO_2 rises in the blood so too does the H^+ concentration. **Chemoreceptors** in the brain detect the rise in H^+ ions and increase the rate of breathing to expel the CO_2. Low levels of CO_2 have the effect of depressing the respiratory system so that H^+ builds up and the pH is once again restored.

The Renal System

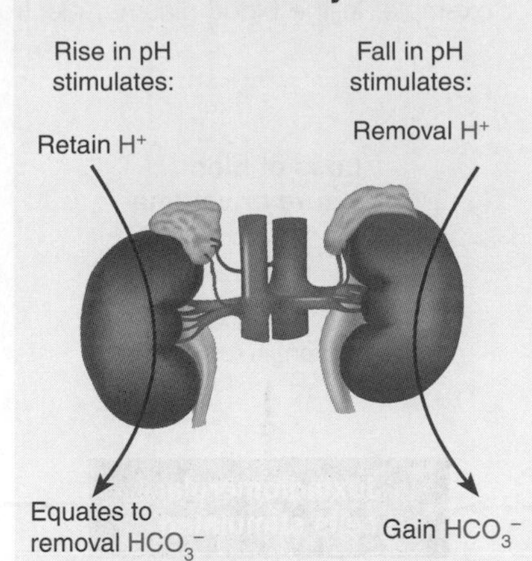

Recall that a net loss of HCO_3^- effectively results in the gain of H^+.

When blood pH rises, bicarbonate is excreted (lost from the body) and H^+ is retained by the tubule cells. Conversely, when blood pH falls, bicarbonate is reabsorbed and H^+ is actively secreted. Urine pH can normally vary from 4.5 to 8.0, reflecting the ability of the renal tubules to excrete or retain ions to maintain the homeostasis of blood pH.

1. Explain why the blood must be kept at a pH between 7.35 and 7.45: ______________________

____

2. A drop in the blood pH to below 7.35 is called metabolic acidosis, even though the blood might still be at pH >7 and not strictly acidic. Describe how metabolic acidosis might arise:

____

3. (a) Describe how the blood buffer system maintains blood pH: ______________________

____

(b) Explain the effects of adding a base (e.g. ingestion of alkaline substances) to the system: ______________________

____

4. (a) Describe the respiratory response to excess H^+ in the blood: ______________________

____

(b) Explain where these H^+ ions come from: ______________________

(c) Describe how **respiratory acidosis** might arise: ______________________

____

5. Explain the role of the renal system in maintaining the pH of the blood: ______________________

____

____

Related activities: *Gas Transport in Humans, Control of Breathing, The Physiology of the Kidney*

ISBN: 978-1-92717357-2

Kidney Dialysis

A dialysis machine is a machine designed to remove wastes from the blood. It is used when the kidneys fail, or when blood acidity, urea, or potassium levels increase much above normal. In kidney dialysis, blood flows through a system of tubes composed of partially permeable membranes. Dialysis fluid (dialysate) has a composition similar to blood except that the concentration of wastes is low. It flows in the opposite direction to the blood on the outside of the dialysis tubes. Consequently, waste products like urea diffuse from the blood into the dialysis fluid, which is constantly replaced. The dialysis fluid flows at a rate of several 100 cm^3 per minute over a large surface area. For some people dialysis is an ongoing procedure, but for others dialysis just allows the kidneys to rest and recover from injury or the effects of drugs or other metabolic disturbance.

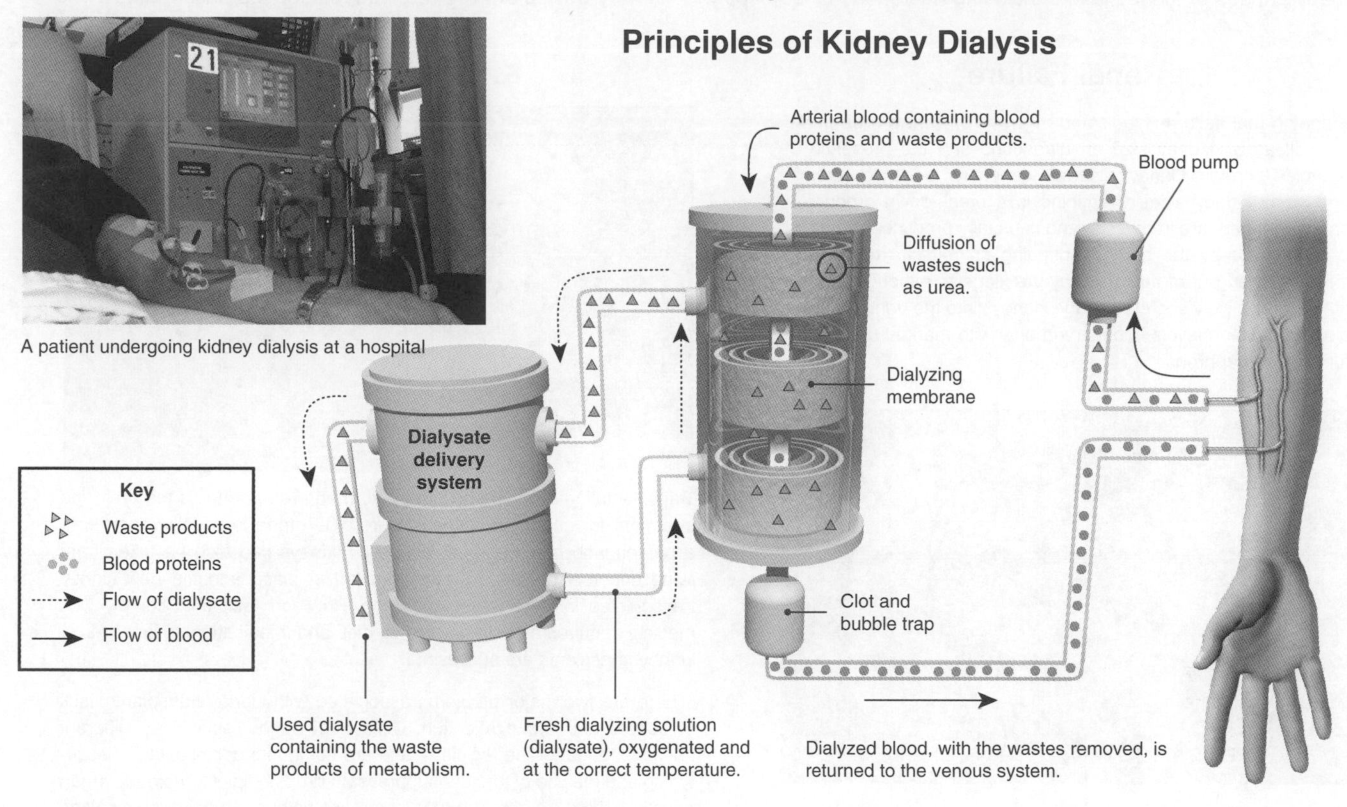

A patient undergoing kidney dialysis at a hospital

1. In kidney dialysis, explain why the dialyzing solution is constantly replaced rather than being recirculated:

2. Explain why ions such as potassium and sodium, and small molecules like glucose do not diffuse rapidly from the blood into the dialyzing solution along with the urea:

3. Explain why the urea passes from the blood into the dialyzing solution:

4. Describe the general transport process involved in dialysis:

5. Give a reason why the dialyzing solution flows in the opposite direction to the blood:

6. Explain why a clot and bubble trap is needed after the blood has been dialyzed but before it re-enters the body:

ISBN: 978-1-92717357-2

Kidney Transplants

Kidney failure (also called renal failure) arises when the kidneys fail to function adequately and filtrate formation decreases or stops. In cases of renal failure, normal blood volume levels and electrolyte balances are not maintained, and waste products build up in the body. Kidney failure is classified as **acute** (rapid onset) or **chronic** (developing over a period of months or years). There are many causes of kidney failure including decreased blood supply, drug overdose, chemotherapy, infection, and poorly controlled diabetic or hypertensive conditions. Recovery from acute renal failure is possible, but chronic renal damage can not be reversed. If kidney deterioration is ignored, the kidneys will fail completely. In some cases diet and medication can be used to treat kidney failure, but when the damage is extensive, **kidney dialysis** or a **kidney transplant** are required to keep the patient alive.

Renal Failure

Kidney (renal) failure is indicated by levels of **serum creatinine**, as well as by kidney size on ultrasound and the presence of anemia (chronic kidney disease generally leads to anemia and small kidney size). Creatinine is a break-down product of creatine phosphate in muscle, and is usually produced at a fairly constant rate by the body (depending on muscle mass). It is chiefly filtered out of the blood by the kidneys, although a small amount is actively secreted by the kidneys into the urine. A rise in blood creatinine levels is observed only with marked damage to functioning nephrons.

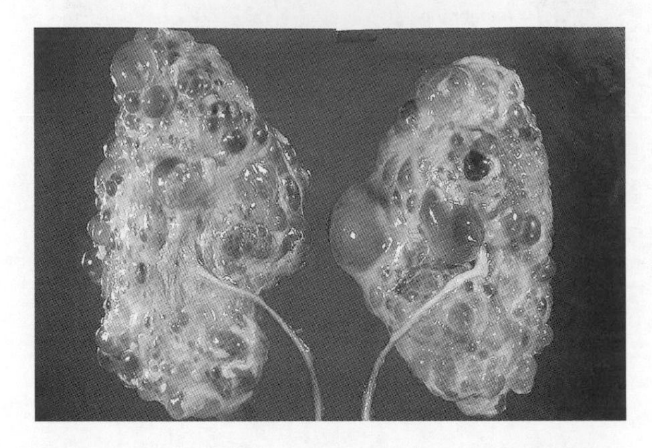

Acute renal failure (ARF) is characterized by decreased urine production (<400mL per day), and commonly arises because of low blood volume (blood loss), dehydration, or widespread infection. In contrast, chronic renal failure, which develops over months or years, is commonly the result of poorly controlled diabetes, poorly controlled high blood pressure, or **polycystic kidney disease**, a genetic disorder characterized by the growth of numerous cysts in the kidneys (above).

Kidney Transplants

Transplantation of a healthy kidney from an organ donor is the preferred treatment for end-stage kidney failure. The organ is usually taken from a person who has just died, although kidneys can also be taken from living donors. The failed organs are left in place and the new kidney transplanted into the lower abdomen. Provided recipients comply with medical requirements (e.g. correct diet and medication) over 85% of kidney transplants are successful.

There are two major problems associated with kidney transplants: lack of donors and tissue rejection. Cells from donor tissue have different antigens to that of the recipient, and are not immunocompatible. Tissue-typing and the use of immunosuppressant drugs helps to decrease organ rejection rates. In the future, xenotransplants of genetically modified organs from other species may help to solve both the problems of supply and immune rejection.

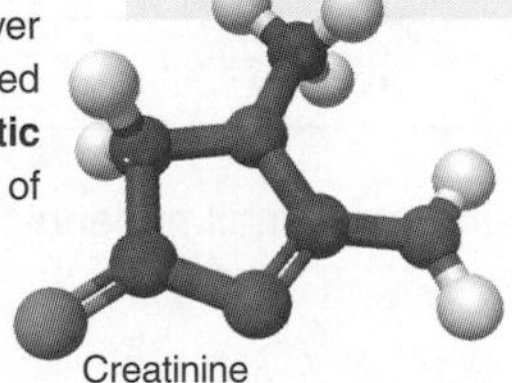

Creatinine

Creatinine levels in both blood and urine is used to calculate the creatinine clearance (CrCl), which reflects the glomerular filtration rate (GFR). The GFR is a clinically important measurement of renal function and more accurate than serum creatinine alone, since serum creatinine only rises when nephron function is very impaired.

1. Distinguish between acute and chronic renal failure and contrast their causes:

2. (a) Explain why a rise in blood (serum) levels of creatinine would indicate a failure of nephron function:

   (b) Explain why a creatinine clearance is a more accurate indicator of renal function than a serum creatinine test alone:

3. Describe some of the advantages and disadvantages of kidney transplantation over a life-time of kidney dialysis:

***Related activities:** Energy for Muscle Contraction, Control of Urine Output, Kidney Dialysis*

ISBN: 978-1-92717357-2

KEY TERMS: Mix and Match

INSTRUCTIONS: Test your vocabulary by matching each term to its definition, as identified by its preceding letter code.

Term		Definition
aldosterone	A	The functional unit of the kidney comprising the glomerulus, Bowman's capsule, convoluted tubules, loop of Henle and collecting duct.
antidiuretic hormone (ADH)	B	The portion of the kidney nephron that leads from Bowman's capsule to the loop of Henle. It is the main site of reabsorption of water and solutes.
bladder	C	Process by which small molecules and ions are separated from larger ones in the blood to form the renal filtrate (in the kidney).
Bowman's capsule	D	Body fluid that is not contained in cells. Examples include the blood plasma and lymph.
collecting duct	E	Part of the kidney nephron between the proximal convoluted tubule and the distal convoluted tubule. Its function is to create a gradient in salt concentration through the medullary region of the kidney.
cortex (of kidney)	F	The ejection of urine from the urinary bladder through the urethra to the outside of the body.
distal convoluted tubule	G	Tubule in the kidney nephron where urine is concentrated as water, and leaves the tubule by osmosis.
electrolyte	H	The fluid contained within cells.
excretion	I	The hormone released in response to low blood volumes, high sodium levels and low fluid intake.
extracellular fluid	J	Sensory organs that detect changes in osmotic pressure.
glomerulus	K	Duct leading from the urinary bladder to the exterior of the body.
intracellular fluid	L	The funnel like upper part of the ureter. It links the collecting ducts to the ureter and drains urine from the kidney to the ureter (and on to the bladder).
kidney	M	A substance that dissociates into ions in solution and can conduct electricity. Examples include sodium, potassium, chloride, and calcium.
loop of Henle	N	The outermost layer of an organ. The portion of the kidney between the renal capsule and medulla.
medulla (of kidney)	O	The portion of the kidney nephron that connects the loop of Henle and the collecting duct.
micturition (= urination)	P	Elimination of waste products of metabolism.
nephron	Q	Hormone produced by the adrenal gland that causes sodium and water to be retained in the body, and potassium to be secreted into the urine.
osmoreceptors	R	The middle region of the kidney.
proximal convoluted tubule	S	Organ of the urinary system that collects urine.
renal corpuscle	T	The filtration unit of the nephron. It consists of the glomerulus and the Bowman's capsule.
renal dialysis	U	Bean shaped organ which removes and concentrates metabolic wastes from the blood.
renal failure	V	A double walled cup, situated in the cortex of the kidney, and surrounding the glomerulus.
renal pelvis	W	Duct that carries urine from the kidney to the urinary bladder.
ultrafiltration	X	A knot of capillaries in the kidney bound within the Bowman's capsule.
urethra	Y	Fluid containing metabolic wastes that collects in the urinary bladder.
ureter	Z	Medical condition in which the kidneys fail to adequately filter toxins and waste products from the blood. Results in edema and disruption of pH and electrolyte balance, and leads to anaemia in the long term.
urine	AA	A medical process designed to remove wastes from the blood when the kidneys have failed.

ISBN: 978-1-92717357-2

Interacting Systems

The Reproductive System

Cardiovascular system

- Estrogens increase blood HDL cholesterol levels and promote cardiovascular health in premenopausal women.
- Pregnancy places extra demands on the cardiovascular system. Blood volume increase 40-50% during pregnancy.
- Local vasodilation is responsible for aspects of the sexual response in men and women.
- Blood transports sex hormones to target tissues.

Respiratory system

- Respiratory system provides O_2 to the reproductive system and disposes of CO_2 produced by cellular respiration.
- Vital capacity and breathing rate increase in pregnancy. Enlarged uterus impairs descent of the diaphragm and can cause shortness of breath late in pregnancy.

Endocrine system

- Reproductive hormones from the ovaries (in females) and testes (in males) are responsible for the development of secondary sexual characteristics. They are regulated via feedback mechanisms to the hypothalamic-pituitary axis.
- Gonadotropins (e.g. LH and FSH) help to regulate gonadal function.
- Placental hormones maintain pregnancy.

Skeletal system

- Sex hormones are responsible for secondary sexual characteristics associated with the skeleton: in males, androgens masculinize the skeleton (broad shoulders and expanded chest) and increase bone density; in females, estrogen causes pelvic widening and maintains bone mass.
- Bony pelvis protect some reproductive organs.

Integumentary system

- In lactating women, milk from mammary glands nourishes the infant.
- Androgens activate oil glands and lubricate skin and hair. Estrogen increases skin hydration and increases skin pigmentation in pregnancy.
- Sex hormones are responsible for secondary sexual characteristics associated with the integument, e.g. appearance of pubic hair and changes in fat distribution associated with male and female body shape.

Nervous system

- Sex hormones masculinize or feminize the brain and influence sex drive.
- The neurohormone, GnRH from the hypothalamus regulates the timing of puberty.
- Reflexes regulate aspects of the sexual response (e.g. orgasm).

Lymphatic system and immunity

- Immune cells protect against pathogens. Regulatory T cells important in the immune tolerance to the developing fetus.
- Lymphatic (and blood) vessels transport sex hormones.
- Maternal antibodies pass to the fetus *in-utero* and are present in breast milk, providing passive immunity.
- Increased abdominal pressure in pregnancy impairs lymphatic return leading to edema.

Digestive system

- Digestive organs are crowded in late pregnancy and constipation and heartburn are common.
- Increased hormone levels result in nausea and vomiting in early pregnancy.

Urinary system

- Increased frequency and urgency of urination in pregnancy as a result of pressure on the bladder and pelvic floor.
- Enlargement of prostate (usually) in older men can impede urination.
- Kidneys dispose of nitrogenous waste and maintain fluid and electrolyte balance of mother and fetus in pregnancy.
- Urethra provides passage for semen.

Muscular system

- Androgens promote an increase in muscle mass in post-pubertal males.
- Pelvic floor muscles provide support for the reproductive organs and are involved in aspects of the sexual response.
- Abdominal, uterine, and pelvic floor muscles are active in childbirth.

General Functions and Effects on all Systems

The reproductive system in adults is responsible for reproduction, i.e. the production of gametes and offspring. Unlike other body systems, which are functioning almost continuously since birth, the reproductive system is quiescent until puberty, at which time it begins development towards maturity.

Disease

Symptoms of disease
- Pain (moderate to severe)
- Abnormal bleeding
- Infertility

Disorders and diseases of the male reproductive system
- Sexually transmitted infections
- Cancers (e.g. prostate cancer)
- Congenital abnormalities
- Functional disorders (e.g. erectile dysfunction, premature ejaculation)

Disorders and diseases of the female reproductive system
- Sexually transmitted infections
- Cancers (e.g. cervical cancer)
- Congenital abnormalities
- Functional disorders (e.g. infertility, ectopic pregnancy, endometriosis)

Medicine & Technology

Diagnosis of disorders
- MRI scan and ultrasound
- Semen analysis
- Laparoscopy
- Blood and DNA (genetic) tests

Treating reproductive disorders
- Drug therapy (e.g. antibiotics)
- Hormone therapy
- Surgery (e.g. hysterectomy)

Treatment of infertility
- Assisted reproductive technologies
- Hormone therapy (e.g. clomiphene)
- Laparoscopic surgery

Contraception
- Physical (barrier) methods
- Hormonal (e.g. oral contraceptive pill)
- Surgical (e.g. vasectomy)

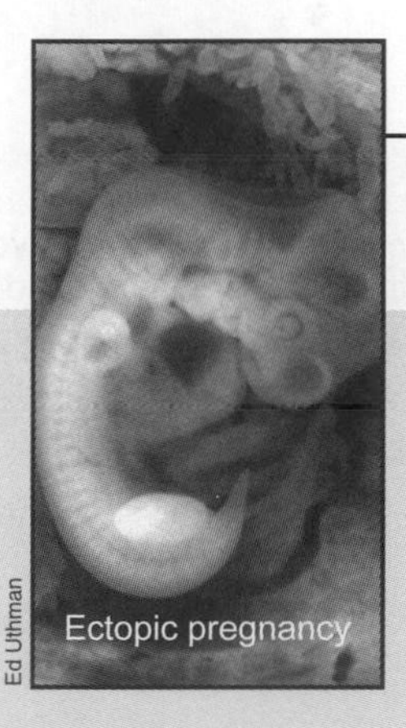

Ectopic pregnancy

Ed Uthman

- Infertility
- Reproductive cancers
- Ectopic pregnancy
- Sexually transmitted infections

- IVF and GIFT
- Ultrasound
- HRT
- Oral contraception
- Pregnancy testing

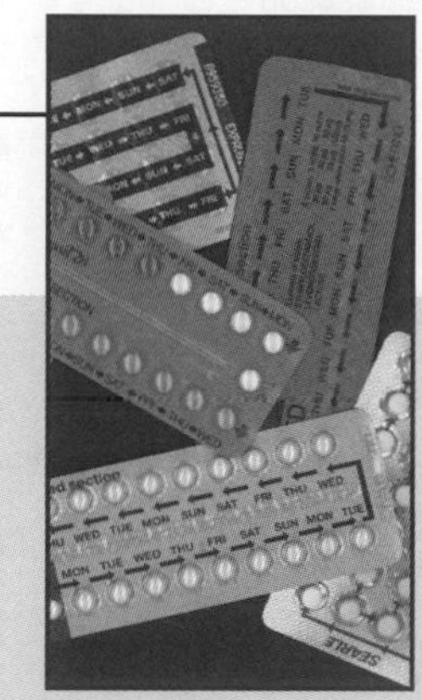

The Human Life Span

The Reproductive System

The reproductive system undergoes marked changes associated with aging and the end of fertility. Disease may affect it both directly and indirectly.

Medical technologies are used to detect, diagnose and treat reproductive disorders and control fertility.

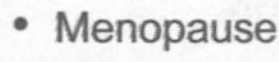

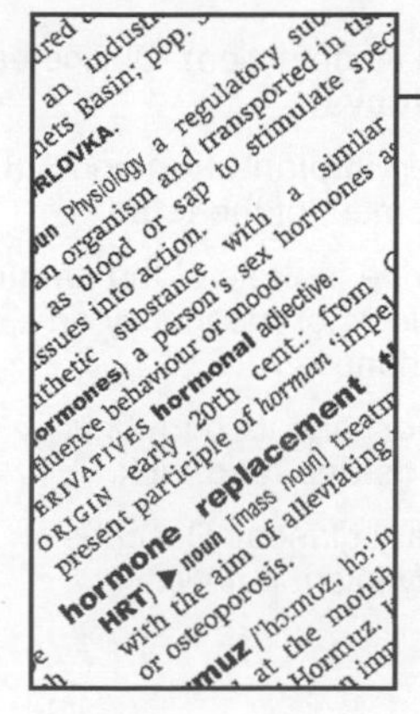

- Menopause
- Decline in hormone levels
- Decline in fertility

Excessive exercise may lead to:
- Exercise induced amenorrhea
- GNRH depression

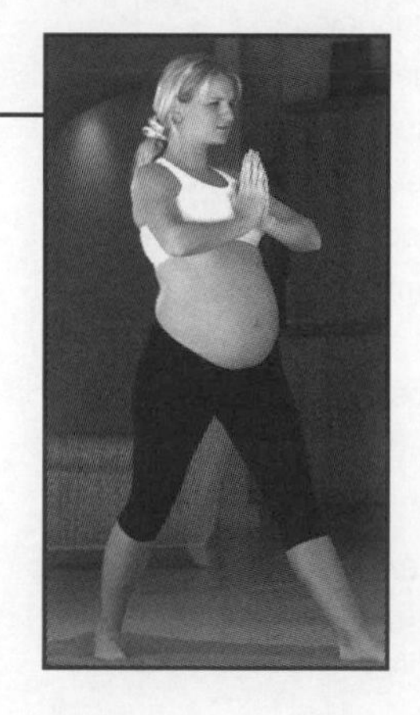

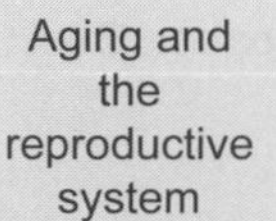

Aging and the reproductive system
- Decline in sperm production and erectile function
- Cessation of menses (women)
- Thinning and prolapse of organs

Effects of exercise on the reproductive system
- Improved muscle tone
- Reduced risk of reproductive cancers
- Heavy endurance exercise can lead to hormonal and menstrual irregularities
- Exercise beneficial during pregnancy

The Effects of Aging

Exercise

Reproduction and Development

Key terms

acrosome reaction
apoptosis
blastocyst
breast
childbirth (=parturition or labor)
cleavage
cortical reaction
degenerative disease
endometrium
estrogen
fertility
fertilization
FSH
gametogenesis
implantation
infertility
lactation
LH
menopause
menstrual cycle
oogenesis
oxytocin
placenta
positive feedback
pregnancy
progesterone
prostaglandins
puberty
reproductive system
spermatogenesis
STI
testosterone

Periodicals:
Listings for this chapter are on page 280

Weblinks:
www.thebiozone.com/weblink/AnaPhy-3572.html

BIOZONE APP:
Student Review Series
The Reproductive System

Key concepts

- The reproductive system is responsible for production of gametes and offspring.
- Its activity is regulated by reproductive hormones from the pituitary and gonads.
- Birth follows a period of gestation in which the fetus is supported by the placenta.
- The human life span is characterized by distinct developmental stages.
- Humans can intervene in their own fertility.

Learning Objectives

☐ 1. Use the **KEY TERMS** to compile a glossary for this topic.

The Reproductive System and Gametogenesis — pages 255-258, 260

☐ 2. Use annotated diagrams to describe the structure and function of the **reproductive system** in male and females, including the roles of the primary sex organs (gonads).

☐ 3. Describe the roles of the sex hormones **estrogen**, **progesterone**, and **testosterone**. Recognize the role of pituitary gonadotropic hormones in regulating gonadal function.

☐ 4. Compare and contrast **gametogenesis** in males and females.

☐ 5. Describe **spermatogenesis** including reference to mitosis, cell growth, meiosis, and cell differentiation. Explain the role of **FSH**, **LH**, and testosterone in this process.

☐ 6. Describe **oogenesis** including reference to mitosis, cell growth, meiosis, the unequal division of the cytoplasm, and the degeneration of the polar bodies.

The Menstrual Cycle, Pregnancy, and Lactation — pages 184, 259, 261-268

☐ 7. Describe the **menstrual cycle** including the development of the ovarian follicles and corpora lutea, the cyclical changes to the uterine **endometrium**, and **menstruation**. Relate the changes in the cycle to the changes in its regulating hormones.

☐ 8. Describe **fertilization**, including the **acrosome reaction**, penetration of the egg membrane, and the **cortical reaction**.

☐ 9. Describe the role of **HCG** in early **pregnancy** (including in its detection). Outline early embryonic development up to the implantation of the **blastocyst**.

☐ 10. Outline the main events in embryonic development between **implantation** and 5-8 weeks. Describe the role of **apoptosis** in the early development of the **fetus**.

☐ 11. Describe the structure and function of the **placenta**. Describe the role of the **amniotic sac** and **amniotic fluid** during pregnancy. Describe the effects of pregnancy on maternal physiology, including anatomical and hormonal changes.

☐ 12. Describe the three phases of childbirth (**parturition** or **labor**) and its hormonal control. Explain how parturition is accomplished through **positive feedback**.

☐ 13. Describe the structure and function of the **breast** (mammary glands). Describe **lactation** and its control and explain its importance to early infant nutrition.

Development and Aging — pages 65, 112-113, 130, 274-276

☐ 14. Describe main events in the growth and development of humans, including changes during infancy, childhood, **puberty**, and adulthood.

☐ 15. Explain the biological basis of **aging**, including reference to **menopause**. Describe some of the **degenerative diseases** of aging.

Reproductive Disease and Technology — pages 269-273

☐ 16. Describe diseases that can affect the reproductive system in males and females, including infectious diseases (such as **STIs**) and non-infectious diseases such as **cancers**. Discuss the implications of reproductive disease on **fertility**.

☐ 17. Identify causes of **infertility** and describe the use of reproductive technologies such as **IVF** to treat female infertility. Describe the control of fertility using **contraception**.

Gametes

Gametes (eggs and sperm) are **haploid** sex cells formed by reduction division in the gonads of females and males. Eggs and sperm differ greatly in their size, shape, and number. These differences reflect their different roles in fertilization and reproduction. Human sperm (or spermatozoa) are highly motile and produced in large numbers. Eggs (or ova) are large, few in number, and immobile in themselves. They move as a result of the wave-like motion produced by the ciliated cells lining the Fallopian tube. Egg cells contain some food sources to nourish the developing embryo. In humans, this food source is very limited because once implantation into the uterus takes place, the fetus derives its nutrient supply from the mother's blood supply.

Egg Structure and Function

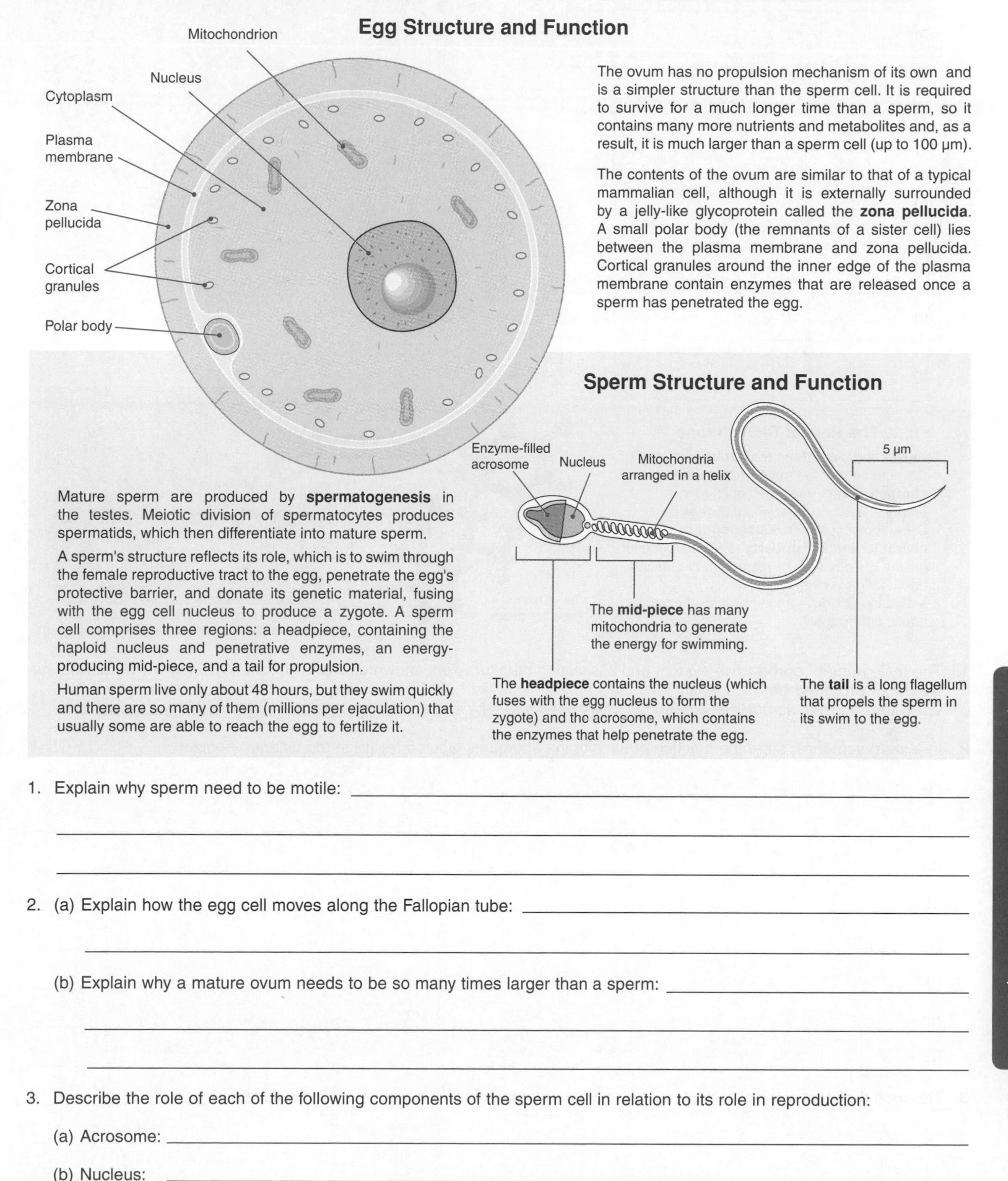

The ovum has no propulsion mechanism of its own and is a simpler structure than the sperm cell. It is required to survive for a much longer time than a sperm, so it contains many more nutrients and metabolites and, as a result, it is much larger than a sperm cell (up to 100 µm).

The contents of the ovum are similar to that of a typical mammalian cell, although it is externally surrounded by a jelly-like glycoprotein called the **zona pellucida**. A small polar body (the remnants of a sister cell) lies between the plasma membrane and zona pellucida. Cortical granules around the inner edge of the plasma membrane contain enzymes that are released once a sperm has penetrated the egg.

Sperm Structure and Function

Mature sperm are produced by **spermatogenesis** in the testes. Meiotic division of spermatocytes produces spermatids, which then differentiate into mature sperm.

A sperm's structure reflects its role, which is to swim through the female reproductive tract to the egg, penetrate the egg's protective barrier, and donate its genetic material, fusing with the egg cell nucleus to produce a zygote. A sperm cell comprises three regions: a headpiece, containing the haploid nucleus and penetrative enzymes, an energy-producing mid-piece, and a tail for propulsion.

Human sperm live only about 48 hours, but they swim quickly and there are so many of them (millions per ejaculation) that usually some are able to reach the egg to fertilize it.

The **mid-piece** has many mitochondria to generate the energy for swimming.

The **headpiece** contains the nucleus (which fuses with the egg nucleus to form the zygote) and the acrosome, which contains the enzymes that help penetrate the egg.

The **tail** is a long flagellum that propels the sperm in its swim to the egg.

1. Explain why sperm need to be motile: ____________________

2. (a) Explain how the egg cell moves along the Fallopian tube: ____________________

   (b) Explain why a mature ovum needs to be so many times larger than a sperm: ____________________

3. Describe the role of each of the following components of the sperm cell in relation to its role in reproduction:

   (a) Acrosome: ____________________

   (b) Nucleus: ____________________

   (c) Mitochondria: ____________________

   (d) Flagellum: ____________________

ISBN: 978-1-92717357-2

Related activities: *Spermatogenesis, Oogenesis*

The Male Reproductive System

The male reproductive system (below) is concerned with producing sperm and delivering them to the female urogenital tract. Mature sperm are ejaculated with fluids from the seminal vesicles and prostate as semen. When a sperm combines with an egg, it contributes half the genetic material of the offspring and, in humans and other mammals, determines its sex.

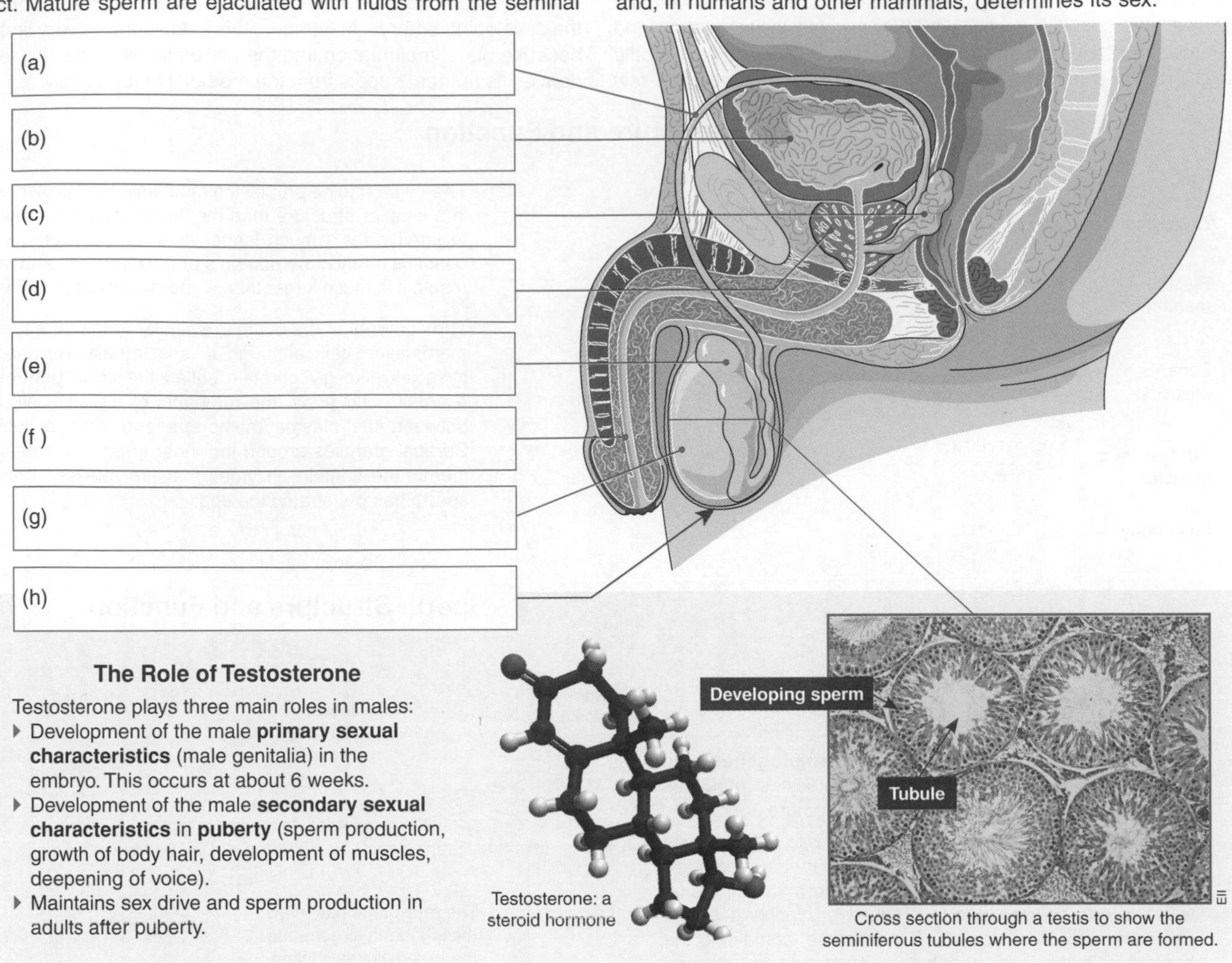

Cross section through a testis to show the seminiferous tubules where the sperm are formed.

The Role of Testosterone

Testosterone plays three main roles in males:

- Development of the male **primary sexual characteristics** (male genitalia) in the embryo. This occurs at about 6 weeks.
- Development of the male **secondary sexual characteristics** in **puberty** (sperm production, growth of body hair, development of muscles, deepening of voice).
- Maintains sex drive and sperm production in adults after puberty.

Testosterone: a steroid hormone

1. The male human reproductive system and associated structures are shown above. Using the following **word list** and the **weblinks provided below**, identify the labeled parts (write your answers in the spaces provided on the diagram).
Word list: *bladder, scrotal sac, sperm duct (vas deferens), epididymis, seminal vesicle, testis, urethra, prostate gland*

2. In a short sentence, state the function of each of the structures labeled (a)-(h) in the diagram above:

(a) ______________________

(b) ______________________

(c) ______________________

(d) ______________________

(e) ______________________

(f) ______________________

(g) ______________________

(h) ______________________

3. Describe the three roles of testosterone in male development and the male reproductive system: ______________________

____

____

4. State the two main roles of the male reproductive system: ______________________

____

Related activities: *Sexual Development*
Weblinks: *Male Reproductive System*

Periodicals:
Spermatogenesis

ISBN: 978-1-92717357-2

Spermatogenesis

Gametogenesis involves meiotic division to produce male and female gametes (sperm and eggs) for the purpose of sexual reproduction. Males produce sperm in the testis by **spermatogenesis**. In humans, sperm production begins at puberty and continues throughout life, although it declines with age. Fluids secreted from various parts of the male reproductive tract support and transport the sperm. Thousands of sperm are produced every second, and take about two months to mature.

Spermatogenesis

Spermatogenesis is the process by which mature spermatozoa (sperm) are produced in the testis. The process is regulated by the hormones **follicle stimulating hormone** (FSH) (from the anterior pituitary) and **testosterone** (secreted from the testes) in response to **luteinizing hormone** (LH) (from the anterior pituitary).

Spermatogonia, in the outer layer of the seminiferous tubules, multiply throughout reproductive life. Some of them divide by meiosis into spermatocytes, which produce spermatids. These are transformed into mature sperm by the process of spermiogenesis in the seminiferous tubules of the testis. Full sperm motility is achieved in the epididymis.

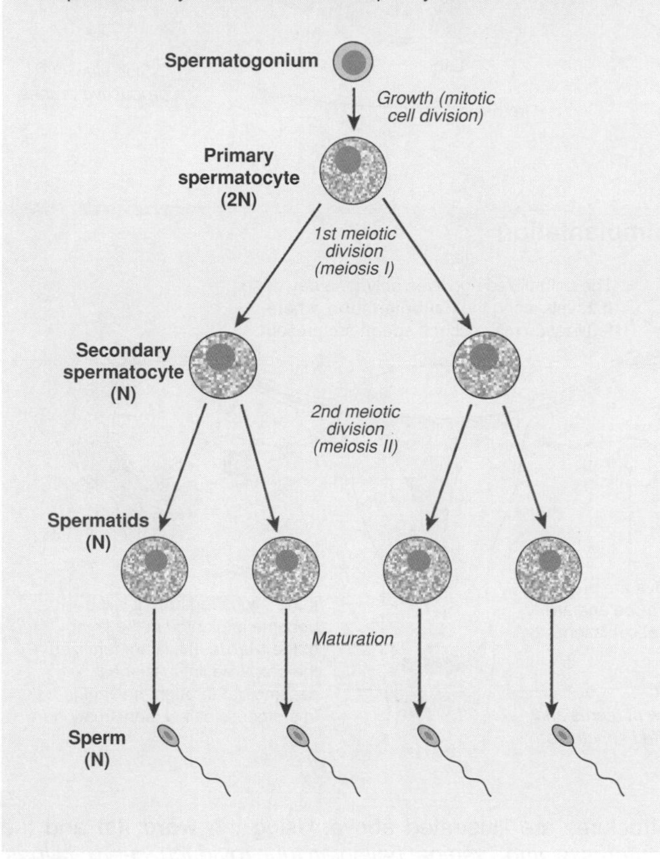

Cross Section Through Seminiferous Tubule

The photograph below shows maturing sperm (arrowed) with tails projecting into the lumen of the seminiferous tubule. Their heads are embedded in the Sertoli cells in the tubule wall and they are ready to break free and move to the epididymis where they complete their maturation. The same cross-section is illustrated diagrammatically (bottom).

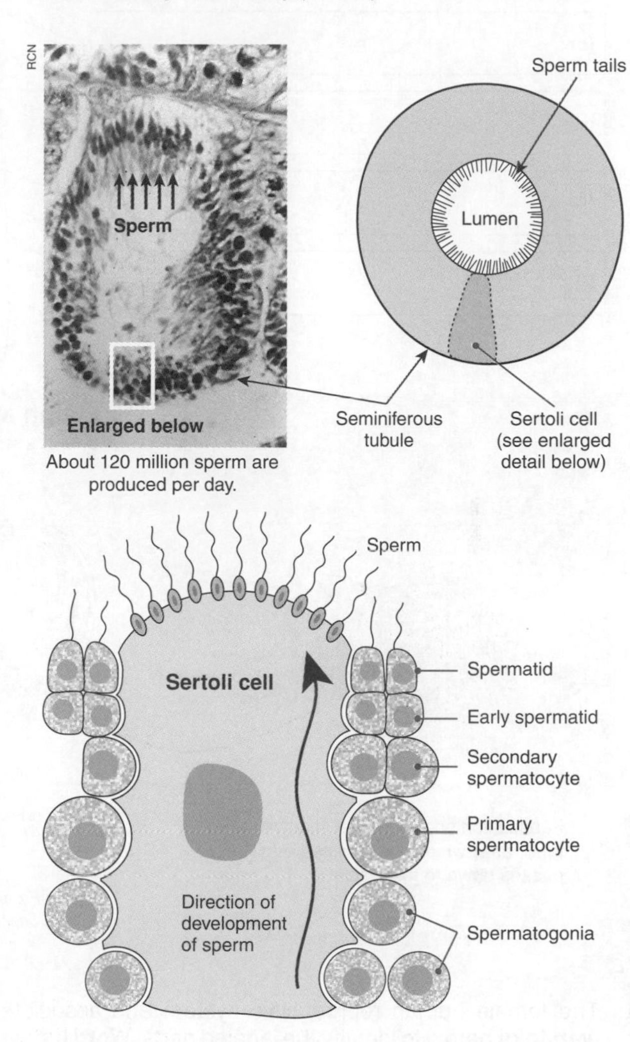

About 120 million sperm are produced per day.

1. (a) Name the process by which mature sperm are formed: ____________

   (b) Identify where this process takes place: ____________

   (c) State how many mature sperm form from one primary spermatocyte: ____________

   (d) State the type of cell division which produces mature sperm cells: ____________

2. Describe the role of FSH and LH in sperm production: ____________

3. Each ejaculation of a healthy, fertile male contains 100-400 million sperm. Suggest why so many sperm are needed: ____________

ISBN: 978-1-92717357-2

Related activities: *Oogenesis, Gametes*

Weblinks: *Spermatogenesis, Comparison of Spermatogenesis and Oogenesis*

The Female Reproductive System

The female reproductive system consists of the ovaries, Fallopian tubes, uterus, the vagina and external genitalia, and the breasts. Although both male and females have breasts, the female breasts (called mammary glands) are modified so that they produce milk after childbirth. The female reproductive system produces eggs, receives the penis and sperm during sexual intercourse, protects and houses the developing fetus, and produces milk to nourish the young after birth.

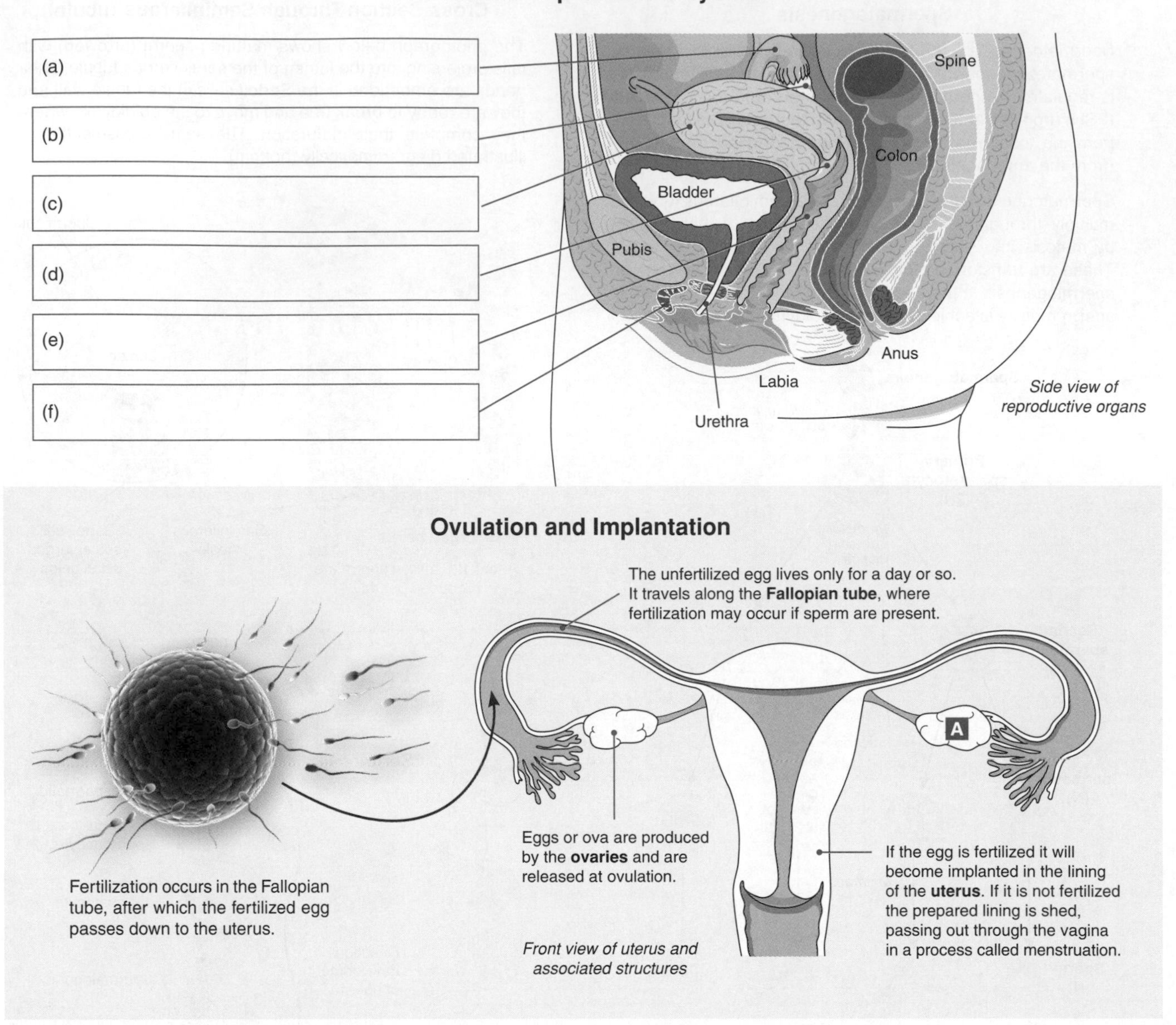

1. The female human reproductive system and associated structures are illustrated above. Using the **word list** and the **weblinks below** to identify the labeled parts. **Word list**: *ovary, uterus (womb), vagina, Fallopian tube (oviduct), cervix, clitoris.*

2. In a few words or a short sentence, state the function of each of the structures labeled (a) - (e) in the above diagram:

   (a) ______

   (b) ______

   (c) ______

   (d) ______

   (e) ______

3. (a) Name the organ labeled (**A**) in the diagram above: ______

   (b) Name the event associated with this organ that occurs every month: ______

   (c) Name the process by which mature ova are produced: ______

4. Where does fertilization occur? ______

Related activities: *The Menstrual Cycle, Oogenesis*
Weblinks: *Female Reproductive System*

ISBN: 978-1-92717357-2

The Menstrual Cycle

Humans are sexually receptive throughout the year and may have sexual intercourse at any time, but fertilization of the ovum is most likely to occur during a relatively restricted period around the time of **ovulation**. The uterine lining thickens in preparation for pregnancy but is shed as a bloody discharge through the vagina if fertilization does not occur. This event, called **menstruation**, characterizes the human reproductive or **menstrual cycle**. The menstrual cycle starts from the first day of bleeding and lasts for about 28 days. It involves a predictable series of changes that occur in response to hormones from the pituitary and ovaries. The cycle is divided into three phases (follicular, ovulatory, and luteal) defined by the events in each phase.

Events in the Menstrual Cycle

Luteinizing hormone (LH) and follicle stimulating hormone (FSH) from the anterior pituitary: FSH stimulates the development of the ovarian follicles resulting in the release of estrogen. Estrogen levels peak, stimulating a surge in LH and triggering ovulation.

Hormone levels: Usually only one of developing follicles (the Graafian follicle) becomes dominant. In the first half of the cycle, estrogen is secreted by the Graafian follicle. The Graafian follicle develops into the corpus luteum (below right) which secretes large amounts of progesterone (and smaller amounts of estrogen).

The corpus luteum: Around day 14, the Graafian follicle ruptures to release the egg (ovulation). LH causes the ruptured follicle to develop into a corpus luteum (yellow body). The corpus luteum secretes progesterone which promotes full development of the uterine lining, maintains the embryo in the first 12 weeks of pregnancy, and inhibits the development of more follicles.

Menstruation: If fertilization does not occur, the corpus luteum breaks down. Progesterone secretion drops, causing the uterine lining to be shed (menstruation). If fertilization occurs, high progesterone levels maintain the thickened uterine lining. The placenta develops and nourishes the embryo completely by 12 weeks.

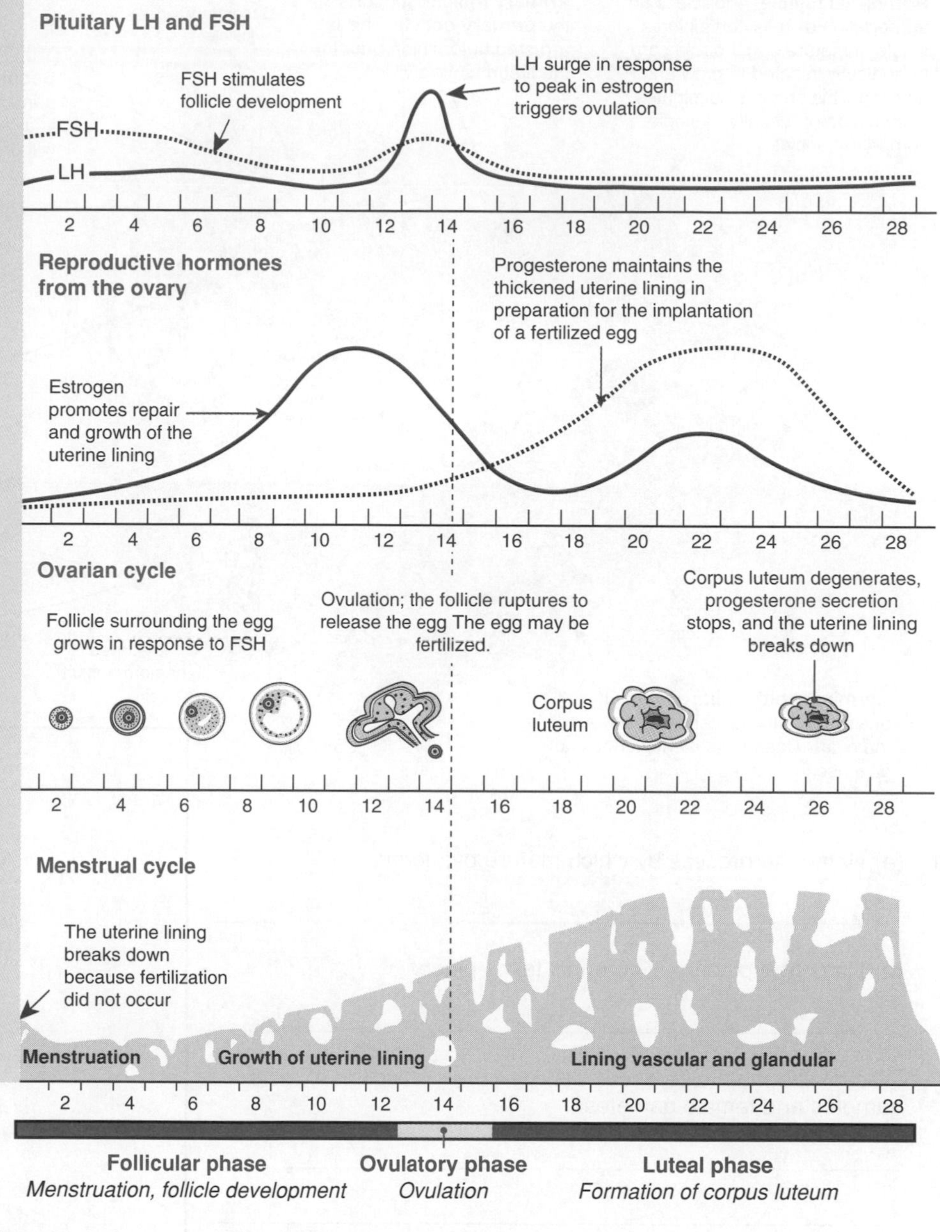

1. Name the hormone responsible for:

   (a) Follicle growth: ______________________ (b) Ovulation: ______________________

2. Each month, several ovarian follicles begin development, but only one (the Graafian follicle) develops fully:

   (a) Name the hormone secreted by the developing follicle: ______________________

   (b) State the role of this hormone during the follicular phase: ______________________

   (c) Suggest what happens to the follicles that do not continue developing: ______________________

3. (a) Identify the principal hormone secreted by the corpus luteum: ______________________

   (b) State the purpose of this hormone: ______________________

4. State the hormonal trigger for menstruation: ______________________

ISBN: 978-1-92717357-2

Related activities: *Control of the Menstrual Cycle*
Weblinks: *The Menstrual Cycle, Ovarian and Uterine Cycle*

Oogenesis

The production of egg cells (**ova**) occurs by **oogenesis**. Unlike spermatogenesis, no new eggs are produced after birth. Instead a human female is born with her entire complement of immature eggs. These remain in prophase of meiosis I throughout childhood. After puberty, most commonly a single egg cell is released from the ovaries at regular monthly intervals (the menstrual cycle). This cell is arrested in metaphase of meiosis II and its second division is only completed upon fertilization. The release of egg cells from the ovaries takes place from the onset of puberty until menopause, when menstruation ceases.

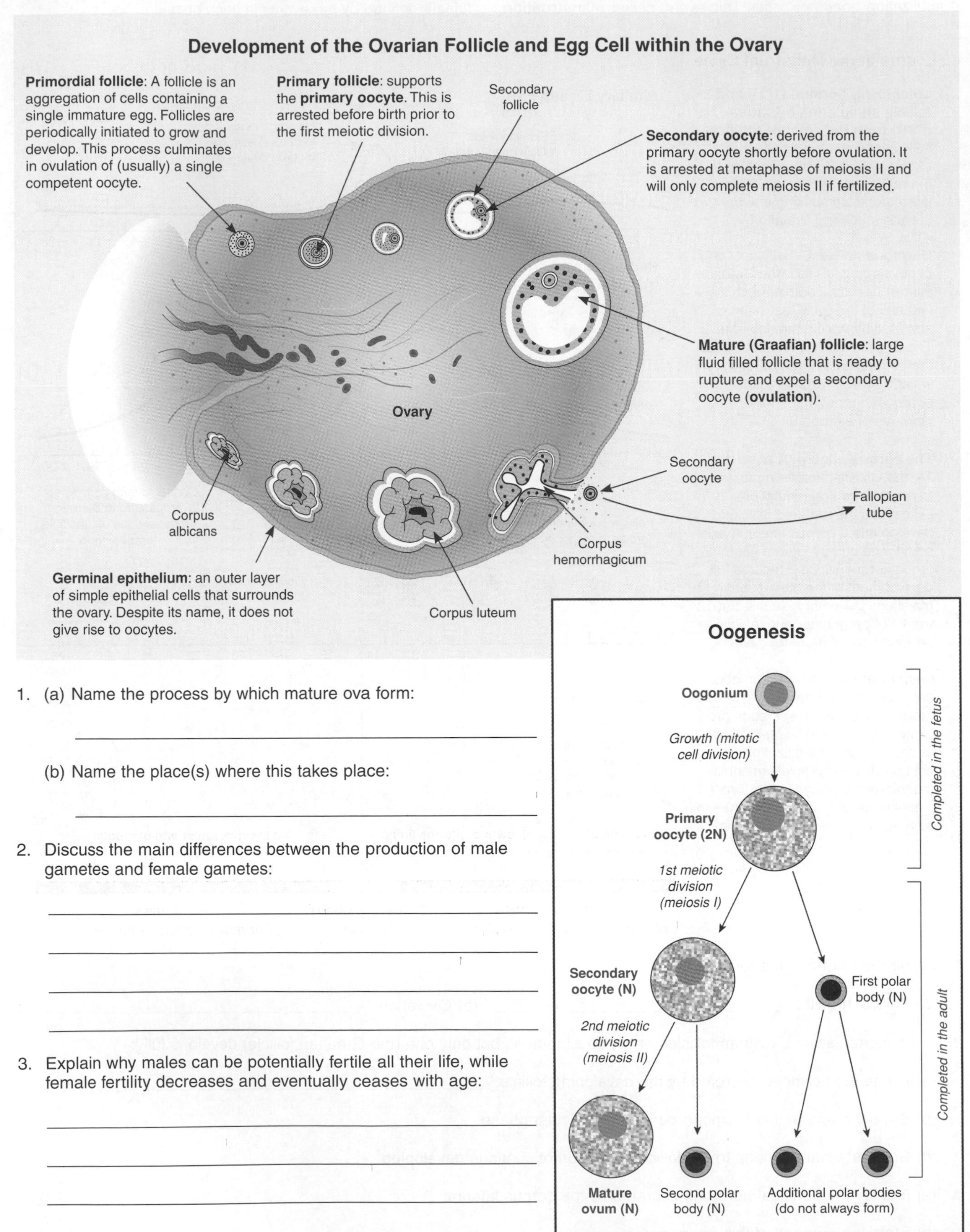

1. (a) Name the process by which mature ova form:

   ____

   (b) Name the place(s) where this takes place:

   ____

2. Discuss the main differences between the production of male gametes and female gametes:

   ____

   ____

   ____

   ____

3. Explain why males can be potentially fertile all their life, while female fertility decreases and eventually ceases with age:

   ____

   ____

   ____

   ____

ISBN: 978-1-92717357-2

Related activities: *Spermatogenesis, Gametes*
Weblinks: *Oogenesis, Comparison of Spermatogenesis and Oogenesis*

Control of the Menstrual Cycle

The female menstrual cycle is regulated by the interplay of several reproductive hormones. The main control centers for this regulation are the **hypothalamus** and the **anterior pituitary gland**. The hypothalamus secretes GnRH (gonadotropin releasing hormone), a hormone that is essential for normal gonad function in males and females. GnRH is transported in blood vessels to the anterior pituitary where it brings about the release of two hormones: follicle stimulating hormone (FSH) and luteinizing hormone (LH). It is these two hormones that induce the cyclical changes in the ovary and uterus. Regulation of blood hormone levels during the menstrual cycle is achieved through **negative feedback** mechanisms. The exception to this is the mid cycle surge in LH, which is induced by the rapid increase in estrogen secreted by the developing follicle.

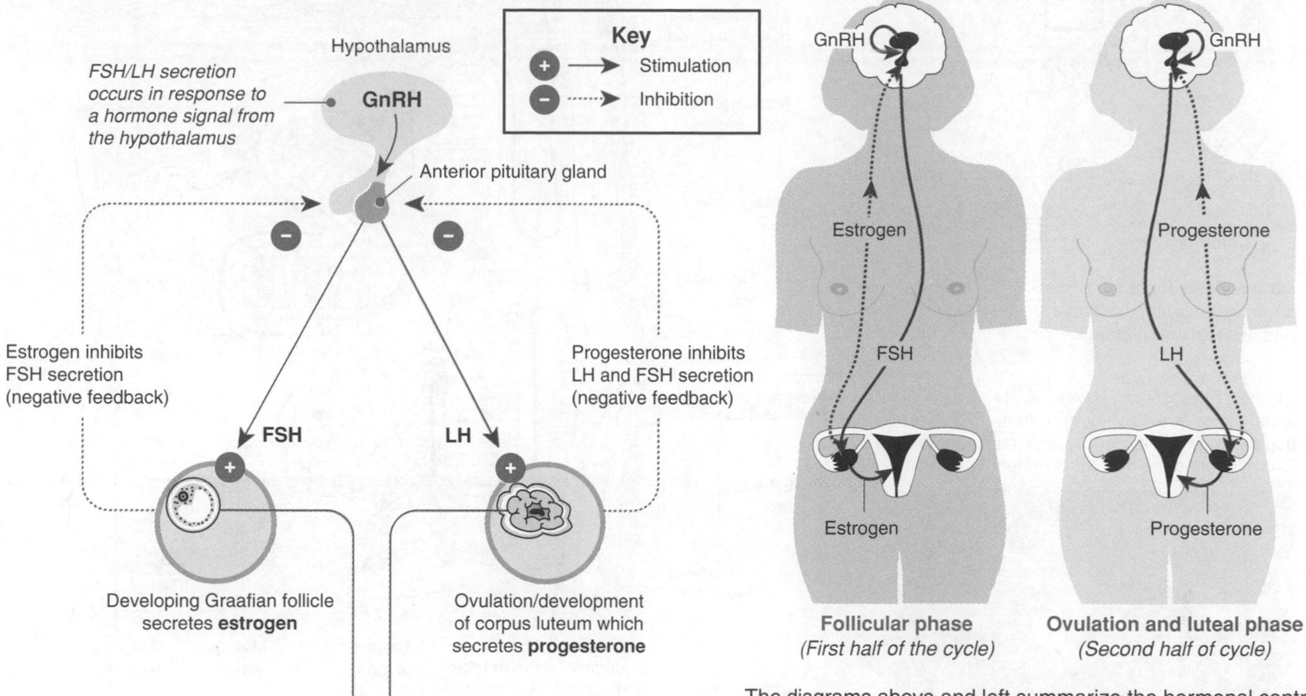

The diagrams above and left summarize the hormonal controls during the menstrual cycle. **In the first half of the cycle**, FSH stimulates follicle development in the ovary. The developing follicle secretes estrogen which acts on the uterus and, in the anterior pituitary, inhibits FSH secretion. **In the second half of the cycle**, LH induces ovulation and development of the corpus luteum. The corpus luteum secretes progesterone which acts on the uterus and also inhibits further secretion of LH and FSH.

1. Using the information above and on the previous page, complete the table below summarizing the role of hormones in the control of the menstrual cycle. To help you, some of the table has been completed:

Hormone	Site of secretion	Main effects and site of action during the menstrual cycle
GnRH		
		Stimulates the growth of ovarian follicles
LH		
		At high levels, stimulates LH surge. Promotes growth and repair of the uterine lining.
Progesterone		

2. Briefly explain the role of negative feedback in the control of hormone levels in the menstrual cycle:

3. **FSH** and **LH** (also known as interstitial cell stimulating hormone in males) also play a central role in male reproduction. Refer to the activity *Male Reproductive System* and state how these two hormones are involved **in male reproduction**:

ISBN: 978-1-92717357-2

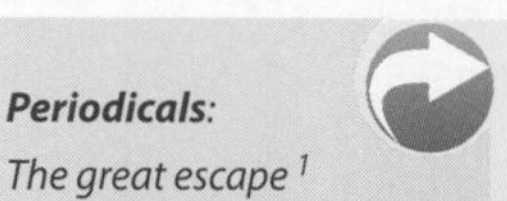

Periodicals: *The great escape* [1]

Related activities: *The Menstrual Cycle*
Weblinks: *Menstrual Cycle Animation*

The Placenta

As soon as an embryo embeds in the uterine wall it begins to obtain nutrients from its mother and increase in size. At two months, when the major structures of the adult are established, it is called a fetus. It is entirely dependent on its mother for nutrients, oxygen, and elimination of wastes. The placenta is the specialized organ that performs this role, enabling exchange between fetal and maternal tissues, and allowing a prolonged period of fetal growth and development within the protection of the uterus. The placenta also has an endocrine role, producing hormones that enable the pregnancy to be maintained.

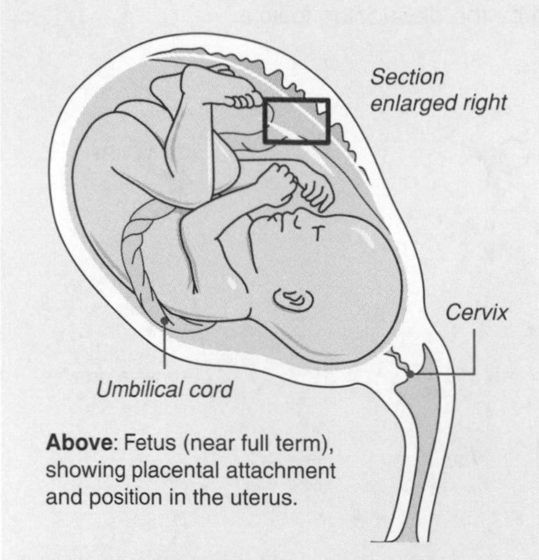

Above: Fetus (near full term), showing placental attachment and position in the uterus.

Below: Photograph shows a 14 week old fetus. Limbs are fully formed, many bones are beginning to ossify, and joints begin to form. Facial features are becoming more fully formed.

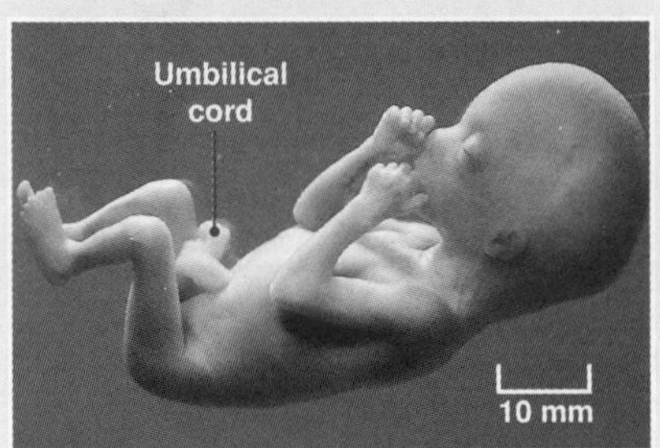

Schematic diagram showing part of the placenta in section

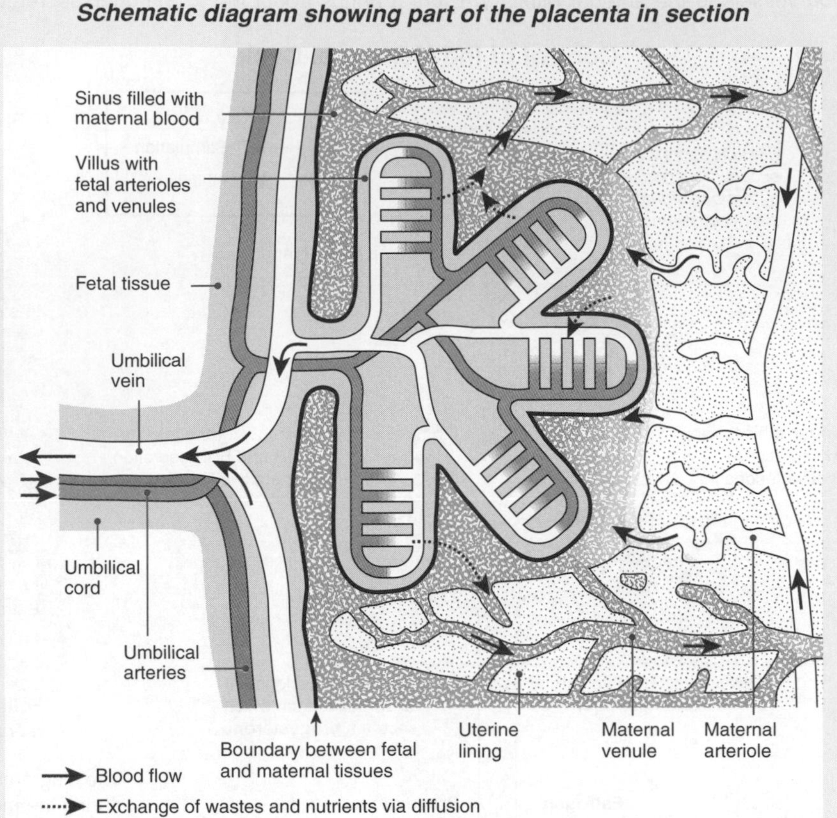

The placenta is a disc-like organ, about the size of a dinner plate and weighing about 1 kg. It develops when fingerlike projections (villi) from the fetal membranes grow into the uterine lining. The villi contain the numerous capillaries connecting the fetal arteries and vein. They continue invading the maternal tissue until they are bathed in the maternal blood sinuses. The maternal and fetal blood vessels are in such close proximity that oxygen and nutrients can diffuse from the maternal blood into the capillaries of the villi. From the villi, the nutrients circulate in the umbilical vein, returning to the fetal heart. Carbon dioxide and other wastes leave the fetus through the umbilical arteries, pass into the capillaries of the villi, and diffuse into the maternal blood. Note that fetal blood and maternal blood do not mix: the exchanges occur via diffusion through thin walled capillaries.

1. In simple terms, describe the basic structure of the human placenta: ______________________

2. The umbilical cord contains the fetal arteries and vein. Describe the status of the blood in each type of fetal vessel:

   (a) Fetal arteries: Oxygenated and containing nutrients / Deoxygenated and containing nitrogenous wastes (delete one)

   (b) Fetal vein: Oxygenated and containing nutrients / Deoxygenated and containing nitrogenous wastes (delete one)

3. Teratogens are substances that may cause malformations in embryonic development (e.g. nicotine, alcohol):

   (a) Give a general explanation why substances ingested by the mother have the potential to be harmful to the fetus:

   ____

   (b) Explain why cigarette smoking is so harmful to fetal development: ______________________

ISBN: 978-1-92717357-2

Related activities: *Fertilization and Early Growth*

Fertilization and Early Growth

When an egg cell is released from the ovary it is arrested in metaphase of meiosis II and is termed a secondary oocyte. **Fertilization** occurs when a sperm penetrates an egg cell at this stage and the sperm and egg nuclei unite to form the zygote. Fertilization is always regarded as time 0 in a period of gestation (pregnancy) and has five distinct stages (below). After fertilization, the zygote begins its **development** i.e. its growth and differentiation into a multicellular organism (see next page).

Fertilization (Time 0)

The stages in fertilization are represented below in a numbered sequence (1-5)

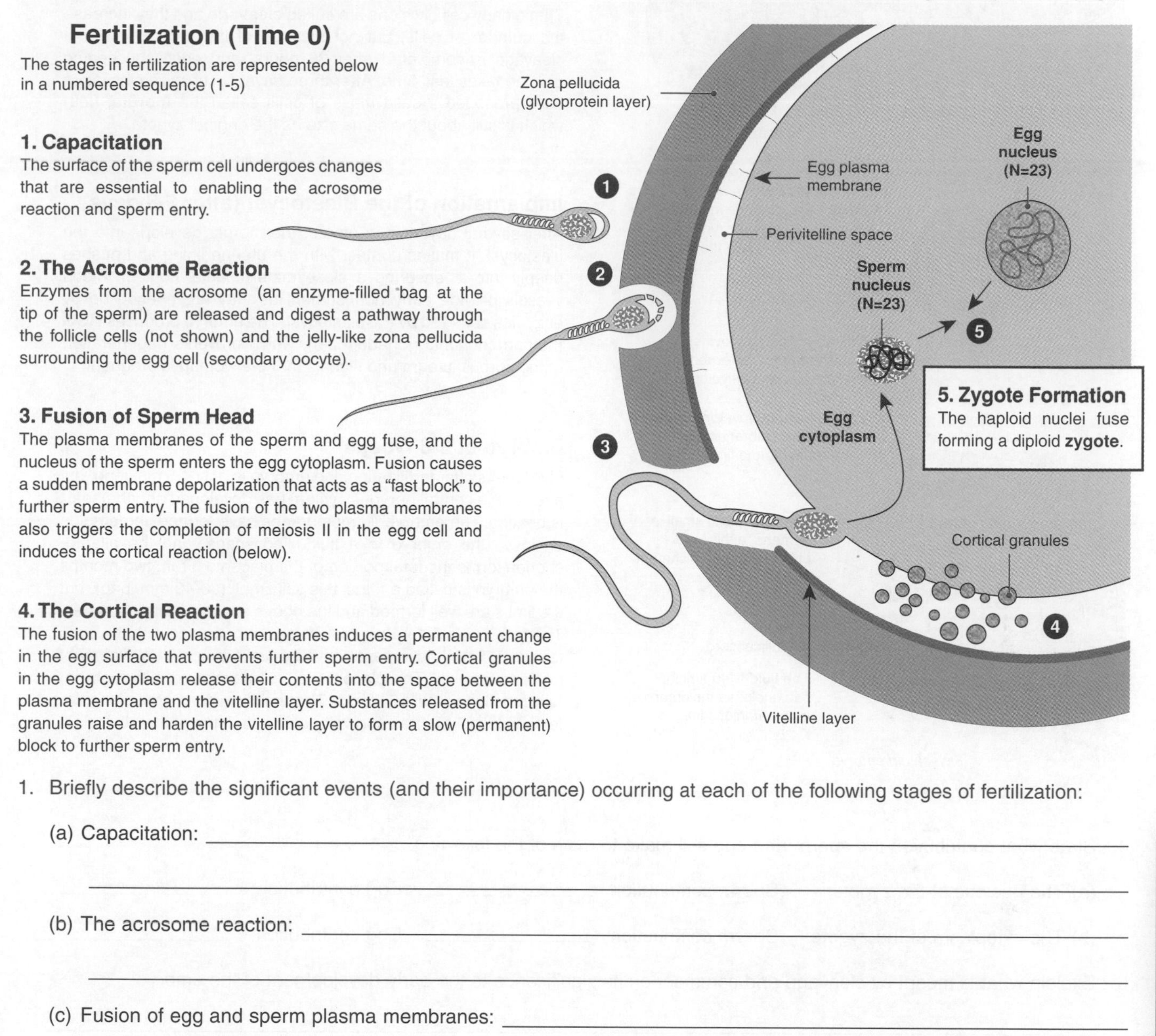

1. Capacitation

The surface of the sperm cell undergoes changes that are essential to enabling the acrosome reaction and sperm entry.

2. The Acrosome Reaction

Enzymes from the acrosome (an enzyme-filled bag at the tip of the sperm) are released and digest a pathway through the follicle cells (not shown) and the jelly-like zona pellucida surrounding the egg cell (secondary oocyte).

3. Fusion of Sperm Head

The plasma membranes of the sperm and egg fuse, and the nucleus of the sperm enters the egg cytoplasm. Fusion causes a sudden membrane depolarization that acts as a "fast block" to further sperm entry. The fusion of the two plasma membranes also triggers the completion of meiosis II in the egg cell and induces the cortical reaction (below).

4. The Cortical Reaction

The fusion of the two plasma membranes induces a permanent change in the egg surface that prevents further sperm entry. Cortical granules in the egg cytoplasm release their contents into the space between the plasma membrane and the vitelline layer. Substances released from the granules raise and harden the vitelline layer to form a slow (permanent) block to further sperm entry.

5. Zygote Formation

The haploid nuclei fuse forming a diploid **zygote**.

1. Briefly describe the significant events (and their importance) occurring at each of the following stages of fertilization:

   (a) Capacitation: ______________________________

   ____

   (b) The acrosome reaction: ______________________________

   ____

   (c) Fusion of egg and sperm plasma membranes: ______________________________

   ____

   (d) The cortical reaction: ______________________________

   ____

   (e) Fusion of egg and sperm nuclei: ______________________________

   ____

2. Explain the significance of the blocks that prevent entry of more than one sperm into the egg (polyspermy):

   ____

   ____

3. (a) Explain why the egg cell, when released from the ovary, is termed a secondary oocyte: ______________________________

   ____

   (b) At which stage is its meiotic division completed? ______________________________

Reproduction and Development

ISBN: 978-1-92717357-2

Related activities: *Female Reproductive System*
Weblinks: *The Acrosome Reaction*

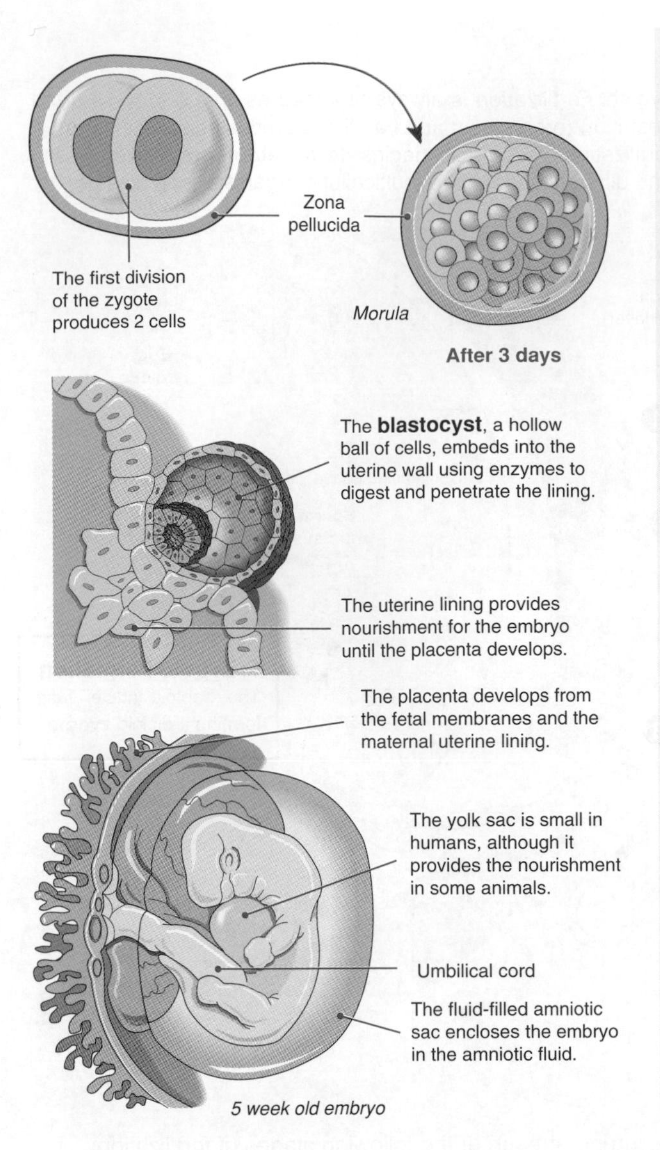

Early Growth and Development

Cleavage and Development of the Morula

Immediately after fertilization, rapid cell division takes place. These early cell divisions are called **cleavage** and they increase the number of cells, but not the size of the zygote. The first cleavage is completed after 36 hours, and each succeeding division takes less time. After three days, successive cleavages have produced a solid mass of cells called the **morula**, (left) which is still about the same size as the original zygote.

Implantation of the Blastocyst (after 6-8 days)

After several days in the uterus, the morula develops into the blastocyst. It makes contact with the uterine lining and pushes deeply into it, ensuring a close maternal-fetal contact. Blood vessels provide early nourishment as they are opened up by enzymes secreted by the blastocyst. The embryo produces **HCG** (human chorionic gonadotropin), which prevents degeneration of the corpus luteum and signals that the woman is pregnant.

Embryo at 5-8 Weeks

Five weeks after fertilization, the embryo is only 4-5 mm long, but already the central nervous system has developed and the heart is beating. The embryonic membranes have formed; the amnion encloses the embryo in a fluid-filled space, and the allanto-chorion forms the fetal portion of the placenta. From two months the embryo is called a fetus. It is still small (30-40 mm long), but the limbs are well formed and the bones are beginning to harden. The face has a flat, rather featureless appearance with the eyes far apart. Fetal movements have begun and brain development proceeds rapidly. The placenta is well developed, although not fully functional until 12 weeks. The umbilical cord, containing the fetal umbilical arteries and vein, connects fetus and mother.

4. State what contribution the sperm and egg cell make to each of the following:

   (a) The nucleus of the zygote: Sperm contribution: ____________ Egg contribution: ____________

   (b) The cytoplasm of the zygote: Sperm contribution: ____________ Egg contribution: ____________

5. Explain what is meant by cleavage and comment on its significance to the early development of the embryo:

   ____

   ____

6. (a) Explain the importance of implantation to the early nourishment of the embryo: ____________

   ____

   (b) Identify the fetal tissues that contribute to the formation of the placenta: ____________

   (c) Suggest a purpose of the amniotic sac and comment on its importance to the developing embryo: ____________

   ____

   (d) Suggest why the heart is one of the very first structures to develop in the embryo: ____________

   ____

7. State why the fetus is particularly prone to damage from drugs towards the end of the first trimester (2-3 months):

   ____

   ____

ISBN: 978-1-92717357-2

Apoptosis and Development

Apoptosis or programmed cell death is a normal and necessary mechanism in multicellular organisms to trigger the death of a cell. Apoptosis has a number of crucial roles in the body, including the maintenance of adult cell numbers, and defence against damaged or dangerous cells, such as virus-infected cells and cells with DNA damage. Apoptosis also has a role in "sculpting" embryonic tissue during its development, e.g. in the formation of fingers and toes in a developing human embryo. Apoptosis involves an orderly series of biochemical events that result in set changes in cell morphology and end in cell death. The process is carried out in such a way as to safely dispose of cell remains and fragments. This is in contrast to another type of cell death, called **necrosis**, in which traumatic damage to the cell results in spillage of cell contents. Apoptosis is tightly regulated by a balance between the factors that promote cell survival and those that trigger cell death. An imbalance between these regulating factors leads to defective apoptotic processes and is implicated in an extensive variety of diseases. For example, low rates of apoptosis result in uncontrolled proliferation of cells and cancers.

Stages in Apoptosis

Apoptosis is a normal cell suicide process in response to particular cell signals. It characterized by an overall compaction (shrinking) of the cell and its nucleus, and the orderly dissection of chromatin by endonucleases. Death is finalized by a rapid engulfment of the dying cell by phagocytosis. The cell contents remain membrane-bound and there is no inflammation.

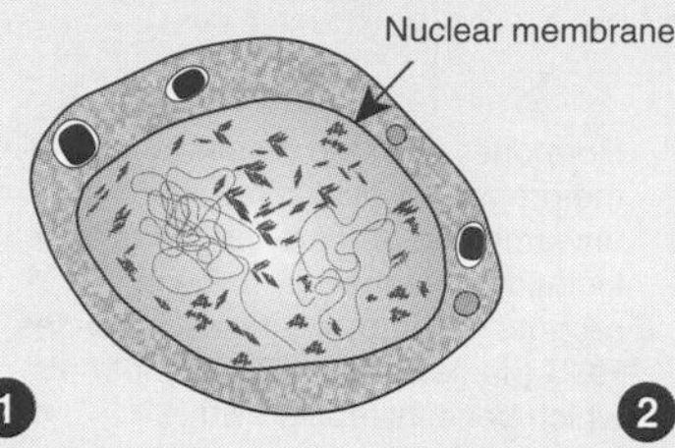

1 The cell shrinks and loses contact with neighboring cells. The chromatin condenses and begins to degrade.

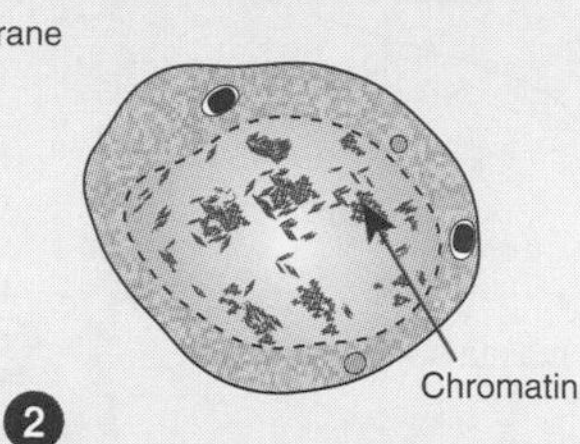

2 The nuclear membrane degrades. The cell loses volume. The chromatin clumps into **chromatin bodies**.

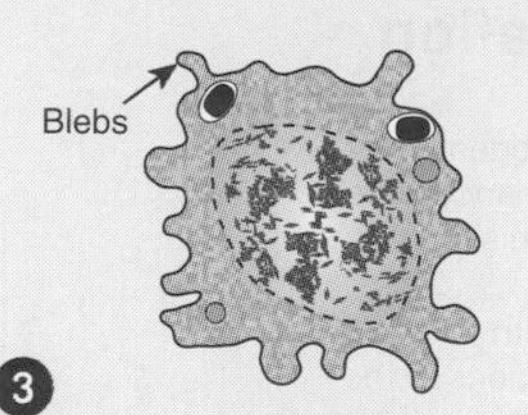

3 **Zeiosis**: The plasma membrane forms bubble like **blebs** on its surface.

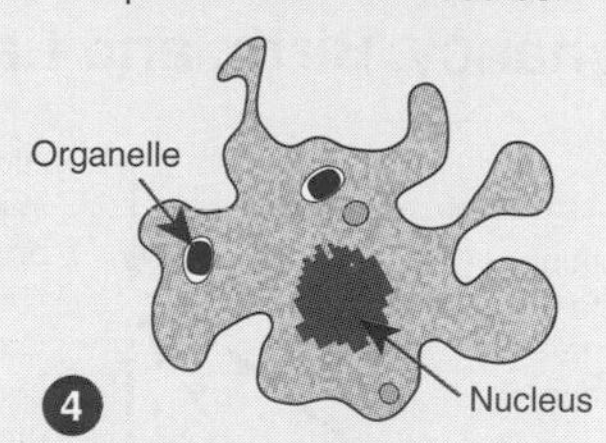

4 The nucleus collapses, but many membrane-bound organelles are unaffected.

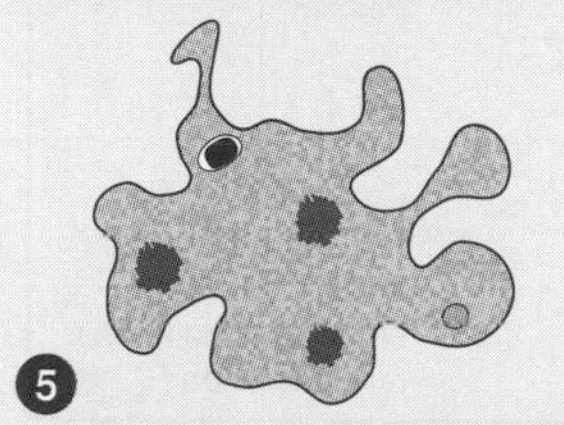

5 The nucleus breaks up into spheres and the DNA breaks up into small fragments.

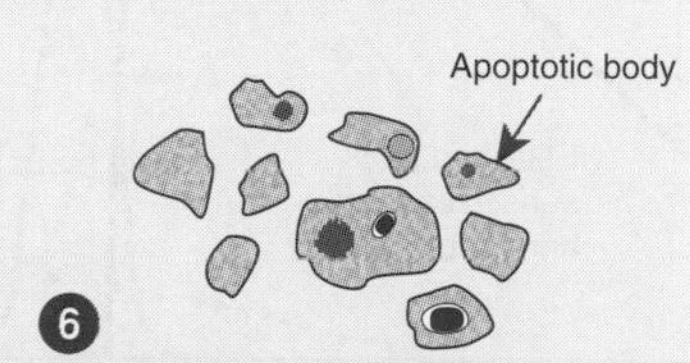

6 The cell breaks into numerous **apoptotic bodies**, which are quickly resorbed by phagocytosis.

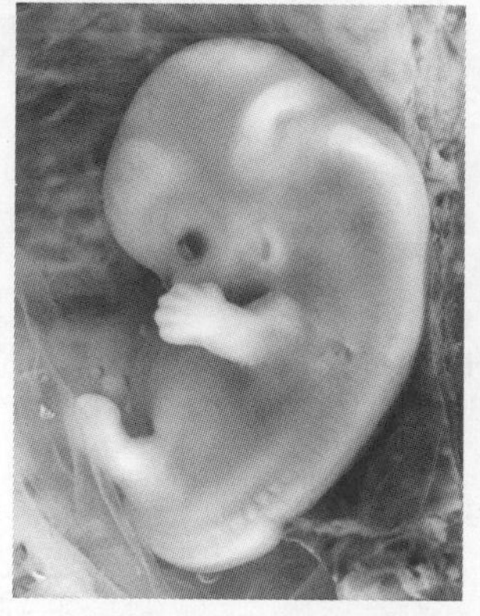

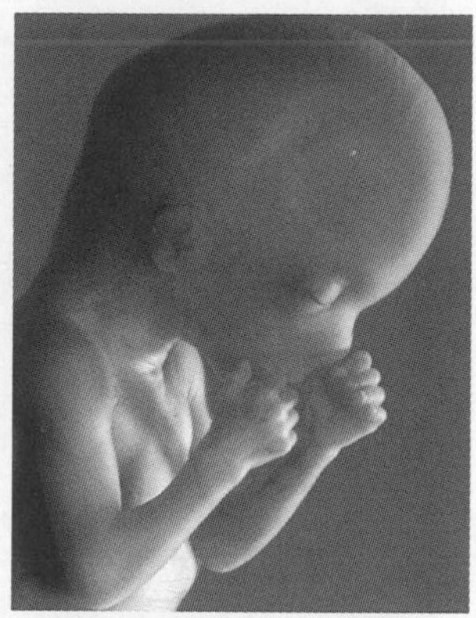

In humans, the mesoderm initially formed between the fingers and toes is removed by apoptosis. 41 days after fertilization (top left), the digits of the hands and feet are webbed, making them look like small paddles. Apoptosis selectively destroys this superfluous webbing and, later in development, each of the digits can be individually seen (right).

Regulating Apoptosis

Apoptosis is a complicated and tightly controlled process, distinct from cell necrosis (uncontrolled cell death), when the cell contents are spilled. Apoptosis is regulated through both:

Positive signals, which prevent apoptosis and allow a cell to function normally. They include:

- interleukin-2
- bcl-2 protein and growth factors

Interleukin-2 is a positive signal for cell survival. Like other signalling molecules, it binds to surface receptors on the cell to regulate metabolism.

Negative signals (death activators), which trigger the changes leading to cell death. They include:

- inducer signals generated from within the cell itself in response to stress, e.g. DNA damage or cell starvation.
- signalling proteins and peptides such as lymphotoxin.

1. The photograph (right) depicts a condition called syndactyly. Explain what might have happened during development to result in this condition:

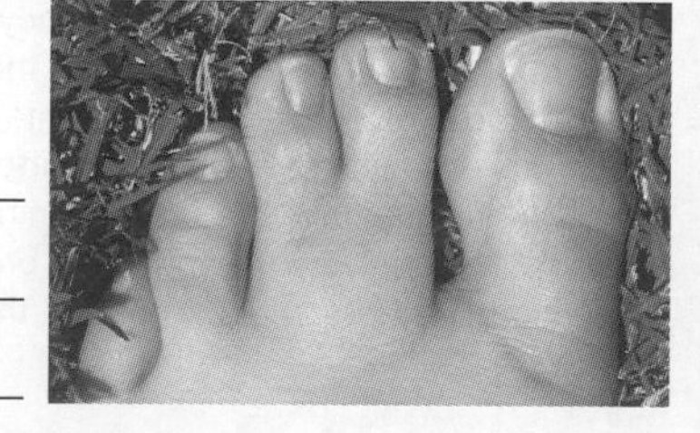

____

____

2. Describe one difference between apoptosis and necrosis: _______________

____

3. Describe two situations, other than digit formation in development, in which apoptosis plays a crucial role:

(a) _______________

____

(b) _______________

____

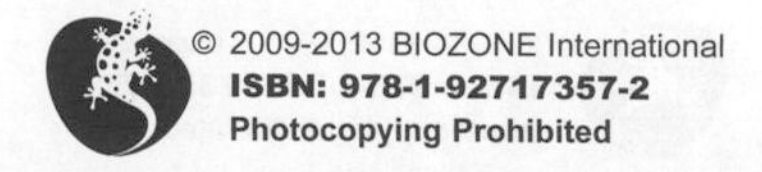

ISBN: 978-1-92717357-2

Related activities: *Growth and Development*
Weblinks: *Apoptosis: The Dance of Death*

The Hormones of Pregnancy

Human reproductive physiology occurs in a cycle (the **menstrual cycle**) which follows a set pattern and is regulated by the interplay of several hormones. Control of hormone release is brought about through feedback mechanisms: the levels of the female reproductive hormones, estrogen and progesterone, regulate the secretion of the pituitary hormones that control the ovarian cycle (see earlier pages). Pregnancy interrupts this cycle and maintains the corpus luteum and the placenta as endocrine organs which maintain the developing fetus for the period of its development. During the last month of pregnancy, the hormone oxytocin (from the posterior pituitary) induces the uterine contractions that will expel the baby from the uterus.

HCG (Human chorionic gonadotropin)
- Secreted by the developing embryo
- Maintains corpus luteum

Progesterone
- Maintains endometrium
- Inhibits uterine contraction

Estrogens
- Maintain endometrium
- Prepare mammary glands for lactation
- High levels induce labor

Human placental lactogen (HPL)
- Stimulates breast growth and development

Relaxin
- Produced by the placenta towards the end of the pregnancy
- Relaxes pubic symphysis at birth
- Helps dilate cervix at birth

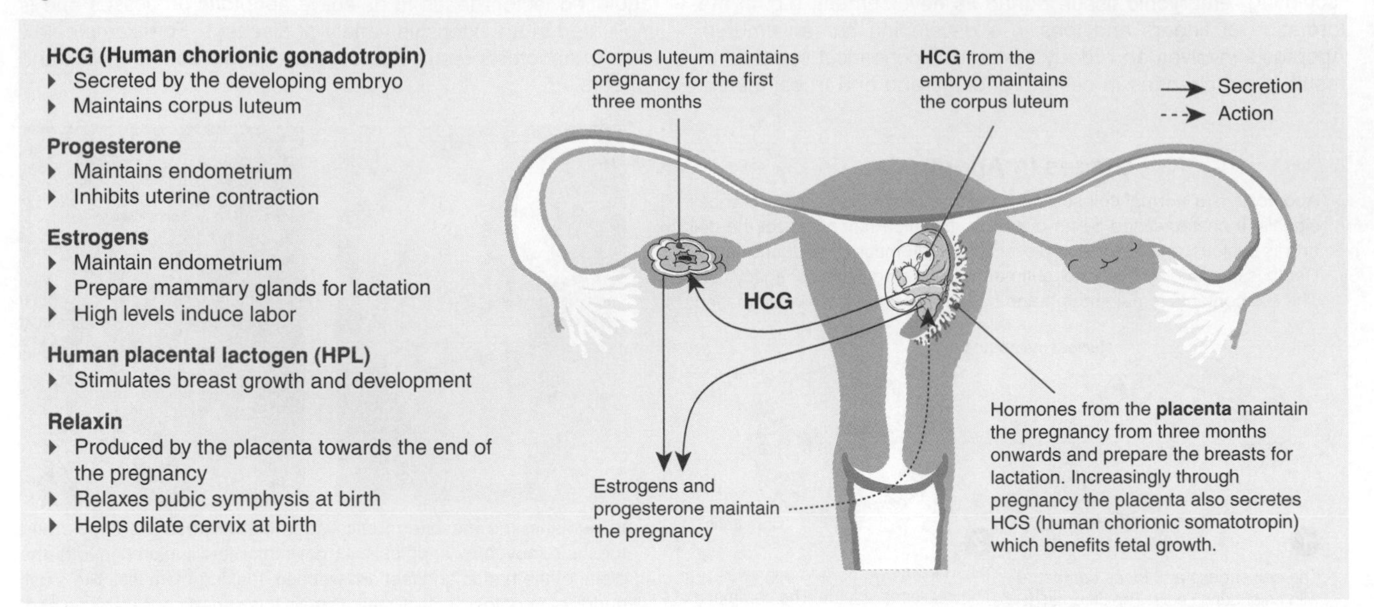

Hormonal Changes During Pregnancy, Birth, and Lactation

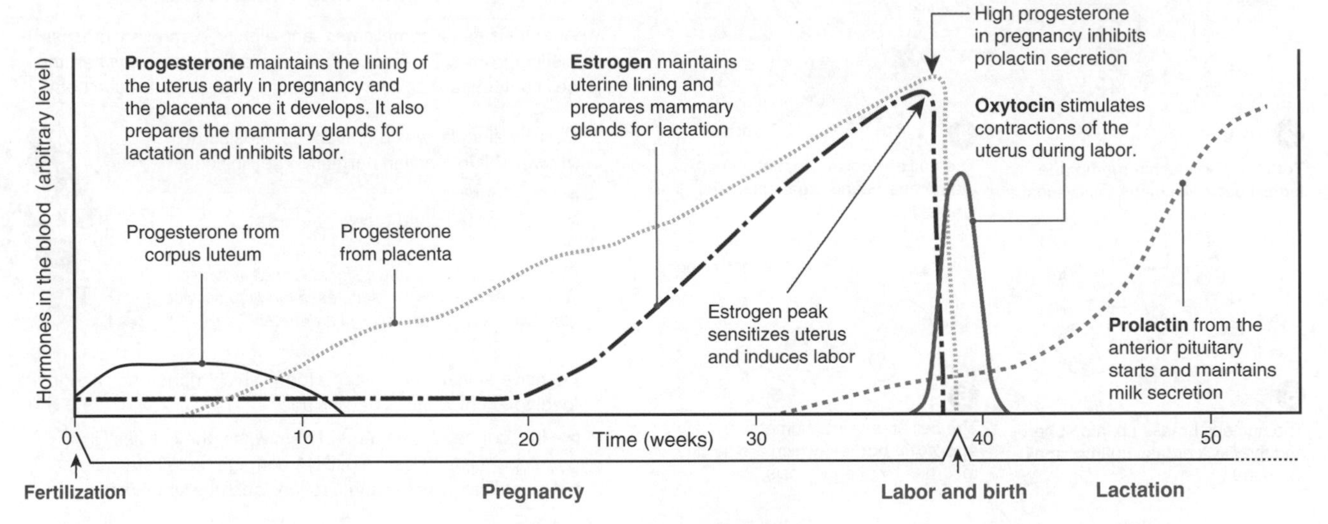

During the first 12-16 weeks pregnancy, the **corpus luteum** secretes enough progesterone to maintain the uterine lining and sustain the embryo. After this, the placenta takes over as the primary endocrine organ of pregnancy. **Progesterone** and **estrogen** from the placenta maintain the uterine lining, inhibit the development of further ova (eggs), and prepare the breast tissue for **lactation** (milk production). At the end of pregnancy, the placenta loses competency, progesterone levels fall, and high estrogen levels trigger the onset of labor. The estrogen peak coincides with an increase in oxytocin. Oxytocin stimulates uterine contractions in a positive feedback loop. The contractions and the increasing pressure of the cervix from the infant stimulate release of more oxytocin, and more contractions and so on, until the infant is born. After birth, prolactin secretion increases. Prolactin maintains lactation during the period of infant nursing.

1. (a) Why is the corpus luteum the main source of progesterone in early pregnancy? ______________________

   ____

   (b) What hormones are responsible for maintaining pregnancy? ______________________

2. (a) Name two hormones involved in labor (onset of the birth process): ______________________

   (b) Describe two physiological factors in initiating labor: ______________________

   ____

Related activities: Control of the Menstrual Cycle, Birth and Lactation

Weblinks: The Chemistry of Pregnancy Tests, HCG Pregnancy Test

ISBN: 978-1-92717357-2

Birth and Lactation

A human pregnancy (the period of **gestation**) lasts, on average, about 38 weeks after fertilization. It ends in labor, the delivery of the baby, and expulsion of the placenta. During pregnancy, progesterone maintains the placenta and inhibits contraction of the uterus. At the end of a pregnancy, increasing estrogen levels overcome the influence of progesterone and labor begins. Prostaglandins, factors released from the placenta, and the physiological state of the baby itself are also involved in triggering the actual timing of labor onset. Labor itself comprises three stages (below), and ends with the delivery of the placenta. After birth, the mother provides nutrition for the infant through **lactation** (the production and release of milk from mammary glands). Breast milk provides infants with a complete, easily digested food for the first 4-6 months of life. All breast milk contains maternal antibodies, which give the infant protection against infection while its own immune system develops.

The Stages of the Birth Process

Stage 1: Dilation

Duration: 2-20 hours

The time between the onset of labor and complete opening (dilation) of the cervix. The amniotic sac may rupture at this stage, releasing its fluid (waters breaking).

The hormone **oxytocin** stimulates the uterine contractions necessary to dilate the cervix and expel the baby. It is these uterine contractions that give the pain of labor, most of which is associated with this first stage.

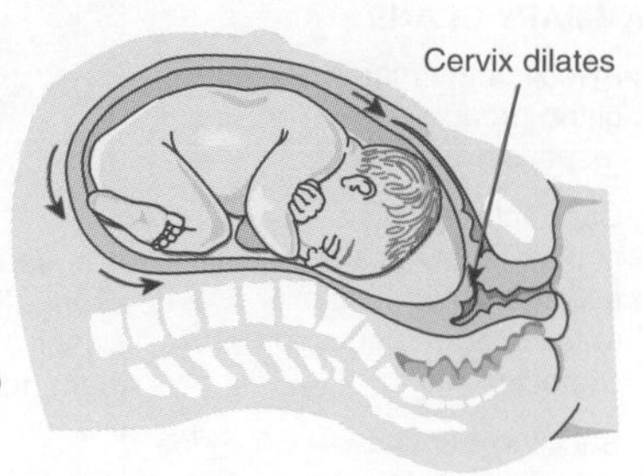

Stage 2: Expulsion

Duration: 2-100 minutes

The time from full dilation of the cervix to delivery. Strong, rhythmic contractions of the uterus pass in waves (arrows), and push the baby to the end of the vagina, where the head appears.

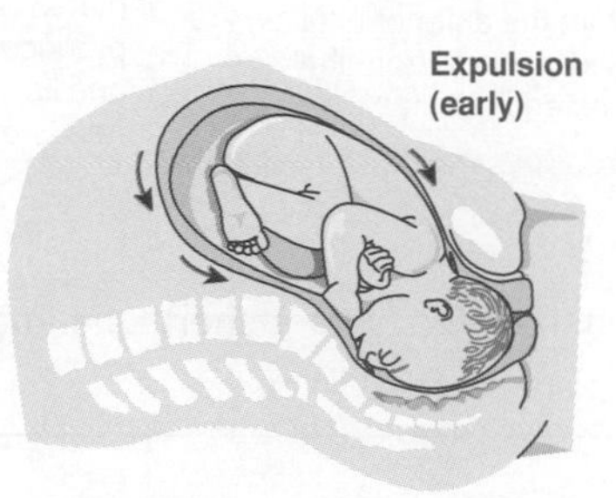

As labor progresses, the time between each contraction shortens. Once the head is delivered, the rest of the body usually follows very rapidly. Delivery completes stage 2.

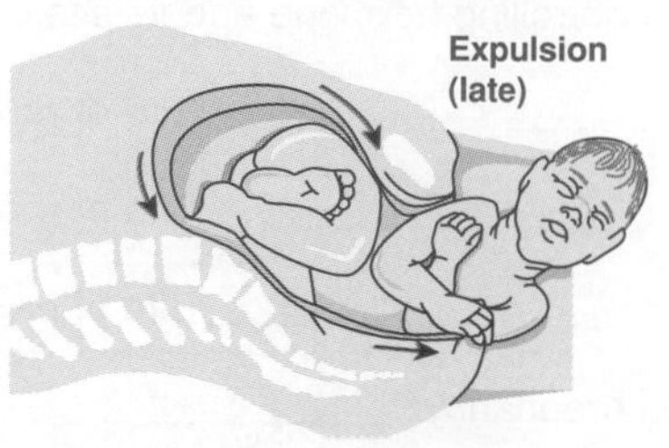

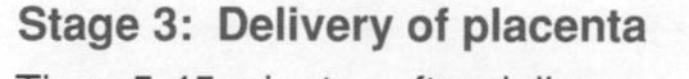

Stage 3: Delivery of placenta

Time: 5-45 minutes after delivery

The third or **placental stage**, refers to the expulsion of the placenta from the uterus. After the placenta is delivered, the placental blood vessels constrict to stop bleeding.

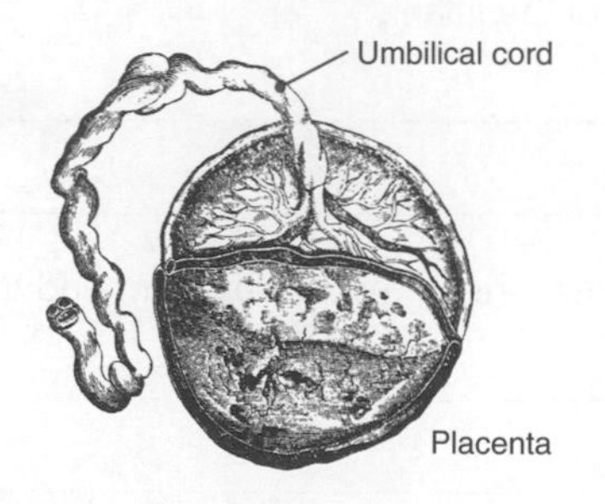

Delivery of the Baby: The End of Stage 2

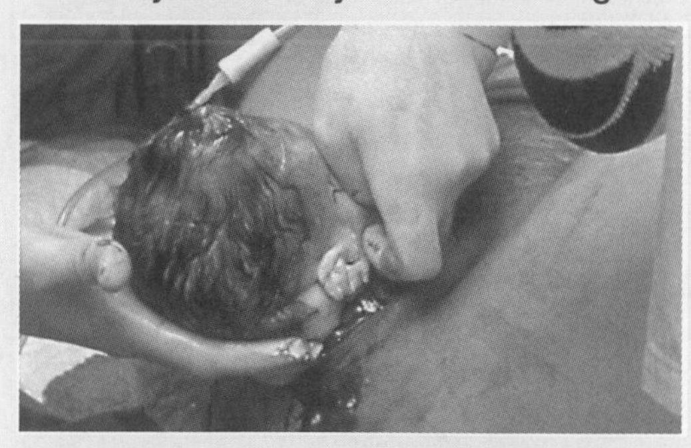

Delivery of the head. This baby is face forward. The more usual position for delivery is face to the back of the mother.

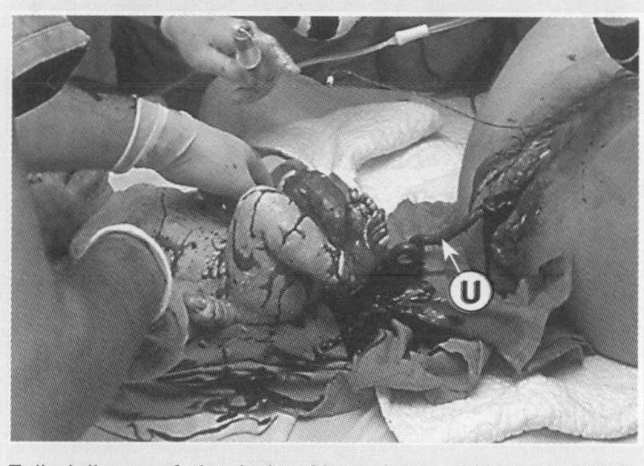

Full delivery of the baby. Note the umbilical cord (U), which supplies oxygen until the baby's breathing begins.

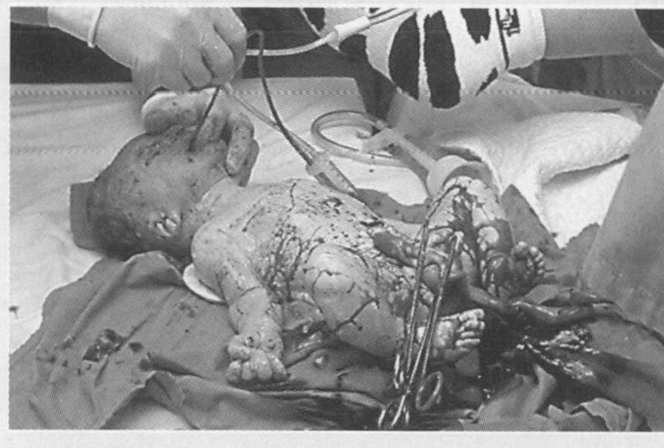

Post-birth check of the baby. The baby is still attached to the placenta and the airways are being cleared of mucus.

1. Name the three stages of birth, and briefly state the main events occurring in each stage:

   (a) Stage 1: ____________________

   (b) Stage 2: ____________________

   (c) Stage 3: ____________________

2. (a) Name the hormone responsible for triggering the onset of labor: ____________________

   (b) Describe two other factors that might influence the timing of labor onset: ____________________

   ____

   ____

ISBN: 978-1-92717357-2

Periodicals:
Let me out!,
The biology of milk

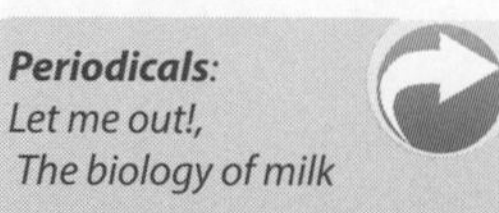

Related activities: *The Hormones of Pregnancy*

Lactation and its Control

After birth, levels of the hormone **prolactin** increase sharply. Prolactin stimulates milk production. **Suckling** maintains prolactin secretion and causes the release of **oxytocin**. Oxytocin induces the milk ducts to contract, resulting in milk release. The more an infant suckles, the more these hormones are produced; an example of positive feedback.

Stimulus to pituitary gland (circled)

Prolactin

Oxytocin

Alveolus

Mammary duct

(+) Symbol indicating stimulation

IN THE LACTATING MAMMARY GLAND:

- Alveoli of the mammary gland produce milk in response to prolactin.
- Contraction of the mammary ducts ejects milk to the nipple in a reflex letdown (induced by oxytocin).
- Suckling stimulates secretion of prolactin from the anterior pituitary and oxytocin from the posterior pituitary.

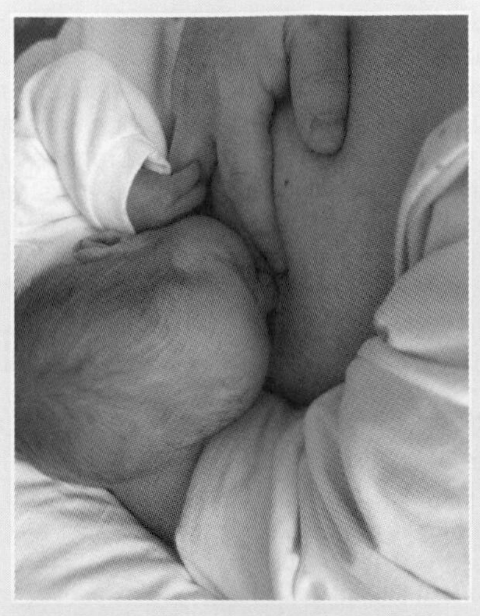

It is important to establish breast feeding soon after birth as this is when infants exhibit the strong reflexes that enable them to learn to suckle effectively. The first formed milk, colostrum, has very little sugar, virtually no fat, and is rich in maternal antibodies. Breast milk that is produced later has a higher fat content, and its composition varies as the nutritional needs of the infant change during growth.

3. Explain why the umbilical cord continues to supply blood to the baby for a short time after delivery: ____________________

____

4. For each of the following processes, state the primary controlling hormone and its site of production:

   (a) Uterine contraction during labor: Hormone: ____________ Site of production: ____________

   (b) Production of milk: Hormone: ____________ Site of production: ____________

   (c) Milk ejection in response to suckling: Hormone: ____________ Site of production: ____________

5. State which hormone inhibits prolactin secretion during pregnancy: ____________________

6. Describe two benefits of breast feeding to the health of the infant:

   (a) ____

   (b) ____

7. (a) Describe the nutritional differences between the first formed milk (colostrum) and the milk that is produced later:

   ____

   ____

   (b) Suggest a reason for these differences: ____________________

   ____

8. Explain why the nutritional composition of breast milk might change during a six-month period of breast feeding:

   ____

9. Infants exhibit marked growth spurts at six weeks and three months of age. At these times, their caloric (energy intake) requirements also increase sharply. With reference to what you know about the control of lactation, suggest how a breast-feeding mother could continue to provide for the increased energy requirements of her infant:

   ____

   ____

ISBN: 978-1-92717357-2

Contraception

Humans have many ways in which to manage their own reproduction. They may choose to prevent or assist fertilization. **Contraception** refers to the use of methods or devices that prevent conception (fertilization of an egg by a sperm). There are many contraceptive methods available including physical barriers (such as condoms) that prevent egg and sperm ever meeting. The most effective methods (excluding sterilization) involve chemical interference in the normal female cycle so that egg production is inhibited. This is done by way of **oral contraceptives** (below, left) or hormonal implants. If taken properly, oral contraceptives are almost 100% effective at preventing pregnancy. The placement of their action in the normal cycle of reproduction (from gametogenesis to pregnancy) is illustrated below. Other contraceptive methods are included for comparison.

Hormonal Contraception

The most common method by which to prevent **conception** using hormones is by using an oral contraceptive pill (OCP). These may be **combined OCPs**, or low dose mini pills.

Combined oral contraceptive pills (OCPs)

These pills exploit the feedback controls over hormone secretion normally operating during a menstrual cycle. They contain combinations of synthetic **estrogens** and **progesterone**. They are taken daily for 21 days, and raise the levels of these hormones in the blood so that FSH secretion is inhibited and no ova develop. Sugar pills are taken for 7 days; long enough to allow menstruation to occur but not long enough for ova to develop. Combined OCPs can be of two types:

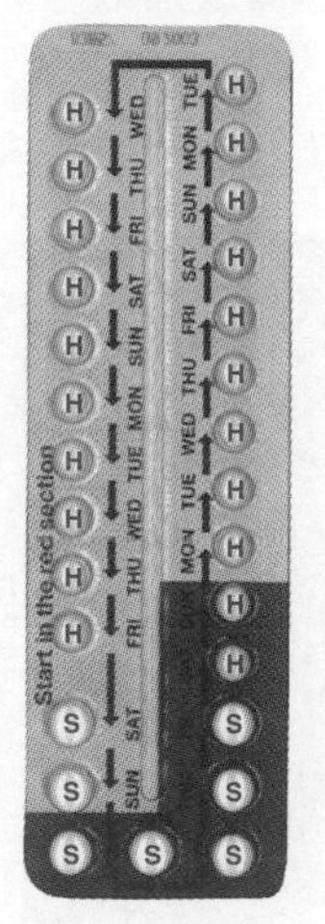

Monophasic pills (left): Hormones (**H**) are all at one dosage level. Sugar pills (**S**) are usually larger and differently colored.

Triphasic pills (right): The hormone dosage increases in stages (**1**,**2**,**3**), mimicking the natural changes in a menstrual cycle.

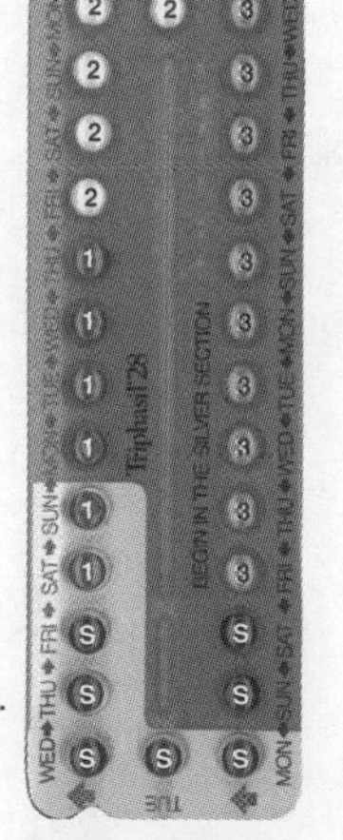

Mini-pill (progesterone only)

The mini-pill contains 28 days of low dose progesterone; generally too low to prevent ovulation. The pill works by thickening the cervical mucus and preventing endometrial thickening. The mini-pill is less reliable than combined pills and must be taken at a regular time each day. However, it is safer for older women and those who are breastfeeding.

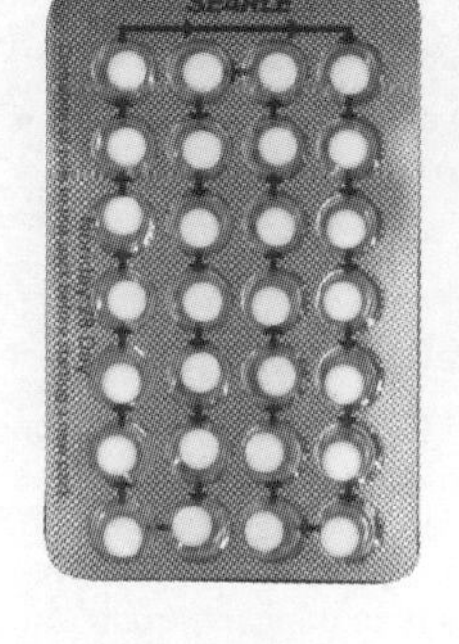

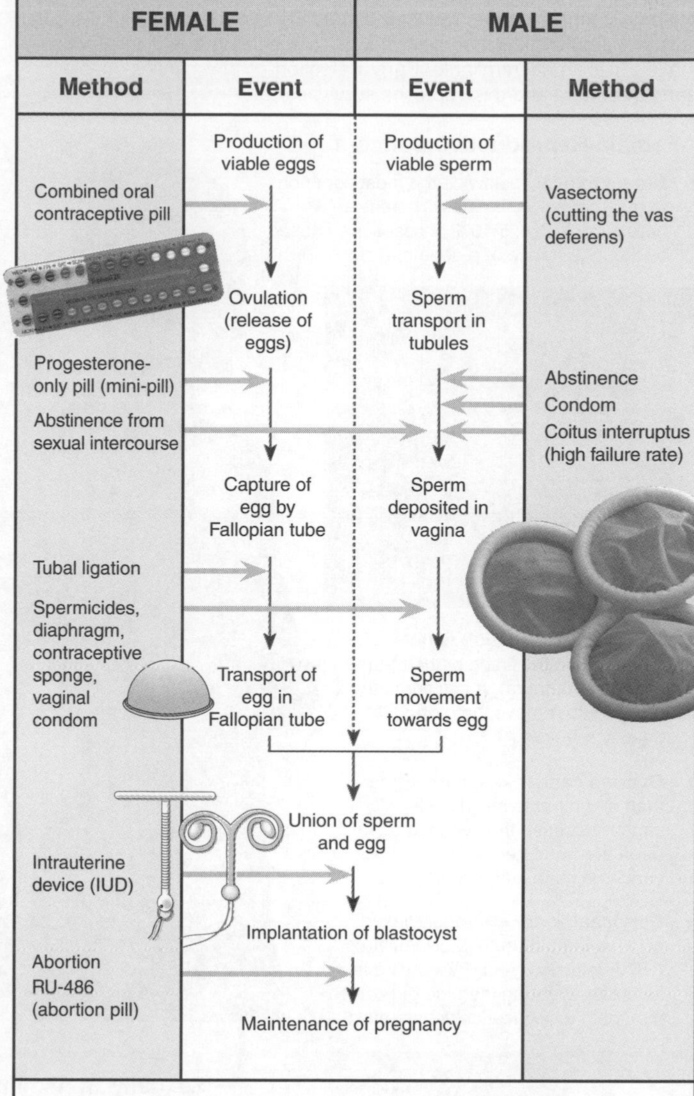

The flow diagram above illustrates where in the reproductive process, from gametogenesis to pregnancy, various contraceptive methods operate. Note the early action of hormonal contraceptives.

1. Explain briefly how the **combined oral contraceptive pill** acts as a contraceptive: ______

2. Contrast the mode of action of OCPs with that of the mini-pill, giving reasons for the differences: ______

3. Suggest why oral contraceptives offer such effective control over conception: ______

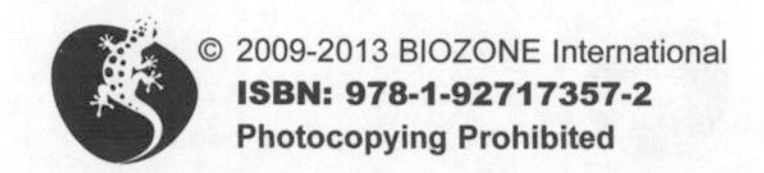

ISBN: 978-1-92717357-2

Periodicals: *Hijacking hormones to regulate fertility*

Related activities: *Control of the Menstrual Cycle*

Diseases of the Reproductive System

A **reproductive system disease** is any condition that affects the male or female reproductive system. There are many causes of reproductive system disease including genetic and congenital abnormalities, abnormal hormone production, functional disorders of the genitalia, infections, and tumors. Some diseases only affect fertility (e.g. erectile dysfunction), while others, such as infections and tumors, can be life threatening. Cancers can affect any part of the reproductive system and may spread (metastasize) from there to other tissues and organs. Some common reproductive system diseases are described below.

Cancers of the Female and Male Reproductive System

Reproductive cancers affect the reproductive organs. The incidence of some cancers can be reduced by making certain lifestyle choices (e.g. not smoking), but some risk factors, including age and genetic makeup) are uncontrollable. Early detection of any cancer enables early treatment and this improves survival rates.

Female Reproductive Cancers

- **Breast cancer** (below) is the most common form of cancer in females. There is a hereditary factor: 5-10% of cases are caused by the inheritance of a gene mutation.

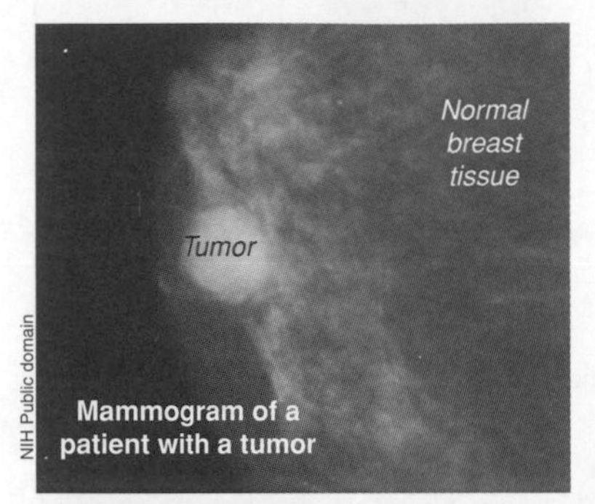

Mammogram of a patient with a tumor

- **Uterine (endometrial) cancer** originates in the lining of the uterus (the endometrium). A hysterectomy (surgery to remove the uterus) is usually required.
- **Ovarian cancer** kills more women than any other gynecological cancer because the symptoms often are not detected until the cancer is quite advanced.
- **Cervical cancer** is strongly linked with having a human papillomavirus (HPV) infection. An HPV vaccine is available to young teenage girls and may help to reduce incidence rates.

Male Reproductive Cancers

- **Prostate cancer** is a slow growing tumor on the prostate gland, and mainly affects men over 40. It can be difficult to detect because it does not produce any symptoms until it becomes large enough to impair the urinary system.
- **Penile cancer**, cancer of the penis, has a survival rate of 65% if detected early. The cause of penile cancer is unknown, but poor genital hygiene and a history of STIs are known risk factors.

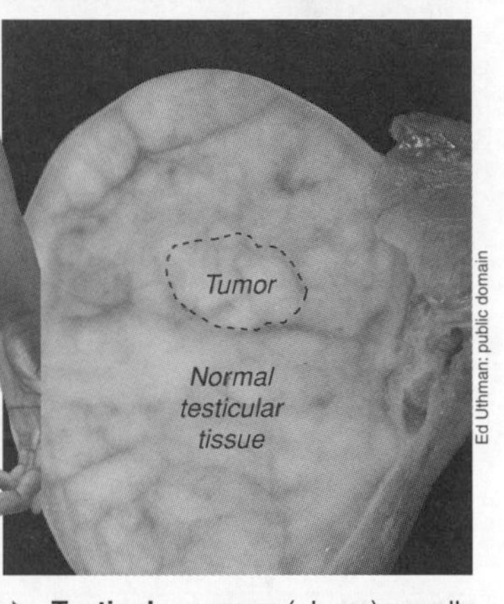

- **Testicular cancer** (above) usually occurs in young men aged 15-35. It is highly treatable and curable.

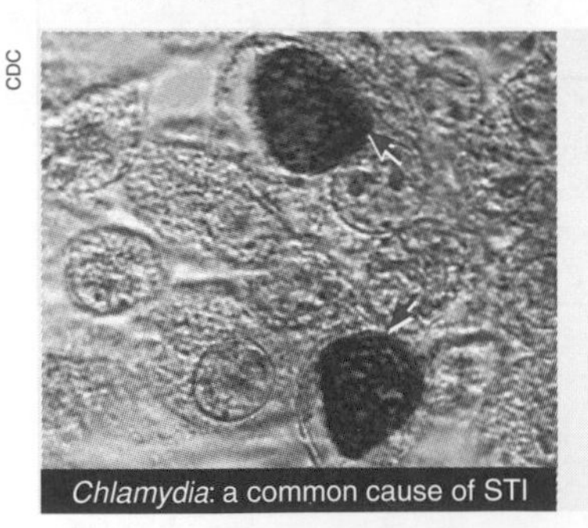

Chlamydia: a common cause of STI

Sexually Transmitted Infections

Sexually transmitted infections (STIs) are passed on by unprotected sexual activity, and infect both male and females. Some STIs are caused by bacteria (e.g. gonorrhoea and chlamydia) and can be treated with antibiotics, but viral infections (such as genital herpes and HIV) have no cure. Some STIs can be difficult to detect because they have no symptoms. If left untreated, STIs can cause a number of related health problems (e.g. cystitis), **infertility** (by damaging the reproductive organs), or death.

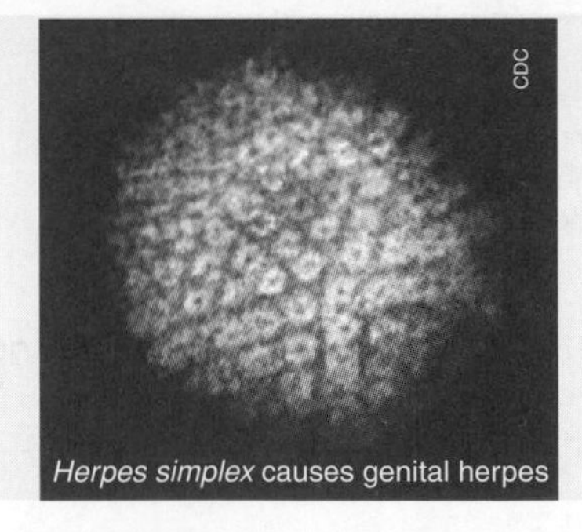

Herpes simplex causes genital herpes

1. Routine screening programs are available for some types of cancers (e.g. cervical cancer and breast cancer). Why is the early detection of cancer important?

____

____

2. Why are ovarian cancer in women and prostate cancer in men more likely to kill than any types of reproductive cancer?

____

____

ISBN: 978-1-92717357-2

Treating Female Infertility

Failure to ovulate is a common cause of female infertility. In most cases, the cause is hormonal, although sometimes the ovaries may be damaged or not functioning normally. Female infertility may also arise through damage to the Fallopian tubes as a result of infection or scarring. These cases are usually treated with hormones, followed by **IVF**. Most treatments for female infertility involve the use of synthetic female hormones, which stimulate ovulation, boost egg production, and induce egg release.

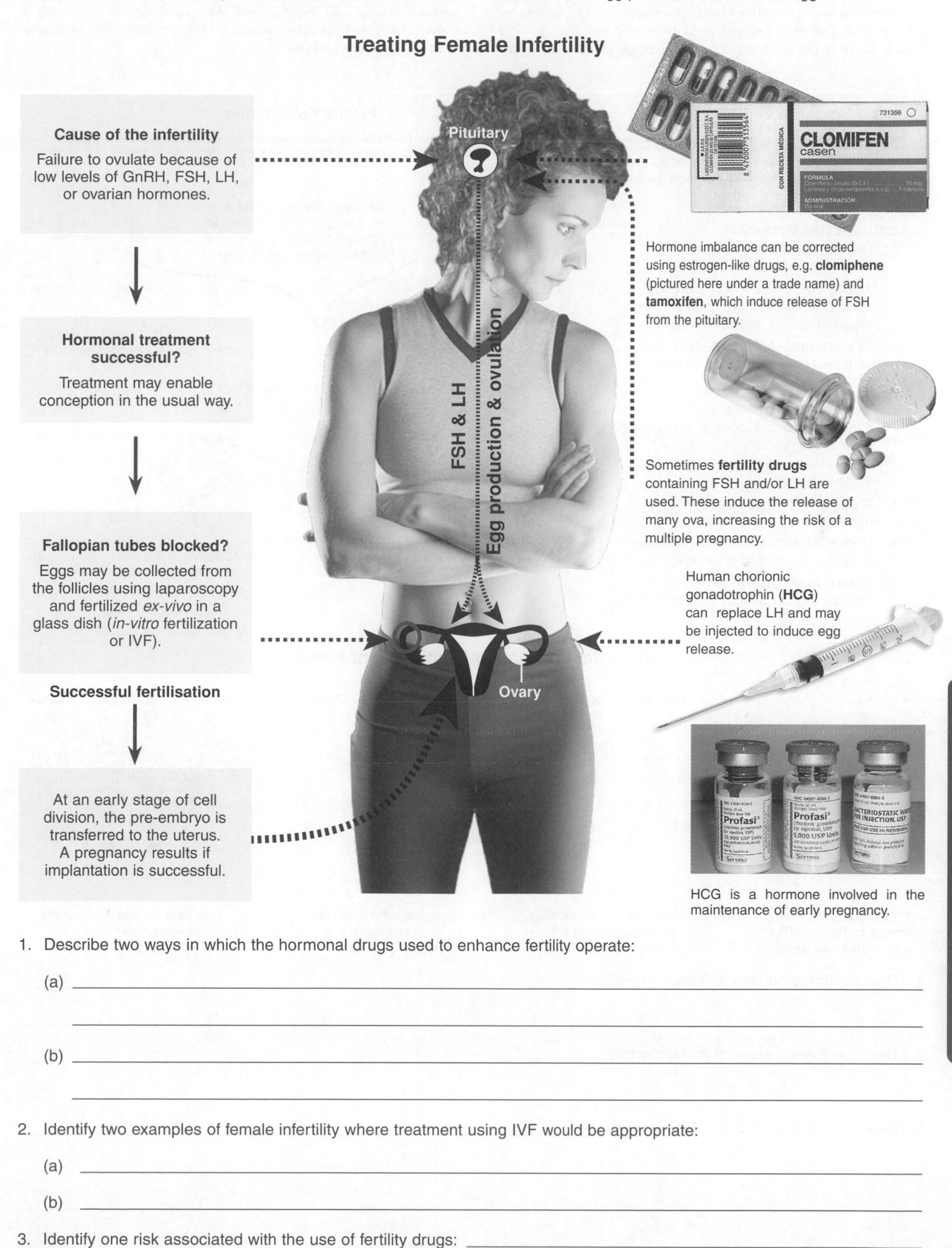

HCG is a hormone involved in the maintenance of early pregnancy.

1. Describe two ways in which the hormonal drugs used to enhance fertility operate:

(a) ____________________

(b) ____________________

2. Identify two examples of female infertility where treatment using IVF would be appropriate:

(a) ____________________

(b) ____________________

3. Identify one risk associated with the use of fertility drugs: ____________________

ISBN: 978-1-92717357-2

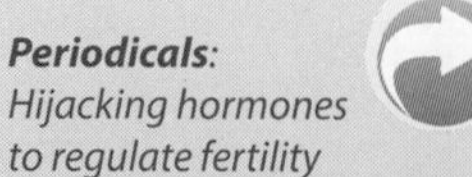

Periodicals: *Hijacking hormones to regulate fertility*

Related activities: *Control of the Menstrual Cycle, In Vitro Fertilization*

In Vitro Fertilization

***In vitro* fertilization** (IVF) may be used to overcome infertility which may result from a disturbance of any of the factors involved in fertilization or embryonic development. Female infertility may be due to a failure to ovulate, requiring stimulation of the ovary, with or without hormone therapy. For couples with one or both partners incapable of providing suitable gametes, it may be possible for them to receive eggs and/or sperm from donors. **Fertility drugs** may be used to induce the production of many eggs for use in IVF, although the natural cycle of ovulation can be used to collect the egg. Fertility drugs stimulate the pituitary gland and may induce the simultaneous release of numerous eggs; an event called superovulation. If each egg is allowed to be fertilized, the resulting embryos may then be frozen after 24-72 hours culture.

Causes of Infertility

Infertility is a common problem (as many as one in six couples require help from a specialist). The cause of the infertility may be inherited, the result of damage caused by disease, or psychological.

Causes of male infertility

- **Penis**: Fails to achieve or maintain erection; abnormal ejaculation.
- **Testes**: Too few sperm produced or sperm are abnormally shaped, have impaired motility, or too short lived.
- **Vas deferens**: A blockage or a structural abnormality may impede the passage of sperm.

Causes of female infertility

- **Fallopian tubes**: Blockage may prevent sperm from reaching egg; one or both tubes may be damaged (disease) or absent (congenital).
- **Ovaries**: Eggs may fail to mature or may not be released.
- **Uterus**: Abnormality or disorder may prevent implantation of the egg.
- **Cervix**: Antibodies in cervical mucus may damage or destroy the sperm.

In Vitro Fertilization

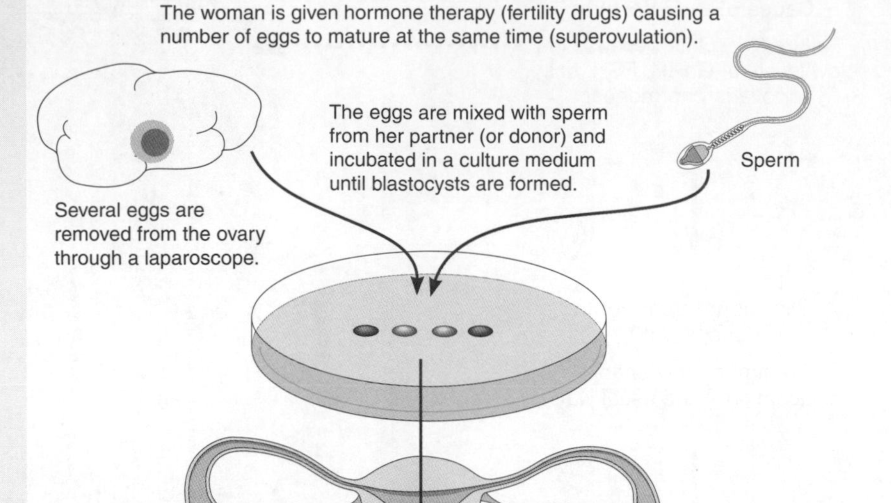

Biological Origins of Gamete Donations

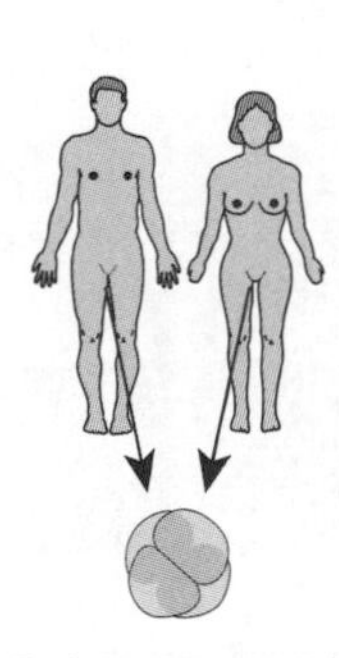

Both partners provide gametes for IVF or GIFT (they donate their own gametes).

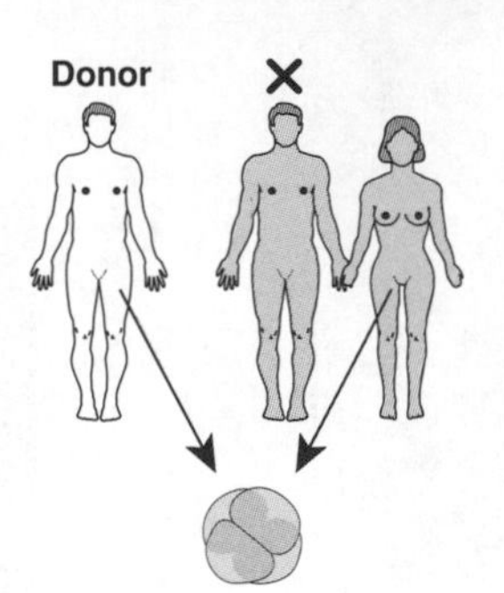

Male partner unable to provide sperm; sperm from male donor.

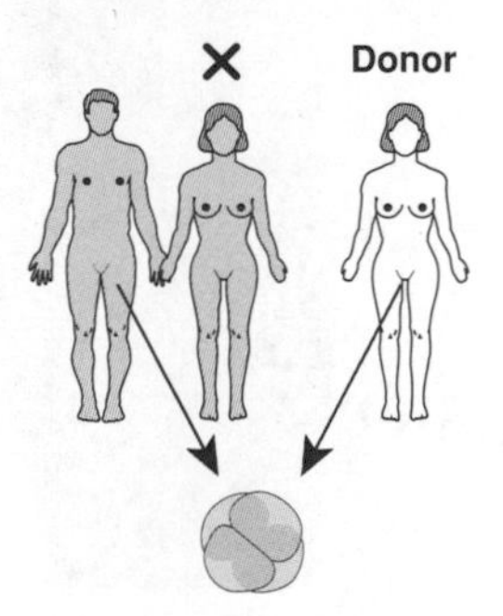

Female partner unable to provide eggs; egg from female donor.

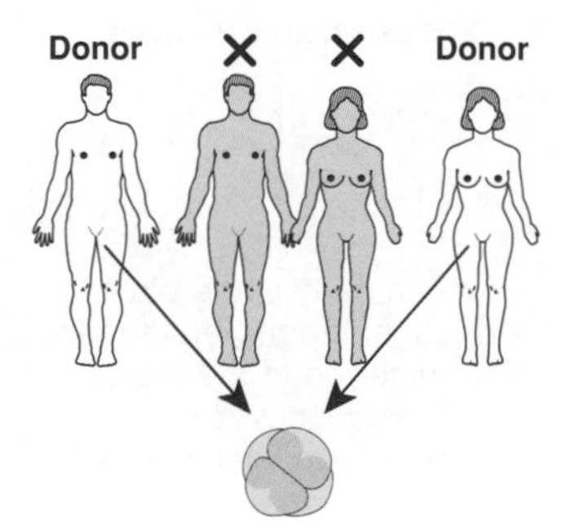

Both partners unable to provide gametes; sperm and egg obtained from donors.

1. Describe three causes of female infertility: ______________________

2. Describe three causes of male infertility: ______________________

3. Describe the key stages of **IVF**: ______________________

Related activities: Treating Female Infertility
Weblinks: In Vitro Fertilization

ISBN: 978-1-92717357-2

IVF raises a number of ethical issues including the concept of personhood, religious beliefs, and the rights and responsibilities of the individual, parents, and community as a whole. It also raises issues over the health and psychological effects off the offspring.

Ethical Issue 1:
The rights of the pre-embryo (blastocyst)

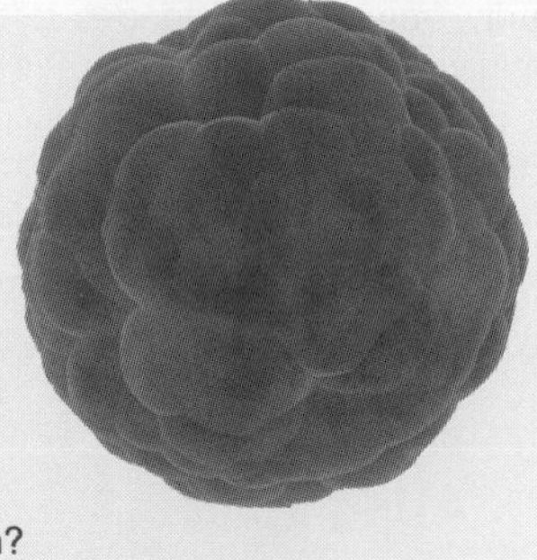

Multiple blastocysts are transferred to a woman's uterus to increase the chances of implantation. After implantation, many of these blastocysts are destroyed by selective pregnancy reduction.

When does personhood or the individual begin?

- If it begins at conception, destruction of the extra embryos technically constitutes murder.
- Many different ideas and definitions exist over the start of personhood or individuality.
  - Can the pre-embryo technically be called an individual during the period of totipotency (about 3 weeks)? During this period, any one of its cells could develop into an individual.

Ethical Issue 2:
Possible wrongs to the couple by the use of IVF

Multiple blastocysts are transferred to a woman's uterus to increase the chances of implantation.

- A multiple pregnancy can have psychological and health effects on the parents.
- A multiple pregnancy can have health effects on the embryos.
- The parents may have to bear the cost of IVF, putting financial strain on them (and possible resentment towards others).
  - Cost may be indirect, because the offspring may have health problems (either caused by IVF or naturally occurring).

Ethical Issue 3:
Possible wrongs to the offspring by the use of IVF

There is some medical evidence to suggest IVF babies have a higher chance of medical problems such as pre-term birth, low birth weight, spina bifida, and heart defects.

- Parents with genetic defects preventing them from conceiving naturally could pass these defects to the offspring via IVF.

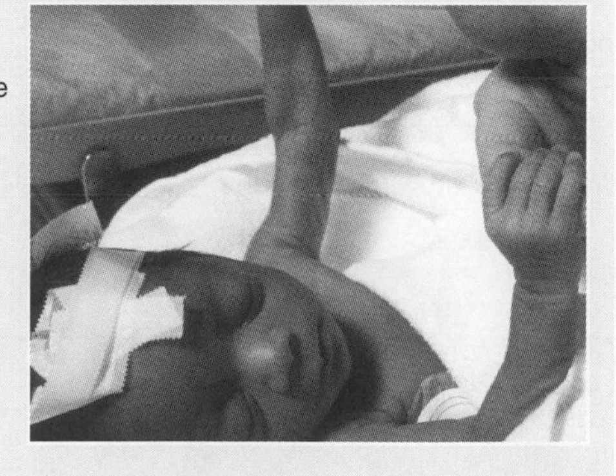

Ethical Issue 4:
Possible wrongs to the community by the use of IVF

IVF is a costly procedure.

- Couples who can afford IVF may be putting money and effort into conception instead of the community.
- The community may have to bear the cost of IVF and welfare for financially struggling parents.
- Offspring with health issues due to IVF may be an ongoing burden to the community.

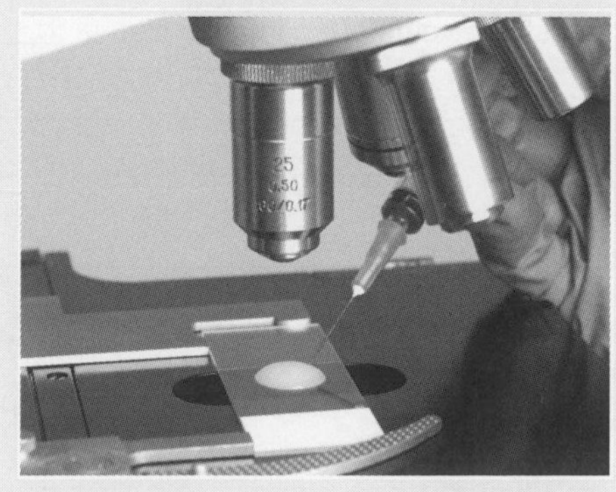

4. Evaluate the ethical and medical issues surrounding IVF treatment. In your opinion what is the best ethical approach to using IVF? Use examples where possible to illustrate your ideas:

ISBN: 978-1-92717357-2

Growth and Development

Development describes the process of growing to maturity, from zygote to adult. After birth, development continues rapidly and is marked by specific stages recognized by the set of physical and cognitive skills present. Obvious physical changes include the elongation of the bones, increasing ossification of cartilage, and changes to the proportions of the body. These proportional changes are the result of **allometric growth** (differential growth rates) and occur concurrently with motor, intellectual, and emotional and social development. These changes lead the child to increasing independence and maturity.

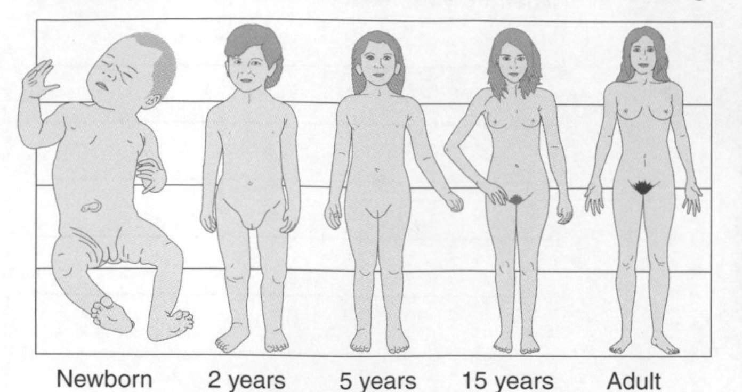

Newborn 2 years 5 years 15 years Adult

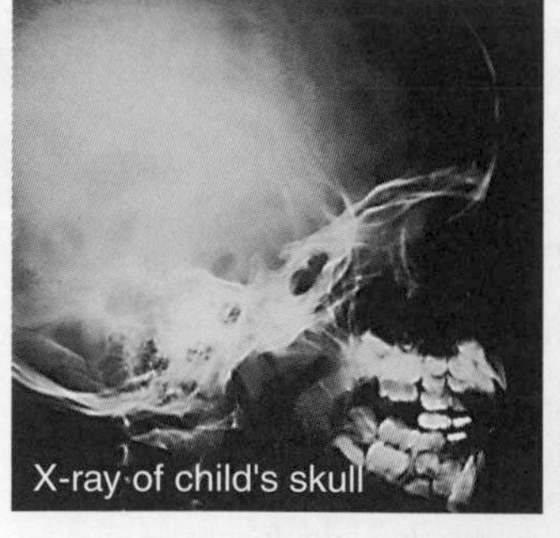

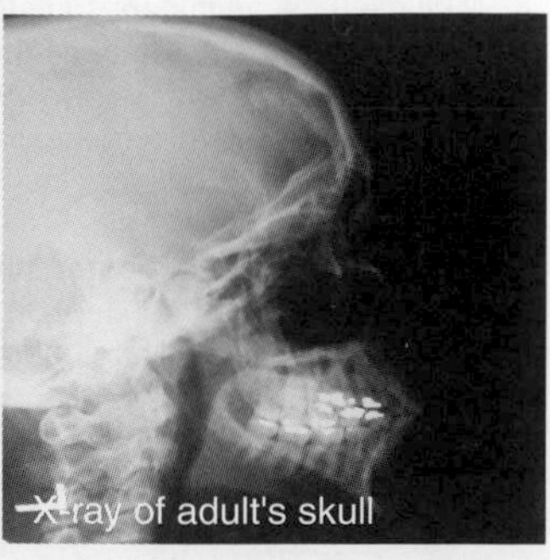

At birth the cranium is very large in comparison to the face and the skull makes up around one quarter of the infant's height. During early life, the face continues to grow outward, reducing the relative proportions of the cranium, while at adulthood the size of the skull in proportion to the body is much less.

By 6 weeks old, a human baby is usually able to hold its head up if placed on its stomach. At 3 months the infant will exercise limbs aimlessly but by 5 months is able to grasp objects and sit up. The infant may be able to crawl by 8 months and walk by 12 months. It is more or less independent by two years and undergoes changes to adulthood at around 11 years of age (puberty).

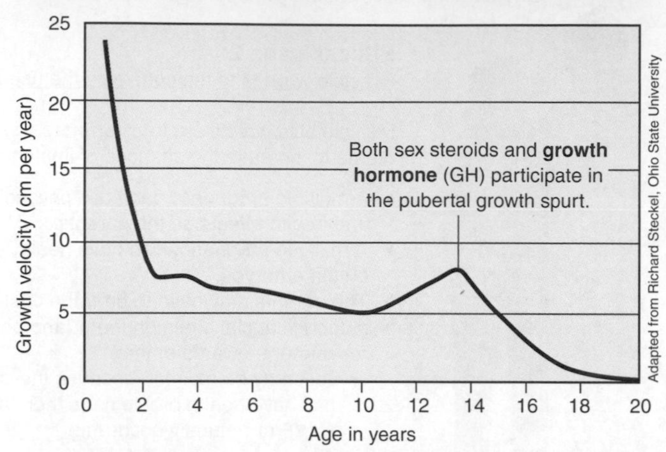

Babies are effectively born premature so that they complete much of their early development in the first two to three years of life. The rate of growth declines slowly through childhood, but increases again to a peak in puberty (the growth spurt). By 20 years of age the cartilage in the long bones has been replaced by bone and growth stops.

1. Describe the most noticeable change in body proportion from birth to adulthood: ______________________

____

2. Describe the changes that occur in the first period of rapid growth in humans: ______________________

____

____

3. Describe the changes that occur in the second period of rapid growth in humans: ______________________

____

4. Answer the following questions with respect to the graph depicting growth rates 0-20 years (above, right):

   (a) Describe what happens to growth velocity in the first two years of an infant's life: ______________________

   ____

   (b) Describe what the graph infers about the rate of growth in the period before birth: ______________________

   ____

   (c) Identify the age range (in years) marking the pubertal growth spurt: ______________________

5. Relate the changes in physical development to the changes occurring in the mental development of an infant:

____

____

ISBN: 978-1-92717357-2

Sexual Development

Humans differentiate into the male or female sex by the action of a combination of different hormones. The hormones testosterone (in males), and estrogen and progesterone (in females), are responsible for puberty (the onset of sexual maturity), the maintenance of gender differences, and the production of gametes. In females, estrogen and progesterone also regulate the menstrual cycle, and ensure the maintenance of pregnancy and nourishment of young.

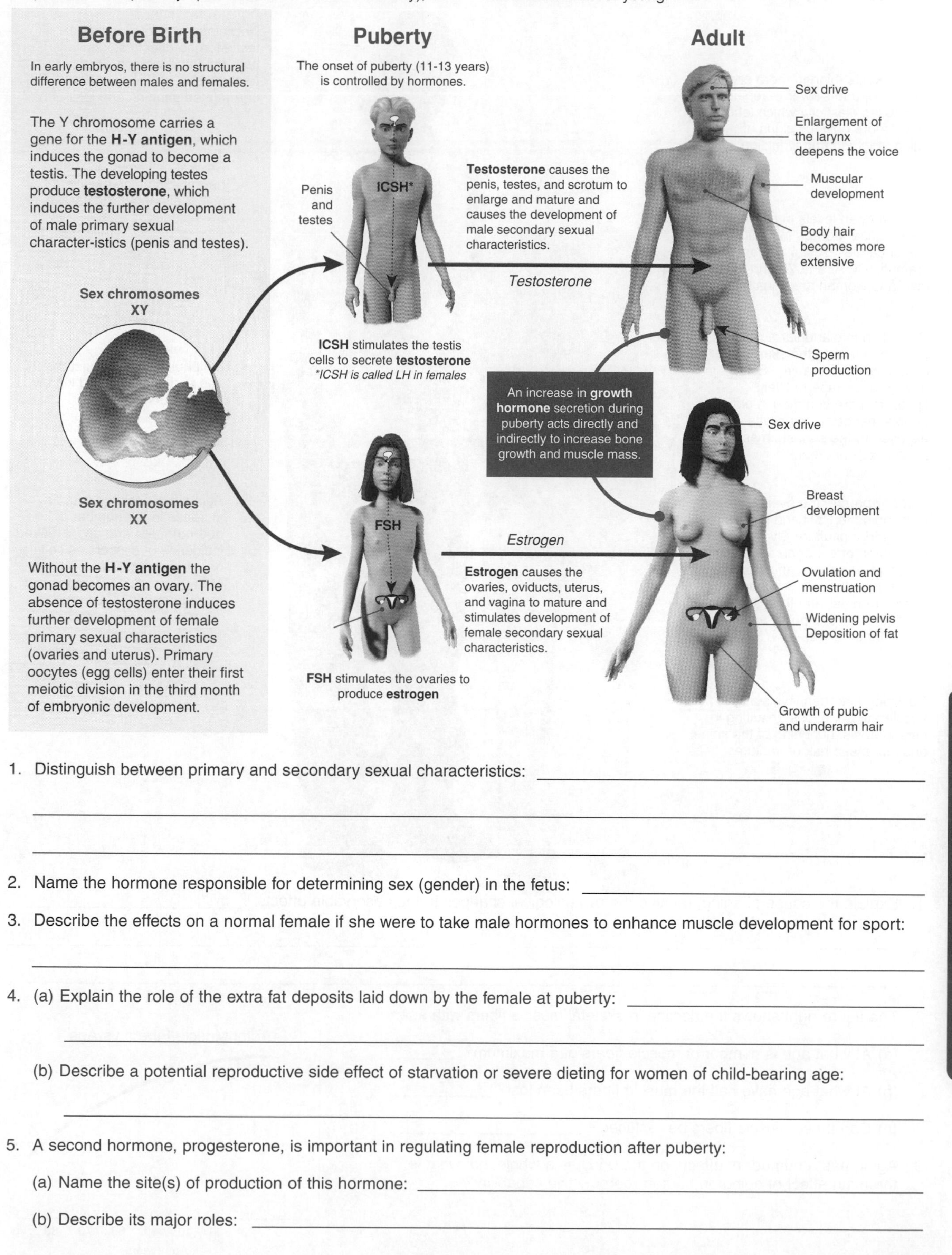

1. Distinguish between primary and secondary sexual characteristics: ______

2. Name the hormone responsible for determining sex (gender) in the fetus: ______

3. Describe the effects on a normal female if she were to take male hormones to enhance muscle development for sport: ______

4. (a) Explain the role of the extra fat deposits laid down by the female at puberty: ______

   (b) Describe a potential reproductive side effect of starvation or severe dieting for women of child-bearing age: ______

5. A second hormone, progesterone, is important in regulating female reproduction after puberty:

   (a) Name the site(s) of production of this hormone: ______

   (b) Describe its major roles: ______

ISBN: 978-1-92717357-2

Related activities: Hormones of the Pituitary, Control of the Menstrual Cycle, The Hormones of Pregnancy

Aging

As soon as we have reached physical maturity, we begin to age. Aging or **senescence** refers to the degenerative changes that occur as a result of cells dying and renewal rates slowing or stopping. It is a general response, producing observable changes in structure and physiology. There is a decline in skeletal and muscular strength, and reduced immune function. Ageing increases susceptibility to stress and disease, so disease and aging often accelerate together.

Older skin is thinner, more easily damaged, and slower to repair. Skin loses elasticity which leads to the more obvious signs of aging, sagging and wrinkling.

Low estrogen levels in post-menopausal women increase the risk of cardiovascular disease (CVD) dramatically. At age 70 and beyond, men and women are equally at risk.

Reduction in glandular activity and blood flow to the skin contribute to skin aging. Skin discoloration (age or liver spots) is more common in older people, especially on areas exposed to the sun, such as the hands, face and neck.

In **menopause**, the ovaries stop responding to FSH and LH from the anterior pituitary. Ovarian production of estrogen and progesterone slows and stops, and the menstrual cycle becomes at first irregular and then stops altogether. At this point, a woman is no longer fertile.

Declining estrogen levels also accelerate bone loss, resulting in osteoporosis, hunching of the spine, and increased risk of fractures.

Declining levels of circulating testosterone affects sex drive in both men and women. Declining testosterone is also associated with age-related thinning and loss of hair.

Several cell types, including neurons, cardiac muscle, and skeletal muscle cannot be replaced.

Metabolic rate decreases with age and digestive and kidney function decline.

Aging is associated with an increase in the number of aberrant cells and an increased incidence of cancers as cellular damage accumulates. Tumors of the reproductive organs are more common in old age.

1. Explain the cause of aging, relating the physiological changes to the observable effects: ____________________

____

____

2. The figure right shows the decline in skeletal muscle fibers with age:

(a) At what age is number of muscle fibers at a maximum? __________

(b) At what age have half the muscle fibers been lost? __________

(c) Can these muscle fibers be replaced? __________

Total Muscle Fibers vs Age

Total number of fibers (x 100): 200, 400, 600, 800

Age (years): 10, 20, 30, 40, 50, 60, 70, 80, 90

3. Aging has a number of effects on the body as a whole, but what is the main effect of aging on human reproductive capability?

____

____

Related activities: *Aging and the Nervous System, Aging and the Endocrine System, Aging and Diseases of the Skeleton*

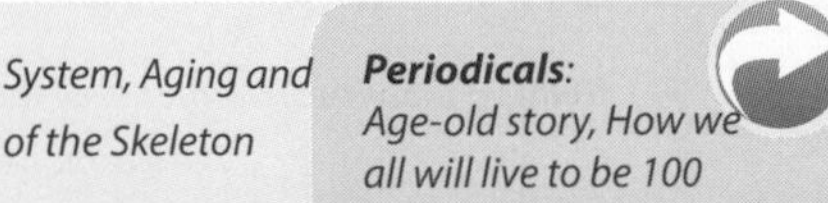

ISBN: 978-1-92717357-2

KEY TERMS: Mix and Match

INSTRUCTIONS: Test your vocabulary by matching each term to its definition, as identified by its preceding letter code.

acrosome reaction ....

apoptosis ....

blastocyst ....

breast ....

childbirth (=parturition or labor) ....

cortical reaction ....

degenerative disease ....

endometrium ....

estrogen ....

fertility ....

fertilization ....

FSH ....

implantation ....

lactation ....

LH ....

menopause ....

menstrual cycle ....

oogenesis ....

oxytocin ....

placenta ....

pregnancy ....

progesterone ....

prostaglandins ....

puberty ....

reproductive system ....

spermatogenesis ....

STI ....

testosterone ....

A Hormone secreted by the anterior pituitary; stimulates ovarian follicular development and estrogen secretions in females and sperm production in males.

B Also called the mammary gland. In females it produces milk after pregnancy.

C The period of physical changes during which a child's body becomes a reproductively capable adult body.

D Capability of producing offspring.

E The system of organs associated with reproduction.

F A hormone with roles in uterine contraction during labor, and milk letdown in lactating females.

G The principal male sex hormone.

H The cycle of changes in reproductive physiology occurring in fertile females. In humans the cycle lasts approximately 28 days.

I The act of giving birth.

J The state of carrying offspring (a fetus), inside the uterus of a female.

K Also called programmed cell death. It is part of normal growth and development, and cellular regulation in multicellular organisms.

L A steroid hormone, which functions as the primary female sex hormone.

M Name given to any disease where the structure or function of tissue progressively deteriorates over time (i.e. with age).

N The permanent change in the surface of the egg cell triggered by the fusion of the sperm plasma membrane and the egg plasma membrane. Contents of the cortical granules are released and harden the vitelline layer of the egg cell.

O Embryos which are about five days old and consist of a hollow ball containing 50-150 cells.

P A steroid hormone involved in the female menstrual cycle, pregnancy, and embryogenesis.

Q An event occurring early in pregnancy in which the blastocyst attaches to the epithelial lining of the uterus.

R The union of male and female gametes to form a zygote.

S The process by which oocytes are produced.

T Cessation of ovulation and menstruation in women.

U The moist mucous membrane lining the uterus.

V Diseases that are passed on by unprotected sexual activity.

W The production of male gametes. Spermatogonia develop into mature sperm cells.

X A process in sperm cells that enables the sperm to break through the egg's coating so that fertilization can occur.

Y Hormone-like substances that are involved in stimulating the contraction of smooth muscle. In labor, they stimulate the uterus to contract.

Z The hormone that stimulates ovulation, luteinization and secretion of estrogen and progesterone in females and stimulates testosterone secretion in males.

AA A specialized organ, characteristic of most mammals, that enables exchanges (via the blood supply) of nutrient, gases, and wastes, between the mother and the fetus.

BB The production and secretion of milk from the mammary glands.

ISBN: 978-1-92717357-2

Direction and Planes

The location and orientation of structures is important when studying human anatomy. Descriptions and locations of structures assume the body is in the **anatomical position** in which the body stands erect facing the observer with feet flat on the floor and hands at the side, palms turned forward. Directional terms describe the orientation and position of organs and structures from this position.

Directional Terms

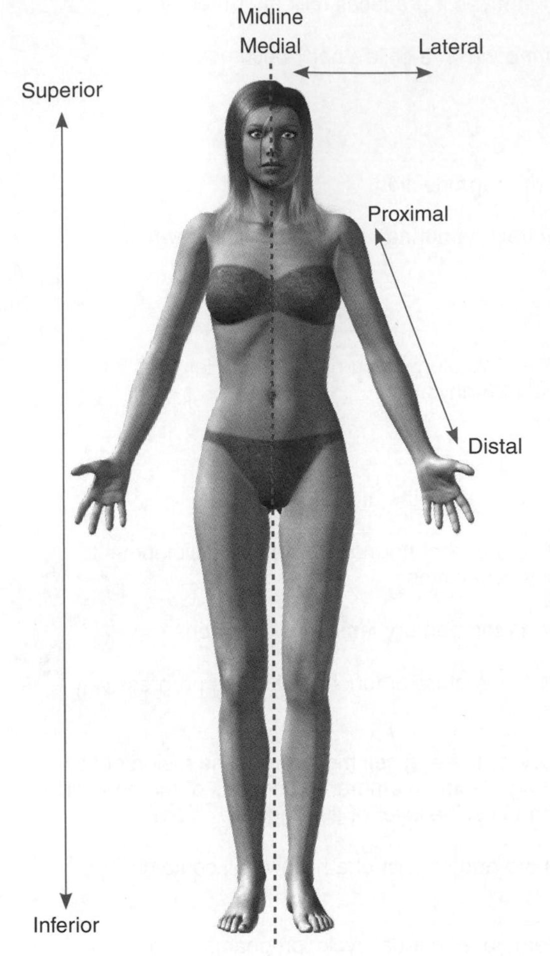

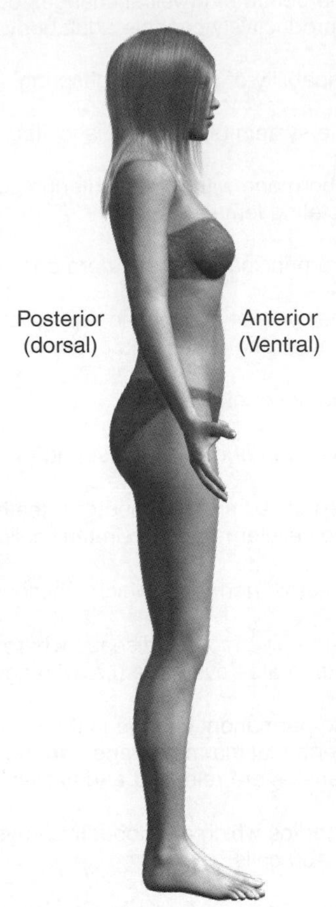

Specific terms are used to give the location and orientation of structures in the body. Many of these terms come in pairs. It is easier to locate and describe structures and movements accurately if you understand these terms.

Proximal: An area towards the attached end of a limb or the origin of a structure, e.g. proximal convoluted tubule.

Distal: An area farthest from the point of attachment of a limb or origin of a structure e.g. distal convoluted tubule.

Superior: Above or over another structure. Towards the head e.g superior vena cava

Inferior: Below or under another structure. Away from the head e.g. inferior vena cava.

Lateral: Away from the midline of the body e.g. lateral collateral ligament.

Medial: Towards the midline of the body e.g. medial collateral ligament

Anterior: Towards the front or ventral surface of the body e.g anterior pituitary gland.

Posterior: Towards the back or dorsal surface of the body e.g. posterior pituitary gland.

Planes and Sections

Planes (flat surfaces) may be cut through a body or organ to a produce section. The plane is named according to the relative direction of the cut surface to the orientation of the organ or structure in the body.

A **frontal plane** (also known as a coronal plane) divides the body into anterior (ventral) and posterior (dorsal) sections.

A **sagittal plane** divides the body into left and right halves. A midsagittal plane does this down the midline of the body (or organ), producing equal sections.

A **transverse plane** divides the body into superior and inferior sections.

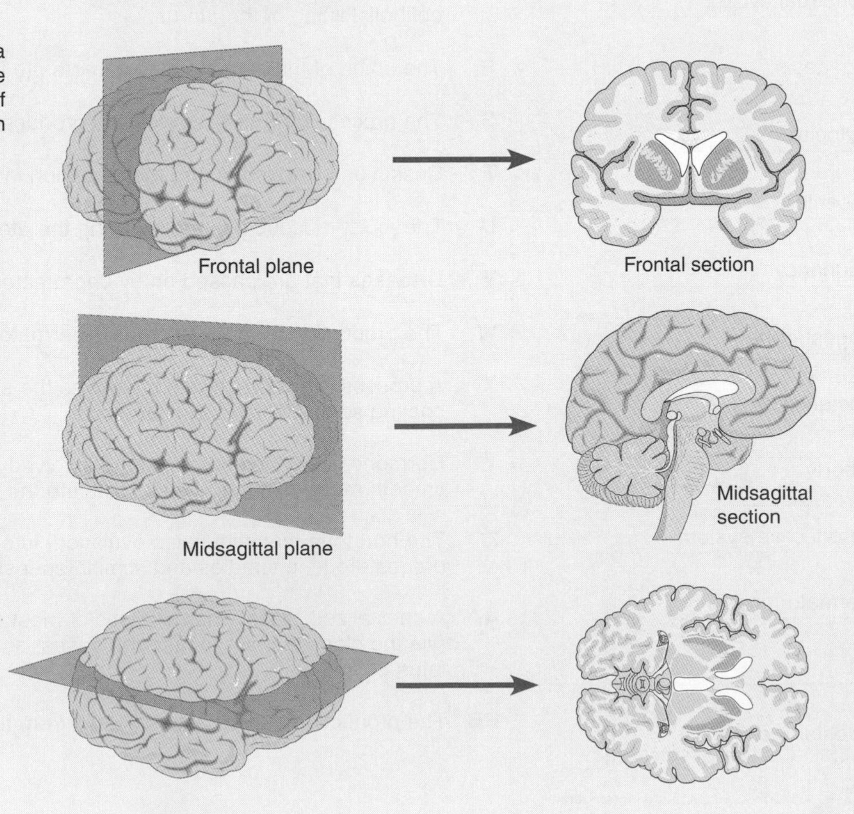

Appendix

PERIODICAL REFERENCES

Cells & Tissues

▶ **Border Control**
New Scientist, 15 July 2000 (Inside Science). *The role of the plasma membrane in cell function: membrane structure, transport processes, and the role of receptors on the cell membrane.*

▶ **The Fluid-Mosaic Model for Membranes**
Biol. Sci. Rev., 22(2), Nov. 2009, pp. 20-21. *Diagrammatic revision of membrane structure and function.*

▶ **Cellular Factories**
New Scientist, 23 November 1996 (Inside Science). *The role of different organelles in plant and animal cells.*

▶ **Getting in and Out**
Biol. Sci. Rev., 20(3), Feb. 2008, pp. 14-16. *Diffusion: some adaptations and some common misunderstandings.*

▶ **How Biological Membranes Achieve Selective Transport**
Biol. Sci. Rev., 21(4), April 2009, pp. 32-36. *The structure of the plasma membrane and the proteins that enable the selective transport of molecules.*

▶ **What is Endocytosis?**
Biol. Sci. Rev., 22(3), Feb. 2010, pp. 38-41. *The mechanisms of endocytosis and the role of membrane receptors in concentrating important molecules before ingestion.*

▶ **The Cell Cycle and Mitosis**
Biol. Sci. Rev., 14(4) April 2002, pp. 37-41. *Cell growth and division, stages in the cell cycle, and the complex control over different stages of mitosis.*

▶ **Living with the Enemy**
New Scientist, 25 Oct. 2008, pp. 26-33. *The sheer diversity of mutations that can turn cells cancerous and drive tumor growth gives endless opportunities to outwit our defense. How can we protect ourselves?*

The Integument & Homeostasis

▶ **Homeostasis**
Biol. Sci. Rev., 12(5) May 2000, pp. 2-5. *Homeostasis: what it is, the role of negative feedback and the autonomic nervous system, and the adaptations of organisms for homeostasis in extreme environments (excellent).*

The Skeletal System

▶ **Aching Joints and Breaking Bones**
Biol. Sci. Rev., 20(2) Nov. 2007, pp. 10-13. *Discusses a number of bone and joint problems which arise with aging. Includes osteoarthritis, osteoporosis, and muscular-skeletal problems.*

The Muscular System

▶ **How Skeletal Muscles Work**
Biol. Sci. Rev., 22(4) April 2010, pp. 10-15. *The structure and function of muscle in humans: contraction, sliding filament theory, and types of muscle.*

▶ **Human Muscle: Structure and Function**
Biol. Sci. Rev., 19(4) April 2007, pp. 25-29. *The structure and function of muscle in humans: contraction and the mechanics of locomotion.*

The Nervous System

▶ **The Autonomic Nervous System**
Biol. Sci. Rev., 18(3) Feb. 2006, pp. 21-25. *Description of the structure and roles of the ANS.*

▶ **Refractory Period**
Biol. Sci. Rev., 20(4) April 2008, pp. 7-9. *The nature and purpose of the refractory period in response stimuli. The principles involved are discussed within the context of the heart's refractory period.*

▶ **Circuit Training**
New Scientist, 9 April, 2011, pp. 35-39. *Discusses methods for treating illness such as depression and Parkinson's disease, and looks if the treatments can be used to enhance everyday performance and feelings.*

▶ **Disco Inferno**
New Scientist, 1 September, 2001, pp. 19. *A half page article describing how ecstasy alters metabolic rate and prevents the body from cooling down.*

▶ **Sense and Sense Ability**
New Scientist, 20 August 2011, pp. 32-37. *Examines the sensory abilities of many different animals, and discusses the components of sensory organs.*

▶ **Infinite Sensation**
New Scientist, 11 August, 2001, pp. 24-28. *The nature of sensory perception. The article has links to web sites where students explore their own responses to sensory inputs.*

▶ **What is a Pacinian Corpuscle?**
Biol. Sci. Rev., 21(3) Feb. 2009, pp. 11-13. *How these pressure receptors in the skin work to convert a mechanical stimulus into a nervous impulse.*

▶ **Generation Specs**
New Scientist, 7 November, 2009, pp. 45-51. *Article covers the human eye and vision, and some conditions such as myopia that can affect the eye.*

▶ **From Genes to Color Vision**
Biol. Sci. Rev., 22(4) April 2010, pp. 2-5. *How do we distinguish color? This article describes the physiological basis of color vision and examines the case for why and how it may have evolved.*

▶ **Alzheimer's: Forestalling the Darkness**
Scientific American, June 2010, pp. 32-39. *Testing for Alzheimer's before symptoms arise can allow drug treatments to be used early. This may increase the chances of the treatment being effective.*

The Endocrine System

▶ **The Ups and Downs of Hormones**
Biol. Sci. Rev., 25(3), February 2013, pp. 30-35. *This accounts discusses cell signaling and hormone action, and explores the changes in hormone activity and disease risk as we age.*

▶ **Food for Thought**
Biol. Sci. Rev., 22(4), April 2010, pp. 22-25. *A clear, thorough account of how the body maintains its supply of glucose long after the nutrients absorbed from a meal have been exhausted.*

▶ **Metabolic Powerhouse**
New Scientist, 11 Nov. 2000 (Inside Science). *The myriad roles of the liver in metabolism, including discussion of amino acid and glucose metabolism.*

▶ **A Diabetes Cliffhanger**
Scientific American, 306(2), Feb. 2012, , pp. 17-19. *Type 1 diabetes is increasing world wide. Hypotheses have been formulated for gluten related diets, fungi, and hygienic living.*

▶ **Glucose: Getting the Balance Right**
Biol. Sci. Rev., 25(1), Sept. 2012, pp. 10-13. *The importance of regulating blood glucose and the consequences of this regulation system failing.*

The Cardiovascular System

▶ **Cunning Plumbing**
New Scientist, 6 Feb. 1999, pp. 32-37. *The arteries actively respond to changes in blood flow, spreading the mechanical stresses to avoid extremes.*

▶ **A Fair Exchange**
Biol. Sci. Rev., 13(1), Sept. 2000, pp. 2-5. *The role of tissue fluid in the body and how it is produced and reabsorbed.*

▶ **The Heart**
Bio. Sci. Rev. 18(2) Nov. 2005, pp. 34-37. *The structure and physiology of the heart.*

▶ **Keeping Pace - Cardiac Muscle & Heartbeat**
Biol. Sci. Rev., 19(3), Feb. 2007, pp. 21-24. *The structure and properties of cardiac muscle.*

▶ **Coronary Heart Disease**
Biol. Sci. Rev., 18(1) Sept. 2005, pp. 21-24. *An account of cardiovascular disease, including risk factors and treatments.*

Appendix

PERIODICAL REFERENCES

▶ **Atherosclerosis: The New View**
Sci. American, May 2002, pp. 28-37. *The pathological development and rupture of plaques in atherosclerosis.*

The Lymphatic System & Immunity

▶ **What is the Human Microbiome?**
Biol. Sci. Rev., 22(2) Nov. 2009, pp. 38-41. *An informative acount of the nature and role of the microbes that inhabit our bodies.*

▶ **Skin, Scabs and Scars**
Biol. Sci. Rev., 17(3) Feb. 2005, pp. 2-6. *The roles of skin, including its role in wound healing and the processes involved in its repair when damaged.*

▶ **Fight for Your Life!**
Biol. Sci. Rev., 18(1) September 2005, pp. 2-6. *The mechanisms by which we recognise pathogens and defend ourselves against them (overview).*

▶ **Inflammation**
Biol. Sci. Rev., 17(1) Sept. 2004, pp. 18-20. *The role of this nonspecific defence response to tissue injury and infection. The processes involved in inflammation are discussed.*

▶ **Lymphocytes**
Biol. Sci. Rev., 12 (1) Sept. 1999 pp. 32-35. *An account of the role of lymphocytes (includes the types and actions of different lymphocytes).*

▶ **Hard to Swallow**
New Scientist, 26 Jan. 2008, pp. 37-39. *Many people fear that vaccines are unsafe and cause health problems. Particular reference to the polio and measles vaccines.*

▶ **Boosting Vaccine Power**
Scientific American, October 2009, pp. 56-59. *This article looks at vaccines and immunization, and how researchers are boosting the effectiveness of vaccines to be even more specific.*

▶ **Monoclonals as Medicines**
Biol. Sci. Rev., 18(4) April 2006, pp. 38-40. *The use and efficacy of monoclonal antibodies in therapeutic and diagnostic medicine.*

▶ **Organ Transplantation**
Biol. Sci. Rev., 25(1) Sept. 2012, pp. 32-35. *The medical and practical issues of organ donation are covered, including tissue matching, organ rejection, and the role of ABO blood groups in tissue typing.*

▶ **Embryonic Stem Cells**
Biol. Sci. Rev., 22(1) Sept. 2009, pp. 28-31. *The future of embryonic stem cell research. Problems and solutions.*

▶ **Genes Can Come True**
New Scientist, 30 Nov. 2002, pp. 30-33. *An overview of gene therapy, and a note about future directions.*

▶ **Gene Therapy: Are We Nearly There Yet?**
Biol. Sci. Rev., 24(4) April 2012, pp. 2-5. *A overview of gene therapy, and recent information*

The Respiratory System

▶ **Gas Exchange in the Lungs**
Bio. Sci. Rev. 16(1) Sept. 2003, pp. 36-38. *The structure and function of the alveoli of the lungs, with an account of respiratory problems and diseases.*

▶ **Humans with Altitude**
New Scientist, 2 Nov. 2002, pp. 36-39. *The short term adjustments and long term adaptations to life at altitude.*

The Digestive System

▶ **The Anatomy of Digestion**
Biol. Sci. Rev., 23 (3) Feb. 2010, pp. 18-21. *The role of each of the components of the human digestive system is described, and explanations are provided about what happens when things go wrong with digestion.*

▶ **The Pancreas and Pancreatitis**
Biol. Sci. Rev., 13(5) May 2001, pp. 2-6. *The structure of the pancreas and its role in digestion.*

▶ **Alimentary Thinking**
New Scientist, 15 Dec. 2012, pp. 38-42. *The role of the enteric nervous system has long been known to control digestion and gut movements, but it seems it has other roles in regulating our general well being as well.*

▶ **Metabolic Powerhouse**
New Scientist, 11 Nov. 2000 (Inside Science). *The myriad roles of the liver in metabolism, including discussion of amino acid and glucose metabolism.*

▶ **The Liver in Health and Disease**
Biol. Sci. Rev., 14(2) Nov. 2001, pp. 14-20. *The various roles of the liver, a major homeostatic organ.*

▶ **Urea**
Biol. Sci. Rev., 17(4) April 2005, pp. 6-8. *An account of nitrogen balance and how and why urea is formed.*

The Urinary System

▶ **Urea**
Biol. Sci. Rev., 17(4) April 2005, pp. 6-8. *An account of nitrogen balance and how and why urea is formed.*

▶ **The Kidney**
Bio. Sci. Rev. 16(2) Nov. 2003, pp. 2-7. *The structure and function of the human kidney, including countercurrent multiplication in the loop of Henle.*

Reproduction & Development

▶ **Spermatogenesis**
Biol. Sci. Rev., 15(4) April 2003, pp. 10-14. *The process and control of sperm production in humans, with a discussion of the possible reasons for male infertility.*

▶ **The Great Escape[1]**
New Scientist 29 Sept 2007, pp. 40-43. *Female reproductive physiology and hormonal control of the menstrual cycle.*

▶ **The Placenta**
Biol. Sci. Rev., 12 (4) March 2000, pp. 2-5. *Placental function and the use of the placenta for prenatal diagnosis and gene therapy.*

▶ **The Great Escape[2]**
New Scientist, 15 Sept. 2001, (Inside Science). *How the fetus is accepted by the mother's immune system during pregnancy.*

▶ **Let me Out**
New Scientist, 10 January 1998, pp. 24-28. *Fetal energy levels have a role in governing the timing of delivery and are communicated via the fetal hypothalamus. An older article but very good in covering general principles.*

▶ **The Biology of Milk**
Biol. Sci. Rev., 16(3) Feb. 2004, pp. 2-6. *The production and composition of milk, its role in mammalian biology, and the processes controlling its secretion.*

▶ **Hijacking Hormones to Regulate Fertility**
Biol. Sci. Rev., 25(1) Sept. 2012, pp. 23-27. *Discusses how female hormones can be manipulated to increase or decrease fertility through IVF and contraception respectively.*

▶ **Age: Old Story**
New Scientist, 23 January 1999, (Inside Science). *The processes involved in ageing. Easy-to-read, but thorough.*

▶ **How We All Will Live to be 100**
Scientific American, Sept. 2012, pp. 40-43. *New approaches in longevity research aim to extend the average life span out to a century or more. Includes a synopsis of causes of mortality: when we die and why.*

Index

Index